STUDENT
DICTIONARY
OF
BIOLOGY

STUDENT DICTIONARY OF BIOLOGY

PETER GRAY

Andrey Avinoff Professor of Biology
University of Pittsburgh

D. Van Nostrand Company
New York • Cincinnati • Toronto • London • Melbourne

Van Nostrand Reinhold Company Regional Offices:
New York Cincinnati Chicago Millbrae Dallas

Van Nostrand Reinhold Company International Offices:
London Toronto Melbourne

Library of Congress Catalog Card Number: 73-742
ISBN 0-442-22816-3

Manufactured in the United States of America.

Published by Van Nostrand Reinhold Company
450 West 33rd Street, New York, N.Y. 10001

Published simultaneously in Canada by Van Nostrand Reinhold Ltd.

15 14 13 12 11 10 9 8 7 6

Library of Congress Cataloging in Publication Data

Gray, Peter, 1908-
    Student dictionary of biology.

    1.   Biology — Dictionaries.   I.   Title.
QH13.G73   574'.03   73-742
ISBN 0-442-22816-3

A BRIEF WORD TO THE STUDENT

If you take ten minutes to read this now it may save you hours later.

HOW TO USE THIS DICTIONARY

There are two ways to use this dictionary. If you want the meaning of a word just look it up. If you are seeking a word you have forgotten, but can remember the root with which it ends or the word with which it is compounded, look under that root or word. There are, for example, numerous words that end in "-genesis" and many kinds of "body"; these words are listed for reference under -genesis or body but are fully defined in their alphabetic position.

SPELLING AND PRONUNCIATION

There is no right or wrong way to spell or pronounce any English word; there is only the custom of the time and place in which you live. Adherence to this custom makes you more comprehensible to your contemporaries.

The trouble about trying to spell or pronounce English is that we use 26 letters, most of Latin origin, to indicate sounds derived from a score of languages. Moreover, two letters are useless ("c", pronounced either "s" or "k": "q", always pronounced "k"), while "s" stands for three sounds (sit, those, leisure) of which the second is adequately represented by "z", and the last by "zh". There is a nasalized and an unnasalized "n" (sin, sing), the latter, very fortunately, always represented by "ng". There is a breathed and an unbreathed "th" (those, this) of which the first requires a special symbol þ because no combination of existing letters can indicate it. The vowels are another matter.

There are four ways of pronouncing "a" (cat, cape, care, calm) which can only be indicated by accents showing that they are respectively short (ă), long (ā), open (â) and long-open (ä). These four words are therefore actually pronounced kăt, kāp, kâr and käm. "e" is either short (met [mĕt]) or long (me [mē]), as is "i" (lit [lĭt], light [līt]). "u" is either short, long or open (cup [kŭp], cute [kyūt], curse [kûrs]). "o" is the most difficult because not only is the single vowel either short, long or open (hot [hŏt], no [nō], order [ôrder]) but the double vowel is either short or long (book [bŏŏk], booze [bōōz]); there are two special combined sounds, the "oi" in boy [boi] and the "ou" in loud or crowd [kroud].

On top of all this, there are the indeterminate vowels which are the sounds you and I make when we pronounce the "a" in acute, the "er" in mother, the "u" in bonus, the "i" in easily, and several others. There is no letter or accent available for these which are all represented in this book by "ə". The words in question are therefore pronounced əkut, muþər, bonəs, ezəle.

It follows that in indicating pronunciations in this book the following symbols, not part of the regular alphabet, must be used*

ă	act, hatch, valve	ĭ	if, women	ōō	ooze, cruise
ā	pay, freight, aid	ī	ice, fright, island	þ	that, those, brother
â	fair, dare	ŏ	hot, collar, comment	ŭ	cup, love
ä	cart, palm	ō	no, croak	ū	use, beauty
ĕ	bet, web	ô	order, ball, pause	û	burn, earn, fern
ē	see, edict, seize	ŏŏ	cook, pull	ə	(indeterminate vowel—see above)

STRESS

Stress is just as, or more important than, pronunciation in spoken English. There are two stresses, a minor and a major, represented in this book by one prime (′) or two primes (″) immediately *following* the syllable to be stressed. An easy example of major and minor stresses is in the words artery [ärt″ərē] and arteriosclerosis [ärtīr′ēōsklərōs″ĭs].

SYLLABIZATION

It is also necessary to indicate in English where the divisions between sounds come. Hothead, for example, could from its spelling alone be pronounced either [hŏth·ēd] or [hŏt·hĕd]; the breaks between sounds are indicated by a dot. Therefore, to pronounce properly the two words given at the end of the last paragraph they should be written artery [ärt″·ər·ē] and arteriosclerosis [ärt·ir′ēō·sklər·ōs″·ĭs].

PRONUNCIATION OF LATIN WORDS

There are many Latin words, and words directly derived from Latin, in this book and there are two ways of pronouncing many of them. Some biologists treat them as English words and put the stress on the last syllable but one; other biologists divide the word so as to indicate its roots. These two methods are often, but not always, the same. Thus everyone says *Eohippus* [ē·ō·hĭp″·əs] (the early horse) and *Mesohippus* [mēs·ō·hĭp″·əs] (the middle horse); but then comes *Merychippos* (the ruminant horse). In English this becomes [mĕr·ē·chĭp″·əs] which gives you a good idea how it is spelled. Originally this was pronounced [mĕr′·ĭk·hip″·əs] to indicate what it meant. In clear cut cases like this I have given both pronunciations prefixing one with *angl.* (for anglicized) and the other with *orig.* (for original). In all cases I have given the Latin termination an indeterminate vowel because, so far as I know, this is all anyone ever uses.

You will find that your instructor has strong, and sometimes violently held, views on which pronunciation to use. You should adapt to these views and be prepared to adapt to the next instructor who may think differently, and finally to make up your own mind when you teach your own classes. As I said at the beginning, neither is right or wrong; it is simply a matter of what is customary in the time and place in which you live.

* This list is similar to that in " The American College Dictionary " (Random House) and is used with the permission of the publisher.

-a- *comb. form* meaning "without". An -n is sometimes appended for the sake of euphony (e.g. anaerobic) or, more rarely, a -p (e.g. aphydrotaxis)

A-discs the darker of the alternate light and dark discs of which striated muscle appears to be composed. They represent the position of the large myosin fibrils which are doubly refracting or anisotropic (*see also* I-disc, Z-band)

A horizon the uppermost level of soil, consisting mostly of decomposing organic detritus (*see also* B, C, and D horizons)

abapical [ăb·ăp″·ic·əl] the antithesis of apical and thus the lower pole of spherical organisms

abdomen [ăb′·dō·man] the posterior region of the animal body; in lower vertebrates it terminates anteriorly in the heart region but in mammals its cavity is separated from the thorax by the diaphragm. In most arthropods it is clearly demarcated from the thorax or cephalothorax (*see also* sessile abdomen, petiolated abdomen)

abdominal rib sternal rib-like bones found in the ventral abdominal wall between the last true ribs in Crocodilia, Sphenodon, and some fossil reptiles (=parasternalia, gastralia)

Abies [ăb′·i·ēz] a genus of coniferales. **A. balsamea** [bôl·săm″·ēə] yields the Canada balsam used by microscopists

abiocoen [ā′·biō·sēn] the sum total of the non-living components of an environment

abiogenesis [ā′·biō·jĕn″·əs·is] the concept that life can arise from non-living material in a relatively short space of time (*see also* biogenesis, neobiogenesis, spontaneous generation)

abioseston [ā′·biō·sĕs″·tən] non-living matter suspended in water (=tripton)

ablastin [ă·blăst″·in] an antibody that inhibits the division of microorganisms

abomasum [ăb′·ō·mās″əm] the fourth division of the artiodactyl stomach (*see also* psalterium, rumen, reticulum) sometimes identified as the "true" stomach

aboral [ăb′·ôr″·əl] pertaining to the surface opposite the mouth, particularly in echinoderms

abortive transduction said of those cases in which the transfused genetic fragment persists without functioning (*see also* transduction)

abscission layer [ăb·sish″·ən] the corky tissue separating the leaf from the branch and at which a deciduous leaf separates from the stem. Also called abscission zone, a term applied to the point of separation of shed animal appendages

abyss- *comb. form* meaning "bottomless" but commonly used for "very deep"

abyssobenthon [ă·bis′·ō·běn″·þŏn] organisms growing on the ocean floor at great depths

abyssopelagic [ă·bis′·ō·pěl·ăg″·ic] pertaining to free living organisms occurring at great depths, restricted by some to those below the 100 fathom line

-ac- *comb. form* meaning a sharp point

-acanth- *comb. form* meaning "thistle head"

acanthella larva [ək̆ăn·thěl′·ə] an acanthocephalan larva developed from an acanthor after the latter has lost its shell and crossed through the gut wall into the intermediate host haemocoel

Acanthobdellida [ək̆·ănþ″·ŏb·děl′·idə] an order of hirudinian annelids erected to contain the single genus Acanthobdella which, though clearly a leech, has chaetae on the second and sixth segment

Acanthocephala [ə·kăn′·thō·sěf″·ələ] a phylum of pseudocoelomate bilateral animals distinguished by an anterior eversible proboscis armed with hooks. Sometimes called the "spiny-headed worms." The genus *Macracanthorhynchus* is the subject of a separate entry.

acanthor larva [ə·kăn″·þôr] a shell-enclosed larva, strongly resembling a hexacanth, which is the infectious stage of acanthocephalans

Acarina [ək̆·ər·īnə] that order of arachnid arthropods which contains the animals commonly called mites and ticks. They are readily distinguished from all other arthropods by the fact that the mouth parts are more or less distinctly set off from the rest of the body on a "false head" or gnathosoma. The division Ixodidae is the subject of a separate entry, as are also the genera *Demodex* and *Sarcoptes*

acceleration the development of a genetically controlled character more rapidly than occurs in the parent or ancestor

accelerator nerve a nerve of the sympathetic nervous system that acts on the heart muscle

accessory gland one that does not directly contribute to the basic function of the organ system with which it

1

is associated. For example, accessory sex glands are all those glands of ·the reproductive system that do not produce gametes

accessory thyroid gland = ultimobranchial body

accidental evolution that which occurs in consequence of a mutation which does not appear to improve survival value

acclimatization [ə·klĭm'·ət·ĭz·ā"·shŭn] the process by which an organism becomes adapted to, or tolerant of, a new environment

accumulator organism one that accumulates a high concentration of an element not known to be essential to its nutrition, as the Brazil nut accumulates barium or the tuna fish mercury

-aceae- suffix indicating familial rank in botanical taxonomy. The zoological equivalent is -idae

acenac- comb. form meaning "a scimitar"

acentric [ā·sĕn'·trĭk] said of chromosomes that lack a centromere

acentric inversion an inversion of any part of the chromosome that does not involve the centromere

acentrous [ā·sĕnt'·rəs] lacking a centrum

acervate [əs"·ĕrv·āt] piled in a heap (see also co-acervate)

Acetabularia [əs·ĕt·ăb"·yūl·âr'·ēə] a genus of Chlorophyceae with an umbrella-shaped thallus containing a single nucleus. Their chloroplasts, alone among green algae, contain the grana typical of higher plants

-acetabuli- comb. form meaning "saucer"

acetabulum 1 [əs·ĕt·ăb"·yūl·əm] (see also acetabulum 2 and 3) the deep socket in the innominate bone at the junction of its three components and into which the head of the femur fits

acetabulum 2 (see also acetabulum 1 and 3) a sucking disc with a raised rim and a central depression, most commonly used of platyhelminths but also applied to many other groups

acetabulum 3 (see also acetabulum 1 and 2) in insects, any cavity into which a joint articulates, particularly the cavity in the coxa ; also, a conical cavity at the front end of some larvae

acetase [əs"·ət·āz'] a general term for enzymes acting on acetate linkages

acetylcholine [ə·sĕt'·ĕl·kō"·lēn] a hormone secreted in the nervous system which affects the conduction of electrical impulses along nerve fibers, and opposes the activity of epinephrin (see also choline)

acetylocholinesterase [ə·sĕt'·ĕl·kō"·lēn·əst"·ər·āz'] catalyzes the hydrolysis of acetylcholine to choline and acetic acid (see also cholinesterase)

achromatic figure [ək·rōm·ăt"·ĭk] all those structures, except chromosomes, pertaining to the process of mitosis

-achyr- comb. form meaning "chaff"

-acr- see -ac-

acicula [əs·ĭk·yūl·ə] needle-shaped objects. In biology usually denotes a size larger than seta or chaeta (e.g. the large aciculum contrasted with the small seta on the parapodium of polychaetes)

acid-fast said of bacteria from which an initial stain is not removed by treatment with acid solutions

acid gland any of numerous invertebrate glands secreting acid, such as the sulfuric acid-secreting salivary gland of shell and limestone boring gastropod mollusca, the formic acid-secreting glands of stinging Hymenoptera, or the HCN-secreting glands of certain diplopods

acid phosphatase [fŏs"·fə·tāz'] an enzyme that catalyzes the hydrolysis of an orthophosphoric monoester in an acid environment to yield alcohol and phosphoric acid

acidophil cell [ăs·ĭd"·ō·fĭl'] any cell which stains readily in an "acid" dye

-acin- comb. form meaning a grape seed

acinar cell [əs·ĭn'·âr] a secretory peripheral cell in the acinus of the pancreas (see also centroacinar cell)

acinous gland = ălveolar gland

acinus [əs·ĭn"·əs] has the same meaning as acinous but is preferred by histologists for the terminal excretory lobes of alveolar glands

Acipenser [ək'·ē·pĕn"·sə] a genus of chondrostean fish called sturgeons, easily distinguished by the heavy, plate-like, ganoid scales along the sides. The eggs are known as "caviar" and the fish was once of great economic importance in both the U.S. and U.S.S.R. It is now almost extinct in the former and rapidly becoming scarcer in the latter

-acm- comb. form denoting a point in the sense of the maximum or minimum of a cyclic curve

Acoela [ā·sēl'·ə] an order of turbellarian platyhelminths distinguished by having a mouth but no intestine. The genus *Convoluta* is the subject of a separate entry

Acoelomata [ā·sēl"·ōm·āt'·ə] a term coined to define those phyla of the animal kingdom which lack a true coelom. This taxon therefore embraces the Platyhelminths and the Nemertea (see also Coelomata and Pseudocoelomata)

acondylose [ā·kŏnd'·il·ōz] lacking joints or nodes

acquired character a character acquired by the individual in the course of its life (cf. Lamarckism)

acquired immunity that which is derived by an individual from a specific immunization through exposure to the infectious agent or a derivative of it

acrasin [ək·rās'·ən] a chemotactic substance, thought to be cyclic AMP, that appears to be secreted by a cell around which other cells congregate

-acro- comb. form meaning "summit", "peak", "point" or particularly when used as a prefix, "apical"

acrocentric chromosome [ək·rō·sĕn'·trĭk] one with a subterminal centromere (see also metacentric chromosome, telocentric chromosome)

acrochordal cartilage [ək·rō·kôrd'·əl] an unpaired cartilage lying immediately above the anterior end of the parachordals in the embryonic chondrocranium

acrodont [ək'·rō·dont] said of a dentition in which the teeth are attached by their sides to a bony ridge in the jaw

Acropora [angl. ə·krŏp'·êr·ə, orig. ăk·rō·pôr"·ə] a common genus of Madreporia, or stone corals. The large branched colonies of several species are common in the waters off Florida and the West Indies.

acrosome [ək·rō·sōm"] a cytoplasmic cap-like structure on the front of a spermatozoan

act- comb. form more properly spelled -akt- meaning a "rocky coast"

ACTH = adrenocorticotropin

-actin- comb. form meaning "ray" more properly spelled -aktin- (cf. -actino-)

actine [ək·tĭn'] a rayed sponge spicule

Actinaria [ăkt"·ĭn·ăr·ēə] a suborder of Zoantharia commonly called sea anemones. They are distinguished from the Antipatharia and Madreporaria by the absence of a skeleton. The genus *Metridium* is the subject of a separate entry

-actino-   *comb. form* from the same root as -actin- but usually taken to mean a "ray of light" and frequently misused in the sense of "star"

actinomere   [ək·tǐn′·ō·mēr] one complete segment of a radially segmented organism

Actinomycetales   [ək·tǐn·ō·mǐ·sět·ǎl″·ēz] an order of Schizomycetes which develop a branching mycelium-like structure, some even bearing conidia. A great majority are found in soil but the group also includes the pathogenic genus *Mycobacterium*

Actinophrys   [ǎkt·ǐn·of′·rǐs] a genus of heliozoan Protozoa with a spherical body. Thin pseudopodia radiate from all the surface and are supported by axial threads that extend to the center of the cell

Actinopterygii   [ǎk·tǐn·op″·těr·ǐj′·ēi] a subclass of Osteichthyes containing most of the boney fishes and distinguished from the Choanicthyes both by their rayed fins and the absence of internal nares. The superorders Chondrostei, Protospondyli and Teleostei are the subject of separate entries

Actinosphaerium   [ǎk·tǐn·ō′·sfēr·ēəm] a genus of heliozoan Protozoa with a sharply defined vacuolated ectoplasm. *A. eichornia* [ik·ôn′·eə] may reach 1 mm in diameter

actinost   [ǎk′·tǐn·ŏst] bones connecting the girdles to the fins in teleost fishes

actinula larva   [ək·tǐn′·yūl·ə] a larva developed from the planula in some hydrozoan coelenterates. It resembles a stalkless polyp and creeps about on its tentacles

activator   an inorganic compound essential component of many enzyme reactions

active immunity   that which is the result of the possession of the appropriate antibodies (*see also* immunity)

active transport   the passage of materials through a differentially permeable membrane against a concentration or electrical gradient

-acule-   *comb. form* meaning a prickle

acut-   *comb. form* meaning sharp

-ad-   *comb. form* indicating "vicinity". It is used in botany principally as a suffix to indicate habitat. In zoology it occurs mostly as a prefix

adanal   [əd·ǎn′·əl] in the vicinity of the anus

-adant-   *comb. form* meaning "attached"

adaptation   the ability of an organism to cope with its environment (*see also* mutual adaption, preadaption)

adaptive race   one which is physiologically, rather than morphologically, distinguished

adaptive radiation   the development, within a taxon, of forms adapted to specific ecological niches

abducens nerve = Cranial VI. Runs from the ventral surface of the medulla to the lateral rectus muscle of the eye

adduction   a moving together

adelo-   *comb. form* meaning "not obvious", "concealed" or, rarely, "unknown"

adelocodonic   [ə·dĕl″·ō·kō·dŏn′·ək] said of medusae which remain attached to the hydroid and therefore lack bells

adelospondylous vertebra   [ə·dĕl′·ō·spŏnd′·əl·əs] a vertebra with a detached neural arch

-adelph-   *comb. form* meaning "brother", but put to a wide variety of uses in biology

adelphogamy   [əd·ĕl′·fō·gǎm″·ē] mating of siblings, though also used in botany for fertilization between neighboring plants

aden-   *comb. prefix* meaning "gland"

adenase   [əd′·ĕn·āz″] an enzyme hydrolyzing adenine to hypoxanthine and ammonia

adenine   [əd′·ĕn·ēn] a 6-amino purine base deriving its name from the original source from which it was isolated

adenohypophysis   [əd·ĕn′·ō·hǐ·pŏf″·ə·sǐs] that portion of the hypophysis which develops from the oral epithelium, originally as Rathke's pouch (*see also* neurohypophysis)

adenoid   [əd″·ĕn·oid′] gland-like. Particularly gland-like masses of lymphoid tissue and specifically such a mass at the back of the pharynx

adenosine   [əd·ĕn′·ō·sēn] a nucleoside the phosphates of which are a primary energy transfer system in living materials

adenosine diphosphate (ADP)   formed in biokinetic systems from ATP

adenosine monophosphate (AMP)   formed in biokinetic systems from ADP

adenosine triphosphate (ATP)   a major energy donor in biokinetic systems

ADH = vasopressin

adhesive gland   any of numerous invertebrate glands which secrete a sticky substance

adhesive papilla   a raised structure, capable of causing adhesion; usually, without qualification, applies to the protuberant ends of the marginal adhesive glands of triclad Turbellaria

adipose fin   a small, posterior, dorsal fin, containing much fatty matter, typical of salmonid fish

adipose tissue   an aggregate of fat-gorged cells

adjective cell   [əd·jěk″·tǐv] (*not* [ǎdj″·ək·tǐv]) in *Chara*, the cell that lies between the oogonium and the stalk cell

adjustor neuron   a neuron which is neither sensory nor motor but which correlates the activities of both

adoption society   an assemblage of one or more organisms living together, though free to dissociate should they wish, and to none of which does the continued association bring any apparent advantage

ADP = adenosine diphosphate

ADPPP = ATP

adradius   [əd′·rād″·ē·əs] the midradius of an interradius

adrenal gland   [əd·rēn′·əl] a gland, adjacent to the kidney in mammals, formed from the fusion of the suprarenal body and the interrenal body of lower forms

adrenaline = epinephrine

adrenocorticotropic hormone = adrenocorticotropin

adrenocorticotropin   [əd·rēn′·ō·kôrt′·ǐk·ō·trōp″·ǐn] a hormone produced by the adenohypophysis that stimulates secretion by the adrenal cortex

adrenoglomerulotropin   [əd·rēn′·ō·glŏm′·ər·yūl·ō·trōp″·ǐn] a hormone secreted by the pineal body which stimulates the secretion of aldosterone by the adrenal cortex

adsere   [ǎd′·sēr] a sere which is going to turn into another sere but not into a climax

-adunc-   *comb. form* meaning "hooked"

advehent veins   [əd·vā′·ənt] veins derived from the posterior cardinals bringing blood to the mesonephoros

adventitia [əd·věn·tĭsh′·eə] the connective tissue sheath of an organ

adventitial wall [əd·věnt′·tĭsh·əl] the outer connective tissue wall of a blood vessel

adventitious embryony a form of apomixis in plants in which an embryo is formed directly by the outgrowth of a cell of the parent sporophyte

adventitious plankton those organisms which have accidentally become planktonic through the action of waves or currents

adventive [əd·věnt′·ĭv] an organism that has been inadvertently introduced to an environment

aecidiospore [ē·sĭd′·ēo·spôr] a teleutospore which always produces a dikaryotic mycelium on germination

aecidium [ē·sĭd′·ē·əm] a cup-shaped sporocarp which produces successive generations of spores from its inner surface

aedeagus [ē·dēg″·əs] the chitinous intromittent organ of male insects (*see also* penis, telopod)

Aedes [ā·ē′·dez] a genus of culicid dipterans containing *A. aegypti* [ē·jĭpt″·ī], the mosquito that transmits yellow fever. This mosquito, though of African origin, thrives in any tropical region including the S.W.U.S.

-aeithal- *comb. form* meaning an "evergreen thicket"

-ael-, aell- *comb. form* meaning a "high wind" or "storm"

Aelosoma [ē·lō·sōm′·ə] a genus of fresh-water oligochaete annelids. *A. tenebrarum* [těn·ē·brâr′·əm] is the green spotted and *A. hemprichi* [hěm·prĭtch′·ī] the red spotted species commonly found among filamentous algae

-aequi- *see* -equi-

-aer- *comb. form* meaning "air"

aeriduct [âr·ĭ·dŭct″] almost any duct concerned with respiration in insects, including the internal trachea of most forms, and the breathing tubes of aquatic larvas

Aerobacter [âr·ō·băkt′·ə] a genus of eubacteriale Schizomycetes widely distributed in the gut of man and other mammals. *A. aerogenes* [âr·oj″·ěn·ēz] is morphologically almost indistinguishable from *Escherichia coli* but, unlike this form, can utilize citric acid

aerobe [âr·rōb′] an organism using air. The adjective, and its derivates, are more frequently used

aerobic [âr·rōb′·ək] said of an organism, or life process, that utilizes, or can only exist in the presence of oxygen (*see also* anaerobic)

aerobic respiration that which requires oxygen (*see also* anaerobic respiration)

aërobiont [âr·rō·bī′·ŏnt] *either* an organism living in air as distinct from water or soil *or* an organism requiring oxygen

aerogen [âr′·rō·jěn] a gas producer, particularly of bacteria

aeroplankton organisms floating in air

aesthete [ēs′·thēt] any invertebrate sense organ, mostly of sensory nerve endings but also applied to sensory hairs and bristles of arthropods

aestivation [ěs′·tĭv·ā·shən] a form of dormancy, not dissimilar from hibernation, in which some organisms pass the summer months

afferent [ăf′·ər·ənt] to convey towards (*see* deferent, efferent)

afferent branchial artery an artery carrying blood to the gill

afferent nerve one which carries impulses to the brain

-aga- *comb. form* meaning "beach" (cf. -psamm-)

agamete [ā·găm″·ēt] any product of reproductive multiple fission that is not a gamete

agamont [ā·găm″·ŏnt] that form of an organism that does not produce gametes; occasionally used as synonymous with schizont

agar-agar [ä″·gə·ä″·gə] a complex polysaccharide derived from several red algae particularly *Gelidium spp.*

agaric [ăg·ăr′·ĭk, ăg′·ər·ĭk] an English word derived from Agaricus and of wide, ill defined, usage though mostly applied to *Agaricus*-shaped forms

Agaricales [ăg·ăr·ĭk′·ăl·ēz] that class of basidomycete fungi which contains all the forms having a fleshy cap (pileus) with gills on the underside of the cap, commonly known as toadstools and mushrooms. The genera *Agaricus*, *Amanita*, *Boletus*, *Coprinus* and *Panaeolus* are the subjects of separate entries

Agaricus [ăg·ăr′·ĭk·əs] the genus of Basidomycetes from which the Agaricales derive their name. *A. campestris* [kăm·pěs′·trəs] is the common or meadow mushroom eaten wherever it occurs. The commercially cultivated mushroom is usually *A. bisporus* [bi·spôr′·əs], considered by many to be inferior in flavor

age the period of time during which an organism, or phenomenon, has existed (*see also* ecological age)

age and area theory the older a species, the larger will be its area of distribution

agglutinated properly "stuck together" but used by entomologists to describe a larva with an unusually heavy chitinous sheath

agglutination the process of sticking, or being stuck together, particularly applied to erythrocytes

agglutinin [əg·glōōt″·ĭn·ĭn] an antibody causing clumping of cells (*see also* autohemagglutinin)

aggregate fruit a fruit derived from many carpels produced by one flower

aggregate plasmodium a myxomycophyte plasmodium congregated without fusion and each cell of which develops separately

aggression the actions of an organism seeking, by physical means, to dominate another organism

aglyph snake [ā·glĭf] one having solid conical teeth (*see also* opisthoglyph and proteroglyph)

Agnatha [ăg·năþ·ə] a class of craniate chordates containing the extinct ostracoderms and the living marsipobranchii (lampreys and hagfish). They are distinguished, as the name indicates, by the absence of a hinged mandible

agouti a pelt color in mammals produced by the fact that each hair is broadly banded in brown and yellow

-agra- *comb. form* meaning "field"

-agrosto- *comb. form* meaning "grass"

agrotype [əg·rō′·tip] an agricultural race or variety

-aigial- *comb. form* meaning "seashore" (cf. psamma)

aileron [āl·ûr′·ŏn] sometimes used as synonymous with alula (q.v.), but more properly a similar structure, actually a large scale, in front of the base of the fore wing

-aio- *comb. form* meaning "eternity" or "everlasting"

air sac 1 (*see also* air sac 2) one of five pairs of outgrowths, derived from the lungs, extending through the body cavity of birds. These are named according to their location: cervical, clavicular, anterior, posterior, thoracic, and abdominal air sacs. These terms are *not* the subject of separate entries

air sac 2 (*see also* air sac 1) the gas containing

portion of the pneumatophore of a siphonophoran coelenterate

-aitio- *comb. form* meaning "cause" or "causation". Usually transcribed "etio"- as in etiology, etc.

Akaryota = Monera

akaryotic [ă·kâr′·ē·ŏt″·ĭk] the condition of lacking a true nucleus, as in bacteria or blue-green algae

akinete [ā′·kin·ēt] a resting algal cell, separated from the thallus, and which will subsequently reproduce

-akm- *comb. form* meaning "point" almost invariably transliterated -acm- (*q.v.*)

akro- *comb. form* meaning "apex", *etc.* almost invariably transliterated -acro- (*q.v.*)

-akti- *comb. form* meaning a "rocky coast", usually transliterated -acti-

-aktin- *comb. form* meaning "ray", usually transliterated -actin- (*q.v.*)

ala temporalis [ălä těmp′·ôr·ăl·ĭs] a cartilage in the embryonic chondrocranium which appears immediately adjacent, and slightly anterior to, the processus alaris

l-alanine [ĕl″·ăl′·ăn·ēn] 1-α-aminopropionic acid CH₃CH(NH₂)COOH. An amino acid not essential in rat nutrition (*see also* phenyl alanine, dehydroxyphenylalanine)

alar [ā′·lä] pertaining to the wing or wing-shaped

alar canal a horizontal channel through the alisphenoid bone of some mammals joining the foramen ovale to the foramen rotundum

alar plate the dorsal half of the side of the brain stem or spinal cord

alarm signal one made by animals denoting the presence of a predator or potential predator. Alarm signals are often shared between different kinds of animals. Thus the alarm signals of a mammal may be comprehensible to birds and *vice versa*

alary zone [āl′·är·ē] that part of the embryonic central nervous system which lies above Monro's sulcus

alate [ā′·lāt] winged

-alb- *comb. form* meaning "white"

albino [ăl·bĭn″·ō] properly an organism lacking all pigment but often applied to partial albinos

albumen [ălb·ū″·měn] any of numerous water-, or saline-, soluble simple proteins. This term, particularly in botany is also used in the sense of "food reserve"

albuminous [ăl·bū″·mĭn·əs] pertaining to, or consisting of, albumen (*see also* exalbuminous)

albuminous cell a parenchymatous cell of gymnosperms analogous to companion cells in angiosperms

albuminous seed one in which some part of the endosperm is retained until germination

Alcyonaria [ăl″·sē·on·âr·ēə] a subclass of anthozoan coelenterates containing, *inter alia*, the "soft corals", horny corals, sea fans, and sea pens. The alcyonaria are clearly distinguished from the Zooantheria by having eight tentacles and eight septa

aldolase [ăl′·dō·lāz″] the term is usually applied to a number of enzymes catalyzing the production of aldehydes from ketose phosphates: more properly called formaldehydelyases or aldehyde lyases

aldosterone [ăl·dō′·stēr·ōn″] a hormone secreted by the adrenal cortex. Active in the control of electrolytes, particularly potassium in blood

alecithal [ā′·lĕs·ĭþ·əl] a term meaning "without yolk" sometimes inaccurately used for isolecithal. No known eggs are completely without yolk

-aleto- *comb. form* meaning "vagrant"

Algae [ăl″·gə] an assemblage of thallophyte plants containing photosynthetic pigments but lacking conducting vessels. They are of very varied morphology and probably of polyphyletic origin. The following subdivisions, each the subject of a separate entry, often regarded as separate phyla: Cyanophyta (blue-green algae), Chlorophyta (green algae), Phaeophyta (brown algae), Rhodophyta (red algae)

algal bloom an explosive growth of planktonic algae

alginase [ăl′·jĭn·āz″] an enzyme catalyzing the breakdown of alginate through the hydrolysis of β-1, 4-mannuronide link

alisphenoid bone [āl′·ē·sfēn′·oid] one of a pair of bones of mixed origin, in the splanchnocranium, lying behind and above the orbitosphenoid and immediately below the lateral edge of the parietal

alitrunk [āl·ē·trungk′] those segments of an insect thorax which bear wings

alkaline phosphatase an enzyme hydrolyzing an orthophosphoric monoester in an alkaline environment to yield an alcohol and phosphoric acid

alkylhalidase [âl·kēl′·hăl·ĭd·az″] an enzyme that catalyzes the hydrolysis of alkylhalides to formaldehyde and the appropriate halide acids

-all- *comb. form* meaning "other" in the sense of "different"

-allag- *comb. form* meaning "exchange", frequently written -allax- in compounds

allantoicase [al·an·tō′·ikza″] an enzyme catalyzing the hydrolysis of allantoate to glyoxylate and urea

allantoinase [āl·ăn·to′·in·āz″] an enzyme catalyzing the hydrolysis of allantoin to allantoic acid

allantois [ăl·ăn·tō″·ĭs] one of the extra-embryonic membranes of amniotes. It is derived as an outgrowth from the posterior end of the embryonic gut. In oviparous animals it presses against the shell as the inner part of the chorioallentoic membrane (*q.v.*) and in mammals contributes to the placenta (*q.v.*)

-allax- the usual *comb. form* of -allag- (*q.v.*). The following terms with this suffix are defined in alphabetic position

MORPHALLAXIS   TROPHALLAXIS

Allee's Law any given habitat has an optimal population level for any given species

allele [ăl·ēl′] alternate forms of genetic characters which occur at the same locus on the chromosome are said to be alleles, or allelic to each other (*see also* allelomorph). The undernoted alleles are the subject of separate entries:

DOMINANT ALLELE   PSEUDOALLELE
ISOALLELE   RECESSIVE ALLELE

allelism [ăl·ēl″·izm] the relationship between two characters which are alleles

-allelo- *comb. form* meaning "mutual"

allelomorph [ăl·ēl″·ō·môrf] two contrasting but closely parallel genetic characters (e.g. smooth or wrinkled skin in peas) are said to be allelomorphs (cf. allele). The term is sometimes, but improperly, used as a synonym for allele

Allen's Law in poikelothermic animals, the relative size of appendages diminishes in a colder environment

alliogenesis [ăl·i·ō·jĕn″·əs·əs] a type of reproduction in which there is an alternation of generations

Allium [ăl·ē·əm] a very large genus of amaryllidaceous plants of great economic importance. *A. cepa* [sē·pə] the onion, *A. porrum* [pŏr″·əm], the leek, *A. ascalonium* [əs·kăl·ōn·ē·əm], the shallot, *A. schoeno-*

prasum [shĕn'·ō·prāz'·əm], chives and *A. sativum* [săt·ĕv'·əm], garlic are only a few of the numerous edible species grown in many parts of the world. There are also a large number of species grown as ornamentals

-allo- *comb. form* meaning "other" in the sense of "different"

allochore [ăl'·ō·kôr'] an organism occurring in two different habitats in the same geographic region

allochronic speciation [ăl·ō·krŏn·ĭk] the production of morphological discontinuity between species thought to be caused solely by the passage of time (*see also* allopatric speciation)

allochronic species species separated in time from a similar, but distinct, species

allochthonous [ăl·ŏk"·þŏn·əs] pertaining to materials entering an aquatic habitat from the external terrain by seepage or drainage. Hence also applied to accumulations of drifted vegetation (cf. autochthonous)

allocryptic [ăl'·ō·kript'·ĭk] said of organisms which hide by covering themselves with other organisms or with inanimate material

allometric coefficient [al'·ō·met'·rĭk] the slope of a curve obtained by plotting the logarithm of some measurement of an organ or part against the logarithm of the whole remainder or another part. This is also termed the heterogonic or heteroausecíc coefficient

allometron [ăl'·ō·mĕt'·rŏn] a genetic change in the proportion of an existing character, such as an increase of depth of color in a flower

allometry [ăl'·ō·mĕt'·ri] disproportionate size, particularly of an organ

allomorphosis [ăl'·ō·môrf·ōs'·is] the rapid development either of specialized organs or of the general specialization of the organism

allomorphotic evolution [ăl'·ō·môr·fŏt'·ĭk] that leading rapidly to specialization

allopatric speciation [ăl'·ō·păt'·rĭk] a morphological discontinuity arising from geographic fragmentation combined with the passage of time (*see also* allochronic speciation)

allopatry the condition of two related populations occupying separate geographic areas and which do not interbreed (*see also* continuous allopatry, disjunct allopatry)

allopolyploid [al"·ō·pŏl·ē·ploid'] a polyploid in which the replicated diploid sets of chromosomes come from genetically different strains

allopreening [al'·ō·prēn'·ing] the mutual preennig of feathers

allosematic [ăl'·ō·səm·ăt"·ik] having coloration mimicing that of another species supposed to be protected by its own coloration

allosomal inheritance [al"·ō·sōm'·əl] the inheritance of characters influenced by genes in an allosome (*e.g.* sex-linked characters)

allosome [ăl"·ō·sōm] a chromosome which is different from the rest, usually the sex chromosome

allosynapsis [ăl'·ā·sin·ăps'·is] pairing at meiosis of chromosomes derived from different ancestors in an amphipolyploid

allotetraploid [ăl'·ō·tet'·rä·ploid] a tetraploid in which one diploid set has been derived from a genetically different parent

-allotri- *comb. form* meaning "unusual", "abnormal"

allotriomorphic [ăl'·ōt'·rē·o·môrf"·ik] having an abnormal or unexpected shape

allotriploid [al"·ō·trĭp'·loid] a triploid with two similar and one dissimilar genomes (*see also* autotriploid)

allotrophic [al"·ō·trôf'·ik] said of an organism dependent on other organisms for its nutrition. Synonymous with parasitic only if all animals are regarded as parasitic on plants

allotype [al"·ō·tip'] a supplementary type described from the specimen of the opposite sex to the original

allozygote [al"·ō·zi'·gŏt] a homozygote with only recessive characters

allomeristic [al"·ō·mēr·ist'·īk] said of organisms which differ in the number of the parts of any organ from that which is customary in the group to which they belong; for example, Pycnogonida commonly have four pairs of legs, but an allomeristic species with twelve pairs has been recorded

Alopex [ăl·ō·peks'] the genus of canid carnivores containing the Arctic Fox (*A. lagopus*) [lā·gō'·pəs]. It differs from other foxes (*Canis, Urocyon*) in many ways, the most obvious being the fully-furred sole of the feet

Alosa [ăl·ō'·sə] a genus of commercially important isospondylous fish. *A. sapidissima* [săp·id·ĭs"·əmə] is the shad

Alouatta [ăl·ōō·ət'·ə] a genus of platyrrhine primates containing the howler monkeys

alp a small pasture in high mountain country. As a well known range has been named for the frequency of such pastures, the term is often misused to describe a high mountain

alpestrine [ăl·pĕst'·rīn] properly applied to plants growing above tree level (but, frequently misused as synonymous with alpine)

apha cell a chromophile cell of the anterior lobe of the pituitary heavily granulated and with a diplosome

alpine [ăl'·pīn] properly applied to organisms occurring in alps (*i.e.* high mountain meadows); it is frequently, however, used as synonymous with alpestrine

alpine tundra a tundra-like habitat occurring in mountainous regions at or around the limit of tree growth

-als- *comb. form* meaning a "grove"

-alt- *comb. form* meaning "high"

-alter- *comb. form* meaning "other"

alternation of generations the condition of an organism in which the diploid generation reproduces asexually a haploid generation that reproduces sexually a new diploid generation and so on

alterne [ôlt'·ûrn] a plant community that is very different from a neighboring community, particularly when such different communities repeat or alternate

-althe- *comb. form* meaning "to increase"

-alto- *comb. form* meaning "high"

altricial [ăl·trish'·əl] said of birds hatched in a condition which requires that they receive parental care

alula [āl'·yōō·lə, ăl·yōō·lə] the thumb-like joint on the wing of the bird, or a posterior basal lobe adjacent to the wing of some Diptera

alveolar [ăl·vē'·ō·lə] adjectival form of alveolus; also applied to that portion of the ectoplasm of a ciliate protozoan which lies immediately beneath the pellicle

alveolar duct a duct connecting alveoli in the lungs to the bronchioles

alveolar gland a much-branched type of gland, with each branch terminating in a secretory capsule

alveolus [ăl·ve"·əl·əs] a terminal cavity of a hollow lobular structure. When used, as is nowadays rarely the case, in distinction from acinus, the alveolus is conical and the acinus more or less spherical or ovate

-ama-   *comb. form* meaning "together"

Amaryllidaceae [ăm'·ĕr·il·ĭd·ās"·ēə] that family of liliflorous monocotyledons which contains not only amaryllis, the tuberose and Narcissus but many forms commonly called "lilies" such as the guernsey lily, the spider lily, etc. Distinguishing characteristics are the inferior 3-celled ovary, the 6-parted perianth and the 6 stamens. The genus Allium is the subject of a separate entry

Amanita [ăm'·ə·nit"·ə] ¹ a genus of agaricale Basidomycetes containing the most deadly of all mushrooms *A. phalloides* [făl·oid'·ēz] and its close relatives *A. verna* [vĕrn'·ə] and *A. sporosa* [spŏr·ōs"·ə]. Only very rarely is ingestion of any of these not followed by death. *A. muscaria* [mŭsk·âr'·ēə] is also poisonous but is used as an intoxicant in parts of Eastern Europe. Many other species are edible and *A. caesarea* [sēz·âr'·ēə] is often found in European markets

-amb-   *comb. form* meaning "around" or "on both sides"

ambiens muscle [amb'·ē·ĕnz] the muscle used to contract the toes by perching birds; it originates on the pelvis and is inserted on the end of the toe

Amblyopsis [ăm·blē·ŏps'·ĭs] a genus of teleost fish, living in caves. The lack of eyes is compensated for by numerous tactile papillae on the head, body and tail

Amblypygi [ăm'·blē·pi"·jē] a small order of arachnid arthropods at one time united with the Uropygi as the Pedipalpi. They are distinguished by the fact that the tarsus of the first walking leg is modified into an extremely long tactile organ

ambon [ăm'·bŏn] the fibrous ring round the socket of a ball-and-socket joint

Ambrosia [ăm·brŏz"·ēə] a genus of flowering plants of the family Compositae. *A. artemsiifolia* (Ragweed) is a notorious producer of hayfever

ambulacrum 1 [ăm'·byūl·ăc"·rəm] (*see also* ambulacrum 2, 3) a walking leg, particularly of insects and insect larvae

ambulacrum 2 (*see also* ambulacrum 1, 3) one of the, usually five, radiating grooves in which the tube feet of echinoderms are placed

ambulacrum 3 (*see also* ambulacrum 1, 2) an adhesive disc, formed of an aggregate of hooks, which terminate the tarsus of Ixodid ticks

Ambystoma [*angl.* ăm·bist"·ōmə, *orig.* ăm·bē'·stōm"·ə] a genus of urodelan Amphibia distinguished from newts by the absence of parasphenoid teeth. There are several American species *A. maculatum* [măk·yūl·ā"·təm] (the Spotted Salamander) being the largest and best known. Several species, or in some cases geographical races of species, exhibit neotony (*i.e.* breed as adult-sized but unmetamorphosed larvas). The best known is *A. mexicanum* [meks·ik·än'·əm] the breeding larva of which was for years regarded as another animal called *Sirenodon* [sĭr·ēn'·ō·dŏn] *mexicanum* (the Axlotl)

ameba [ə·mē"·bə] an English word derived from the Latin *Amoeba* but of less restrictive use in such phrases as "the amebas living in man" or "the limax amebas"

ameboid [ăm·ē"·boid] pertaining to, or resembling, any rhizopod protozoan, particularly *Amoeba*

ameboid motion movement with the aid of blunt pseudopodia

ameiotic [ā·mī·ŏt'·ĭk] said of a maturation division of a gamete in which the diploid number of chromosome is not reduced to the haploid

ameiotic parthenogenesis parthenogenesis from a gamete which has become haploid through a single mitotic division

ametabolous [ā·mĕt'·ăb"·əl·əs] said of those wingless insects the eggs of which hatch into a nymph, differing from the adult principally in size and in the life history in which there is thus no metamorphosis

Amia [ām"·ēə] a genus of protospondylous osteichthyes. *A. calva* [kăl'·və], the Bowfin, is the only extant species

amictic egg [ā'·mĭkt·ĭk] an egg that cannot be fertilized, and therefore can only develop parthenogenetically into a female (*see also* mictic egg)

amidase [ăm'·ĭd·āz"] a general term for an enzyme catalyzing the hydrolysis of numerous carboxylic acid amides to the corresponding carboxylic acids and ammonia

aminase [ăm'·in·āz"] a general term for enzymes catalyzing reactions that produce urea or ammonia

aminidase [ăm·in'·id·āz"] a general term for an enzyme catalyzing the hydrolysis of terminal links in various mucopolysaccharides

amino acid [ə·mēn"·ō] an organic acid with an amino group at one end and therefore possessing both acidic and basic linkages. Many but not all amino acids are known to take part in protein synthesis

amitosis [ā·mĭt·ō'·sĭs, əm·ĭt'·ōs·ĭs] cellular division in which mitotic activity has not been observed

amixis [ā·mĭks'·ĭs] the absence of fertilization; occasionally extended to mean specifically by reason of the absence of gonads

ammocoetes larva [ăm·ō·sē'·tēz] the freshwater larva of the sea lamprey *Petromyzon marinus*

ammonifier [əm·ōn"·ē·fi·ər] an organism, particularly a bacteria, that liberates ammonia

Ammon's horn [əm"·ənz] an enlarged, and inwardly rolled, portion of the hippocampus in mammals

amnion 1 [ăm"·nē·ən] (*see also* amnion 2, 3) that embryonic membrane (*q.v.*) which encloses the embryos of reptiles, birds and mammals. In telolecithal, and similar forms, it arises as folds over the head and tail from the extra embryonic blastoderm (*see also* proamnion)

amnion 2 (*see also* amnion 1, 3) the thick outer wall of the invaginated ectodermal disk in Nemertea

amnion 3 (*see also* amnion 1, 2) a membrane resembling an amnion and enclosing the embryo of many insects and other invertebrates

amniote [ăm"·nē·ōt] said of an organism, or group of organisms, which possesses an amnion. Frequently used in the sense of "higher" vertebrates, that is the reptiles, birds and mammals (*see also* anamniote)

amniotic fluid [ăm'·nē·ŏt'·ĭk] the contents of the amniotic pouch

amniotic pouch the sac which contains the amniotic fluid

Amoeba [ə·mē"·bə] a genus of amoebid protozoans. The species *A. proteus* [prō'·tē·əs] is commonly used for class teaching. The correct name of this form is probably *Chaos diffluens* [kā"·ŏs dĭf'·lyū·ĕnz]

Amoebida [ə·mē"·bĭd·ə] that class of sarcodinous protozoa which contains the amebas. The method of locomotion is characteristic. The following genera are the subject of separate entries: *Amoeba, Chaos, Endolimax, Entamoeba, Hydra, Naegleria, Pelomyxa*

amoebocyte [əm·ē″·bō·sĭt] any coelomocyte or hemocyte which exhibits amoeboid movement

AMP adenosine monophosphate (*see also* acrasin, nucleosidase)

-amphi- *comb. form* meaning "around", or "double"

amphiapomict [ăm′·fĭ·ăp″·ō·mĭkt] an organism which reproduces both sexually and parthenogenetically

amphiarthrosis [ăm′·fĭ·är·þrō″·sĭs] a joint capable of flexion, but not free movement. In human anatomy a fibrocartilagenous joint

amphiaster [ăm′·fĭ·ăst·ər] the two asters, one at each end of the cell, from which the spindle fibers diverge in cell division

Amphibia [ăm·fĭb′·ēə] a class of chordates commonly called amphibians. They are distinguished by the absence of scales, hairs, or feathers in the skin. Among the frogs, salamanders, toads and the like, enough are amphibious to justify the name. The orders Apoda, Laybrinthodonta, Salienta and Urodela are the subject of separate entries

amphiblastula [ăm′·fē·blăst″·yū·lə] a blastula in which the cells of one pole differ markedly either in size or shape from those of the other pole

amphiblastula larva the free-swimming larva of some calcareous sponges consisting of an anterior hemisphere of flagellated micromeres and a posterior hemisphere of non-flagellated macromeres

amphicoelous vertebra [ăm′·fē·sēl″·əs] one in which the centra are biconcave (*see under* vertebra for other types)

amphigean [*angl.* əm·fĭj·ē·ən, *orig.* ăm′·fē·jē″·ən] pertaining both to the Old and New Worlds

amphigenesis [ăm′·fē·jĕn″·əs·ĭs] development inaugurated by the fusion of two unlike gametes

amphigenous castrate [*angl.* əm·fĭj′·ən·əs, *orig.* am′·fē·jĕn″·əs] the condition of a male that shows an intermingling of female characters or *vice versa*

amphikaryon [ăm′·fē·kăr″·ē·ŏn] a nucleus containing two haploid genomes

amphimict [ăm′·fē·mĭkt″] an organism which reproduces by amphimixis

amphimixis [ăm′·fē·mĭks″·ĭs] the union of two gametes from separate parents

amphimorula [ăm′·fē·mŏr″·yūl·ə] a morula derived from an amphiblastula

Amphineura [ăm′·fē·nyōŏr″·ə] a class of symmetrical Mollusca either lacking shells (Aplacophora) or with 8 transverse dorsal calcareous plates (Polyplacophora). All are commonly called chitons in English though this term is properly applied only to the order Polyplacophora

Amphioxus [ăm′·fē·ŏks″·əs] a name almost universally misused for *Branchiostoma* (*q.v.*)

amphiplatyan vertebra [ăm′·fē·plăt″·ē·ən] one in which both posterior and anterior surfaces of the centra are flat

amphiploid [ăm′·fē·ploid] a type of polyploid characterized by the addition of two sets of chromosomes from each of two species

amphipneustic [ăm′·fē·nōōs″·tĭk, ăm′·fē·nyōō″·stĭk] having two methods of respiration, as amphibia having both gills and lungs or aquatic insect larvas having functional spiracles at both the anterior and posterior ends

Amphipoda [am′·fē·pŏd″·ə] an order of Crustacea in which there is no distinct carapace and the first thoracic somite is coalesced with the head. They are distinguished from the closely allied Isopoda by the laterally compressed body, the presence of a distinct telson, and the biramous first antennae. The group contains those forms commonly known as sand fleas, sand hoppers, and scuds or side swimmers. The genera *Caprella* and *Gammarus* are the subject of separate entries

Amphisbaena [ăm′·fĭs·bēn″·ə] the genus containing the Worm Lizards. They are burrowing, totally subterranean, worm-like forms without limbs and with only one lung. Some place them in a distinct order (Amphisbaenea) separate from the Squamata

Amphitrite [*angl.* ăm·fīt″·trĭt·ē, *orig.* ăm′·fē·trī″·tē] a genus of Polychaetae Sedentaria, with a cylindrical, tapering body, living in sand tubes. The tentacles are very numerous and contractile and there are 3 pairs of branching gills on the peristomium

amphitrophic growth [*angl.* ăm·fĭt″·rŏf·ĭk, *orig.* ăm′·fē·trŏf·ĭk] that in which the lateral shoots and buds of a plant grow more rapidly than the terminal

Amphiuma [ăm′·fē·yūm·ə] a genus of eel-like urodele Amphibia with minute lidless eyes and vestigeal limbs. *A. means* [mē″·əns] is a common American species growing to 3 ft long

amplexus [ăm·plĕks″·əs] an embrace, particularly a sexual embrace, and specifically the sexual embrace of frogs and toads (Salienta)

-ampull- *comb. form* meaning a "narrow-necked vessel"

ampulla [ăm·pŭl″·ə] literally a small, sub-spherical, flask. Used in biology for almost any sub-spherical hollow body including the muscular sac at the base of an echinoderm tube foot, the cavity in the coenosteum in which develops the gonophore of a millopore hydrozoan, the flotation bladders of brown algae, etc., etc. (*see also* Lorenzini's ampulla, Savi's ampulla, Vater's ampulla)

Ampullaria [ăm·pŭl′·âr·ēə] a common and widely distributed genus of freshwater gastropods usually called pond snails. The entire family Ampullariidae [ăm′·pŭl·âr·ĭ′·ĭ′·dē] are sometimes referred to as the apple snails

-amyl- *comb. form* meaning "starch"

amylase [ăm′·əl·āz″] an enzyme hydrolyzing 1,4-glucan links, in starch, glycogen and related polysaccharides (*see also* glucoamylase)

amyloplast [əm′·ĭl″·ō·plăst, əm′·əl·ō·plăst″] a plastid modified for the storage of starch grains

-an- *see* -a-

-ana- *comb. form* confused between two roots and therefore meaning both "without" and "up"

anabiosis [ăn′·ă·bi·ōs″·ĭs] the condition of an organism which has passed into a resting stage that is cyclic or seasonal, and produced by a change in the environment, such as the loss of moisture

Anaboena [ăn·ă·bēn″·ə] a genus of cyanophyte Algae, closely allied to *Oscillatoria*, and the members of which form simple filamentous colonies

anabolic [ăn·ă·bŏl″·ĭk, ăn·ăb′·əl·ĭk] pertaining to those metabolic activities which are concerned with the breakdown of larger molecules to smaller ones

Anacanthini [ăn·ă·kănth′·inī] an order of osteichthyes containing the codfish and their allies. They are easily recognized by the fact that the ventral fins are anterior to the pectoral fins. The genera *Gadus* and *Melanogrammus* are the subject of separate entries

Anacardiaceae [ăn·ə·kärd′·ē·āsē] that family of sapindalous dicotyledons which contains, *inter alia*, the mangos, the poison oaks, and the cashews. It is distinguished from the closely related Sapindaceae by the presence of resins. Many also contain irritant oils. The genus *Rhus* is the subject of a separate entry

anadromous [*angl.* ăn·ăd″·rōm·əs, *orig.* ăn·ă·drōm′·əs] said of marine fish that enter fresh waters to spawn (*cf.* catadromous)

anaerobic [ăn·âr·ō″·bĭk] said of an organism, or life process, that does not utilize, or cannot exist in, the presence of oxygen

anaerobic metabolism that which does not utilize oxygen (*see also* fermentation)

anaerobiont [ăn·âr″·ō·bĭ·ŏnt] an organism capable of anaerobic existence

Anagasta see Ephestia

anal [ā″·nəl] pertaining to the anus in all the meanings of this word (*see also* adanal)

anal angle the angle of the hind wing of an insect that lies nearest to the abdomen

anal fin a ventral, unpaired, usually posterior, fin of a fish

anal gland a term loosely applied to any gland adjacent to or associated with the anus, in many fish and birds, and often used where rectal gland would be better. In insects, a gland opening in an area adjacent to the anus

anal pit an invagination precursor to the anus in embryos

anal plate any plate abutting on or surrounding the anus. The term is common in arthropods and is used also for the transversely enlarged ventral scale that covers the anus in lizards and snakes

anal siphon the breathing tube of aquatic dipterous larvae

analogous [ăn·ăl′·əg·əs] used in anatomy to describe structures of similar function but different phylogeny (e.g. the vertebrate and invertebrate eye). Also used in ecology to describe organisms of similar habitat or distribution (*see also* homologous)

anamniote [ăn·ăm′·nē·ōt] a term used, in contrast to amniote, to describe those craniate vertebrates (fish and Amphibia) by which no amnion is produced in the course of their development

anamorphosis [ăn′·ă·mŏrf·ōs″·ĭs] a process of slow, steady evolution not involving immediately apparent gross mutant variations (*cf.* saltation)

anaphase [ăn″·ə·fāz′] the third stage of mitotic division in which the chromosomes split and start moving in the direction of the poles (*see under* -phase for other terms of this series)

Anaplura [ăn″·ə·plŏŏr″·ə] that order of insects which contains the parasitic lice of mammals, allied to, and once united with, the Hemiptera

anaspid [ăn·ăp″·sĭd] pertaining to a reptilian skull with no temporal openings

Anaspida [ăn·ăp″·sĭd·ə] a subclass of reptiles without temporal openings in the skull. It includes both the very primitive Permian forms like *Seymouria* and all of the extant Chelonia (turtles). The genera *Captorhinus* and *Seymouria* are the subject of separate entries

anastomosis [ăn′·ăst′·ō·mō″·sĭs] a union between two things, originally between two seas. Now used principally of blood vessels or nerves

ancestrula [ăn′·sĕs″·trōō·lə] a colony of animals produced by asexual reproduction from a metamorphosed sexually produced larva: particularly the original zooid from which a colony of Ectoprocta is derived

Anchetherium [ănk′·ē·þēr·eəm] a genus of Miocene equids. It appears to have been the first equid to migrate from the New World to the Old

Anchylostoma [ăn·kil·ôst·ōmə] a genus of anchylostomatid nematodes best known for the human parasite *A. duodenale* [dyōō·ō·dĕn′·äl·ē], endemic to most northern countries. Infection is through skin contact with feces-contaminated soil or water. *A. braziliense* [brăz·il·ĕn″·sē] is the cat hookworm and *A. caninum* [căn·in″·əm] the dog hookworm

anci- *see* anko-

Ancistrodon [ăn·sĭst″·rō·dŏn] a genus of pit vipers of which there are two species in the U.S. *A. piscivorus* [pĭ·sĭv′·ər·əs] is the Water Moccasin and *A. contortrix* [kŏn·tôr′·trĭks] is the Copperhead

Ancylostomidae [ăn·kil″·ō·stōm′·ĭd·ē] a family of nematode worms, commonly called hookworms, distinguished by the large mouth opening into a buccal cavity and by the possession of cutting teeth or plates on the anterior margin of the mouth. The genera *Anchylostoma* and *Necator* are the subject of separate entries

-andr- *comb. form* meaning "male" (*see* -androus, -aner)

androchore [ăn′·drō·kôr″] a plant dependent on man for its distribution

androconium [ăn′·drō·kōn″·eəm] a patch of modified scales, thought to function as an odor-producing organ, on the wings of male butterflies

androecium [ăn′·drēs″·ēəm] the male portion of a flower consisting of the stamens and their associated parts

androgen [ăn″·drō·jen″] a general term for male hormones

androgenesis [ăn′·drō·jen″·əs·ĭs] development of an egg having only a paternal nucleus i.e., male genes

-androus pertaining to a male, or male part. The substantive form terminates in -andry and the alternative adjectival form -andric is occasionally used. Compounds so formed are not listed separately (*see* protandrous)

androsterone [*angl.* ăn·drŏs″·tə·rōn′, *orig.* ăn′·drō·ster″·ōn] a hormone secreted by the testes having the same activity as testosterone

androsynhesmia [ăn′·drō·sĭn·hĕz″·mēə] a group of males gathered together during the breeding season (*see also* synhesmia, gynosynhesmia)

-anem-, -anemo- *comb. forms* meaning "wind"

aner [ā″·nə] a male ant

anesthesia the condition of not perceiving

anestrous cycle [ăn′·ēs″·trəs] the period between estrous cycles or the non-breeding period (*see also* estrous cycle, diestrous cycle)

aneucentric translocation [ăn′·yōō′·sĕn″·trĭk] one which involves the centromere

aneuploid [ăn·yōō′·ploid] an irregular polyploid in that the nuclei do not contain a multiple of the haploid number of chromosomes

-ang- *comb. form* meaning a "hollow container"

-angio- *comb. term.* meaning "vessel" or "container"

angioblast [ăn′·jē·ō·blăst] mesenchyme cells precursor to blood vessels and cells

angiospermae [an′·jē·ō·spûrm·ē] those Spermatophytae (q.v.) in which the seeds are contained in an ovary (cf.) Gymnospermae

angium [ăn·jē″·əm] any container, particularly, in botany those concerned with reproduction. The following terms using -angium as a suffix are defined in alphabetic position:

GAMETANGIUM   MICROSPORANGIUM
GONIANGIUM   SPERMATANGIUM
GONIDANGIUM   SPORANGIUM
GYNOGAMETANGIUM   SYNANGIUM
MEGASPORANGIUM

Anguilla [an·gwil″·ə] a genus of freshwater apodid teleost fish, or eels. *A. anguilla*, the European Eel, migrates from ponds to rivers (traveling over wet grass by night) down rivers to the Atlantic, and then to the Sargossa Sea. Here they breed and die. The eggs hatch into leptocephalus larvas. These start to migrate back to Europe, a three-year trip, in the course of which they metamorphose into elvers that reascend the rivers. The American eel, *A. rostrata* [rŏs·trä′·tə, rŏs·trā′·tə] also goes to the Sargossa Sea to breed, but it only takes the larvas one year to return as elvers

Anguilliformes = Apodes

Anguillula [ăn·gwil·yūl″·ə, ăn·gwil″·yūl·ə] a genus of ascaroid nematode worms of which *A. aceti* [ăs·ē′·ti] the "Vinegar Eel" is known to every student of biology

Anguina [ăn·gwin′·ə] a genus of nematode worms parasitic on plants. *A. tritici* [trit′·ik·ē] is a major pest of wheat in every country of the world

angular bone one of a pair of chondral bones, forming the surface of the inner angle of the lower jaw between the dentary and the prearticular, in many vertebrates other than mammals (*see also* multangular bone, supraangular bone)

angular process a process on the dentary bone posterior, (or ventral) to the articular process. In the marsupials, the angular process is directed medially

animal any organism that lacks chlorophyll and that is not thought, as are fungi and some other parasitic plants, to be immediately descended from chlorophyll bearing ancestors. Most animals have locomotor responses to external stimuli.

animal pole the upper, yolk-free, pole of an egg (*see also* vegetal pole)

Animalia [ăn′·im·ăl″·ēə] that kingdom of organisms which contains the animals. There is no single characteristic distinguishing animals from plants; several are given under the entry "animal"

-aniso- *comb. form* meaning "unequal"

anisoplanogamete [ăn·is′·ō·plăn·ō·găm″·ēt] a motile gamete which differs in size from another planogamete produced by the same organism

Anisoptera [ăn·is·ŏp″·tər·ə] a sub-order of odonatan insects in which the hind wings are wider than the front wings and the wings are held horizontally when at rest. Commonly called dragonflies

anisospore [ăn′·is·ō·spôr″] a spore which is produced in two types and is presumably therefore a form of sexual reproduction, commonly contrasted with isospores produced by the same form as in the "sexual" and "asexual" reproduction of radiolarian protozoa (*see also* isospore)

anlage [ăn·läg·ə] a German word of numerous meanings used by some embryologists in the sense of "precurser" or "rudiment"

Annelida [ăn′·ĕl″·id·ə] a phylum of bilateral coelomate animals containing the earthworms, bristleworms and leeches. They are distinguished from other worms by their metameric segmentation. The classes Archiannelida, Hirudinea, Oligochaeta and Polychaeta are the subject of separate entries

annual ring a growth ring in organisms, such as trees, that have annual periodicity (*see also* false annual ring)

annular cartilage the main supporting structure of the buccal funnel of Cyclostomes

annulus [ăn″·yūl·əs] literally a "ring" and used in biology for numerous annular structures particularly a thin membrane extending from the stalk to the rim of the cap in many agaricale basidomycete fungi

Anodonta [ăn′·ō·dŏnt″·ə] a genus of pelecypod mollusks distinguished from all other freshwater clams by the absence of teeth from the hinge of the shells. There are many species living in mud

anoestrous *see* anestrous

Anopheles [ăn·ŏf′·ĕl·ēz″] a genus of culicid dipterans containing many mosquitoes acting as human disease vectors. About 20 species can carry malaria of which *A. quadrimaculatus* [kwăd′·rē·măk·yūl·ă″·təs] is the most important North American representative

Anolis [ăn·ōl′·is] a genus of lizards (Squamata) mistakenly called Chameleons, a term properly reserved for the old world genus *Chamaeleo*. *A. carolinensis* [kâr·ō·lin·ĕns·is] is a familiar object of laboratory study

anomocoelous [*angl.* ăn·ə·mŏs″·əl·əs, *orig.* ăn′·ăm·ō·sĕl·əs] having different kinds of centra in various parts of the vertebral column

Anomura [*angl.* ən·ŏm″·ə·rə, *orig.* ăn′·ō·myūr″·ə] a division of reptant decapod Crustacea with the abdomen bent under itself though less so than in the Brachyura. This division includes, *inter alia*, the hermit crabs and their allies. They lack a chela on the third pair of legs. The genus *Pagurus* is the subject of a separate entry

-anoplo- *comb. form* meaning "unarmed"

Anoplura [*angl.* ăn·ŏp″·lōōr·ə, *orig.* ăn′·ō·plyūr″·ə] the order of insects that contains the forms commonly called sucking lice. They are characterized as wingless mammalian parasites with mouth parts adapted to sucking

Anostraca [ăn·ŏst″·rə·kə] an order of branchiopod Crustacea commonly called fairy shrimps. They are distinguished by their elongate body lacking a carapace and their stalked eyes. They also swim upside down. The genera *Artemia* and *Eubranchipus* are the subject of separate entries

anovulatory cycle [ăn′·ŏv·yūl·ă″·tôr·ē] the occurrence of menstruation or estrous without ovulation

anoxybiont [*angl.* ăn·ŏks′·ē·bi·ŏnt, *orig.* ăn·ŏks′·ē·bi″·ŏnt] an organism incapable of using oxygen as distinct from one which is aerobic, frequently used of facultative anaerobes

Anseriformes [ăn′·sĕr·ē·fôrm″·ēz] that order of birds which contains the ducks, geese, and swans. They are distinguished by the flattened bill and webbed feet

l-anserine [ĕl·ăn′·sĕr·ēn] (N-methylcarnosine). A little known amino acid isolated from some birds including, as the name indicates, geese (*see also* carnosine)

ant- *comb. prefix* meaning "against", "instead", or "opposite". Takes the form "anti-" before consonants (*cf.* ante-)

ant popular name of hymenopterous insects of the suborder Formicoidea

ant lion any of several species of plannipenidan insects, the larvae of which trap ants in funnel-shaped excavations in loose soil

antagonistic symbiosis a condition in which one symbiont seeks to establish domination over the other. Occasionally used as synonymous with parasitism

antarctalian [ănt″·ärk·tāi′·ēən] the south polar equivalent of arctalian (*q.v.*)

antarctic [ănt·ärk″·tĭk] that part of the planet Earth which lies within 23° 30′ of latitude from the "south", or lower, pole (*see also* arctic)

antarctic tundra resembles the arctic tundra save that there are many fewer shrubs

ante- *comb. prefix* meaning "before"

Antedon [ănt′·ē·dŏn″] a genus of stalked crinoids found below the 1000 meter level in most seas

-antenn- *comb. form* properly mean the "yard" or "spar" of a sailing vessel though more often used in biology in the derivative sense of "antenna"

antenna [ăn·tĕn″·ə] an anterior sensory appendage. The term is not confined to the jointed appendage of arthropods but is used also for analogous, unsegmented, structures in polychaete worms and rotifers. The term is also used for antenna-like processes arising from the rostellum of some orchids.

antennal segment that segment of the arthropod head from which the antennae arise, usually considered to be segment two

antennula [ăn·tĕn″·yūl·ə] the second, of two, antennae

antennule [ăn·ten″·yūl] the smaller pair when there are two pairs of antennae in an arthropod

anterior abdominal vein a vein formed in some amphibia, reptiles and birds, from the fusion in the midline of the two epigastric veins

anterior cephalic duct a lymph duct, parallel to the jugular vein

anterior chamber that portion of the cavity between the cornea and the lens of the eye which is exterior to the iris

anterior horn one of the dorsal (anterior) arms of the H-shaped mass of gray matter seen in a transverse section of the spinal cord

anterior peduncle [pĕd′·ŭnkl] the fiber tracts connecting the cerebellum with the mid-brain

anterior tectal cartilage one of a pair of cartilages forming the anterior portion of the orbit and anterior roof of the cyclostome chondrocranium

-anth- *comb. form* meaning "flower". The *adj. terms.* -anthic and -anthous (*q.v.*) are interchangeable but compounds are in this dictionary recorded only as the latter. A distinction is sometimes, but rarely, made between -anthous, -anthemous (*q.v.*) and -antherous (*q.v.*)

anther [ăn·thûr′] that expanded portion of a stamen in which the pollen is formed but frequently used as though synonymous with stamen

-anthemous pertaining to a flower as a whole (*see* calycanthemous)

antheridium [ăn″·thĕr·ĭd″·ēəm] that structure which bears the male gametes in cryptogams

antherozooid [ăn″·þĕr·ō·zō″·ĭd] a motile male, plant gamete

Anthoceros [*angl.* ăn·þŏs″·ə·rŏs, *orig.* ăn′·thō·sĕr″·ŏs] a typical genus of Anthocerotae with the characteristics of the class

Anthocerotae [*angl.* ăn·þŏs·ə·ō·tē, *orig.* ăn′·thō·sĕr″·ō·tē] a class of bryophyte plants commonly called hornworts. They differ clearly from the mosses and liverworts in the absence of leaf-like structures and in the fact that the chloroplasts of the thallus cells have conspicuous pyrenoids. The archegonia and antheridia are developed within the thallose body but the sporophytes arise as horn-like structures from the gametophytes

anthocyanin [ăn′·thō·si″·ən·ən] a general term for blue and red glycoside pigments found in plant cells

anthogenesis [ăn·thō·jĕn″·əs·ĭs] the production of both males and females by parthenogenesis

anthotropism [ăn′·thō·trōp″·ĭzm] the movement of a flower in response to stimuli

-anthous pertaining to flowers. Frequently misused for -antherous

Anthozoa [ăn′·thō·zōə″] that class of coelenterates which contains the sea anemones, sea pens, sea fans and corals. The Anthozoa are clearly distinguished from the other classes of Coelenterata by the lack of a medusoid generation and the strongly developed stomodaeum. The orders Alcyonaria and Zoantharia are the subjects of separate entries

-anthrop- *comb. form* meaning "man"

Anthropithecus = *Pan*

Anthropodea [ăn′·prō·pōd″·ēə] a sub-order of primate mammals containing the tailless forms most commonly referred to as apes and great apes. The family Pongidae is the subject of a separate entry

anthropoid ape a term applied by some to all anthropoids but by others confined to the great apes in contrast to the lesser apes

Anthropoidea [ăn′·thrō·poid·ēə] an order of primate mammals containing the apes and monkeys. They are distinguished from the Lemuroidea and Lorisoidea by an expanded brain case and an abbreviated facial skeleton in which the eyes are set close together and face forward. The suborders Catarrhyna, Platyrrina and Anthropodea are the subjects of separate entries

-anthy *subs. form* derived from -anth (q.v.). Most words listed under -anthous (q.v.) can be turned into nouns by substituting -anthy for -anthous

-anti- *comb. form* meaning "against"

antiaposematic color [ănt′·ē·ăp′·ō·sĕm·ăt″·ik] either coloration which disguises a predator or one which is used as a threat (*see also* sematic color, aposematic color, pseudepismatic color)

antibiosis [ănt·ē′·bĭ·ōs″·ĭs] the method of existance of a microorganism which secretes a substance destroying or inhibiting other microorganisms

antibody the specific substance produced in blood serum by the injection of a specific antigen

anticryptic color [ăn·tē·krĭpt′·ĭk] a color or color pattern used for concealment by a predator (*see also* cryptic color, procryptic color)

antimere [ăn′·tē·mēr″] either of the two halves of bilaterally symmetrical object, or a homologous part repeated in segments arranged round the axis of a radially symmetrical organism

antimorph [ănt'·ē·môrf"] a mutant allele which inhibits the production of an ancestral or wild type structure

Antipatharia [ănt'·ē·păth·är"·ēə] a suborder of Zoantharia distinguished by a black horn-like central axis. Usually mistaken for Alcyonarians and popularly called black corals

antipathetic said of two organisms, the parts of which do not graft or transplant easily to each other, but which are not specifically antagonistic to each other

antiserum [ănt'·ē·sēr"·əm] a serum containing antibodies

antithetic generation [ănt·ē·thĕt'·ĭk] said of those alternations of generations in which the alternates are fundamentally different in appearance and origin

antitropism said of a plant that twines against the direction of the sun

antrorse [ăn·trôrs"] directed frontwards or upwards (see also detrorse)

antrum [ăn"·trəm] either a cavity in a bone or that portion of the gonoduct which, when not otherwise distinguished, lies immediately adjacent to the gonopore

Anura [ə·nyōōr"·ə] a term once used for tailless Amphibia (frogs and toads) now replaced by Salientia. However the adjective anuran [ə·nyōōr"·ən] is widely current

anus [ā"·nəs] the orifice which terminates the alimentary canal (see also suffix "-proct")

aorta [ā·ôrt"·ə] any large arterial vessel communicating directly with the heart (see also dorsal aorta, ventral aorta)

aortic arch [ā·ôrt"·ĭk] one of up to six pairs of blood vessels which rise from the dorsal aorta along the sides of the pharynx through the gill arches

aortic root one of a pair of dorsal blood vessels into which the aortic arches open

apatetic color [ăp'·ă·tĕt'·ĭk] those colors which enable an organism to mimic either its environment or another organism or, by extension, any color thought to have a protective function

aphid [ā"·fĭd] anglicized form of the homopteran family Aphidae. The generic term Aphis [ā"·fis] is frequently misused as a substitute. A few psyllids are popularly called "aphids"

aphotic zone that zone of water which lies at such a depth that it receives no significant amount of light

Aphrodite [ăf'·rō·dit"·ē, ăf'·rə·dit·ē] a genus of Polychaetae Errantia, mostly very large and commonly called "Sea Mouse". A. hastata, [hăs·tät·ə] of the Atlantic coast is typical, being about five inches by two inches with the back covered with iridescent felted setae

Aphyllophorales [ā·fĭl'·ō·fôr·āl"·ēz] an order of basidiomycete fungi of varied appearance and characteristics. It includes the polypores like Fomes as well as the chanterelles. Sometimes fused with the Agaricales but differing from them in lacking the typical pileum ("cap") and stipe ("stem") of the mushrooms and toadstools. The genera Clavaria, Fomes, and Polyporus are the subject of separate entries

aphytal zone [ā·fit"·əl] those waters in which the penetration of light is too little to support photosynthesis

apical literally pertaining to the apex. It is often, particularly in invertebrate anatomy and embryology, used to indicate the uppermost point of a spherical or conical organism

apical cell a single cell from which, in some plants, all derivative meristem is clearly produced and which is therefore responsible for continued growth

apical initial one of a group of cells no one of which can be specifically recognized as a specific apical cell (see also subapical initial)

apical meristem the meristem at a growing tip, either shoot or root

apices plural of apex

Apis [āp"·is] the type genus of the apoid hymenoptera. A mellifica [mĕl·if'·ik·ə] is the Honey Bee and is the only species found outside southeast Asia

Aplacophora an order of amphineuran mollusca distinguished by the absence of a shell and by their worm-like shape. They are marine, often deep-water, forms

aplanogamete [ā·plăn"·găm·ēt'] a non-motile gamete

aplanospore [ā·plăn'·ō·spôr] a non-motile spore

-apo- comb. form meaning "away from" or "separate"

apocrine gland [ăp·ō·krin"] a gland in which part of the cytoplasm of a cell, as well as the secretion, is lost

apocrine sweat gland any sweat gland producing a secretion other than sweat. These glands are produced from, or associated with, hair follicles (see also sweat gland, eccrine sweat gland)

Apocrita [ăp·ō·krī"·tə] a suborder of hymenopteran insects in which the basal segment of the abdomen is fused with the thorax and separated from the rest of the thorax by a constricted area often referred to as a wasp-waist. This group contains all of the hymenoptera except for sawflies and horntailes (cf. Symphyta). The family Braconidae is the subject of a separate entry

Apocynaceae [ăp'·ō·sin·ā"·sē] the dogbane family of plants, mostly twining shrubs. The genus Rauvolfia is the subject of a separate entry

Apoda 1 [ā·pŏd"·ə] (see also Apoda 2) an order of holothurioid echinodermata distinguished from all others by lacking both ambulacral feet and respiratory trees. The genus Leptosynapta is the subject of a separate entry

Apoda 2 (see also Apoda 1) an aberrant order of worm-like amphibia distinguished by the absence of limbs.

apodeme [ăp"·ō·dēm'] those hollow inwardly projecting portions of an arthropod exoskeleton which form an internal framework for the support of muscles (see also apophysis)

Apodes [ā·pŏd"·ēz] an order of osteichthyes containing the eels. They are distinguished by the absence of pelvic or ventral fins and by the fact that the air bladder has an open duct to the throat. The genus Anguilla is the subject of a separate entry

Apodiformes [ā·pŏd'·ē·fôrm"·ēz] that order of birds which contains the swifts and hummingbirds. They are distinguished, amongst other things, by their very small feet which makes it impossible for them to take flight from flat surfaces

apodome [ăp"·ō·dōm'] the internal portions of an arthropod skeleton and therefore comprising both apodeme (q.v.) and apophysis (q.v.). Used by many as synonymous with apodeme apoenzyme [ăp'·ō·ĕn·zīm"] an enzyme that cannot function without a coenzyme (see also holoenzyme)

apogamete [ăp'·ō·găm"·ĕt] one which is formed by apomixis (q.v.)

apomeiosis [ăp"·ō·mī'·ōs·ĭs] the production of a reproductive cell (e.g. a spore) without reduction of the chromosome number

apomict [ap"·ō·mĭkt] an organism produced by apomixis

apomixis [ăp'·ō·mĭks"·ĭs] the replacement of sexual reproduction by any form of propagation which, by avoiding meiosis and syngamy, permits the reproduction unchanged of a given genotype

apophysis [ə·pŏf"·ə·sĭs] literally "an outgrowth away from" specifically an outgrowth of a bone from a vertebra, outgrowths or ingrowths from the exoskeleton of an insect or other arthropod, a basal swelling beneath the capsule of some mosses and the basal swelling on the cone scale of some pines (see also anterior apophysis, gonapophysis, parapophysis, zygapothesis)

apopyle [ăp"·ō·pīl'] the opening of the flagellated chamber of the sponge into an excurrent canal

aposematic color [ăp'·ō·sĕm·ăt"·ĭk] colors supposed to repel predators (see also sematic color, antiaposematic color, pseudoepisematic color)

aposeme [ap"·ō·sĕm'] a population, all the individuals in which, though taxonomically distinct, share the same aposematic coloration

appendage a subordinate part which is appended to a major part. Most frequently used in biology for the articulated structures (e.g. limbs) of arthropods and vertebrates. The undernoted terms are defined in alphabetic order:

BUCCAL APPENDAGE   INTERCALARY APPENDAGE

Appendicular muscle a muscle moving a limb

Appendicularia = Larvacea

appendix anything appended to a regular structure as the anal cerci of insects. Without qualification usually refers to the vermiform appendix (see also hydatid appendix)

apposition [ap'·ə·zĭsh"·ən] the growth of a cell wall, or other structure, by the successive deposition of layers on its outside (cf. intersusception)

-apsid- comb. form meaning arch or vault. The following terms with this suffix are defined in alphabetic position:

ANAPSID   EURYAPSID
DIAPSID   PARAPSID

apterium [ăp·tēr"·ēəm] the space of bare skin between the tracts of contour feathers in birds

Apus [ā·pəs] the only genus of notostracan Crustacea to occur in N. America. They are readily recognized by the large oval carapace in front of a many segmented body terminating in a pair of segmented anal cerci

apyrase [āp'·ir·āz] a plant enzyme which, when activated by Ca, catalyzes the hydrolysis of ATP to ADP (cf. ATPase)

aqueous humor the fluid filling the anterior and posterior chambers of the eye

Arachnida [ə·răk"·nĭd·ə] a large class of the phylum Arthropoda containing the forms commonly known as scorpions, spiders, mites, ticks, harvestmen, and some other lesser known groups. They are distinguished by the division of the body into a cephalothorax and abdomen, the former bearing six pairs of appendages the first of which are chelicerae and the second palpae. The remaining four pairs of appendages are walking legs. The orders Acarina, Araneae, Palpigrada, Pedipalpi, Phalangida, Pseudoscorpionida, and Scorpionida are the subject of separate entries. The Tardigrada, Linguatulida, and Pynogonida are thought by some also to belong in the Arachnida

Arachnodiscus [ər·ăk'·nō·dĭsk"·əs] a genus of disk-shaped marine planktonic Bacillariophyceae ("diatoms"). Many are relatively large (0.2 to 0.5 mm)

arachnoid [ər·ăk·noid"] properly resembling a member of the Arachnida but used in anatomy in the sense of "spider web-like"

arachnoid membrane an inner subdivision of the pia mater

Araneida [ər·ăn·ā'·ĭd·ə] that order of arachnid arthropods which contains the spiders. In general they are distinguished from other arachnids by the possession of non-chelate chalicerae, bearing at their tips the opening of poison glands, and by the production of several kinds of silk from spinnerets on the rear of the abdomen. The genus Latrodectus is the subject of a separate entry

Aranzio's duct = ductus venosus

Arbacia [är·bā'·sē·ə] a genus of centrachinoid echinoderms. A. punctulata [pŭnk·tyōō·lā'·tə] is the common reddish brown sub spherical sea urchin of the Atlantic coasts

-arbor- comb. form meaning "tree"

Arcella [är·sĕl'·ə] a genus of testaceous sarcodine protozoans with a test shaped like a mushroom with an aperture where the stalk would be

arch 1 (see also -arch- 2) a curved structure, or a pair of curved structures, which maintains an aperture. In anatomy extended to structures with the shape, but not the function, of an arch. The undernoted are the subject of separate entries:

AORTIC ARCH   MANDIBULAR ARCH
BRANCHIAL ARCH   NEURAL ARCH
EXTRAHYOID ARCH   OCCIPITAL ARCH
GILL ARCH   SUBOCULAR ARCH
HEMAL ARCH   VISCERAL ARCH
HYOID ARCH   ZYGOMATIC ARCH

-arch- 2 (see also arch 1) comb. form meaning "primitive" or "beginning" but by extension "origin", particularly as applies to seres (q.v.)

archaeocytes [är'·kē·ō·sĭtz] large, wandering amoebocytes in the wall of sponges. They are thought by some to give rise to sex cells, or to control regeneration

Archaeopteryx [ärk'·ē·ŏp"·tĕr·ĭks] a genus of fossil birds from the upper Jurassic known from two specimens of A. lithographica [lĭth·ō·grăf'·ikə]. It differed from all modern birds in possessing a long feathered tail. Archaeornis [ärk'·ē·ôrn"·ĭs], from the same period, is now thought to be identical

Archaeornis see Archaeopteryx

Archaeornithes [ärk'·ē·ôr·ni"·thēz] a doubtfully valid taxon erected to containing Archaeopteryx in contrast to all other birds which are placed in the Neornithes

archegonium [ärk·ē·gōn"·ēəm] that structure which bears the female gamete in lower plants

archenteric pouch [angl. ärk·ĕnt"·ər·ĭk, orig. ärk·ĕn·tēr"·ĭk] one of the paired, segmented, dorsoventral protuberances of the archenteron from which the mesoderm derives

archenteron [ärk·ĕnt′·ər·ŏn] the cavity, precursor to the gut, formed by the invagination of the blastula and thus the cavity of the gastrula

Archeozoic era [ärk·ē·ō·zō′·ik] a geologic era extending from about 2 billion years ago to 1 billion years ago. It was followed by the Proterozoic and preceded by the Azoic. It is presumed that organized life evolved in this epoch

-archi- = -arch-

Archiannelida [ärk′·ē·ǎn·ĕl″·id·ə] a small class of annelid worms having many of the characteristics of polychaete larvae. The genus *Polygordius* is the subject of a separate entry

archibenthic [ärk′·ē·bĕn″·thĭk] pertaining to the zone of the ocean between approximately fifty fathoms and five hundred fathoms

archicentrous [ärk′·ē·sĕnt″·rəs] said of those vertebrates in which the base of the neural arch contributes to the substance of the centrum

archigastrula [ärk′·ē·gǎs″·trōō·lə] that type of gastrula in which the endoderm is produced by invagination (emboly). This is usually what is meant by the term gastrula when used without a qualifying adjective

archipallium = olefactory cortex

Architeuthis [ärk′·ē·tyū·thĭs] a genus of cepalopod Mollusca containing the "giant squids". These abyssal forms are the largest invertebrates extant, some reaching a total length of more than 50 feet.

architomy [*angl.* ärk·it′·əm·ē, *orig.* ärk′·ē·tŏm″·ē] the breaking of an annelid into two individuals at a position where a new head structure has been reorganized (*cf.* paratomy)

-archo- *comb. form* meaning "chief" or "principal"

Archosauria [ärk′·ō·sôr″·ēə] a subclass of reptiles distinguished by the fact that the postorbitosquamosal arch separates the temporal openings. Only the Crocodilia are extant. All the huge extinct dinosaurs belong in this subclass.

arctalian [ärk·tāl′·ē·ən] pertaining to Arctic waters the southern limits of which are defined by the presence of floating ice (*see also* antarctalian)

arctic that area of the planet Earth which lies within 23° 30′ of latitude of the N. pole (*see also* antarctic, holarctic, nearctic, palaearctic)

arctic tundra that area of northern tundra which is north of the tree zone but on which there still occur a few stunted shrubs

arctogaea [ärk′·tō·jēə″] one of the primary zoogeographic zones comprising most of the northern hemisphere and those parts of Asia that lie to the north and west of Wallace's Line (*q.v.*)

arcualia [är·kwäl′·əa, ärk′·yū·äl″·ēə] cartilaginous precursors of the neural arch of the vertebra which persist as simple rods in Cyclostomes

area pellucida [âr″·ēyə pə·lyōō′·sid·ə] the clear embryonic area that overlies the blastocoel in the early development of a telolecithal egg

area vasculosa [vǎs·kyūl·ōs″·ə] the extraembryonic blood- and blood vessel-forming area (ring) in the telolecithal blastoderm

aren- *comb. form* meaning "sand" (*cf.* -psam-)

Arenicola [ăr·ĕn·ik′·ōl·ə] a genus of marine, mud-dwelling polychaete annelids with external gills and in which the parapodia are greatly reduced. They are also unusual among annelids in possessing otocysts

areola 1 [är·ē·əl·ə] literally, a small area, and variously used in biology to refer to limited areas of surfaces, small cells, or even tesselated patterns. Specifically the area of the iris that immediately borders on the pupil of the eye, the interstices between the veins of a leaf, the naked tracts between scales of the feet of birds, the bare area surrounding the boss of an echinoderm tubercle and a large pseudoporein Ectoprocta

areola gland the area immediately surrounding the nipple of a mammary gland (*see also* Montgomery's gland)

areolar tissue the loose connective tissue between the basement membrane of an epithelium and the underlying tissues

argentaffine cell [är·jĕnt″·ə·fĭn] any cell, particularly one in the intestinal glands, which stains readily by silver techniques

-argill- *comb. form* meaning "clay"

arginase [är′·jĭn·āz″] an enzyme catalysing the hydrolysis of arginine to ornithine and urea

l-arginine [ĕl·är′·jĭn·ēn] 1-Amino-4-guanidovaleric acid. $N_2CH_3NHCH_2CH_2CH_2CH(NH_2)COOH$. An amino acid known to be essential in rat nutrion

arginine cycle *see* urea cycle

-argo- *comb. form* meaning "passive"

Argonauta [är′·gŏn·ôrt″·ə] a genus of cephalopod Mollusca in which the mantle is not united to the head and in which there is a spiral shell with septa. The shell-like egg case is often called the "paper nautilus" shell, since it is similar in shape to that of *Nautilus*

Arguloidea [är′·gyū·loid″·ēə] a group of parasitic Crustacea lacking egg-sacs and with a strongly depressed body bearing sucking discs in front. They are variously regarded as a separate order of Crustacea or a suborder of Copepoda

aril [âr″·il] an outgrowth of the funiculus which envelops the integument of plant ovule. It may be fleshy, as in the "fruit" of Taxus, or fibrous as the aril that coats the peach stone

Arion [âr″·ēən] a genus of pulmonate gastropods. *A. circumspectus* is the commonest garden slug of northeastern America. Many species of *Arion* are predatory on earthworms

Armadillidium [är′·mǎ·dil″·id·ēəm] a genus of terrestrial isopod Crustacea, commonly called "Pill Bugs". *A. vulgare* [vŭl·gär′·ā], the well known garden pest, is of world-wide distribution

-arrhe- *comb. form* meaning "male"

arrhenogeny [*angl.* ər·ĕn·ŏj″·ən·ē, *orig.* ăr′·ĕn·ō·jĕn″·ē] the condition of producing exclusively male offspring

arrhenotoky [*angl.* ə·rĕn·ŏt″·ə·kē, *orig.* ăr′·ĕn′·ō·tō″·kē] the parthenogenetic production of male offspring

-arsen- *comb. form* meaning "male" or "masculine" (*see* monarseny)

Artemia [är·tĕm″·ēə] a genus of anostracan Crustacea containing *A. salina* [sə·lēn·ə], the "brine shrimp". This form can live in sea water evaporated down to the point of crystallization. Many primitive peoples rely on it to remove the insoluble materials in the course of the preparation of edible salt from sea water

arterial trunk the main arterial vessel leaving the ventricle

arteriole [är·tēr′·ē·ōl, ärt·ə·rē·ōl′] diminutive of artery

artery   a vessel which conducts blood away from the heart. The names of specific arteries are given in alphabetic position:

-arthro-   *comb. form* meaning "joint"

Arthropoda   [är·thrŏp″·ə·də, är·þrō·pō″·də]   an enormous phylum of the animal kingdom distinguished, in all classes except the Onycophora, by a chitinous segmented exoskeleton and jointed appendages. The body cavity is a haemocoel. The following classes are the subject of separate entries: Arachnida, Chilopoda, Crustacea, Diplopoda, Eurypterida, Insecta, Onycophora, Pauropoda, Pycnogonida, Symphyla, Trilobita and Xiphisura

arthrosis   [är·thrō′·sis]   a joint. The following types of arthrosis are listed in alphabetic position:

AMPHIARTHROSIS   SYNARTHROSIS

DIARTHROSIS

arti-   *comb. form* meaning "complete"

articular   [är·tik″·yūl·ə]   pertaining to a joint

articular bone   one of a pair of membrane bones lying at the inner posterior angle of the lower jaw in vertebrates other than mammals (*see also* retroarticular bone, prearticular bone)

articular cartilage   the hyaline cartilage of a joint

articular membrane   the flexible membrane between the segments of arthropods and the joints of arthropod appendages

articular process   a process from the dentary bone, posterior to the coronoid process (where there is one) which articulates the dentary with the zygomatic arch

artifact   [är′·tə·făkt′]   an appearance, or structure, produced by treatment of material in the course of preparation and that is not present in the original material before the manipulation

artificial hibernation   a condition showing physical and physiological resemblances to hibernation induced in a form that does not normally hibernate

artificial parthenogenesis   the laboratory-induced development of an egg using a stimulus other than a spermatozoa

-artio-   *comb. form* meaning "even"

Artiodactyla   [*angl.* ärt′·ē·ō·dăkt″·ələ, *orig.* ärt·ē·ō·dăk·til·ə]   an order of placental mammals once fused with the Perissodactyla into the order Ungulata. The Artiodactyla are commonly referred to as the cloven-hoofed animals and therefore contain the cattle, sheep, antelopes, deer, and similar forms. They are distinguished by having an even number of digits with the axis of symmetry passing between these digits. The genera *Camelus, Lama* and *Odocoileus* are the subject of separate entries

artioploid   [*angl.* ärt′·ē·ŏp″·loid, *orig.* ärt′·ē·ō·ploid″]   having chromosomes multiplied by even numbers

arytenoid   [ə·rit′·ən·oid]   in the shape of a pitcher

arytenoid cartilage   the posterior of the three laryngeal cartilages

-asc-   *see* -asco-

Ascaphus   [*angl.* əsk″·ə·fəs, *orig.* ăs·kăf″·əs]   a very primitive frog (Salientia) misleadingly known as the tailed frog. The "tail" is a permanently everted cloaca

Ascaris   [ăsk″·ər·is]   a genus of ascaroid nematodes. *A. lumbricoides* [lŭm·brik·oid′·ēz] is the largest nematode parasite of man, sometimes attaining a length of 14 inches; infection is by the ingestion of egg-containing fecal material. The pig *Ascaris*, often used in class teaching, is a variety *A. lumbricoides* var. *suum*

[syōō·əm′].   "*A. megalocephala*"   [mĕg·əl·ō·sĕf″·ələ], the horse ascaris, is properly *Parascaris equoreum*

Ascaroidea   [ăsk′·ə·roid″·ēə]   a suborder of Nematoda distinguished by the three prominent lips and the spirally curled hinder end of the male. The genera *Anguillula, Ascaris* and *Parascaris* are the subject of separate entries

ascending colon   that portion of the colon which curls forward on the right side of the body in some mammals. The term "ascending" derives from human anatomy

ascending tract   the dorsal funiculus of the spinal cord which carries impulses towards the brain

Aschelminthes   [ăsk′·hĕl·min″·thēz]   a taxon of the animal kingdom which may either be regarded as a superphylum containing the phyla Rotifera, Gastrotricha, Kinorhyncha, Priapulida, Nematoda, Nematomorpha or as a phylum containing these taxa as classes

-ascia-   *comb. form* meaning "hatchet"

-ascid-   *see* -asc-

Ascidiacea   [ăs·id′·ē·ās″·ēə]   a class of tunicate chordata containing the sea squirts. They are mostly sedentary forms with a recurved gut and a pharynx modified|to form a ciliary-mucoid filter-feeding apparatus The product of sexual reproduction is a tadpole larva. The genera *Molgula* and *Ciona* are the subject of separate entries

-asco-   *comb. form* literally meaning "wine skin" but extended in biology to many bladder-like structures. Frequently abbreviated to -asc-

ascocarp   [ăs·kō′·kärp″]   the fruiting body of ascomycete fungi. It consists of an outer layer of interwoven hyphae and an inner layer of parenchymatous fibers in which the asci develop

ascogonium   [ăs·kō·gō′·nəm]   that structure which gives rise to the sexual reproductive organs in ascomycete fungi

Ascomycetes   [ăs′·kō·mī·cē″·tez]   a phylum of fungi distinguished by the presence of an ascus. The group is divided into the Hemiascomycetidae (yeasts, and similar forms) with single asci and Euascomycetidae with a fruiting body (the ascocarp) bearing many asci. The following orders of Ascomycetes are the subject of separate entries Mucorales, Endomycetales, Taprinales, Plectascales, Sphaeriales, Clavicipitales, Helotiales, Pezizales and Tuberales

asconoid   [ăs′·kŏn·oid″]   a type of sponge structure consisting of a simple vase-like shape lined with choanocytes (*cf.* leuconoid, syconoid, sylleibid)

ascospore   [ăs″·kō·spôr]   one of usually eight spores produced in the ascus of an ascomycete fungus

ascus 1   [ăs″·kəs]   the spore sac of ascomycete fungi, typically containing eight ascospores within a tubular, or oval, case produced either from the ascogonium or from specialized hyphae

Aselloidea   [ăs′·ĕl·oid″·ēə]   a suborder of free living isopod crustacea having the uropods terminal and the first pair of pleopods modified to form a thin opercular plate which covers the other pleopods. They are further distinguished from the Oniscoidea by their aquatic habit. The very similar Bopyroidea are parasitic. The genus *Asellus* is the subject of a separate entry

Asellus   [əs·ĕl′·əs]   a typical genus of asselloid Isopoda, with both marine and fresh water species. *A.*

communis [kŏm′·yūn·ĭs] is the commonest U.S. fresh water isopod

asexual reproduction   the⟋replication of an individual of and by itself

-asko-   *comb. form* meaning ''wine-skin'', usually transliterated -asco-

asparaginase   [ăs·păr′·ə·jĭn·āz″]   an enzyme catalyzing the hydrolysis of asparagine to aspartate and ammonia

l-asparagine   [ĕl·ăs·păr·ə·jĭn, ăs·păr·ə·jĭn]   α-aminosuccinamic acid. A common amino acid not essential in rat nutrion

Aspergillus   [ăs′·pĕr·jĭl″·əs]   a genus of plectascale ascomycete fungi. *A. niger* [nĭ′·jər], a common black mold, has been used in the production of both citric and tartaric acids. *A. fumigatus* [fyōō·mĭg·ät″·əs, fyōō·mĭg·āt″·əs] can cause tubercular cysts in the lungs of man and of birds

-aspid-   *comb. form* meaning ''shield''

aspidospondylous   vertebra   [ăs′·pĭd·ō·spŏn″·dəl·əs]   one in which the centra are derived from the ossification of cartilaginous arches

assemblage   any group of organisms taken together without, unless further defined, any other connotation

assembly   the smallest community recognized in ecology

assimilation   the basic power of living matter to change other things·into its own substance (*see also* genetic assimilation)

association 1   so many ecologists have applied so many meanings to this word that it is now valueless as a specific term; in general, it usually indicates a large assemblage of organisms in a specific area with one or two dominant species (*see also* faciation, lociation, sociation, associes, consocies and, in another sense, heterogenetic association). The following derivative terms are defined in alphabetic position:

CHIEF ASSOCIATION
CLOSED ASSOCIATION
COMPLEMENTARY ASSOCIATION
HOMOTYPICAL ASSOCIATION
INTERMEDIATE ASSOCIATION
MIXED ASSOCIATION
MÜLLERIAN ASSOCIATION
OPEN ASSOCIATION
PASSAGE ASSOCIATION
PROGRESSIVE ASSOCIATION
PURE ASSOCIATION
RETROGRESSIVE ASSOCIATION
STABLE ASSOCIATION
SUBASSOCIATION
SUBORDINATE ASSOCIATION
SUBSTITUTE ASSOCIATION
TRANSITIONAL ASSOCIATION
UNSTABLE ASSOCIATION

associes   [ə″·sō·shēz]   a term only slightly more specific than association. It refers in general to a transitory or intermediate stage in the development of an association taken as a whole. It has also been defined as a developmental unit of a consocies

assortive mating   the system under which mates are selected on the basis of a character thought to be desirable and which therefore appears with increasing frequency in the population

assortment   the separation of genes at meiosis (*see also* independent assortment)

astaxanthin   [*angl.* əs·tăks·ăn″·þĭn, *orig.* ăs·tə·zăn″·þĭn]   green or blue oxygenated carotenoid biochrome commonly found in marine crustacea. Boiling liberates the red carotenoids

-aster-   *comb. form* meaning ''star'' (*see also* -astr-)

aster   [ăs″·tə″]   mitotic figures that contain star-like structures

Asterias   [ăs·tēr′·ē·əs]   a cosmopolitan genus of asteroid echinoderms, so common that they are what most people mean when they say ''sea star'' or ''star fish''. They are a devastating predator on oyster beds

Asteroidea   [ăs′·tĕr·oid″·ēə]   that class of echinodermata containing the sea stars and starfish. They are distinguished by their flattened star-like shape with five (rarely more) radiating arms. The genera *Asterias, Goneaster* and *Solaster* are the subject of separate entries (*see also* Phanerozoa)

Asteroxylon ⸝ [*angl.* əs·tĕr″·ŏks″·ə·len, *orig.* ăs′·tĕr·ō·zi″·lŏn]   a palaeozoic plant probably belonging in the Lycopsida but also having affinities with the Zosterophyllophytina

-asthes-   *comb. form* meaning ''perception'' or ''sense''. Properly, but very rarely, spelled -aesthes-

-astr-   *comb. form* meaning ''star'' (*see also* -aster-)

astragaloid   [əs·trăg′·əl·oid]   dice-shaped, for the reason that the original dice were knuckle bones

astragalus bone   the tarsal bone that lies at the base of the tibia

astral ray   one of the streaks seen in the cytoplasm that radiate from each centriole during cell division and are thus the basis of the spindle

astrocyte   [ăs′·trō·sit″]   *either* a star-shaped neuroglia cell *or* any star-shaped cell particularly, those in stroma tissues

astroglia   [ăs′·trō·glē″·ə]   macroglial cells with large nuclei and many ramifying fibers (=astrocyte)

-atacto-   *comb. form* indicating irregularity

Ateles   [*angl.* ət·ēl′·ēz, *orig.* ā·tēl·ēz]   a genus of platyrrhine primates containing the long-legged, long-tailed spider monkey

atlas   the vertebra next to the skull, articulating with the occipital condyles (*see also* proatlas)

atoke   [ā″·tōk″]   the anterior sexless portion of the body of some polychaete worms

ATP = adenosine triphosphate

ATPase   an enzyme, otherwise called myosin, which, when activated by Ca, acts on ITP, CTP, GTP, and UTP and ATP to yield the corresponding diphosphate. Another form activated by Mg acts only on ITP and ADP

atretic follicle   [ə·trĕt·ĭk]   a primordial follicle of an ovary which degenerates *in situ*

atrial cavity   [āt″·rē·əl]   the cavity that, in prochordates, lies between the pharynx and the body wall

atrial groove   an ill-defined term often used for the epibranchial groove of *Branchiostoma* in contrast to the endostyle

atriopore   [āt″·rē·ō·pôr′]   the opening of the atrium particularly in *Branchiostoma*

Atrium   [āt″·rē·əm]   properly ''chamber''. In anatomy refers to the cavity of the auricle of the heart or to the tympanic cavity of the ear. In zoology refers to any of numerous cavities, particularly the cavity between branchial pouch and the body wall of prochordates

atrocha larva   [ā′·trōk″·ə]   a type of trochophore larva uniformly ciliated and without a preoral band

Atropa   [ət′·rō·pə]   a genus of solanaceous flowering

plants. *A. belladonna* [běl'·ə·dŏn"·ə] (deadly nightshade) yields the alkaloid atropine

atrophic [ā·trŏf"·ĭk] said of materials that cannot be used as nutrients

atropine [ăt"·rō·pēn] an alkaloid widely used in medicine occurring in many plants of the family Solanaceae. *Atropa belladonna* is the principal commercial source

attenuation used specifically of the gradual reduction in virulence of a microorganism

auditory nerve Cranial VIII. Running from the inner ear to the upper surface of the anterior region of the medulla

auditory vesicle the embryonic vesicle that gives rise to the inner ear of the adult

Auerbach's plexus [ou·ər·bakh'] an autonomic plexus in the muscle layers of the intestine

-aur-, -auro- *comb. form* confused from three roots and therefore meaning "ear", "gold" or "air"

Aurelia [ô·rēl'·eə] a genus of scyphozoan coelenterates. *A. aurita* [ôr·ət·ə] is probably the commonest jellyfish in the world and is almost universally used in class teaching. It is peculiar, compared to most scypozoans, in having such very short marginal tentacles

auricle [ôr"·ĭk·əl] almost any ear-shaped structure but most frequently the derivative of the atrium of the heart. Also swollen lateral appendages or parts of appendages in many insects. One of a pair of prominent lateral ciliated projections posterior to the corona on some rotifers. An internal projection from the perignathic girdle of some echinoderms. Lateral sensory projections from the head of some turbellaria, and a spirally coiled process arising, in some ctenophora, from the comb row

auricularia larva [ôr·ĭk'·yūl·âr"·ēə] the larva of a holothurian echinoderm

auroral [ôr·ôr"·əl, ər·ôr·əl] pertaining to the dawn (*cf.* crepuscular)

australopithecine [ô·strāl"·ō·pith"·ə·sĭn] a term coined to cover many races of *H. erectus* (*see Homo*) found in East Africa. Many (*Australopithecus, Zinjanthropus, etc.*) were at one time accorded separate generic rank

autecology [ôt"·ē·kŏl"·əjē] the relation of an individual organism, as distinct from the association of organisms to their habitat

autocentrous [*angl.* ôt'·os"·ən·trəs, *orig.* ôt'·ō·sĕn"·trəs] said of those vertebrates in which the neural arches chondrify separately from the centrum

autochthonous 1 [ôt'·ok"·thĕn·əs] (*see also* autochthonous 2) pertaining to organic materials produced within an aquatic habitat

autochthonous 2 (*see also* autochthonous 1) aboriginal

autoecious parasite [ôt'·ēs"·ēəs] one which passes its entire life on an individual host

autogamy 1 [*angl.* ôt'·ŏg"·əm·ē, *orig.* ôt'·ō·găm"·ē] the fusion of two reproductive nuclei within a single cell derived from a single parent. Particularly in ciliate Protozoa in some of which two of the eight or more divisions of the micronucleus fuse to produce a new macronucleus

autogenic succession [*angl.* ôt·ŏj"·ən·ĭk, *orig.* ôt'·ō·jēn"·ĭk] one resulting from biotic, as distinct from climatic, causes

autogenous = endogenous

autohemaglutinin [ôt'·ō·hēm"·ə·glōō·tən·ĭn] an antibody causing clumping of red cells in the organism which produces it

autologous graft [ôt'·ŏl"·ə·gəs] one of which donor and recipient are the same individual

autolysis [*angl.* ôt·ŏl"·əs·ĭs, *orig.* ôt·ō'·lī"·sĭs] the degradation after death of cell contents by the contained enzymes

automictic parthenogenesis [ôt·ō·mĭk·tĭk] the condition when a haploid gamete divides without reduction thus giving a diploid offspring

automixis [ôt·ō·mĭks·ĭs] = automictic parthenogenesis

autonomic ganglion [*angl.* ôt'·ŏn"·əm·ĭk, *orig.* ôt'·ō·nŏm"·ĭk] one which sends the axons of its cells to cardiac and smooth muscles and to glandular epithelium. They lie in a paired segmental chain along the lumbar spinal cord

autonomic nervous system that part of the vertebrate nervous system that controls involuntary muscles, and is therefore the sum of the sympathetic and parasympathetic nervous systems

autopelagic plankton that which lives continuously on the surface without seasonal or diurnal migrations

autophyte [ôt'·ō·fīt"] a plant that does not require decomposed organic matter for its nourishment and is therefore capable of synthesising its food from inorganic salts (*cf.* saprophyte)

autoploid [*angl.* ôt·ŏp"·loid, *orig.* ôt'·ō·ploid] = autopolyploid

autopolyploid [ôt'·ō·pŏl"·ē·ploid] a polyploid in which all the diploid sets have come from the same parent species

autosegregation [ôt'·ō·sĕg·rəg·ā"·shŭn] gene rearrangement in diplospores resulting from the failure of meiosis through the formation of a restitution nucleus

autosomal linkage [ôt'·ō·sōm"·əl] the linkage of alleles on the same autosome

autosome [ôt"·ō·sōm] any chromosome except a sex chromosome

autostyly [*angl.* ôt·əst"·əl·ē, *orig.* ôt'·ō·stil"·ē] the condition of having the lower jaw attached directly to the skull

autotomize [*angl.* ô·tŏt·əm·iz, *orig.* ôt'·ō·tōm"·iz] to shed a part intentionally as many crustacea shed limbs or some lizards shed tails

autotriploid [*angl.* ô·tŏt'·trə·ploid, *orig.* ôt'·ō·trĭp"·loid] a triploid (*q.v.*) in which the three diploid sets are identical (*see also* allotriploid)

autotrophic [ôt'·ō·trôf"·ĭk] said of an organism capable of utilizing inorganic materials in the synthesis of living matter *or*, more rarely, of an organism capable of securing its own food as opposed to a parasite (*see also* photosynthetic autotroph)

autozooid [ôt'·ō·zō"·ĭd, ôt'·ō·zōō·ĭd] the feeding member of a polymorphic hydrozoan colony

-auxes- *comb. form* meaning "growth"

auxesis [ôks'·ēs"·sĭs] literally, growth, but used specifically for an increase in cell size without cell division or the transformation of a small cell into a large one; it has also been used as an expression for the maximum size to which a cell may grow before cell division is inevitable. Sometimes used as synonymous with "swelling" (on a plant stem) or "dilation" (of a diatom trustule)

auxiliary cell   the algal cell that unites with the conjugation tube

auxins   [ôks"·ĭnz]   a group of plant hormones principally known for their function of inducing roots in cuttings but also responsible for the dominance of vertical over lateral growth and the formation of wound tissue. Indole acetic acid and indole butyric acids are well known examples (*see also* cytokinins, gibberellins)

-auxo-   *comb. form* meaning "increase"

auxotrophic   [*angl.* ôks·ŏt"·trŏf·ĭk, *orig.* ôks'·ō·trŏf·ĭk]   said of microorganisms requiring certain specific nutrients (*cf.* prototrophic)

-av-   *comb. form* meaning "bird"

Avena   [ə·vēn"·ə]   a large genus of graminaceous plants. *A. sativa* [săt·ēv'·ə], the Oat, is possibly derived from *A. fatua* [fât'·yū·ə]. The coleoptile is widely used in experiments on auxins

Aves   [ā"·vēz]   the class of gnathostomatous craniate chordates that contains the birds. They are distinguished by the presence of feathers on the skin and the habit of laying eggs with calcified shells; most have functional wings

avicularium   [ə·vĭk'·yōōl·âr·ēəm]   a modified zooid in the shape of a bird beak found in ectoproct bryozoans

awn 1   [ôn]   the bristle which terminates the bract of some grasses, of which barley is a typical example

-ax-   *comb. form* meaning "axle" or "axis"

axenic   [ăks·ēn'·ĭk, əks·ĕn·ĭk]   said of organisms isolated in pure culture from their normal environment or animals sterile both internally and externally (*see also* gnotobiotic, monaxenic)

axial skeleton   that part of the vertebrate skeleton which runs down the dorso-median line and therefore comprises the skull and vertebral column

-axilla-   *comb. form* meaning "arm-pit"

axis 1   (*see also* axis 2, axis 3)   the second cervical vertebra on which the atlas pivots

axis 2   (*see also* axis 1, axis 3, plane)   the point, or line, around, along or across which symmetry is established or gradients measured. The following terms are defined in alphabetic position:
LONGITUDINAL AXIS   TRANSVERSE AXIS
SAGITTAL AXIS

axis 3   (*see also* axis 1 and axis 2)   the main part of an elongate body, particularly a plant (*see* tropaxis)

axolotl   [ăks"·ə·lŏtl]   the unmetamorphosed, but sexually mature, larvae of some salamanders, particularly *Ambystoma*

axon   [ăk"·sŏn]   those nerve fibers that pass from a nerve cell to the site where the impulse is to be conducted (*cf.* dendrite)

axon hillock   a prominence on a nerve cell from which the axon arises

axopod   [ăks"·ō·pŏd]   a permanent pseudopodium, stiffened by an axial filament

Aysheaia   [āsh·ē'·ĭ·ē·ə]   a genus of Cambrian marine fossils so strongly resembling Peripatus as to be placed in the Onycophora and to suggest an ancient marine origin for that group

Azoic era   [ā·zō"·ĭk]   a geological era covering the first four or five billion years of the Earth's existence. It is followed by the Archeozoic. There are no records of organized living forms

Azolla   [ā·zŏl"·ə]   a genus of ferns of the family Salviniaceae, forming small floating fronds of alternating leaves with aerial and submerged lobes. *A. filiculoides* [fil·ĭk'·yūl·oid'·ēz] is well known to aquarists

Azotobacter   [ā'·zō·băk"·tə, ăz'·ō·băk"·tə]   a genus of relatively large ovoid eubacteriale schizomycetes. They are nitrogen fixing soil bacteria

azurophilic granule   [ăz·yōōr"·ō·fil'·ĭk]   one that selectively stains with azure dyes in azure-eosin combinations (*cf.* eosinophilic granule)

B

B-horizon the layer of soil immediately under the A-horizon containing considerable organic matter carried to it by rainfall from above or ground water from below

Babesia [băb·ēz'·eə] a genus of hemosporidean protozoans. *B. bigemina* [bī·jĕm'·in·ə] is the cause of Texas cattle fever (*see Boophilus*). There are many other species causing diseases of ungulates, mostly in the Old World. All are conveyed by ticks. Most of the diseases are called piroplasmosis because *Piroplasma* was once used for *Babesia*

-bac- *comb. form* meaning "berry." The form -bacc- is common in compounds

-bacill- *comb. form* meaning a "little staff" or "little rod" but frequently used in biology in the derivative sense of bacillus (*cf.* bacul)

Bacillariophyceae [bəs'·il·âr·ē·ō·fis"·ē·ē] a class of chrysophyte Algae containing the forms known as diatoms. They are distinguished by the possession of a silicified skeleton composed of two overlapping frustules. The genus Arachnodiscus is the subject of a separate entry

bacillary dysentery *see Shigella*

Bacillus [bə·sil"·əs] a genus of spore-forming aerobic eubacteriale Schizomycetes closely allied to the anaerobic *Clostridiums*. Many are pathogens of insects, *B. popilliae* [pŏp·il"·iē] causing the milky disease of Japanese beetles and commercially available for the control of that pest. Many are thermophilic and *B. stearothermophilus* [stēr'·ō·bĕrm·ŏf"·il·əs] is responsible for much food spoilage. The only dangerous human pathogen is *B. anthracis* [ăn·thrā"·sis], causative agent of anthrax

bacillus [bə·sil"·əs] literally, a rod or staff, and applied not only to bacteria but also to diatoms and some other organisms of that shape

back-mutation the mutation of a mutant back to the form from which it was derived

bacteria [băk·tēr"·ēə] a term loosely applied to a group of minute (of the order of $1\mu \times 10\ \mu$), plants, lacking chlorophyll and closely allied to the fungi. The "true bacteria" include the orders Pseudomonadales and Eubacteriales but the Actinomycetales, Chlamydobacteriales, Beggiatoales, Myxobacteriales and Spirochaetales are usually thought of as "bacteria."

bacterial capsule a thickened layer, usually polysaccharide, around some bacteria

bacterial dissociation the appearance of types differing from the original but which lack the genetic stability that would be indicated by the use of the word mutant

bacteriocin [băk·tēr"·ē·ō'·sin] a protein produced by one bacteria that is lethal to another

bacteriophage a virus infesting, and usually lysing, bacteria. Bacteriophages are of many shapes from a sphere to a rod but many of the most widely studied species are tadpole-shaped. All can alter the heritable characters of an infected cell

-bacul- *comb. form* meaning "rod" or "staff" (*cf.* -bacill-)

Baillarger's line a thickened layer of tangential fibers along the boundary between the superradial and interradial networks in the cortex of the forebrain

Balanomorpha [băl·ən·ō·môrf'·ə] a division of cirripede, commonly called rock barnacles. They all have the same general form of a shrimp attached by the small of its back, protected by moveable calcareous plates. The genus *Balanus* is the subject of a separate entry

Balantidium [băl·ən·tĭd'·ēəm] a genus of holotrichan ciliate protozoans parasitic in the colon of many amphibians. *B. entozoon* [ĕn·tō·zō'·ən] is found in *Rana* where it is often confused with *Nyctotherus*. Many are found in invertebrates

Balanus [băl'·ən·əs] the commonest genus of cirrepedes. They attach to any surface they can find in all the coastal waters of the world and are a pestilential fouler of ships' bottoms

balsam a sticky exhudate of numerous pinaceous trees consisting essentially of one or more resins dissolved in one or more essential oils. The term is also applied to balsam producing trees (*see also* Canada balsam)

band 1 (*see also* disc) a narrow strip. The following derivative terms are defined in alphabetic position:

CASPARIAN BAND	LATERAL BAND
FLAGELLATED BAND	MARGINAL BAND
GERM BAND	MESODERMIC BAND

band cell a granulocyte the nucleus of which is not divided into parts

-bar- *comb. form* meaning "weight" or "heavy" (*cf.* -bary-)

barbel diminutive of barb, particularly as applied to sensory organs of fish

barbiturase [bär·bĭt"·chûr·āz'] an enzyme catalyzing the hydrolysis of barbiturates to malonates and urea

bare name (*nomen nudum*) [nō'·měn·nyūd·əm] a published binomial not supported by evidence sufficient to permit its official adoption

bark A loosely defined term, usually taken to mean all tissues external to the vascular cambium of a woody stem

baroceptor [băr·ō·sěpt'·ôr] an organ perceiving weight

barophilous [*angl.* bə·ŏf"·əl·əs, *orig.* băr'·ō·fil"·əs] tolerant of weight, said particularly of bacteria growing at great depths in the ocean

barren a tract of land, either in high altitudes or high latitudes on which there are typically shrubs, with occasional stunted trees, but no regular trees

Bartholin's glands mucous secreting glands in the vestibule of the vagina of some mammals, probably homologous with Cowper's glands

-bary- *comb. form* meaning "weight" or "heavy" (cf. -bar-)

-bas- *comb. form* meaning "bottom"

basal pertaining to, forming, or found at, the base. Also, by a confusion of meaning, pertaining to basic dyes

basal body = basal granule

basal granule a tubular swelling on the base of a cilium. It is usually considered that new cilia arise by the division of basal granules (= basal body, kinetosome)

basal plate 1 (*see also* basal plate 2) a cartilage lying immediately below the foramen magnum in a chondrocranium

basal plate 2 (*see also* basal plate 1) the ventral half of the lateral side of the brain stem or spinal cord

basal zone that part of the embryonic central nervous system that lies below Munro's sulcus

basement membrane the layer that lies between epithelial cells and the underlying connective tissue but in insects applied also to the inner surface of the eye

basidial layer [bās·ĭd"·ēəl] that portion of the mycilium in fungi from which basidia arise

basidium [bās·ĭd"·ēəm] the condiophore of a basidiomycete fungus

Basidiomycetes [bās·ĭd·ēō·mi·sĕt·ēz] a class of fungi distinguished by a unique type of spore, the basidiospore, of which four are formed on each basidium. The following orders are the subject of separate entries: Ustalaginales (the smut fungi), Uredinales (the rust fungi), Tremallales (the jelly fungi), Aphyllophorales, Agaricales (mushrooms and toadstools), and Gasteromycetales (puffballs and their allies)

Basilona [băz·əl·ōn'·ə] a genus of citheroniid moths containing *B. imperialis* [ĭm·pēr"·ē·āl·ĭs] (the imperial moth) that rivals *Citheronia regalis* [rĕg·āl'·ĭs] in size and in the frequency with which it is miscalled *Cercropia*

basimetrical [bās·ē·mĕt·rĭk·əl] the distribution of organisms on the sea-bottom, either vertically or horizontally

basioccipital bone [bās·ē·ŏk·sip"·ət·əl] one of a pair of chondral bones of the skull lying immediately below the foramen magnum (*see also* occipital bone, exoccipital bone, supraoccipital bone)

basipodite [*angl.* bās·ĭp·əd·ĭt, *orig.* bās·ē·pōd"·ĭt] the joint of a crustacean appendage that lies between the ischiopodite and the coxopodite

basisphenoid bone [bās·ē·sfēn"·oid] one of a pair of chondral bones on the base of the skull immediately anterior to the basioccipitals (*see also* sphenoid bone,

alisphenoid bone, laterosphenoid bone, orbitosphenoid bone, parasphenoid bone)

basitrabecular plate [bās·ē·trəb·ĕk"·yūl·ə] the anterior region of the base of the chondrocranium of cyclostomes

basiventral cartilage [bās·ē·vĕnt"·rəl] the cartilaginous precursor of the lateral wall of the centrum in vertebrates

basket cells stellate nerve cells, the axons of which envelope Purkinje's cells

basophil cell a descendant of a basophilic metamyelocyte in which the nucleus is scarcely apparent owing to the dense mass of basophilic granules

basophilic erythroblast the descendant of a proerythroblast and the direct precursor of a polychromatophilic erythroblast; distinguished from a proerythroblast by the greater basophilic staining capacity of the nucleus, the smaller size, and the absence of clearly visible nucleoli

basophilic metamyelocyte a descendant of a basophilic myelocyte in which the nucleus is U-shaped

basophilic myelocyte a descendant of a promyelocyte in which the granules are apparent, but the nucleus is still ovoid

bast [băst] the phloem of plants. The term is commonly used for dried, fibrous, strips of phloem

bastard wing a modified digit, representing a thumb, on the wing of a bird

Batesian mimicry the mimicry of an edible form by an inedible one

-bath- *comb. form* = -bathy- The following terms with this suffix are defined in alphabetic position:
EURYBATH   STENOBATH

bathile [băth·il] pertaining to the bottoms of deep lakes (*see also* chilile, pythmic)

-bathy- *comb. form* meaning "low", "deep" "depth" and "broad"

bathylimnetic [băth'·ē·lĭm·nĕt"·ĭk] the condition of bottom-dwelling plants which may either float or be rooted

bathypelagic plankton that which shows a diurnal vertical migration

bathyphilous [*angl.* băth·əf"·ə·ləs, *orig.* băth'·ē·fil"·əs] dwelling in lowlands or in the depths of the ocean

Batoidea [bət·oid"·ēə] an order of elasmobranch Chondrichythes containing the rays and skates. These are superficially distinguished from the Selachii (sharks) by their flattened form and long tail. The genera *Dasyatus, Pristis, Raja* and *Torpedo* are the subject of separate entries

-bdell- *comb. form* meaning "leech"

Bdelloidea [dĕl·oid'·ēə, dəl·oid"·ēə] a class of Rotifera, with completely retractile anterior ends, distinguished by the leech-like creeping motion that is used as an alternative to swimming

Bdelloura [dĕl·yōōr'·ə] a genus of marine tricladid Turbellaria with a glandular adhesive disc. All members of the genus are parasitic or commensal on *Limulus polyphemus*

beach that area on the shores of an ocean which lies between high and low water mark. The term is also applied to sandy areas, subject to wave action, on the shores of fresh-water lakes

beak the horny mouth parts of birds and reptiles, and the very similar mouth parts of some cephalopod mollusks. The term is also loosely applied to any hardened prominence associated with the mouth parts of any animal and with many beak-like structures such as the gnathosome of acarines and the pedicel valve of brachiopod valves

beard   any tuft of filaments on any part of an animal, but in vertebrates usually the head or breast. Widely used of the barbels of catfish and pendant tufts of feathers in birds. Not infrequently misused as though synonymous with byssus and sometimes even applied to the gills of pelycepod mollusks. In botany sometimes used for awn

Beggiatoales   [běg'·ē·ăt"·ō·ăl·ēz]   an order of Schizomycetes occurring as trichomes which frequently show bending or flexing, or in single cells which may glide over the surface of the substrate. They are by some regarded as Cyanophyceae which have lost their pigment

Bellini's duct   collecting tubules of the metanephros

-benth-   comb. form meaning "deep"

benthic   [běn"·thǐk]   pertaining to depths (see also archibenthic, eurybenthic)

benthon   [běn"·thǒn]   organisms growing on the bottom or living in deep water. In botany sometimes used as synonymous with benthophyte (see also plankton, nekton)

benthos   [běn'·thǒs]   depths, or deep zones of water

Bergh's Theory   see gonocoel theory

Bergman's Law   the average size of individuals of a given species is larger in those inhabiting colder climates than those inhabiting warmer ones

Beroë   [bē·rō'·ē]   a cosmopolitan, genus of flask-shaped Ctenophora. The body is a compressed ovoid with the mouth extending across the whole of one end. *B. cucumis* [kyū·kyū"·mǐs], about 10 cm long by 9 cm wide, is common off the northern parts of both the Atlantic and Pacific coasts

berry   a fruit in which all the ground tissue, either of the ovary wall, placenta, or both, forms a fleshy mass. In compound names the term applies equally to the plant bearing the fruit

Bertin's columns   the connective tissue lying between the lobes of the medulla in polylobular kidneys

beta cell   a chromophile cell of the anterior lobe of the pituitary showing less granulation and fewer mitochondria than the alpha cells

Betz's cell   one of the large pyramidal cells in the motor region of the cerebral hemisphere

-bi-   comb. form meaning "two"

bicornuate uterus   the type of uterus in which both uteri are fused, but have short lateral extensions

bicuspid   having two sharp points

bicuspid tooth = premolar

bidder's organ   the remnants of the testes in some female anuran amphibia

biennial   an event which occurs once every two years or a plant which blooms two years after the seeds are sown and then dies

bilateral cleavage   that in which the blastomeres exhibit marked bilateral symmetry

bilateral symmetry   a type of triaxial symmetry in which all three axes differ from each other, i.e., in which the dorsal and ventral surface are distinguished

Bilateralia   [bī'·lăt·ěr·āl"·ēə]   a division of the animal kingdom erected to contain all those forms which show bilateral symmetry. It therefore contains all the phyla of the animal kingdom except Protozoa, Coelenterata, Porifera, and Ctenophora

bile   the secretion of the gall bladder of mammals. It is colored dark green by the breakdown products (biliverdin, bilirubin) of hemaglobin and, as an alkaline wetting agent, assists in the emulsification of fat

bile duct   the duct leading from the gall bladder to the small intestine

Bilharzia = Schizostoma

bilirubin   [bĭl'·ē·rōō"·bĭn]   the breakdown product of biliverdin

biliverdin   [bĭl'·ē·vâd"·ĭn]   the primary breakdown product of hemoglobin; it is itself converted to bilirubin

binary fission   division into two

-bini-   comb. form meaning twin

binomial, binomial name   the correct Latin name of an organism consisting of the genus name followed by the species name

-bio-   comb. form indicating life

biochore   [bī'·ō·kôr]   a group of similar biotopes so large as to form a recognizable habitat; thus forests, deserts and prairies are biochores; a tree with its associated organisms is a biotope

biochrome   [bī'·ō·krōm]   any pigment found in a living organism (see also indigoid biochrome, quinone biochrome)

biocommunication   the process of transferring information between non-human organisms

biocoen   [bī'·ō·sēn]   the sum total of the living components of an environment

biocycle   the biosphere is divided into three environmental biocycles known as the marine, the fresh-water and the terrestrial

biogenesis   [bī·ō·jĕn'·əs·əs]   the production, or origin, of life (see also abiogenesis, neobiogenesis)

biogenetic law   the concept of Haeckel that "ontogeny recapitulates phylogeny"; that is, that successive stages of development approximate evolutionary ancestral stages

biological control   a reduction in the population of an unwanted species by the intentional introduction of a predator, parasite or disease

biological productivity   the increase in biomass, usually measured in protein-time units

bioluminescence   luminescence produced by living organisms (see also luciferin, luciferase)

biomass   [bī'·ō·măs]   the total weight of all living organisms on the planet or in specific habitats on the planet, or of a specific taxon

biome   [bī"·ōm]   a large community of living organisms having a peculiar form of dominant vegetation and associated characteristic animals

biomonad   [bī'·ō·mŏn"·əd]   a symbiotic system of biomores (cf. biomolecule, biomore)

biomore   [bī'·ō·môr]   an aggregate of similar biomolecules (cf. biomolecule, biomonad)

Biomphalaria   see Planorbis

bion   [bī'·ŏn]   a living unit including not only cells but also viruses. By some the term is considered synonymous with "individual" and by others as a variant spelling of biome. There is a further confusion in the literature with "biont"

biont   [bī"·ŏnt]   a living thing (but see remarks under bion). Also widely used in the sense of a member of a biome. The specific condition of a -biont is -biosis

bioseston   [bī'·ō·sĕs"·tŏn]   the living component of the seston

biosis   [bī·ōs'·ĭs]   *either* the condition of being alive *or* the conditions pertaining to a specific type of life.

An individual in a condition of -biosis is a -biont, or -bion and it is said to be -biotic. The following terms with this suffix are defined in alphabetic position:

ANABIOSIS EPIBIOSIS
ANTIBIOSIS GNOTOBIOSIS
CLEPTOBIOSIS PARABIOSIS
CRYPTOBIOSIS SYMBIOSIS

biosphere that portion of the planet earth (i.e., the upper soil and the lower air) that contains living organisms

biota [bī·ōt″·ə] the sum total of the living organisms of any designated area

-biotic see -biosis

biotic potential the possibility of a specific organism surviving in a specific environment, particularly an unfavorable one

biotic province a biogeographical division of a land mass

biotope [bī′·ō·tōp″] an ecological niche, or restricted area, the environmental conditions of which are suitable for a certain fauna and flora. A tree with its associated organisms is a biotope; a forest is a biochore

bipartite uterus a type where paired, tubular uteri fuse at the point of junction with the vagina

bipinnaria larva [bī′·pĭn·âr″·ēə] the free-swimming larva of asteroid Echinodermata which precedes the brachiolaria

bipolar cell a nerve cell in the retina with one axon and one dendrite connecting the photoreceptors with the ganglion cells

biradial cleavage that in which the tiers of blastomeres are symmetrical with regard to the first cleavage plane

biradial symmetry a type of triaxial symmetry in which only two axes differ from each other. That is, in general terms, there is no difference between the ventral and dorsal surface (in which it differs from bilateral symmetry)

Biston [bĭs″·tən] a genus of geometrid moths. *B. betularia* [bĕt′·yūl·âr·ēə] occurs in light and dark forms, the latter becoming increasingly numerous in industrial areas where pollution has darkened the bark of trees and killed the lichens against which the "peppered" wings of the light form are camouflaged. This is often used as an example of "survival of the fittest"

bivalent [bī·văl″·ənt] one of the two pairing homologous chromosomes in the prophase of meiosis (*see also* diachistic bivalent)

bladder any thin walled, expansible structure either plant or animal. Without qualification, usually refers to the urinary bladder of mammals. The following derivative terms are defined in alphabetic position

GALLBLADDER URINARY BLADDER
SWIM BLADDER

-blast- (*see also* -cyte-) *comb. form* meaning "bud", or "shoot" but also used for any developing or developmental structure, particularly cells, or for cells or structures which are producing something. There is much confusion, apparently not entirely due to misprints, with -plast-. For example, "blepharoblast" and "blepharoplast" are both used for the same structure but "bioblast" and "bioplast" appear to be distinguishable. The following terms with the suffix -blast are defined in alphabetic position:

ANGIOBLAST CHONDROBLAST
BLEPHAROBLAST CNIDOBLAST
ENDOBLAST NEMATOBLAST
ERYTHROBLAST NEOBLAST
FIBROBLAST NEUROBLAST
HEMATOBLAST NORMOBLAST
HISTIOBLAST ODONTOBLAST
HYPOBLAST OSTEOBLAST
LEMNOBLAST PERIBLAST
LIPOBLAST PLASMABLAST
MEGAKARYOBLAST SCLEROBLAST
MEGALOBLAST SPONGIOBLAST
MESOBLAST STATOBLAST
MONOBLAST TELOBLAST
MYELOBLAST TROPHOBLAST

blastema [blăst·ēm″·ə] literally a sprout. In zoology an area in an embryo of segregated cells which will subsequently develop into a specific organ. In botany the radicle and plumule of an embryo before the appearance of the cotyledon

-blastic pertaining to developing structures or parts of developing structures. The form -blastous is occasionally seen, as is the substantive -blasty, particularly in botany. The following terms with this suffix are defined in alphabetic position:

DIPLOBLASTIC
GYMNOBLASTIC
HOLOBLASTIC
TRIPLOBLASTIC

Blastocladiales [blăst′·ō·klăd·ē·âl″·ēz] an order of uniflagellate Chytridiomycete Fungi distinguished by isogamous sexual reproduction. The genus *Allomyces* is the subject of a separate entry

blastocoel [blăst′·ō·sēl″] the segmentation cavity or cavity of the blastula

blastocone [blăst′·ō·kōn″] a segment of a cleaving blastodisc which is not yet separated from the yolk, and cannot, therefore, be referred to as a blastomere

blastocyst [blăst′·ō·sist″] the early mammalian embryo following cleavage and during implantation

blastokinesis [blăst′·ō·kīn·ēs″·əs] activity resulting in the reorientation of an embryo within an egg

blastospore [blăst′·ō·spôr″] an ascomycete fungal spore produced by budding and which can itself reproduce by budding

-blastous see blastic

blastozooid [blăst′·ō·zō″·ĭd] an individual resulting from asexual reproduction

blastula [blăs″·tyū·lə] the stage that terminates the cleavage of many animal eggs and which usually consists of a hollow sphere of cells. The following derivative terms are defined in alphabetic position

AMPHIBLASTULA DISCOBLASTULA
COELOBLASTULA PERIBLASTULA

-blasty see under -blastic

-blephar- *comb. form* meaning "eyelid" or "eyelash" and sometimes used as an adjectival form of flagellum

blepharoblast [blĕf″·ə·rō·plăst′] the granule formed at the base of the axoneme of a flagellum

Blochmann's body any intracellular organism in the egg of an arthropod. Most are bacteria, supposedly symbiotic

block mutation a mutation of a number of neighboring loci

blood a fluid connective tissue used for the transport of dissolved or dispersed materials in animals. In vertebrates the oxygen carrying compounds (hemoglobins) are localized in erythrocytes but in other groups these pigments (hemoglobins, hemocyanins, hemovanadins) are dispersed in the plasma

blood island an extraembryonic aggregate of blood-forming cells

bloom a sudden increase in the number of Algae in lakes. The term is often modified by seasonal adjectives

body 1 (see also body 2, corpus, corpuscle) any organ, or organelle, not possessing a specific name. The following derivative terms in this sense are defined in alphabetic order:

BASAL BODY	MAMMILLARY BODY
BROWN BODY	NISSL BODY
CENTRAL BODY	PERIBILIARY BODY
COCCYGEAL BODY	POLAR BODY
GOLGI BODY	REDUCTION BODY
HABENULAR BODY	SUPRARENAL BODY
INTERRENAL BODY	ULTIMOBRANCHIAL B.
INTERVERTEBRAL BODY	WOLFFIAN BODY
LATERAL GENICULATE B.	

body 2 (see also body 1) a whole organism, or the major part of an organism. The following derivative terms in this sense are defined in alphabetic position: BLOCHMANN'S BODY FOREBODY HINDBODY

-bol- comb. form originally meaning "throw" but extended first to meaning any kind of "movement" and later "change". The biological compounds are very confused because the substantive terminations -boly and -bolism and the alternative adjectival terminations -bolous and -bolic have all developed distinct meanings

Boletus [bŏl·ēt″·əs] a large genus of agaricale basidiomycete fungi. They are typically mushroom-shaped but the large fleshy cap has pores in place of gills. *B. edulis* [ĕd′·yūl·ĭs] occurs both in Europe and the U.S. The French name *cêpe* [sāp] is frequently used in English

-bolic (see also -bolous, -bolism, -boly) adjectival form from -bol limited in biology to the meaning "change". The undernoted compounds with this suffix are defined in alphabetic position:
ANABOLIC CATABOLIC METABOLIC

-bolism (see also -bolic, -bolous, -boly) substantive termination pertaining either to the adjectival terminations -bolic, or -bolous even though these differ in meaning. Thus metabolic and metabolous have, by usage, developed quite different meanings but both refer to "metabolism" (i.e. a "condition of change"). To avoid repetitive definitions, only the adjectival forms are defined in this dictionary

-bolous (see also -bolic, -boly, -bolism) adjectival form from -bol- but, unlike -bolic, limited in biology to "change" in the sense of metamorphosis (see metabolous)

-boly (see also -bolism, -bolic, -bolous) substantive termination derived from -bol- with no apparent purpose except to avoid -bolism when this, in combination with the desired prefix (e.g. ana-, meta-) is pre-empted. The following terms with this suffix are defined in alphabetic position:
EMBOLY EPIBOLY

Bombyx [bŏm′·bĭks] the genus of moths containing *B. mori* [môr′·ĭ], the larva of which is the silkworm. Originally Chinese, it has been introduced into most civilized countries

bone 1 (see also bone 2) a calcified connective tissue forming the principal portion of the skeleton of most vertebrates. The following types of bone are defined in alphabetic position:

CANCELLOUS BONE	LAMELLAR BONE
CARTILAGE BONE	MEMBRANE BONE
CHONDRAL BONE	SPONGY BONE
COMPACT BONE	TRABECULAR BONE
DERMAL BONE	

bone 2 (see also bone 1) a particular mass of bone. Specific bones are defined in alphabetic position

Bonellia [bŏn·ĕl″·ēə] a genus of Echiurida remarkable both for the length of the female's proboscis and for extreme sexual demorphism. In *B. viridis*, for example, a three inch female can extend the proboscis for more than a yard. The male of this species is a ciliated ovoid about 3 mms long which lives as a commensal in the uterus or coelom of the female.

book lung a series of leaf-like respiratory plates found on the postero-dorsal surface of some arachnids, particularly scorpions (see also gill book)

Boophilus [bō·ŏf·əl·əs] a genus of ixodid acarines *B. annulatis*, the "Texas Cattle Tick", carries the dreaded Texas cattle fever (see Babesia). It is the only known tick that is confined to one host and, for this reason, is the only tick to be exterminated in the United States. The tick, and its accompanying disease, survive in other parts of the world

bordered pit a plant pit in which the secondary cell wall arches over the pit cavity

-borea- comb. form meaning northwind

-bostrych- comb. form meaning "a ringlet"

Botallo's duct = ductus arteriosus

-bothr- comb. form meaning "hole"

bothrenchyma [bŏth·rĕn″·kə·mə] plant tissue composed of pitted cells or ducts

-bothro- comb. form meaning a "hole" or "groove"

-botry- comb. form meaning a "bunch of grapes"

Botrychium [bŏt·rĭ″·kē·əm] one of the three extant genera of Ophioglossales. They are distinguished from other ferns by their leaf-like frond terminating either in a leaf-like frond or a fertile spike

botryoidal [bŏt′·rē·oid″·əl] in the form of a bunch of grapes

bottom recessive an individual possessing all the recessive alleles of a group

bouquet stage early leptotene stage of meiosis in which chromosomes are arranged in a bouquet pattern

Bowman's capsule = glomerular capsule

Bowman's glands small, scattered, sensory glands in the nasal cavity of mammals

Bowman's membrane a clear membrane immediately underlying the cornea of the eye

Boyden's sphincter that part of Oddi's sphincter which lies around the preampullary portion of the bile duct

brachial canal [brāk″·ēəl] the canal in the oral arm of Schyphozoa

-brach- comb. form meaning "arm" but frequently confused in compounds with "-brachy-"

brachiation to progress by swinging from hand to hand from one branch of a tree to another

brachiocephalic artery [brāk′·ēō·sĕf″·əl·ĭk] the common trunk of the subclavian and corotid arteries

brachiolaria larva [brāk′·ēō·lâr″·eə] the free-swimming larva of asteroid Echinodermata which develops from the bipinnaria larva

Brachiopoda [brăk′·ē·ŏp″·ō·də] a phylum of lophophorian coelomate animals commonly called lampshells. They are distinguished from other lophophorians by the possession of a bivalved shell. The classes Echardines and Testicardines are the subject of separate entries

Brachiosaurus [brăk′·ēō·sôr″·əs] a giant herbivorous, semi-aquatic, saurischian dinosaur of the late Jurassic

brachium [brăk″·ēəm] literally "arm" but applied less to these structures in vertebrates, than to the arms of such forms as crinoid echinoderms

-brachy- *comb. form* meaning "short" but frequently confused in compounds with -brach-

brachymeiosis [brăk′·ē·mi″·ōs·ĭs] meiosis resulting in half the usual haploid number of chromosomes

Brachystola [*angl.* brăk′·ĭs·tə·lə, *orig.* brăk·ē·stŏl″·ə] a genus of orthopteran insects. *B. magna* [măg″·nə] is the lubber grasshopper usually miscalled *Romalea*

Brachyura [brăk′·ē·yûr″·ə] a section of reptant crustacea distinguished by the greatly reduced abdomen flexed beneath the thorax and the heavy chelae. The whole group in general is frequently referred to as the "true" crabs. The genera *Uca* and *Cancer* are the subject of separate entries

Braconidae [brăk·ŏn′·id·ē] a large family of apocritan hymenopteran insects closely resembling the Ichneumonidae but differing from them in certain details of the wing and, in general, having thicker bodies. *Habrobracon junglandis* is properly *Bracon hebetor* [brăk′·ən heb′·ət·ôr]

bract 1 (*see also* bract 2) a small leaf on a floral stem often closely pressed to the base of the flower and sometimes, as in Poinsettia and Bougainvillea, brilliantly colored and frequently mistaken for a petal

bract 2 (*see also* bract 1) a leaf-like, protective, zooid of a siphonophoran hydrozoan

-brady- *comb. form* meaning "slow"

brain the large anterior end of the central nervous system. The term was at one time confined to vertebrates but is nowadays often used for the cerebral ganglion of invertebrates, particularly insects. The following derivative terms are defined in alphabetic position (*see also* cephalon) :

FOREBRAIN   HINDBRAIN   MIDBRAIN

brain stem that portion of the brain which remains after the various evaginations forming the cerebrum and cerebellum, etc., have been produced

-branch- *comb. form* meaning a "fin" or "gill"

branch 1 [brănch, brănch] (*see also* branch 2) any offshoot leaving a mainstem or trunk

branch 2 [brănk] (*see also* branch 1) a gill. The following terms with this suffix are defined in alphabetic position :

HEMIBRANCH   HOLOBRANCH   PSEUDOBRANCH

branchial arch [brănk′·ē·al] there is great confusion in the literature as to whether this term is synonymous with visceral arch, whether it applies to all the visceral arches except the maxilla, or whether it should be confined to those visceral arches which bear gills in fish and to their homologues in higher forms

branchial artery an artery associated with a gill and derived from the aortic arch

branchial net the basket of anastomizing blood vessels that forms the pharynx of ascidians

branchial pouch a series of respiratory pouches in marsipobranchia corresponding to the gills of higher forms

Branchiobdellidae [brănk′·ē·ŏb·dĕl″·id·ē] a family of parasitic oligochaete annelids sometimes still classed with the *Hirudinea*; they lack setae and have a sucker at the posterior end of the body but not at the anterior

branchiomere [brănk′·ēō·mēr] that division of the mesoderm from which the mesodermal components of a visceral arch are developed

Branchiopoda [brănk·ēō·pŏd·ə] a subclass of crustacean arthropods containing the animals called fairyshrimps, tadpole-shrimps, clam-shrimps and waterfleas. They are distinguished from other crustacea by the leaf-like appendages on the thorax that serve both for locomotion and respiration

branchiostegal ray [brănk′·ēō·stĕg″·əl] one of a series of bony rays projecting backwards from the hyoid pouch of fish. They support the floor and sides of the gill chamber

branchiostegite [*angl.* brănk′·ē·ŏst″·əg·ĭt, *orig.* brănk′·ēō·stĕg″·ĭt] an extended fold of the cephalothoracic pleura that covers the gills in some Crustacea

Branchiostoma [brănk′·ē·ŏst″·əm·ə] the best known genus of cephalochordates, with numerous species all looking much alike. *3. lanceolatus* [lăns′·ēō·lā″·təs], which almost everything shown to students is called, is a European species. There are half a dozen American species of which *B. virginiae* [vûr·jin′·iē] is commonest Atlantic and *B. californiensis* [kăl′·ē·fôrn″·ē·ĕn·sĭs] the Pacific form

Brassica [brăs′·ĭk·ə] a genus of cruciferous plants that has furnished man with food since the dawn of history. There is some argument as to whether the cole crops are descended from *B. oleracea* [ōl′·ē·ās″·ēə] some deriving the root forms (turnip, rutabaga), etc. from a postulated *B. rapa* [răp″·ə]. *B. chinensis* [chi′·nĕn″·sĭs] (Chinese Cabbage) is also known only as a cultivar but *B. nigra* [nĭg″·rə] (Black Mustard) and *B. hirta* [hĭrt′·ə] (White Mustard) are widespread weeds

break a spontaneous, but not of necessity heritable, change appearing in an F-1 generation of plants (*see also* chromatid break)

breed a variety or race of constant genetic character. The term is mostly used of domestic animals and is by some considered synonymous with race

-bri- *comb. form* meaning "to acquire strength"

brille [bril] a transparent scale fitting closely over the eye of a snake

-brious *adj. term.* from -bry-

-bronch- *comb. form* meaning "wind pipe"

bronchiole [brŏnk′·ē·ōl] the terminal divisions of second order bronchi

bronchus [brŏnk′·əs] properly, the two tubes into which the trachea divides. The following derivative terms are defined in alphabetic order :

DORSAL BRONCHUS   SECOND ORDER BRONCHUS
MESOBRONCHUS

Brontosaurus [brŏnt′·ō·sôr″·əs] a genus of extinct diapsid reptiles of the Jurassic. They were very similar to *Diplodicus* but had spatulate teeth

brood an assemblage of individuals all hatched at the same time from eggs laid by a single parent or any similar conglomeration of synchronously produced offspring

brown body   *either* brownish colored fatty masses found in hibernating mammals *or* brownish masses produced by the degeneration of an ectoproct polyp which subsequently regenerates into a new polyp *or* a small clump of cells found in the coelom of holothuria

Brucellaceae [brōō'·sĕl·ās·ē] a family of gram negative, coccoid, eubacteriale Schizomycetes the great majority of which are pathogenic to man and domestic animals. The genus *Pasteurella* is the subject of a separate entry

Brunner's glands   enzyme secreting glands in the submucosa of the anterior region of the duodenum

brush border   the appearance of epithelial cells in which microvilli are of a size visible by optical microscopy

-bry-   *comb. form* meaning to "swell", "grow" or "burst forth", often transliterated -bri-

Bryophyllum [*angl.* bri·of'·əl·əm, *orig.* bri'·ō·fil"·əm] a large genus of crassulaceous plants most of which can produce plantlets from the edge of the leaf

Bryophyta [bri'·ō·fit"·ə] a subkingdom of plants containing the liverworts, hornworts, and mosses. They are distinguished by the absence of a vascular system, of true roots and of true leaves. The alternate generations are distinct

Bryozoa [bri'·ō·zō"·ə] a no longer acceptable, but once widely used, taxon composed of the Ectoprocta and Endoprocta

buccal [bŭk'·əl] pertaining to the mouth or mouth cavity (*see also* peribuccal)

buccal appendage   any mouth part of an arthropod which is articulated and moveable

buccal tube   that portion of the alimentary canal, particularly in invertebrates and hemichordates, which lies between the mouth and the pharynx

bud   a protruberance on an organism that will subsequently grow into an organ, or another organism. Most widely used in botany for the precursors of flowers or shoots. The following derivative terms are defined in alphabetic position:

IMAGINAL BUD	PERIOSTEAL BUD
LIMB BUD	TASTE BUD
LUNG BUD	

buffer species   a form eaten by a predator, but which is not the natural prey of that predator and, therefore, "buffers" the effect of the predator on its normal prey

buffering gene   modifier gene

Bufo [bū"·fō"] a very large genus of terrestrial Salientia universally called toads. They are protected against predators by an intensely irritating, often highly toxic, secretion of the parotid glands and skin glands. The South American giant toad (*B. marinus* [mə·rēn'·əs]) is such a prodigeous eater of insects that it has been introduced into all the sugar-growing areas of the world where it has proved the only effective control against the Sugar Cane Beetle

Bugula [byōō'·gyū·lə] a genus of marine Ectoprocta the branches of which carry the zooecia in from 2 to 6 rows; there is no operculum. There are many species mostly forming colonies from one to three inches high. The aviculariae are usually pedunculate

bulb 1   (*see also* bulb 2) in zoology any hollow globose organ, frequently with the connotation that a hollow open shaft is attached to the bulb. In this sense, the following derivative terms are defined in alphabetic order:

DUODENAL BULB	FLAME BULB
EJACULATORY BULB	OLFACTORY BULB
END BULB	

bulb 2   (*see also* bulb 1) in botany a globose mass of fleshy leaves which serves as the dormant phase of many flowering plants. The term is often, in error, applied to the corm. In this sense the following derivatives are defined in alphabetic position:

PLUMULE BULB	PSEUDOBULB	TUNICATED BULB

bulb scale   one of the modified leaves forming a bulb

bulbil 1 [bŭl'·bil] (*see also* bulbil 2) a small bulb, produced above ground

bulbil 2   (*see also* bulbil 1) a resting vegetative gennule produced by some lower plants

bulbourethral gland [bŭl'·bō·yōōr'·ēth"·rəl] = prostate gland

bulbus arteriosus [bŭlb"·əs ärt·ēr'·ē·ōs"·əs] a strongly muscularized region immediately anterior to the conus arteriosus in some fish

-bull-   *comb. form* meaning "blister"

bulla [bŏŏl'·ə] any hollow knob or disc. Also an unveined patch on the wing of an insect (*see also* tympanic bulla)

bullar recess   a cavity in the tympanic bulla of some mammals, separated from the ossicular cavity by a septum derived from the tympanic bone

bundle 1   (*see also* bundle 2) in botany a common abbreviation for vascular bundle, that is a group of plant conducting vessels, frequently formed by the breakup of the vascular cylinder

bundle 2   in zoology a large mass of nerve, or other fibers (*see* His's bundle)

bundle cap   a group of parenchyma cells no longer concerned with conduction

bundle sheath   cells surrounding the vascular fascicles in a leaf

bunodont [byōō'·nō·dŏnt"] having tubercules on the crown of the molar teeth

bursa [bûrs"·ə] literally a purse, but used in biology for any pouch or sac. Particularly (1) an eversible pouch used to grasp the rear end of the female in copulating Acanthocephalans (2) a pouch alongside each side of each arm on the oral surface of ophiuroid echinoderms (3) a copulatory expansion at the hinder end of strongyloid nematodes supported by expanding rays. The following derivative terms are defined in alphabetic position:
COPULATORY BURSA
GENITAL BURSA

Busycon [byōō'·sik·ŏn] a genus of large marine gastropod mollusks. The common American Whelk is *B. caniculatum* [kăn·ik'·yūl·āt"·əm] with a 5" shell. The helicoid masses of eggs of this form are a common classroom object. The "whelk" eaten in Europe is *B. undatum* [ŭn·dāt'·əm]

byssus [bis'·əs] literally, freshly combed flax and therefore applied to any loose, long, tuft. Particularly a series of branched projections at the poles of some nematode eggs, a loose stipe of some basidomysetes and the attaching fibers of sessile pelecypod mollusks

C

C cell = chief cell

C-horizon the soil layer under the B-horizon consisting principally of unconsolidated inorganic matter

C mitosis one which has been arrested by the action of colchicine or some similar reagent

-caco- *comb. form* meaning "bad"

caconym [kăk″·ō·nĭm] a name which is linguistically impossible

Cactaceae [kăk·tās″·ē] the only family in the order Opuntiales. The thick fleshy leaves and spines of the cactus distinguish the family from most groups except some of the Euphorbiaceae from which, however, they may be clearly distinguished by the berry like fruit. The genera *Opuntia* and *Lophophora* are the subject of separate entries

caddis larva the larva, usually case-bearing, of a trichopteran insect

caecum [sē″·kəm] any blindly ending pouch. Unless qualified, usually refers to the pouch coming off, in mammals, from the junction of the ileum and the colon, at the end of which in some forms, there is the vermiform appendix. The following derivative terms are defined in alphabetic position:

HEPATIC CAECUM   PYLORIC CAECUM

-caeno- *see* -kaino-

Caiman [kā″·mən] a genus of Central and Southern American crocodilid reptiles. The word is interchangeable in Latin or English

Calamophyta [*angl.* kăl·ə·mŏf″·ĭt·ə, *orig.* kăl′·ăm·ō·fĭt″·ə] a phylum of pteriodophyte plants containing the horsetails. They are distinguished from other pteriodophytes by the presence of a whorl of leaves at each node. The majority of groups are known only as fossils

Calanus [kăl′·ə·nəs] a genus of Copepoda with numerous species and nearly related genera that form a significant fraction of the zooplankton of northern oceans

-calath- *comb. form* meaning "vase-shaped"

calcaneum bone [kăl·kăn″·ēəm] the tarsal bone that lies between the head of the fibula and the cuboid bone

Calcarea [kăl·kâr′·ēə] a class of Porifera distinguished by the presence of a calcareous skeleton. The genera *Grantia*, *Leucosilenia* and *Sycon* are the subject of separate entries

-calce- *comb. form* meaning "slipper"

calces plural of calx

calciferous gland [kăl·sĭf″·ə·rəs] one of several oesophageal out-foldings in many oligochaete annelids. They control the calcium content of the blood (cf. chalk glands)

calcified cartilage cartilage permeated, or even replaced, by calcium salts, but without any organized structure as in bone

Caligus [kăl″·ĭg·əs] a large genus of marine parasitic Copepoda. They have a wide, ovoid, flattened body and attach themselves as temporary ectoparasites to numerous species of fishes

Callinectes [kăl′·ĭn·ĕkt″·ēz] a genus of crabs with a markedly serrated carapace. *C. sapidus* is the common "blue crab" of the Atlantic coast

calotte [kə·lŏt′] the polar cap of a holoblastically cleaving egg

calvarium [kăl·vâr′·eəm] the dome of the anthropoid cranium

calx [kălks] the heel, or the portion of a limb corresponding to the heel in other forms

-calyb- *see* kalyb-

-calyc- *comb. form* meaning a "shallow cup" (cf. -cyath-)

calyoptis larva [kăl′·ĕ·opt″·is] the zoaea larva of euphausiadid crustacea in which the carapace overlaps and conceals the eyes

calypter [kăl·ĭp″·tə] the dipterous alula when it is sufficiently large to cover the halter

-calyptre- *see* kalyptr-

calyx 1 [kā″·lĭks] (*see also* calyx 2) any cup-like or funnel-shaped structure (cf. infundibulum) particularly the area immediately surrounding the renal papilla in many mammalian kidneys. Also that portion of an Entoproct that contains the viscera in distinction to the stalk, the aboral cup of a crinoid echinoderm, and a spicule-containing basal portion of the anthocodium of some Alcyonaria

calyx 2 (*see also* calyx 1) the outer cycle of the appendages of a flower, usually consisting of sepals

Camarina [kăm″·âr·ēn″·ə] a genus of fossil Foraminifera. Some were nearly a foot in diameter and were therefore probably the largest Protozoa that ever existed

Cambarus [kăm′·bär″·əs, kăm′·bə·rəs] a genus of fresh water decapod crustaceans, commonly called crayfish. *C. bartoni* [bär·tōn′·i] is the common crayfish of the east

cambial zone [kăm′·bē·əl] that region of the cambium containing the initials and their immediate derivatives

cambium [kăm″·bē·əm] (*see also* cambium 2) that plant tissue which adds new tissues to the thickness of a plant stem. The following derivative terms are defined in alphabetic order:

PROCAMBIUM   VASCULAR CAMBIUM
PSEUDOCAMBIUM

cambium 2 (*see also* cambium 1) the inner layer of the periosteum

Cambrian period the oldest of the Paleozoic periods, extending from about 500 million years ago to about 400 million years ago. It is preceded by the Proterozoic and followed by the Ordovician. The only plants in this epoch were Algae but protozoans, sponges and mollusks were well developed. Trilobites were the dominant form

Camelus [kăm·ēl′·əs, kăm′·əl·əs] the genus of artyodactyl mammals containing the Camel (*C. bactrianus* [băkt′·rē·än″·əs], with the two humps) and the Dromedary (*C. dromedarius* [drŏm′·ĕd·är″·ēəs], with one hump)

-camp- *comb. form* hopelessly confused between two Greek and one Latin roots and therefore variously meaning "caterpillar", "marine" and "plain" in the sense of a flat area of ground

campt- *see* kampt-

Canada balsam the natural exudate of the pinaceous tree *Abies balsamea*. It was once the principal mountant used in histological technique but is now often replaced by synthetic substitutes

canal a term commonly used in the biological sciences to denote a narrow tube. The following derivative terms are defined in alphabetic position:

ALAR CANAL   NEURAL CANAL
CONJUGATION CANAL   NEURENTERIC CANAL
COPULATION CANAL   PODIAL CANAL
CUVIER'S CANAL   RADIAL CANAL
GYNECOPHORIC CANAL   RING CANAL
HAVERSIAN CANAL   SEMICIRCULAR CANAL
KUPFFER'S CANAL   STONE CANAL
LAURER'S CANAL   VOLKMANN'S CANAL
MARGINAL CANAL

1-canaline [kăn′·ə·lin] an amino acid NH_2OCH_2-$CH_2CH(NH_2)$ COOH most commonly known as a hydrolysis product of canavanine

l-canavanine [ĕl·kăn·ə·văn′·in] an amino acid $NH_2C(=NH)-NH-OCH_2-CH_2CHNH_2-COOH$ first isolated from jackbean meal

cancellous bone [kăn″·sĕl·əs] bony tissue in which the trabeculae are joined as in a scaffolding. Sometimes called "spongy bone"

Cancer [kăn″·sə] a genus of brachyuran decapod Crustacea with many species. *C. magister* [măj′·ist·ə] is the common edible crab of the California coast but the Atlantic species are rarely eaten

Canidae [kăn″·id·ē] a family of carnivorous mammals containing the dogs, wolves, jackals and their immediate allies; the family is distinguished by the large smooth, auditory bulla and the long and prominent paraoccipital process; all are furnished with short, thick, non-retractile claws. The genera *Alopex, Canis, Curon,*

Dusicyon, Lucaon, Thos, Urocyon and *Vulpes* are the subject of separate entries

canine tooth a conical tooth (which, when enlarged, is frequently called a fang) between the last incisor and the first premolar of mammals. The term is also applied to large, conical, teeth in the jaws of other vertebrates

Canis [kān′·is] the type genus of the Canidae containing the true dogs. The domestic dog is of unknown, polyphyletic origin. There are several wolves variously called the European wolf, Timber wolf, Afghan wolf, Japanese wolf, etc. but it seems most likely that they are all geographic races of *C. lupus* [lōō′·pəs]. The coyote is *C. latrans* [lā″·trănz]. The foxes, once considered *Canis* are now in the genera *Valpes, Urocyon* and *Alopex*, and the jackals in the genera *Thos* and *Dusicyon*

Cannabis [kăn″·ə·bis] a genus of moraceous plants *S. sativa* [săt·ēv′·ə] was once widely cultivated for the production of the fiber hemp and also as a decorative garden plant. This is now discouraged since the dried leaves and stems are "marijuana" while the dried gummy flowers and seedpods are "hashish"

canopy tree one having its largest branches at the crown

Cantharellales [kăn′·thə·ĕl·āl″·ēz] a class of basidomycete fungi characterized by a rather stiff upright growth which may be branched like a coral in the Clavariaceae or cup-shaped in the Cantharellaceae

cap 1 (*see also* cap 2, 3) a cap-shaped mass of cells (*in this sense see* root cap)

cap 2 (*see also* cap 1, 3) a cap-like mass of protoplasm (*in this sense see* polar cap)

cap 3 (*see also* cap 1, 2) a cap-like organ (as, for example, knee-cap [= patella bone])

capillary [kăp·il″·ə·ē] those minute blood vessels, lacking any muscular wall, which connect the arteries and veins

-capit- *comb. form* meanning "head"

capitate bone [kăp′·it·āt] the carpal that lies at the base of the third metacarpal, between this and the lunate bone

capitulum [kăp·it′·tyū·ləm] the upper of the two articular processes of the rib

-capnod- *see* -kapnod-

Caprella [kăp·rĕl″·ə] an aberrant genus of marine amphipod crustaceans. The thoracic segments are thin and elongate, the abdomen being reduced to a knob with a pair of rudimentary legs

Caprimulgiformes the order of birds that contains the goat-suckers and their allies. They are distinguished by their short, weak, somewhat hooked bill with imperforate nostrils and by their loose fluffy plumage

-caps- *see* -kapsa-

capsaicine [kăp·sā″·is·in] the pungent principal of the fruits of the genus *Capsicum*. A solution of 10 parts per million causes a burning sensation on the tongue and a 0.1% solution causes severe blistering of the skin of the hands

Capsicum [kăp′·sik·əm] a genus of solanaceous plants of great economic importance. *C. annulum* [an″·yōō·əm], *C. frutescens* [frōōt″·ĕs·ĕnz] and their numerous hybrids furnish the Sweet Peppers, Paprikas, Tabasco Peppers, etc., of commerce. The pungent principal is capsaicine

capsule 1 (*see also* capsule 2–4) a fruit resembling a legume but differing from it in that the seeds are in more than one compartment and the fruit strips along more than two lines when ripe

capsule 2 (*see also* capsule 1, 3, 4) spherical, or sub-spherical, portions of the skull. In this sense, the following derivative terms are defined in alphabetic order:

CRANIAL CAPSULE   OLFACTORY CAPSULE
NASAL CAPSULE   OPTIC CAPSULE

capsule 3 (*see also* capsule 1, 2, 4) any thin-walled globose structure. In this sense the following are defined in alphabetic order:

BACTERIAL CAPSULE   GLISSON'S CAPSULE
BOWMAN'S CAPSULE   GLOMERULAR CAPSULE
CARTILAGE CAPSULE   UTERINE CAPSULE

carapace [kăr'·ə·păs, kăr'·ə·păs] any dorsal skeletal shield over an animal. It is applied equally to the chitinous structures of arthropods, the horny structures of armadillos, or the bony structures of turtles

Carassius [kăr·ăs"·ēəs] a genus of teleost fish containing the Goldfish *C. auratus* [ôr·ā'·təs]. This interbreeds freely with the carp (*Cyprinus cyprinus*) from which it differs only in lacking barbels

carbhemoglobin (carbaminohemoglobin) [kärb'·hēm·ō·glōb"·ĭn, kärb'·əm·ēn"·ō·hēm'·ō·glōb"·ĭn] the compound formed betweeen hemoglobin and carbon dioxide

carbon cycle the cycle through which carbon, changed to a carbohydrate by synthesis, ultimately reappears in the atmosphere as carbon dioxide

carbonic anhydrase [kär·bŏn'·ĭk ăn·hīd'·rāz] an enzyme catalyzing the breakdown of carbonic acid to carbon dioxide and water

Carboniferous period one of the Paleozoic periods extending from about 250 million years ago to 200 million years ago. It was preceded by the Devonian and followed by the Permian. There were extensive forests of giant spore-bearing plants (ferns and club mosses). It saw the beginnings of insects and reptiles

carboxylase ' [kăr·bŏks"·ĭl·āz'] a group of ligase enzymes catalyzing the attachment of carbon dioxide using energy derived from the breakdown of ATP, and therefore, more properly called ligases (*see also* decarboxylase)

Carcharodon [kär·kăr'·ō·dŏn] a genus of selachian elasmobranchs containing *C. carcharias* [kär·kăr"·ē·əs] the Great White Shark and one of the few indubitable man-eaters. They are mostly about 20 feet long, weighing more than a ton.

-card- *comb. form* confused between a Latin root meaning "hinge" or "pivot" and a Greek root meaning "heart"

cardiac ganglion a prevertebral ganglion near the origin of the pulmonary artery

cardiac muscle the peculiarly branched striated muscle of which the chordate heart is composed. At one time thought to be entirely syncytial but now known to be composed of multinucleate cells

cardiac stomach the region of the stomach nearest to the esophagus (cf. pyloric stomach)

cardinal sinus the principal blood-collecting vessel of vertebrate embryos running laterally along the body from front to back. The anterior and posterior sini unite to form the Cuverian sinus

cardines [kär·dĭn'·ēz] plural of cardo

cardiogenic plate [kärd'·ēō·jĕn"·ĭk] the early primordium of the heart of the human embryo

cardium [kärd"·ēəm] heart, but used mostly in compounds, among which the following are defined in alphabetic position:

ENDOCARDIUM   MESOCARDIUM
EPICARDIUM   MYOCARDIUM

carina [kăr·ēn"·ə, kär·īn"·ə] literally, a keel; particularly, in botany, keeled petals of flowers and in zoology the ridge or keel along the bottom of a bird's sternum

carmine [kär'·mēn, kär'·mĭn] a dye derived from the cocineal insect (*Coccus cacti*). It is widely used to stain wholemounts

carnassial tooth [kär·năs'·ēəl] the posterior, cutting premolar in carnivores

Carnivora an order of placental mammals distinguished by their carnivorous habit and the adaptation of the teeth for this purpose. Among the better-known are the dogs, cats, weasels, bears, walruses and seals. The suborders Fissipedia and Pinnipedia are the subjects of separate entries

carnosine [kär"·nō·sin] β-alanylhistidine. [N≡NH —CH₂CH]NHCOCH₂CH₂NH]—COOH an amino acid common in muscular tissue but not essential to rat nutrition

carotenoid [kə·rŏt'·ən·oid] a large class of oil soluble yellow, orange, or red biochromes

carotid artery [kə·rŏt'·ĭd] one of two arteries arising from the first aortic arch. The internal carotid goes to the brain; and the external carotid supplies the superficial parts of the head

carotid gland a spongy swelling at the base of the internal and external carotid artery in some Amphibia

-carp- *comb. form* meaning "wrist", or "wrist joint". A frequent, and very confusing, transliteration of "-karp-"

carp 1 (*see also* carp 2) properly "karp", a fruit of a vascular plant, or part of a fruit. Any of the derivative substantives can be rendered as adjectives terminating in -carpic, or -carpous. In this sense the following terms using this suffix are defined in alphabetic position:

ENDOCARP   EXOCARP   MESOCARP

carp 2 (*see also* carp 1) properly "karp," a fruit and hence a reproductive ("fruiting") body of a lower plant. In this sense the following terms using this suffix are defined in alphabetic position:

ARCHICARP
ASCOCARP
SPOROCARP

carpal [kär"·pəl] pertaining *either* to the wrist *or* to the angle of the last joint of a bird's wing

carpal bone one of a series of small bones lying between the radius and ulna and the metacarpals. They are usually called wrist bones (*see also* metacarpal bone)

-carpho- *see* -karph-

carpogonium [kärp'·ō·gōn"·əm] the organ in the lower plants, particularly Algae, that gives rise to a female gamete

carpopodite [*angl.* kär·pŏp"·əd·ĭt, *orig.* kärp'·ō·pŏd"·ĭt] the joint of a crustacean appendage that lies between the propodite and the meriopodite

carposperm [kärp"·ō·spĕrm'] the female gamete of Algae after fertilization

carpospore 1 [kärp'·ō·spōr"] (*see also* carpospore 2) a non-motile, unicellular spore of Rhodophyceae

carpospore 2 (*see also* carpospore 1) a haploid

spore produced by meiosis from a zygote. Diploid carpospores are produced by direct proliferation of the zygote

carrying capacity   the potential of a given habitat for the support of a population, a level beyond which no major increase in the population can occur

cartilage 1   [kårt"·əl·əj] (*see also* cartilage 2) a connective tissue, usually skeletal, which consists of a matrix of glycoproteins and chondroitin sulfate containing the scattered, usually paired, cells that secreted it. In this sense the following derivative terms are defined in alphabetic position:

ARTICULAR C.	FIBROCARTILAGE
CALCIFIED C.	HYALINE CARTILAGE
ELASTIC C.	PROCARTILAGE

cartilage 2   (*see also* cartilage 1) a cartilaginous skeletal element. Specific cartilages are defined in alphabetic position.

cartilage bone   a bone formed by the replacement of cartilage

cartilage capsule   the layer of matrix immediately surrounding a cartilage cell

-cary-   *comb. form* meaning "nut", properly and frequently transliterated -kary-

caryopsis   [kår'·ē·ŏps"·ĭs] a one-seeded dry fruit with an adherent seed coat. Most grains (corn, wheat) are of this type

cascade determination   the concept that embryos or regenerating tissues depend on determinants that become serially independent

Casparian strip   a strip of lignified, and often suberized material in the radial and transverse walls of plant endoderm cells. Also known as Casparian band

cast   [kăst, kåst] anything which is shed in a form resembling the original, as the skin of snakes or the molted exoskeletons of arthropod instars

caste   [kåst] a group of specialized function in a social order. Among insects they are often morphologically distinguished

Castor   [kăst'·ôr] the type genus of the Castoridae containing the American Beaver (*C. canadenis* [kăn'·ə·děn"·sĭs]) and the Eurasian Beaver (*C. fiber* [fĭb'·ə])

Castoridae   [kăst·ôr'·ĭd·ē] that family of sciuromorph rodents which contains the beavers. They are distinguished by their heavy build and modifications for aquatic life including short ears and valvular nostrils

castration cell   a signet-ring type of gonadotroph cell in a castrated animal

Casuariiformes   [kăz'·yōō·âr·ē·fôrm"·ēz] an order of birds containing the Australasian casowaries and emus. They are distinguished from the other flightless birds by the fact that the feathers have an aftershaft as large as the main shaft

cata-   *see* kata-

catabolic   [kăt'·ə·bŏl"·ĭk] pertaining to those metabolic activities which are concerned with synthesis

catadont   [kăt'·ə·dŏnt] having teeth only in the mandible

catadromous   [*angl.* kăt·ăd"·rəm·əs, *orig.* kăt'·ə·drōm"·əs] descriptive of fish which pass from freshwater to salt water for reproductive purposes

catalase   [kăt'·əl·āz"] a hemoprotein enzyme, or group of hemoprotein enzymes, which catalyzes the production of oxygen and water from hydrogen peroxide or water and aldehyde from hydrogen peroxide and alcohol

cataphyll   [kăt'·ə·fĭl] leaf-like structures such as bud-scales, rhizome scales, etc.

Catarrhina   a division of primates embracing the Old World monkeys and apes which are distinguished from the Platyrrhina by the fact that the nostrils point downwards; this taxon therefore contains the Simioidea and Anthropodea. The genera *Macaca, Parapithecus* and *Proconsul* are the subject of separate entries

-caten-   *comb. form* meaning "chain"

-caud-   *comb. form* meaning "tail"

caudad   [kôr"·dăd"] moving or turning in the direction of the tail

caudal   [kôr'·dl] pertaining to the tail

caudal fin   the terminal posterior unpaired fin of the fish, commonly called the tail or tail fin

caudal glands   literally, glands of the tail region, but usually applied without qualification, to the adhesive tip of Turbellaria, and some other invertebrates

caudal peduncle   the tapered posterior end of the body of elasmobranch fish

caudal sclerotomite   the posterior of the two sclerotomites (q.v.) which compose a somite, and which therefore forms all the anterior portion of a vertebra except the anterior zygapophysis (*see also* cranial sclerotomite)

caudal somite   one of the posterior divisions of a metamerically segmented form, particularly used of vertebrates in the embryonic condition

caudal vertebra   one of the vertebra of the tail, posterior to the pelvic symphysis

Caudata = Urodela

-caudic-   *comb. form* meaning the "trunk" of a tree

-caul-   *comb. form* meaning "stem" or "stalk"

caulescent   [kôl·ĕs'·ənt] a condition intermediate between sessile and stalked

caulome   [kôl'·ōm] the "stalk" of a hydroid hydrozoan coelenterate (*see also* hydrocaulome, rhizocaulome)

-caus-   *see* -kaus-

cavity   any hollow space, not otherwise designated, in an organism. The following derivative terms are defined in alphabetic position:

ATRIAL CAVITY
GASTROVASCULAR CAVITY
PERICARDIAL CAVITY

CDP   cytidine diphosphate

Ceboidia   [sē·boid'·ēə] a sub-order of primate mammals containing all of the New World forms except the Hapaloidea from which they differ in having nails rather than claws, though the nails of the family Pithecidae are notably claw-like

-ceci-   *see* -keki-

cecum   *see* caecum

Cedrus   [sē"·drəs] the genus of Old World Coniferales containing the cedars. The leaf arrangement resembles the larches but the leaves are persistent

cell   this word originally meant one of the smallest rooms of a large house used for the storage of food or, later, prisoners. The reticulated structure seen by early microscopists in plant tissue seemed to be clearly analogous to the rooms in a house and were therefore called cells. In contemporary biology the word has come to mean a living unit consisting of a nucleus, cytoplasm and numerous specialized organelles. The term is also used for the clear space between the veins of insect wings. Specific cells are defined in alphabetic position. The root -cyte, meaning a "pouch", is regarded by many biologists as synonymous with cell and numerous terms will be found under this suffix

cell division frequently taken as synonymous with mitosis even though this term applies properly only to the nucleus

cell membrane the external limiting membrane of the cytoplasm of the cell

cell plate the first appearance of the future cell wall between two cell walls at the conclusion of mitosis (cf. phragmoplast)

cell wall a non-living protective coat around a cell, particularly a plant cell (*see also* primary cell wall, secondary cell wall)

cellulase [sĕl'·yōō·lāz"] an enzyme catalyzing the breakdown of cellulose through the hydrolysis of β-1,4 glucan links

cement sac that portion of an invertebrate oviduct in which eggs are coated with a capsule or shell

cementum [sə·mĕn'·təm] the material cementing a bony tooth into its socket

-cen- *see* -coen-

cenogenous [*angl.* sĕn·ŏj"·ən·əs, *orig.* sĕn'·ō·jĕn"·əs] said of an organism which is sometimes oviparous and at other times viviparous

cenosis | [sēn·ōs"·is] a community dominated by two distinct, but not of necessity mutually antagonistic, species. Used by some as synonymous with community and by others as synonymous with association

cenospecies [sēn'·ō·spēs·əz] species that can interbreed with each other

Cenozoic era [sēn'·ō·zō"·ĭk] a geologic epoch lasting from about 60 million years ago to 1 million years ago. It is divided into the Quaternary and Tertiary periods

-centr- *comb. form* meaning "middle" (*but see* -kentro-)

central body the nuclear material in a blue-green alga

central cell that cell in the archegonium in which the oosphere and ventral canal-cell arise

central cord cells resembling vessels in moss stems

central mass the lump of cells remaining at one pole when the morula of a mammal develops an asymmetrical cavity. It produces both blastodisc and endoderm

central nervous system the brain and spinal cord

centriole [sĕnt'·rē·ōl] two minute areas of differentiated cytoplasm which arise from the centrosome at the beginning of mitosis. Each forms one pole of the mitotic figure

centroacinar cells [sĕnt'·rō·ăs·ĭn"·är] an internal, non-secretory cell in the acinus of the pancreas (*see also* acinar cell)

centrolecithical [sĕnt'·rō·lĕs·ĭth"·ĭk·əl] said of a heavily yolked egg, such as that of an insect, in which the yolk is concentrated at the center

centromere [sĕn"·trō·mĕr'] the point at which two chromatids are fused along the course of a chromosome, usually indicated by a transparent constriction from which the spindle fibers usually originate during mitosis

centrosome [sĕn"·trō·sōm'] a minute area of differentiated cytoplasm which gives rise to two centrioles at the beginning of mitosis

centrum the basal portion of a vertebra formed as a

replacement of the original notochord. The following derivative terms are defined in alphabetic position:

ECTOCHORDAL CENTRUM   PLEUROCENTRUM
HOLOCHORDAL CENTRUM   STEGOCHORDAL CENTRUM
HYPOCENTRUM

-cephal- *comb. form* meaning "head". The adjectival forms -cephalous and -cephalic appear synonymous but -cephaline has developed a distinct meaning as has the substantive cephalon

cephalic lobe the expanded anterior region of a nemertine

cephalization [sĕf'·ăl·ĭz·ā"·shŭn] the process by which the highest degree of specialization became localized in the anterior end of animals

Cephalochordata [sĕf"·əl·ō·kôrd'·āt·ə] a subphylum of acraniate chordata which contains *Branchiostoma* ("Amphioxus") (the lancelet). It is distinguished from other acraniates by retaining throughout life the notochord, nerve chord, and metamerically segmented muscles

Cephalodiscida [sĕf"·əl·ō·dĭsk'·ĭdə] an order of pterobranchiate hemichordates distinguished by the fact that the so-called colony is really an aggregate of discontinuous individuals

Cephalodiscus [sĕf"·əl·ō·dĭsk'·əs] the best known genus of pterobrachian Hemichordata. The individual zooids inhabit a relatively large coenecium. Each zooid has a trunk sac which rests on a stalk and from which arise arms bearing tentacles. Most are Antarctic or sub-Antarctic, usually living at depths of from 200 to 2000 feet

cephalomere [sef"·əl·ō·mĕr'] those segments of a metamerically segmented animal, particularly an arthropod, which are considered to belong to the head

cephalon [sĕf"·əl·ŏn'] brain or head as indicated by the nature of the compound. The following terms using this suffix are defined in alphabetic position:

DIENCEPHALON   PROSENCEPHALON
MESENCEPHALON   RHOMBENCEPHALON
METENCEPHALON   TELENCEPHALON
MYELENCEPHALON   THALAMENCEPHALON

Cephalopoda | [sĕf"·əl·ō·pōd'·ə] a class of mollusca containing the squids, cuttlefish, and octopus. This class is characterized by the fact that the foot is drawn out into tentacles surrounding the head. The following genera are the subject of separate entries: *Architeuthis, Loligo, Sepia*

cephalothorax [sĕf'·əl·ō·thôr"·ăks] the anterior body division of many arthropods consisting of the head fused with one or more of the thoracic segments

cephalotroch larva [sĕf'·əl·ō·trōk"] a trochophore-like stage in the development of polyclad turbellaria distinguished from a trochophore by the possession of one preoral and one postoral ciliated band

cephalous [sĕf'·əl·əs] pertaining to the head

-ceptor *abbr. term* for "receptor". The following terms using this suffix are defined in alphabetic position:

BAROCEPTOR   NOCICEPTOR
ENTEROCEPTOR   PROPRIOCEPTOR
EXTEROCEPTOR   TELOCEPTOR
INTEROCEPTOR

-cera- *comb. form* meaning "horn"

ceras (*plur.* cerata) [sēr'·əs, sēr·āt'·ə] a dorsal process on a nudibranch mollusk which contains a diverticulum from the digestive gland

Ceratium [*angl.* sə·rāsh'·ē·əm, *orig.* sĕr·āt'·ē·əm] a

genus of Dinophyceae with an ornamented test of ten plates frequently extended in spines. Common in both fresh and salt water. *C. tripos* [trī'·pŏs] is the best known marine, and *C. hirudinella* [hir'·ōō·dĭn·ĕl"·ə] the commonest fresh water species

Ceratocystis [*angl.* sĕr·ə·tŏs"·əs·ĭs, *orig.* sĕr'·ə·tō·sĭst"·ĭs] a sphaeriale ascomycete fungus of great economic importance as the causative agent of Dutch Elm disease

Ceratodus [sĕr·ăt'·ō·dəs] a genus of fossil Dipneusti found in the Triassic and Jurassic. The extant form is *Neoceratodus*

Ceratotherium [sĕr·ā'·tō·thēr"·eəm] a monotypic genus of perissodactyla containing *C. simus* [sī"·məs], the African horned "white rhinoceros". The term "white" is a corruption of the Dutch "weit" (wide) and has no reference to color

-cerc- *comb. form* meaning tail

cercal [sĕr"·kəl] properly pertaining to the tail but used in biology principally in relation to the caudal fin of fish. The following terms with this suffix are defined in alphabetic position:

GEPHYROCERCAL	HYPOCERCAL
HETEROCERCAL	ISOCERCAL
HOMOCERCAL	PROTOCERCAL

cercaria larva [sə·kâr"·ēə] the free-swimming larva of trematode Platyhelminthes (flukes), which develops from the redia

-cercid- *see* kerkid-

Cercopithecoidea [sə·kō'·pĭth·ē"·koid·ē] that super family or primate suborder which contains the Old World monkeys. They are distinguished by their downwardly directed nostrils (*see* catarrhine) and by the absence of a prehensile tail

cercopod [sə'·kō·pŏd] a posterior, segmented projection from an arthropod abdomen; the term includes the anal cerci of insects as well as analogous structures in some crustacea

Cercropia [sə·krōp'·ēə] a "genus" of moths fairly indiscriminately used for *Citheronia, Basilona,* various Saturniidae and *Bombyx.* The "cercropia moth" is actually the saturniid *Samia cercropia*

cerebellar hemisphere [sĕr'·ə·bəl"·ə] a large hemisphere on the cerebellum developed in mammals between the vermis and the flocculus

cerebellum [sĕr·ə·bĕl'·əm] one of a pair of outgrowths from the latero-dorsal wall of the mesencephalon. In birds and mammals these outgrowths form deeply convoluted lobes

cerebral aqueduct [sĕr'·ə·brəl, sĕr'·ē"·brəl] the narrow canal through the mesencephalon of the embryonic brain

cerebral eye an eye embedded in the substance of the brain of rotifers

cerebral ganglion the dorsal (i.e. supra-esophageal) ganglion of invertebrates

cerebral hemisphere one of a pair of thickened evaginations from the dorsal wall of the diencephalon. The cerebral hemispheres form the major part of the adult brain of birds and mammals

Cerebratulus [sĕr'·ə·brăt"·yū·ləs] the best known genus of nemerteans. *C. lacteus* [lăk'·tē·əs], of the Atlantic coast, is a flesh colored worm about six feet long (a twenty foot specimen has been recorded) by an inch wide. *C. albifrons* [ăl·bē'·frŏnz] of the Pacific

coast is rarely more than a foot long, dark, brown with a white head

-cerebri- *comb. form* meaning "brain"

cerebrospinal fluid [sər'·ə·brō·spīn"·əl, sər·ē"·brō·spīn"·əl] the fluid in the cavity of the central nervous system. It has essentially the composition of slightly diluted lymph and is excreted by the choroid plexus

cerebrospinal ganglion a ganglion either in the brain or spinal cord which contains the cell bodies of peripheral nerve fibers

cerebrum [sə·rēb"·brəm] that portion of the brain which consists of paired lobes derived from the diencephalon. It may be a scarcely noticeable swelling, as in fish, or the major portion of the brain, as in mammals. The term is sometimes loosely used as synonymous with brain

Ceriantharia [sər·ē·ăn·þâr·ēə] an order of zoantharian coelenterates distinguished as elongate anemone-like forms lacking a pedal disc

cervical [sər·vik"·əl, sûrv"·ik·əl] pertaining to the neck and also the neck of the uterus. These are sometimes distinguished by using the first pronunciation given above for the neck and the second for the neck of the uterus

cervical line the line or junction between the enamel and cementum in a mammalian tooth

cervical shield that part of the exoskeleton which lies immediately above the upper side of the head in many insect larvae

cervical sinus a depression in the pharyngeal region of an embryo

cervix [sûr'·viks] the junction of the vagina and uterus in many mammals

-cest- *comb. form* meaning "ribbon" or "girdle"

Cestoda [sĕs·tōd"·ə] the class of parasitic platyhelminthes containing the tapeworms. It is distinguished by the complete absence of a digestive system or mouth, and the division of the body into proglottids. The genera *Diphyllobothrium, Echinococcus, Hymenolepsis* and *Taenia* are the subjects of separate entries

Cestus [sĕs'·təs] a genus of tentaculate Ctenophora and one of the most beautiful animals in the world. The glass clear body is three feet long by two inches wide with four longitudinal rows of iridescent combs. It has borne the popular name of "Venus Girdle" since Roman times

Cetacea [sêt·ăs"·eə] the order of placental mammals that contains the whales and porpoises. They are clearly distinguished by the modification of the limbs for swimming

Cetorhinus [sêt'·ə·rīn"·əs] a genus of selachian elasmobranchs containing the basking shark *C. maximus* [măks'·əm·əs]. This huge beast, which may reach a length of 50 feet and a weight of four tons is commercially fished for its liver oil. It has numerous gill rakers and is a plankton feeder

cH a symbol sometimes used for the pH of soil

-chaen- *comb. form* meaning gape

-chaet- *comb. form* meaning "bristle"

chaeta [kē"·tə] the chitinous structures projecting from some annelid worms and most arthropods. It usually designates a stouter structure than a seta and in insects refers specifically to a jointed outgrowth from the epidermis

Chaetogaster [kēt'·ō·găst"·ər] a genus of small freshwater Oligochaetae distinguished by the lack of dorsal chaetae. It has been found free-living and also parasitic in the hepatopancreas of the snails *Lymnaea* and *Planorbis*

Chaetognatha [kēt'·ŏg·nāth"·ə] a group of coelomate bilateral animals comprising the planktonic organisms called arrowworms. They are mostly torpedo-shaped with large grasping spines at the head and a horizontal tail fin at the posterior end. The genus *Sagitta* is the subject of a separate entry

Chaetonotus [kēt'·ō·nōt"·əs] a genus of Gastrotricha with close to a hundred species. Both surfaces are covered with scales and spines and the caudal forks are short. *C. brevispinotus* [brĕv'·ē·spin"·ōt·əs] is very frequent in protozoan cultures

Chaetopterus [*angl.* kēt·ŏpt"·ər·əs, *orig.* kēt'·ō·tĕr"·əs] a weirdly aberrant species of Polychaeta sedentaria. The worm lives in large, U-shaped, parchment-like tubes in sand. The parapodia of the middle region are expanded into huge wings, or paddles, that are used to maintain a flow of water through the tube. The common Atlantic *C. pergamentaceus* [pĕr'·găm·ĕnt·âr"·ēəs] is brilliantly phosphorescent

chain ganglion one of a series of metameric ganglia lying near the aorta in vertebrates, communicating both with each other and with the visceral branches of the spinal nerve

chalaza [kə·lā·zə] one of the two apical twists of thick albumen which holds the yolk of a bird's egg in place

chalk-gland a gland depositing and voiding excess calcium in some plants

chalone [kā"·lōn"] a hormone exercising an inhibitory action on a metabolic process

-chamae- *comb. form* meaning "on the ground"

Chamaeleo [kăm·ēl"·ē·ō] a genus of Old World lizards well-known both for their ability to change color and for the enormously long tongue with which they catch insects

chamber a cavity in an animal or animal organ. The following derivative terms are defined in alphabetic order:

ANTERIOR CHAMBER
PERIBRANCHIAL CHAMBER
POSTERIOR CHAMBER

Chaos [kā"·ŏs"] a genus of amebid protozoans best known for the giant ameba *Chaos chaos* which may reach 2 mms in length. The form commonly called *Amoeba proteus* is better called *Chaos diffluens* [dif'·lyū·enz]

chaparral [shăp'·ə·əl] vegetation, primarily broadleaf, evergreen, thorny shrubs growing in regions with prolonged dry periods particularly in S.W.U.S.

Chara [kâr"·ə] a genus of Charaphyta in which some "branches" arising from the nodes adhere longitudinally to the intermodal cells to form a cortex

character a specific, determinate attribute. The following derivative terms are defined in alphabetic position:

ACQUIRED CHARACTER	SEX-INFLUENCED C.
CHARACTERISTIC C.	SEX-LINKED CHARACTER
DOMINANT CHARACTER	SEXUAL CHARACTER
EPIGAMIC CHARACTER	TAXONOMIC CHARACTER
RECESSIVE C.	

character displacement the situation which results when the differences between two species are exag-

gerated by natural selection in the areas where they overlap but not in areas where each exists alone

characteristic character the ultimate phenotypic expression in the adult of a gene or group of genes

Charaphyta [kâr'·ə·fit·ə] a group of plants variously regarded as a family of green algae or as a separate phylum. They are called stoneworts because the cell walls are strengthened with deposits of calcium carbonate. They differ from green algae by having apical growth with internodal, and nodal, portions from the latter of which whorled branches arise. The "branches" and internodal portions are large multinucleate cells. The genera *Chara* and *Nitella* are the subject of separate entries

-chas- *comb. form* meaning "branching" or "separation"

-chasko- *comb. form* meaning open, usually transliterated -chasco-

cheilostom [kil'·ə·stōm] the anterior of the three regions into which the buccal capsule of a nematode may be divided

-cheimo- *comb. form* meaning "winter" frequently transliterated chimo-

-cheir- *comb. form* meaning "hand" frequently transliterated -chiro-

Cheiropsis see Hippopotamus

chela [kē'·lə] an arthropod limb in which the ultimate joint is articulated with the base of the penultimate so as to form a claw, like that of the lobster

chelicera [kē·lis'·ər·ə] one of the first pair of appendage of arachnids. They are chelate in scorpions but modified as poison fangs in spiders and as biting mouth parts in ticks

cheliped [kēl'·ə·pēd] the principal chela-bearing limb of an arthropod such as the first periopod of a lobster

Chelonea [kēl·ōn'·ēə] the only living order of anapsid reptiles. This order contains the tortoises and turtles which are clearly distinguished from other reptiles by the possession of a bony or leathery shell. The genera *Chelonia, Chelydra, Dermochelys, Eretmochelys, Macrochelys, Pseudomys, Testudo* and *Tryonyx* are the subject of separate entries

Chelonia [kē·lōn'·ēə] the genus that has given its name to the order Chelonia. *C. mydas* [mi"·dəs] is the Green Turtle, once widely hunted for food

Chelydra [kē·li"·diə] a genus of Chelonian reptiles containing the Snapping Turtles. *C. serpentina* [sĕrp'·ĕnt·ē"·nə] is found in all states east of the Rockies (*see also Macrochelys*)

chemolithotrophic bacteria [kĕm'·ō·lith'·ō·trōf"·ik] those bacteria which are capable of synthesizing all their protoplasmic constituents from inorganic materials and which derive the energy to do this from the oxidation of inorganic materials

chemoreceptor [kĕm'·ō·rē·sĕpt"·ə] a receptor capable of perceiving dissolved or dispersed substances foreign to the normal environment

chemosynthetic autotroph [kĕm'·ō·sin·thĕt"·ik] an organism, such as bacteria of the genera *Nitrosomas* and *Thiobacillus* that derive energy solely from inorganic nutrients (*see also* photosynthetic autotroph)

chemotaxis [kĕm'·ō·taks"·is] movement in response to inorganic substances

-chers- *comb. form* meaning "dry land"

chevron bones [shĕv"·rən] one of the small

V-shaped bones which replace the hemal arch in higher vertebrates

-chiasm-   comb. form meaning "a cross" or, more properly, the shape of the Greek letter "Chi"

chiasma 1   [kī·ăz″·mə]   (see also chiasma 2) a point of contact between homologous paired chromosomes at which there is an exchange of material (see distal chiasma, proximal chiasma)

chiasma 2   (see also chiasma 1)   pertaining to other structures (see optic chiasma)

chiasma terminalization   the movement of chiasmata at the end of the meiotic prophase

chief association = stable association

chief cell   a chromophobe cell in the anterior lobe of the pituitary

-chil-   comb. form meaning "lip" or "margin" more correctly, but rarely, transliterated -cheil-

chilile   [kil′·il]   the bottoms of shallow lakes or of the shallow portions of lakes (see also bathile, pythmic)

Chilopoda   [angl. kil·ŏp″·əd·ə, orig.   kil′·ō·pŏd″·ə] the only class of opisthogoneate arthropods, containing the centipedes. The flattened shape and widely separated legs distinguish them clearly from the millipedes with which they were once associated. In many centipedes the first pair of walking legs is modified as a poison jaw. The genera Geophilus, Lithobius, Scolopendra and Scutigera are the subject of separate entries

-chim-   comb. form meaning "winter" or, by extension, "cold"

chimaera   [kī·mēr″·ə]   an organism containing tissues derived from two genetically distinct parents; most chimaeras are known only from the plant kingdom

Chimaera   a genus of holocephaline Chondrichthyes often called Rabbitfish. C. monstrosa [mŏn·strōs′·ə], adequately described by its name, reaches a length of five feet

chimpanzee   the simiid primate of the genus Pan now considered to have only the one species P. troglodytes [trŏg′·lō·dit″·ēz]. It is distinguished from the gorilla by its smaller size, friendly disposition, and unpigmented skin; and from the orang-utan by its short hair

-chion-   comb. form meaning snow

chir-   see cheir-

Chiroptera   [kir·ŏp″·tər·ə]   the order of mammals that contains the bats. The adaptation of the limbs for flight is characteristic of the group. The genera Myotis and Phyllostomus are the subject of separate entries

chitin   [kīt′·ĭn]   a substance used in the exoskeleton of all arthropods, in the skeletal elements of many other invertebrate animals and in the cell wall of many fungi. It is a high molecular weight polymer composed of n-acetylglucosamine residues joined together by β-glycosidic linkages (see also pseudochitin)

chitinase   [kīt·ĭn·āz]   an enzyme catalyzing the breakdown of chitin through the hydrolysis of α-1,4 acetylamino-2-deoxy-D-glucoside links

Chiton   [kīt′·ŏn]   a genus of polyplacophoran amphineuran mollusks. They are mostly Pacific forms though C. tuberculatus [tyū′·bĕrk·yōōl·āl″·əs] is found in the West Indies. They are so typical of the amphineurans that this order is often known as "the chitons"

-chlamy-   comb. form meaning "cloak", but extended to the mantle of mollusks and the perianth of flowers

Chlamydobacteriales   [klăm′·ĭd·ō·băk′·tēr·ē·āl″·ēz] an order of gram-negative schizomycetes found in both fresh and salt water. They occur in trichomes, sometimes with false branching, and the sheaths frequently contain iron or manganese deposits

-chlor-   comb. form meaning "grass-green"

Chlorella   [klôr·ĕl′·ə]   a genus of unicellular, non-motile, planktonic chlorophyte algae. Species of the genus have been widely used in experiments on photosynthesis and attempts have also been made to culture them on a commercial scale as a food source

DDT-dehydrochlorinase   [dē·dē·tē′·dē·hi′·drō·klôr″·ĭn·āz] an insect enzyme that detoxifies DDT

Chlorococcum   [klôr′·ō·kŏk″·əm] a genus of chlorophyte Algae with a relatively large non-motile vegetative form and biciliate zoospores

chlorocruorin   [klôr′·ō·krōō·ôr″·ĭn]   a green respiratory pigment found in some polychaete annelids. It resembles hemoglobin in having iron linked to a porphyrin prosthetic group

chlorophyll   [klôr′·ə·fil, klôr·ō·fil]   the photosynthetic pigment of plants. It is a magnesium centered porphyrin containing, as its most distinctive constituents, a hydrophilic 5-membered carbocylic ring and a lipophilic phytol tail

chlorophyllase   [klôr′·ō·fil′·āz]   an enzyme that catalyzes the hydrolysis of chlorophyll to phytol and chlorophyllide and also the transfer of chlorophyllide

Chlorophyta   a sub-phylum of Algae known as green algae, a name both descriptive and distinctive. The genera Acetabularia, Chlorella, Chlorococcum, Fritschiella, Gonium, Oedogonium, Spirogyra Tetraspora and Vaucheria are the subject of separate entries as is also the order Volvocales

chloroplast   [klôr″·ō·plăst′]   an organnelle present in photosynthetic plants containing chlorophylls and, sometimes, other pigments (see also grana)

Choeropsis   [kēr·ōp′·əs]   a monotypic genus of Hippopotomidae containing C. liberiensis [li·bēr′·ē·ĕn″·sĭs], the Pigmy Hippopotamus

Choloepus   [kōl·ēp′·əs]   the genus of xenarthran mammals containing the Two-toed Sloth C. didactylus [di·dăct″·il·əs]. It has actually five toes but only two fingers

chondriomere   [kŏnd′·rē·ō·mēr]   that section (the middle section) of a spermatozoa in which the mitochondria are found

Chondrus   [kŏn″·drəs]   a genus of rhodophyte algae. C. crispus [krĭsp′·əs] is "Irish Moss" or "Carragaen" widely used for food

Choanichthyes   [kō′·ən·ĭk″·thēz]   a subclass of Osteichthyes distinguished from all other boney fishes by the presence of internal nares. The orders Crossopterygii and Dipneusti are defined in alphabetic order

choanocyte   [kō″·ən·ō·sit]   a cell typical of sponges which possesses a thin protoplasmic collar round a flagellum. A similar type of cell is found among some flagellate protozoa

cholecystokinin   [kōl·ĕ·sĭs″·tō·kīn″·ĭn]   a hormone secreted in the duodenum that stimulates the gall bladder to release bile

choline   [kō′·lēn]   (β-Hydroxyethyl)-tri-methylammonium hydroxide. A water soluble nutritional factor frequently classed as a vitamin (see also acetylcholine)

cholinergic fiber   [kōl′·ən·ərj″·ik]   a nerve fiber the termination which liberates acetylcholine when activated

cholinesterase   [kōl′·ən·ĕst″·ĕr·āz]   an enzyme that catalyzes the hydrolysis of choline esters into choline

and the appropriate acid (*see also* acetylcholinesterase)
-chondr- *comb. form* meaning both "granule" and "gristle" ("cartilage")

chondral bone [kŏn"·drəe] a bone that arises in or around cartilaginous precursors (*see also* endochondral bone, perichondral bone)

Chondrichthyes [kŏn·drĭk'·thēz] that class of gnathostomatous craniate chordates which contains those fishes which differ from the Osteichthyes in lacking a bony, as distinct from an occasionally calcified, skeleton. The subclass Elasmobranchii and Holocephali are defined in alphabetic position

chondroblast [kŏn"·drō·blăst'] a cell which secretes the matrix of cartilage

chondroclast a cell the secretions from which lyse the matrix of cartilage

chondrocranium [kŏn·drō·krān"·ēəm] a skull, or skull rudiment, consisting entirely of cartilage. In some elasmobranchs, it may be calcified but is never ossified

chondrocyte [kŏn"·drō·sit'] any cell in a cartilagenous matrix

chondrosis [kŏn'·drē·ōs"·ĭs] a cartilagenous joint (*see also* synchondrosis)

Chondrostei [kŏn·drŏst·ēi] a superorder of osteichthyes retaining the heterocercal tail but having the skeleton largely reduced to cartilage. The genera *Acipenser* and *Polyodon* are the subject of separate entries

-chord- *comb. form* meaning "cat gut", or by extension, any straight cylindrical form. Now thoroughly confused with -cord- which is the English form of the same word. The following terms using the suffix -chord are defined in alphabetic position

HYPOCHORD   NOTOCHORD   STOMOCHORD

chordal usually, in biology, means pertaining to the notochord

Chordata [kôr·dāt'·ə] a phylum of bilateral coelomate animals containing the vertebrates and their allies. The phylum is characterized by the possession of a hollow, dorsal nerve cord, a notochord, and gill slits in the anterior region of the alimentary canal at some stage of development. In some groups all of these characters occur only in embryonic or larval stages

-chore- literally "a place", thus giving rise to two common suffixes

-chore 1 (*see also* -chore 2) a suffix indicating the agent of plant dispersion. The very numerous compounds using the suffix in this sense are not defined in this dictionary since the meanings are self evident from the roots. Typical examples are anemonochore [ən·ĕm'·ŏn·ō·kôr"] (wind dispersed) and zoochore [zō"·ō·kôr'] (dispersed by animals)

-chore 2 (*see also* -chore 1) a suffix indicating a place or part. The following terms using the suffix in this sense are defined in alphabetic position:

ALLOCHORE   EURYCHORE
BIOCHORE

chori- *comb. form* meaning "separate"

chorion [kôr'·ēən] the outer layer of the embryonic sac of a mammal consisting of both trophoderm and mesoderm elements. The outermost extraembryonic membrane in birds and reptiles. The vitelline membrane in tunicates and fishes, the shell of an insect egg, a membrane round the egg outside the vitelline membrane and produced by the follicle cells in the ovary

choroid plexus [kôr'·oid] an extension of a plexus

of blood vessels into the cavity of the brain. It secretes the cerebro-spinal fluid

-chrom- *comb. form* meaning "color". In biology the substantive "chrome" is usually confined to pigments and "chromatin" for those portions of a nucleus which stain readily after histological fixation. "Chromatism" and "chromatic" usually refer to pigments while the termination "-chromous", though often used as synonymous with -chromatic, pertains to color as distinct from the causative agent of color

chromaffin cell [krōm'·ə·fĭn] a cell first identified as staining heavily in histological techniques using chromium. These cells secrete adrenalin and are largely concentrated in the adrenal medulla. They originate from sympathetic ganglia and are also known as chromaphile cells

chromatic figure the arrangement of the chromosomes in a mitotically dividing cell (cf. amphiastral figure)

chromatid [krōm"·ə·tid] one of the pair of elongate structures which together form a chromosome

chromatin [krōm"·ə·tĭn] those parts of the nucleus which stain densely in nuclear stains, mostly DNA and RNA

chromatophorotropic hormone [krōm"·ə·tō·trŏp'·ik] = melanocyte-stimulating hormones

chrome in biology a pigment (*see* remarks under -chrom-). The following derivative terms are entered in alphabetic position:

BIOCHROME	OMMOCHROME
CYTOCHROME	PHYTOCHROME
ENDOCHROME	QUINONE BIOCHROME
INDIGOID BIOCHROME	SCHEMOCHROME

chromocenter [krōm"·ō·sent·ər] the area of a chromosome immediately on each side of the centromere

chromomere [krōm"·ō·mēr] deeply staining structures found along the length of a chromosome, particularly in the leptonema stage. The term is not specific but is applied to any division recognizably constant in size and position, such as the bands on the salivary gland chromosomes of diptera, or the lumps from which loops project in amphibian bottle-brush chromosomes

chromonema [krōm'·ō·nē"·ma] coiled intertwined threads of which a chromosome is constituted

chromophobe cells [krōm"·ō·fōb'] cells which do not stain under specific circumstances, particularly those cells of the pituitary which are thus contrasted with the acidophyles and the basophils

chromoprotein [krōm'·ō·prō"·tēn] a compound protein which contains a pigment, usually metal containing, but sometimes a carotenoid. Many (e.g. hemoglobin) function as oxygen carriers

chromosomal sterility a form of sterility, present in many hybrids, resulting from structural differences between chromosomes

chromosome [krōm"·ə·sōm'] thread-like bodies into which the hereditary material of the nucleus is organized. They usually become apparent only in the course of cell division. The following derivative terms are defined in alphabetic position: (*see also* allosome, androsome, autosome, zygosome)

ACROCENTRIC C.	HETEROCHROMOSOME
DAUGHTER CHROMOSOME	HETEROPYCNOTIC C.
GIANT CHROMOSOMES	HOMOLOGOUS C.
HETEROBRACHIAL C.	ISOCHROMOSOME

LAMP BRUSH C.
METACENTRIC C.
POLYTENE C.
SEX CHROMOSOME
TELOCENTRIC C.

TELOMITIC CHROMOSOME
W CHROMOSOME
X-CHROMOSOME
Y-CHROMOSOME
Z-CHROMOSOME

-chron- comb. form meaning "time"

chronic pertaining to time. -chronous, not used in this work, is synonymous.

-chrys- comb. form meaning "gold"

chrysalis [krĭs'·ə·lĭs] a popular name for the pupa of lepidoptera

Chrysomonadales [krī''·sō·mŏn·əd·āl''·ēz] an order of chrysophycean chrysophytes containing those forms in which the vegetative phases are motile (but see next entry)

Chrysomonadida [krī'·sō·mŏn''·ĭd·ə] an order of mastigophorous protozoa distinguished by the presence of yellow or brown chromoplasts. Most biologists regard these forms as algae, in which case the Chrysomonida of the zoologists corresponds to the Chrysomonadales of the botanists

Chrysophyta [angl. krĭs·ŏf''·ət·ə, orig. krī'·sō·fĭt''·ə] a phylum of Algae distinguished by the fact that the chromatophores are golden since carotene and xanthophylls are sufficiently predominant to mask the color of the chlorophyll

Chrysops [krĭs''·ŏps] a large genus of tabanid Diptera called "Horse Flies" in the U.S. and "Mango Flies" in those parts of the world in which they transmit *Loa*

chyle [kĭl] either lymph rendered milky by emulsified fat or partially digested nutrients in the alimentary canal

chyme [kĭm] the partially digested contents of the stomach

-chyme termination indicating animal tissue—the botanical equivalent is -enchyma (see collenchyme, mesenchyme)

-chytr- comb. form meaning a small pot

-cil- comb. form meaning "hair" (see cnidocil)

ciliary [sĭl'·yə·ē] pertaining to the eyelid, to the eyebrow or to cilia

ciliary glands modified sweat glands in the eyelids

ciliary muscle the muscle in the ciliary process of the eye which moves the lens

ciliary process the free, anterior edge of the choiroid in the eye

Ciliata [sĭl'·ē·āt''·ə] a class of Protozoa distinguished by the possession of numerous cilia at some stage of their life history and a sub-pellicular system of kinetosomes even when cilia are temporarily absent. Most have polymorphic nuclei. The subclasses Holotricha and Spirotricha are the subject of separate entries (see also Protociliata)

ciliated epithelium columnar, cuboidal, or pseudostratified columnar epithelium, the free surface of which bears cilia

cilium [sĭl''·ē·əm] a prominent vibratile ectoplasmic process found on many cells throughout the animal kingdom. Each cilium consists of nine pairs of peripheral filaments wrapped round a central pair, the whole embedded in a matrix

Cimex [sī''·mĕks] a genus of blood sucking hempteran insects. *C. lectularius* [lĕk'·tyūl·âr''·əs] is the human parasite commonly called the bed-bug

Cinchona [sĭn·kōn'·ə] a genus of South American rubiaceous trees, several species of which yield quinine from the bark

-cincinnal- comb. form meaning curly or curled

-cinct- comb. form meaning "girdle"

-cine- comb. form meaning "move" or "movement" frequently, and properly, transliterated -kine-

cingulum [sĭng''·gyōō·ləm] literally, a girdle and used for almost any structure having this appearance including the ciliary zone on the disc of a rotifer, the clitellum of oligochaete annelids, the connecting edges of the frustules of diatoms, a circular ridge on the crown of some teeth, the junction between the stem and the root of a plant and a colored band on gastropod shells

Ciona [sī·ōn'·ə] a monotypic genus of ascidians of wide distribution. *C. intestinalis* [in·tĕst'·in·āl''·is] is four or five inches long by two thick. Its shape and white, almost transparent appearance make it easy to identify

circadian [angl. sûr·kād'·ēən, orig. sûr'·kə·dē''·ən] repeated more or less daily—i.e. on a 23 hr to 25 hr cycle (cf. diurnal). The anglicized pronunciation, now almost universal, is unfortunate since it conceals the meaning

circular ocellus a closed ocellus with a definite lens

circulation is primarily applied to the movement of blood. The following derivative terms are defined in alphabetic position:

PULMONARY C.   RESPIRATORY C.   SYSTEMIC C

circulatory system the sum total of the vessels that convey fluid, usually blood, throughout organisms (see also closed circulatory system, open circulatory system)

-circum- comb. form meaning "round" in the sense of surround

circumfluence [sûr·kŭm'·flōō·ĕns] a flowing together of pseudopodia around the food particle (cf. circumvallation)

circumscript muscle used as the antithesis of diffuse muscle in those forms in which the latter occurs

circumvallate papilla [angl. sûr·kŭm''·vəl·āt, orig. sûr'·kŭm·văl''·āt] one of many large, club shaped, sensory papillae in the tongue, sunk deeply into the mucous membrane and surrounded by a circular furrow.

circumvallation [sûr'·kŭm·văl''·ă·shŭn] the capture of prey by a protozoa, which throws a large cup round the object and the adjacent water (cf. circumfluence)

-cirr- comb. form meaning a "curl" of hair. Very commonly misspelled -cirrh-

Cirripedia [sĭr'·ē·pēd''·ēə] a subclass of crustacean arthropods containing the forms called barnacles as well as the parasitic Rhizocephala. This last group is regarded by many as a separate class leaving the Cirripedia to consist only of the order Thoracida. The divisions Balanomorpha, Lepadomorpha and Rhizocephala are the subject of separate entries

cirrus [sĭr''·əs] in zoology any flexible, often tactile projection. In general a cirrus is shorter than a flagellum, and longer than a papilla. In many invertebrates synonymous with penis and used also for projections from the stalks of crinoid echinoderms

cis configuration [sĭs] the condition in which two mutants at different sites within a cistron are located on the same chromosome

cis-trans effect the condition in which two mutants, at different sites within a cistron may act as alleles in the trans configuration or as a single gene in the cis configuration

cistron [sǐs·trən] a functional genetic unit thought to control the synthesis of a single product (*see also* recon, muton, configuration)

Citheronia [sìth'·ər·ōn"·ēə] a genus of large moths. *C. regalis* [rə·gāl"·ìs] is the largest N. American moth, having a wing spread of up to six inches. This is the "Royal Moth" commonly miscalled *Cercropia*, The genus *Basilona* is the subject of a separate entry

citric cycle *see tricarboxylic cycle*

l-citrulline [ĕl·sìt·trŭl"·ēn] α-amino-δ-ureidovaleric acid. $H_2NCONH(CH_2)_3CHNH_2COOH$. An amino acid, not essential for rat nutrition which, though first indicated from the watermelon, appears most commonly to be synthesized from ornithine by microorganisms

-clad- *comb. form* meaning "branch"

Cladocera [klə·dŏs'·ər·ə, klä·dŏs'·ə·rə] an order of branchiopod crustacea containing the animals called water-fleas. They are distinguished from the closely allied Conchostraca in being laterally compressed and enclosed within a bivalved shell and by the biramous second antennae which serve as organs of locomotion. The genus *Daphnia* is the subject of a separate entry

cladode [klǎd'·ōd] a flattened portion of a stem, resembling a leaf in function and structure (cf. phyl-

Cladonia [klǎd·ōn'·ēə] a genus of lichens. The species *C. rangiferina* [rǎnj·ìf'·ə·in"·ə] is a "reindeer moss", a major food of reindeer and caribou

-cladous pertaining to a branch. The very numerous compounds with this suffix are not defined since the meaning is self evident from the roots employed (e.g. brachycladous = short branched)

clamp cell hyphal connections, or bridges that, in the basidiomycete fungi, permit the migration of nuclei and thus the maintenance of the dikaryotic condition

clan in ecology, a local group of organisms of a restricted number of types intermediate between a colony and a society

clandestine evolution [klǎn·dĕst"·ìn] that which takes place in the larva and is transmitted by neoteny but which subsequently forms part of the genetic composition of the adult

clas- *see klas-*

clasper 1 (*see also* clasper 2) a modified portion of the pelvic girdle of male elasmobranch fish, serving as an intromittent organ

clasper 2 (*see also* clasper 1) a modified appendage, not functioning as a penis, used by a male to clasp a female, particularly in insects

class a taxon of either the plant or animal kingdom ranking immediately below phylum and above order. The Insecta, for example, are a class of the phylum Arthropoda and the Coleoptera (beetles) are an order of Insecta (*see also* subclass, superclass)

-clast- *comb. form* meaning to shatter (*see* chondroclast, osteoclast)

-claus- *comb. form* meaning "closed"

-clav- *comb. form* meaning "club"

Clavaria [klə·vâr'·ēə, klăv·âr·ēə] a genus of aphyllophorale basidiomycete fungi often called coral fungi, a term descriptive both of the general shape and the brilliant color of many

clavate [klǎv'·āt] club-shaped

Claviceps [klăv"·ē·sĕps'] a genus of clavicipitale ascomycete fungi best known for *C. purpurea* [pŭr·pyōōr'·ēə], the dried sclerotium of which is "ergot of rye" containing numerous poisonous compounds. These were at one time principally used as abortifacients. It was later used as a hallucinogen since it contains large quantities of lysergic acid some of which may be naturally transformed into LSD

Claviciptales [klǎv'·ē·sìp·ìt·ăl"·ēz] an order of Ascomycetes distinguished by having the asci in small perithecia in a stalked stroma. Most are plant parasites but a few parasitic on insects. The genus *Claviceps* is the subject of a separate entry

clavicle bone a bone in the pectoral girdle which articulates with the scapula at one end and usually lies free at the other. In Crossopterygian fish it is fused to the anterior end of the cleithrum (*see also* interclavicle bone)

clay soil particles, or soil primarily composed of particles, less than 2 microns in diameter

cleavage the process by which a fertilized egg divides into blastomeres. The following derivatives are defined in alphabetical order:

BILATERAL CLEAVAGE	INDETERMINATE C.
BIRADIAL CLEAVAGE	MEROBLASTIC CLEAVAGE
DETERMINATE C.	RADIAL CLEAVAGE
DISCOIDAL CLEAVAGE	SPIRAL CLEAVAGE
HOLOBLASTIC C.	SUPERFICIAL CLEAVAGE

cleavage cell = blastomere

-cleid- *see* -kleid-

cleidoic egg one, like that of a bird, which is enclosed in a more or less moisture-proof container or shell

cleithrum bone a double bone lying alongside and fused with the chondral bones of the pectoral girdle in some Amphibia and lying between the pectoral girdle and the opercular complex in teleost fish. See *also* postcleithrum bone, supracleithrum bone)

-clem- *comb. form* meaning "branch" (cf. -clad-) but by many writers used only in the sense of "twig"

-clept- *comb. form* meaning "thief" (*see* myrmecoclepty)

cleptobiosis [klĕpt'·ō·bi·ōs"·ìs] a form of symbiosis based on theft, best known in ants in which one species will systematically steal the collected food of another laboring species

-cles- *comb. form*, properly transliterated -kles- meaning "closed"

-climac- *comb. form* meaning "ladder"

climate the sum total of the meteorological phenomena of any given area of the planet. The following derivative terms are listed in alphabetical order:

BIOCLIMATE	MICROCLIMATE

climatic ecotype an ecotype effected by climatic as distinct from soil conditions (*see also* ephaphic ecotype)

climatic race one which is adapted to a different climatic environment from that in which its nearest relatives live

climax a plant association which has reached its full development and is likely to remain stable unless disturbed by climatic or other environmental changes. The following derivative terms are defined in alphabetic position (*see also* unit, zone):

DISCLIMAX	PRECLIMAX
EDAPHIC CLIMAX	PREVAILING CLIMAX
MONOCLIMAX	QUASICLIMAX
PANCLIMAX	SERCLIMAX
POLYCLIMAX	SUBCLIMAX
POSTCLIMAX	TEMPORARY CLIMAX

climax community a stable climax

climax unit a group of associated species contributing to a climax

climax some local minor changes in the composition of a climax due to local changes in environment

-clin- *comb. form* hopelessly confused between two roots meaning "bed" or "slope" and "bend", the last often extended to "tend towards". The adjectival form in the "bend" sense is -clinal and that of the "tend" sense -clinic. The ecological "cline" is presumably from "slope" but the connection is not clear

cline [klīn] in the sense of a slope or gradient and particularly, in ecology, a group, within a population of one species, which shows differing characters reflecting a geographic or ecological situation; particularly used when such characters grade gradually from one to the other. The following derivative terms are defined in alphabetic position:

ECOLINE   HIBRID CLINE
GENOCLINE   THERMOCLINE
GEOCLINE   TOPOCLINE

clinoid process a process on the orbitosphenoid bone

Clione [klī·ōn"·ē] a genus of scaphopod mollusks without shells and with triangular paddles. *C. limacina* [lim"·ə·sīn"·ə], a pale translucent blue slug-shaped form, an inch and a half long, is an important component of Arctic plankton, sometimes drifting as far south as the New England coast. In the Arctic they are so numerous as to be an important constituent of the food of balaenid whales

clisere [klī"·sēr] a condition intermediate between two seres (*see also* preclisere, postclisere)

clitellum [klət'·ĕl·əm, klī'·tĕl·əm] a glandular annular swelling anterior to the genital apertures of land-dwelling oligochaete worms. It is used for the secretion of the egg case

-clito- *comb. form* meaning a "hillside" or "slope"

clitoris [klit'·ər·ĭs] the vestigial penis of a female mammal

cloaca [klō'·ək·ə, klō·āk·əˌ] a posterior involution of the body into which open both the anal and urinary apertures and, in many forms, the sex ducts

cloacal membrane a temporary ectodermal-endodermal plate closing the opening of the cloaca into the proctodeum of the embryo

clone [klōn] a group of organisms descended asexually from a single ancestor

Clonorchis [klōn"·ôrk·ĭs, klŏn"·ôrk·ĭs] a genus of trematodes of which the best known is *C. sinensis* [sĭn·ĕn'·sĭs], the Chinese liver fluke. This human parasite is distributed all over the far east. The miracidia develop in water snails, the cercariae liberated from the sporcysts form metacercariae in cyprinid fish. When these are eaten raw, human infection occurs

closed association one in which there is no room for further growth

closed circulatory system one in which arteries join veins through arterioles and venules, and capillaries (*see also* open circulatory system)

closed community one in which no further species can find lodgement, either because of unfavorable environmental conditions or because all ecological niches are already occupied

closed fertilization the condition of a plant which is fertilized by its own pollen

closed formation an assemblage of plants so closely placed in the soil, that others cannot obtain a foothold among them

closed society a group of animals, human or not, which will not admit any addition to their ranks

closing plate a membrane formed by the junction of a pharyngeal pouch with a visceral groove in the embryo

-closter- *comb. form* meaning "spindle"

Clostridium [klŏs·trĭd'·ēəm] a genus of spore forming anaerobic soil Bacillaceae. *C. botulinum* [bŏt'·yū·lĭn"·əm] produces the deadliest toxin known in imperfectly sterilized canned and bottled food. *C. tetani* [tĕt'·ən·ī] causes tetanus, and *C. septicum* [sĕpt'·ĭk·əm] gas gangrene in imperfectly treated wounds. Many, however, are harmless and *C. acetobutylicum* [ăs·ēt'·ō·byūt·ĭl"·ĭk·əm] is used in the commercial production of butyl alcohol

Clupea [klōō·pē"·yə] a commercially important genus of teleost fishes. *C. harengus* [hăr·ĕn'·gəs] is the atlantic herring and *C. pallasi* [păl·ăs"·ī] the pacific herring

-clyp- *comb. form* meaning a round shield, larger than -pelt- (cf. -pelt- and scut-)

clypeus [klip"·ēəs] the shield shaped median anterior plate on an insect head

-clys- *comb. form* meaning "wave action" or "tide"

-clyst- *comb. form* meaning "pipe" (*see* physoclystous)

CMP cytidine 5'—phosphate

-cnem- *comb. form* meaning "knee". Frequently confused with -nema- (*see* metacneme, protocneme)

cnemial pertaining to the tibia

cnemidium [nē·mĭd"·eəm] that portion of a bird's leg which bears scales and not feathers

-cnid- *comb. form* meaning "nettle" and by extension "sting"

Cnidaria [nī·dâr"·ēə] a term used to denote a subphylum of coelenterata when the ctenophores are included, the latter then being placed in the Acnidaria

cnidoblast = nematocyte

cnidocil [nīd"·ō·sĭl'] a projection on a nematocyte which triggers the discharge of the nematocysts

cnidocyst [nĭd"·ō·sĭst'] the rigid oval capsule containing the eversible thread in the cnidoblast

CNS a widely used abbreviation for "central nervous system", meaning, in general, the brain and spinal cord

co- *see* -con-

co-acervate [kō·ăs'·ər·vāt] aggregates of colloidal particles in suspension

co-enzyme a non-protein component of many enzymatic reactions to which it is essential, but from the other components of which it may be separated (cf. activator)

cocaine a drug once derived by extraction from the leaves of *Erythroxylon coca*. Now synthesized and largely replaced by other compounds

-cocc- *comb. form* meaning "grain" or "kernel" and, by extension, "nucleus"

-cocco- *comb. form* from same root as -cocc- but in biology usually meaning "berry" or "chamber"

Coccus [kŏk"·əs] a genus of coccid insects containing *C. cacti* [kăk"·tī] from which the dye carmine is extracted

coccygeal body [kŏk'·sĭj"·ēəl] a small, heavily vascularized body situated just in front of the apex of the coccyx of great apes; the term is sometimes applied to the coccygeal gland

coccygeal gland properly, the uropygeal gland but sometimes applied to a coccygeal body

-cochl- *comb. form* meaning either "spoon" or "spiral shell"

cochlea [kŏk″·lēə] the spiral ("shell-like") half of the labyrinth of the inner ear

cocoon 1 [kə·kōōn″] (*see also* cocoon 2) the woven silk case which surrounds the pupa of many Lepidoptera, or the eggs of many spiders

cocoon 2 (*see also* cocoon 1) any invertebrate structure in which numerous eggs, capsulated or uncapsulated, are enclosed in a common capsule, or shell such as that secreted from the clitellum of some oligochaete annelids

-cod- *comb. form* meaning "head" in the sense of a lump on the end of a stalk but hopelessly confused in compounds with -codi- and -codo-

codon 1 [kō″·dŏn″] (*see also* codon 2) a unit in the cistron controlling the synthesis of a single polypeptide unit of the genetic code

codon 2 (*see also* codon 1) the umbrella of a hydrozoan medusa (*see also* entocodon)

codonic [kō·dŏn′·ik] pertaining to codon 2 (*see* adelocodonic, phanerocodonic)

Codosiga [*angl.* kō·dŏs″·ig·ə, *orig.* kō′·dō·sig″·ə] a genus of Mastigophora with the form of a stalked colony of collared cells, each like a choanocytes of a sponge. It has been suggested that Protozoa of this type may have been ancestral to the Porifera

-coel- *comb. form* meaning "heaven" hence, following the Greek view of this space, "a cavity". Properly, but very rarely, transliterated -koelo-

coel [sēl] (*see also* enteron) a cavity. The following terms with this suffix are defined in alphabetic position:

AXOCOEL	METACOEL
BLASTOCOEL	MYOCOEL
ENTEROCOEL	NEPHROCOEL
ENTEROHYDROCOEL	NEUROCOEL
EPICOEL	PSEUDOCOEL
EXOCOEL	SCHIZOCOEL
GONOCOEL	SCLEROCOEL
HAEMOCOEL	SOMATOCOEL
MESENCOEL	SPONGOCOEL
MESOCOEL	TELOCOEL

Coelenterata [sēl·ĕnt″·ər·āt·ə] a phylum of the animal kingdom distinguished by the presence of nematocysts, a diploblastic body wall, and a single internal cavity opening to the exterior only at the oral end. So defined the phylum excludes the Ctenophora (comb-jellies) which lack nematocysts. The classes, Anthozoa, Hydrozoa and Scyphozoa are the subject of separate entries

coelenteron [sēl·ĕnt″·ər·ŏn] the single cavity of the coelenterates

coeliac artery [sēl′·ē·ăk] an artery arising from the dorsal aorta near the radix, and dividing into the gastric, splenic and hepatic arteries

coeliac ganglion a prevertebral ganglion connected with the chain ganglia of the thoracic region

coeloblastula [sēl′·ō·blast″·yūl·ə] a blastula having a central cavity. This is what is usually meant by the use of the word blastula without qualification

coelogastrula [sēl′·ō·găst″·rōō·lə] a gastrula derived from a coeloblastula

coelom [sē″·lōm, sē″·ləm] (*see also* coel) the body cavity that is limited on all surfaces by the mesoderm. This is also called the "true coelom" to distinguish it from other types of body cavity

Coelomata [*angl.* sēl·ŏm″·ə·tə, *orig.* sēl′·ō·māt″·ə] a term coined to define those phyla of the animal kingdom which possess a true coelom. This taxon, therefore, embraces the Chaetognatha, Hemichordata, Pogonophora, Phoronidia, Ectoprocta, Brachiopoda, Siphunculida, Echinodermata, Prochordata, and Chordata

coelomic funnel [sēl·ŏm″·ik] a ciliated funnel, draining fluid from the coelom (= nephrostome)

coelomocyte [sēl·ōm′·ō·sit, sēl·ŏm·ō·sit] cells analogous to hemocytes, but found free in coelomic fluid (*see also* pseudocoelomocyte)

coelomoduct [sēl·ōm′·dŭkt, sēl·ŏm·ō·dŭkt] any duct which connects the coelom to the exterior, usually applied to the terminal tubule of nephridia

coelomostome [sēl′·ō·stōm, sēl′·əs·tōm] a primary coelomic funnel or nephrostome derived directly from an expansion of the coelom, in distinction to other metanephridia in which the connection to the coelom is secondary

-coelous possessing, or in the form of, a cavity, particularly that of the centrum of a vertebra. The following terms with this suffix are defined in alphabetic position:

AMPHICOELOUS	DICOELOUS
ANOMOCOELOUS	SCHIZOCOELOUS

-coen- *comb. form* properly from a Latin root meaning "filthy" but widely used as the transliteration of a Greek root, (properly -koin-) meaning "sharing" or "togetherness"

coen the sum total of the components of an environment. Occasionally used as an abbreviation for coenosis (*see also* abiocoen, biocoen)

coenocyte [sēn″·ō·sit′] an organism consisting of many plant cells within a single cell wall. The botanical equivalent of the zoologist's syncytium

coenosarc [sēn″·ō·särk] the non-living portion of a hydrozoan hydroid colony

coenurus larva [sēn·yūr′·əs] a type of cysticercus in which the inner wall of the bladder proliferates groups of other cysticerci which do not become detached

coenzyme A [kō′·ēn·zim″] the most important coenzyme in biokinetic systems. The systematic name (3′-phosphoadendine diphosphate-pantoyl-β-alanilecysteamine) is very rarely used

cohesion-tension theory the theory that trees can raise a column of liquid to a great height, owing to the cohesion of water molecules that produces tensile strength in thin columns

-col- *comb. form* meaning "to dwell". The *adj. terms* -coline and -colous are both used but the former appears to be obsolescent

colchicine [kŏl″·chə·sēn, kōl″·chə·sēn] an alkaloid, derived from the liliaceous herb *Colchicum autumnale* and which inhibits the formation of the mitotic spindle

-cole- *comb. form* meaning "sheath" better transliterated -kole-

Coleoptera [kōl′·ē·ŏpt″·ərə] the order of insects that contains the beetles. They are distinguished by the metamorphosis of the anterior wings into strongly chitonized elytra, or wing covers. There are more kinds of beetle than of any other animal

coleoptile [kōl′·ē·ŏpt″·il] the first true leaf of a monocotyledon

colicin [kŏl'·is·ĭn] any of several surface produced substances from bacteria which serve to inhibit or kill other bacteria

-coline adj. term. from -col- referring to habitat. For compounds see under -colous

coliphage [kŏl'·ē·fāj, kŏl'·ē·fâj] a bacteria phage specific to Escherichia coli

-coll- comb. form confused from one Latin and one Greek root and therefore meaning either "neck" or "glue". The prevalence of collagenous connective tissues in animals has led to a further meaning "connective tissue", even though, in plants, these may be cellulosic

collagen [kŏl·ə·gĕn] literally, that which yields glue. It is a protein, usually occurring in fibrous form and is the chief constituent of white connective tissue and a major constituent of fibrous cartilage

collar cell = choanocyte

collaterial gland [kŏl'·ə·tēr"·ēəl] glands in the wall of the oviduct of some arthropods which secrete a sticky material holding the eggs in masses

collecting cell a round cell lacking chlorophyll, immediately beneath the palisade cells in the leaf

collecting hair a hair on a stile, intended to entrap pollen

collecting tubule a functional unit of a compound kidney into which several convoluted tubules open and which itself opens into the ureter

collector nerve a nerve found in some fish, connecting additional anterior segmental nerves to an appendage

Collembola [kŏl·ĕm'·bŏl·ə] an order of wingless insects distinguished from all other insects by the presence of nine postcephalic segments the last six of which bear ventral appendages. These characteristics are considered by some to distinguish the Collembola so clearly from all other insects that they should be placed in a separate class of the Arthropoda

collenchyma [angl. kŏl·ĕn'·kəm·ə, orig. kŏk'·ĕn·kim"·ə] elongated prismatic cells in seed plants which, though they function as the supporting structures of young leaves and stems, are still capable of growth

collenchyme [angl. kŏl·ĕn'·kim, orig. kŏl'·ĕn·kim"] an embryonic tissue of cells loosely embedded in a gelatinous matrix

colon [kŏl'·ən] the terminal section of the alimentary canal of vertebrates. The term is often extended to corresponding regions of the invertebrate gut. The following derivative terms are defined in alphabetic position:

ASCENDING COLON   MESOCOLON
DESCENDING COLON   SIGMOID COLON
ILEOCOLON   TRANSVERSE COLON

colony 1 (see also colony 2) a natural community of two or more species. The term does not designate any degree of mutual interdependence

colony 2 (see also colony 1) a visible, laboratory-induced, growth of microorganisms usually on a solidified medium; many colonies are true clones; most descriptive terms of colonies are self-explanatory (e.g. dull colony, lobate colony); specific colony designations are curled colony (one in the form of parallel, wavy chains), erose colony (one having an irregularly shaped border), punctiform colony (one which is only just apparent to the naked eye), raised colony (one with a raised edge), daughter colony (a colony derived, either by transfer or outgrowth from the bacterial colony)

color a sensation induced in the occipital region of the cerebrum by the differential effect of various wavelengths of photon energy on the retina. These sensations are vocalized in conventional terms (red, blue, etc.) established by custom and correlated with wavelength. The term is used in biology in the sense of pigment (see pigment, chrome), as descriptive of a particular color (color 1, below), as descriptive of the method by which a color is produced (color 2, below), and the supposed function of color (color 3, below)

color 1 (see also color 2, 3) as descriptive of a particular color

color 2 (see also color 1, 3, pigment, chrome) the method by which a color is produced (in this sense see structural color, Tyndall color)

color 3 (see also color 1, 2) color as a functional phenomenon. In this sense, the following terms are defined in alphabetic position:

ANTICRYPTIC COLOR   EPIGAMIC COLOR
APATETIC COLOR   EPISEMATIC COLOR
APOSEMATIC COLOR   PROCRYPTIC COLOR
CONFUSING COLOR   PSEUDOPISEMATIC COLOR
CRYPTIC COLOR   SEMATIC COLOR
DYMANTIC COLOR

colostrum [kŏl·ŏst"·rəm] the milk produced by the mammary glands immediately following birth

-colous pertaining to the location, or habit, of growth. The very numerous compounds using this suffix are not defined since their meaning is obvious from the roots (e.g. arenicolous = sand dwelling)

-colp- comb. form meaning "bosom", better transliterated -kolp-

columnar epithelium epithelium of which each cell is in the form of a simple elongate column

comb 1 (see also comb 2) an erect pectinate caruncle on the head of a bird

comb 2 (see also comb 1) any structure which has closely apposed teeth along one side (see pollen comb, tactile comb)

comb rows the band of ctenes on a ctenophore

combat dance an exhibition between two males either of which may desire to establish dominance over the other, without causing him bodily harm

-comma- comb. form confused between a Latin and a Greek root and therefore variously meaning "ornamentation" (better -komma-) and "septum"

Commelinaceae that family of monocotyledons which contains the spiderworts. Distinguishing characteristics are the transformed anthers and stamen hairs together with the complete differentiation of the perianths into calyx and corolla. The genus Tradescantia is the subject of a separate entry

commensalism the condition of two organisms living together for the purpose of sharing each other's food

commiscuum an assemblage, either real or postulated, of all members of a taxonomic group which are capable of interbreeding (cf. convivum)

commissure 1 (see also commissure 2) literally, a joint or seam. Particularly, in botany, where plant organs unite

commissure 2 (see also commissure 1) a bundle of nerve fibers connecting two ganglia (see hippocampal commissure, pallial commissure)

common carotid artery the portion of the ventral aorta between the third and fourth arches, that carries blood for the carotids alone, in those tetrapods in which the radix disappears between the third and fourth arches

common oviduct the terminal duct resulting from the union of two or more oviducts or oviductules

common ventricle the cavity of the diencephalon

community a group of organisms which may be a clan, a cenosis in the ecological sense, or a climax. The following derivatives are defined in alphabetic position : CLIMAX COMMUNITY OPEN COMMUNITY CLOSED COMMUNITY

-como- *comb. form* meaning "hair" better transliterated -komo-

compact bone a bone in which the bony structure is predominant as distinct from cancellous bone

companion cells specialized parenchyme-type cells in the phloem of angiosperm plants. They arise by division from the sieve tube mother cell and are thought to be connected by plasodesmata to the sievecells themselves

comparium [kŏm·pâr·ēəm] a biosystematic unit composed of one or more cenospecies that are able to intercross. Distinct comparia are unable to intercross

compensatory hypertrophy the rapid overgrowth of a smaller existing organ to take the place of one which is lost Thus in many arthropods with a large and small chela the result of the loss of the large chela is an enlargement of the small chela and a regeneration of a new small chela on the limb from which the large chela was lost

competence reactivity of a tissue or cells to an inductor in development

complement a component of blood which reacts with sensitized cells to cause lysis

complemental male a dwarf, frequently bizarre, male usually associated with, and sometimes parasitic on, a larger female

complementary association an association of plants which do not compete either because they root at different depths or have their principal above-ground forms at different seasons

complementary gene a gene which is unable to produce a phenotypic expression in the absence of one or more other genes

complementary society = complementary association

complete metamorphosis in insects, a metamorphosis in which eggs, larva, pupa and adult are completely polymorphic

complete penetrance said of a gene which always produces an effect

complex formation a group of plants which is approaching a climax

Compositae an enormous family of campanulalous flowers. The typical involucrate head with the gamopetalous flowers and one-seeded dry fruits are typical. It is the largest family of flowering plants containing more than eight hundred genera and close to fifteen thousand species. It would be misleading to pick out from this mass "typical" flowers though it must be pointed out that the daisies, thistles and sunflowers include as well as the less obvious goldenrods and marigolds. The genus *Ambrosia* is the subject of a separate entry

compound eye an eye typically found in arthropods consisting of numerous, discrete, small eyes (ommatidia) joined together

compound gland any gland of which the duct is branched and terminates in separate glandular elements

compound leaf one which possesses several blades

compound ocellus an ocellus with annular divisions

compound protein one which has one or more prosthetic groups in addition to the basic amino acid structure

-con- *comb. form* meaning "cone", sometimes transliterated -kon- or a prefix meaning "with". In the latter meaning the "n" is frequently dropped and sometimes changed to "r"

concentrated nervous system one in which at least some of the nerve cells are concentrated in the ganglia, in contrast to the nerve nets of simple invertebrates

conceptacle a word which has been applied to almost any type of receptacle in a plant or animal, but which is generally used to refer to a superficial cavity opening outwards, and frequently to one which is concerned in reproduction

-conch- *comb. form* meaning "shell"

conch 1 [kŏnch] (*see also* conch 2) the external ear of mammals

conch 2 (*see also* conch 1) in the meaning of shell in general. In this sense see dissoconch and protoconch

Conchostraca [kŏn'·kŏs·trək"·əl] an order of branchiopod crustacea commonly called clam shrimps. They are distinguished by being laterally compressed and entirely enclosed within a bivalved shell

concrementation the process of getting rid of nitrogenous wastes by converting them to insoluble substances. Melanin and guanin are frequent end products

conditioned dominance the situation in which a dominant gene can be modified by another gene

condominant species two or more, not necessarily mutually antagonistic, species that dominate a community (*cf.* cenosis)

-condyl- *comb. form* meaning "a knuckle" but extended to "knobbly joint". Better transliterated -kondyl-

condyle 1 (*see also* condyle 2) the surface at the posterior end of the skull which articulate with the anterior vertebra (*see* occipital condyle)

condyle 2 (*see also* condyle 1) in insects, any process which articulates an appendage to the body and particularly that which articulates the base of the mandible to the head

Condylura [kŏn'·dil·yūr"·ə] the genus of insectivores that contains the star-nosed mole *C. cristata* [krĭst·ät·ə]. The English name is adequately descriptive

cone 1 (*see also* cone 2–4) in botany cone-shaped groups of scales, or leaves, particularly those which constitute the gamete bearing organs of coniferous trees

cone 2 (*see also* cone 1, 3, 4) a photoreceptor in the retina functioning principally under conditions of bright illumination in the perception of color

cone 3 (*see also* cone 1, 2, 4) conical prominences in general. In this sense the following derivative terms are defined in alphabetic position :
BLASTOCONE
FERTILIZATION CONE

cone 4 (*see also* cone 1–3) one of the projections from the surface of a mammalian tooth

configuration in biology used particularly of chromosomes (*see* cis-configuration, trans-configuration)

confusing color sematic color which gives a different appearance when the animal is at rest than when it is moving

congenital immunity that which results from the transfer of antibodies from the mother to the offspring

congestin [kŏn·jĕst'·ĭn] a toxin obtained by the

glycerine extract of nematocyst bearing tentacles (cf. hypnotoxin and thalassin)

-coni-   *comb. form* (better -koni-, -konid-) meaning "duct"

conid   [kŏn′·ĭd]   the cusps on a lower molar (*see also* cone 4)

conidiophore   [kŏn′·ĭd·ēō·fôr″]   an aerial hypha that bears conidia

conidium   [kŏn′·ĭd·ēəm]   the asexual reproductive body constructed from the tips of special hyphal branches in ascomycete fungi. The term is also used as synonymous with gonidium

Coniferales   [kŏn′·ĭf·ĕr·āl″·ēz]   that order of Gymnospermae which contains the yews and pines. The name, which means conebearers, is descriptive. The genera *Abies, Cedrus, Cupressus, Larix, Picea, Pinus, Taxus* and *Tsuga* are the subjects of separate entries

Conium   [kōn″·ēəm]   a genus of umbelliferous plants. *C. masculatum* [măs′·kyūl·ā″·təm] is the Poison Hemlock used to execute Socrates in 399 B.C.

conium   [kōn″·eəm]   the Latin form of cone used in the following terms that are defined in alphabetic position:
ANDROCONIUM   OCTOCONIUM   STATOCONIUM

conjugant   one of a pair of conjugating protozoans, or a protozoan about to conjugate. The following derivative terms are defined in alphabetic position:
EXCONJUGANT   MICROCONJUGANT
MACROCONJUGANT

conjugate   coupled together as protozoa in conjugation, or paired as two leaflets

conjugated protein = compound protein

conjugation   the joining together of organisms, as in the conjugation of *Paramoecium*; the joining together of living entities such as gametes; the joining together of parts of entities such as chromosomes (*see also* deconjugation, scalariform conjugation)

conjugation canal   a connection between algal gametes

conjunctiva 1   [kŏn′·jŭnkt′·ĭv″·ə]   (*see also* conjunctiva 2) the epithelium of the inner surface of the eyelid and outer surface of the eye

conjunctiva 2   (*see also* conjunctiva 1) the membranous connection at the joint between the segments of an insect (= coria)

conjunctive symbiosis   the condition, as in lichens, where two symbionts appear to form a single individual

connecting tissue   cells lacking chlorophyll lying adjacent to the veins of leaves

connective tissue   those mesodermal tissues which bind and connect other tissues together. Blood, for want of a better classification, is usually regarded as a connective tissue (*see also* bone, cartilage, ligament, periportal connective tissue, tendon, white fibrous connective tissue, yellow elastic connective tissue)

-conop-   *comb. form* meaning "gnat", better transliterated -konop-

consociation   [kŏn′·sŏs·ē·ā″·shŭn]   an association having a single dominant and distinguished from other types of associations by this single characteristic

consocies   [kon′·sŏ·shēz]   a portion of an association lacking one or more of its dominant species, or a portion of an association characterized by one or more of the dominants of the association; the term is also used of a seral community with a single dominant (*see also* socies, associes, isocies, subsocies)

consors   [kŏn·sôrz′]   one of a group of associated organisms which cannot be characterized by such customary terms as commensal, symbiont, etc.

consortes   [kŏn·sôrt′·ēz]   plural of consors

consortium   [kŏn·sôrt′·ēəm, kŏn·sôr″·shŭn]   a somewhat ill-defined term more or less meaning symbiosis, sometimes reserved for the type of mutualism found in lichens

-cont-   *see* -kont-

contingent symbiosis   the condition of an organism that lives within another, apparently only for shelter and without either causing damage or receiving nutrients. In this sense synonymous with endobiosis

continuum   [kŏn·tĭn′·yū·əm]   the area in which two contiguous populations overlap

contour feather   a complete feather which lies on the outside of the feather mass of a bird and serves to establish the apparent shape of the animal

contractile vacuole   a vacuole found in many Protozoa, which rhythmically collects and then discharges, its contents. The primary function appears to be the maintenance of osmotic equilibrium with the environment

conus arteriosus   [kō″·nəs ärt′·yēr·ē·ōs″·əs]   the part of the truncus arteriosus that contains valves

convergent evolution   the condition where two forms of widely different ancestry arrive at a similar adaptive form. The metatherian and eutherian moles are a well known example

converse ocellus   one in which the distal ends of the retinal cells receive the light

convivium   [kŏn·vĭv′·ēəm]   those members of a commiscuum that are prevented from interbreeding by geographical barriers

Convoluta   [kŏn′·vŏl·yōō″·tə]   a genus of acoelan turbellaria. *A. roscoffensis* [rŏs′·kŏf·ĕn″·sĭs], a species with symbiotic green algae, is found off the Brittany coast. It has a built in biological clock the mechanism of which is still unexplained. The animal normally rises through the sand into pools at low tides and descends into the sand at high tide. This rise and fall, synchronous with the tide, is maintained in aquaria, even at some distance from the coast

convoluted tubule   that portion of the kidney unit which connects Henley's loop to the main collecting duct

Copepoda   [kōp·ē″·pŏd·ə]   a subclass of crustacean arthropods showing striking polymorphism. All save a very few aberrant parasites may be distinguished by bearing the eggs in egg masses paired, or unpaired, attached to the outside of the body through the genital aperture. The free-living planktonic forms are easily recognized by the long first antennae which is used either for swimming or as a balancing organ. The genera *Calanus, Caligus* and *Lernaea* are the subject of separate entries

Coprinus   [kə·prĭn″·əs]   a genus of agaricale basidiomycete fungi. *C. comatus* [kōm·ā′·təs] (Shaggy-mane) is a well known edible form

coprodeum   [*angl.* kə·prō′·dēəm, *orig.* kŏp′·rō·dē″·əm]   the anterior portion of the cloaca into which the anus opens

coprozoic   [kŏp′·rō·zō″·ĭk]   pertaining to organisms living on the excreta of other forms

copula   [kŏp″·yōōl·ə]   literally a connection, specifically an unpaired cartilage at the anterior end of the hyoid, joined by the first ceratobranchials behind and the hypohyals in front

copulation canal   a canal used for copulation in some hermaphrodite invertebrates, but which is not a vagina in the sense of being confined to the female part of the reproductive system

copulation path   the route followed by the sperm from the end of the penetration path to the female nucleus

copulatory bursa   a seminal recepticle which houses sperm for a brief space of time

-corac-   comb. form meaning "raven" better transliterated -korak-

coracidium larva   [kôr'·ə·sĭd"·ēəm]   the membrane-enclosed onchosphere of cestodes

coracoid   [kŏr'·ə·koid"]   in the shape of a crow's beak

coracoid bone   a bone of the pectoral girdle articulating with the scapula and contributing to the glenoid cavity (see also procoracoid bone)

-corb-   comb. form meaning "basket"

corbiculum   [kôr'·bĭk"·yūl·əm]   a basket of high, curved, setae rising from the dilated hind tibia of bees; often called pollen basket (see also scopa)

-cord-   comb. form meaning "heart" but also used in the sense of "cord" (properly -chord-)

cord   in the sense of string, or rope. The following derivative terms are defined in alphabetic position:
CENTRAL CORD   SPERMATIC CORD
FLAGELLAR CORD   UMBILICAL CORD
PULP CORD   VOCAL CORD

Coregonus   [kə·rĕg"·ən·əs]   a commercially important genus of fresh water teleost fish. *C. clupeaformis* [klōō'·pēə·fôrm"·ĭs], and several other species, are the Whitefish of commerce. The European anadromous species *C. oxyrhynchus* [oks'·ē·rĭn"·kəs] is sometimes locally known as Sea Trout, a term used in the U.S. for *Cynocion* and properly in Europe for a sea-run *Salmo trutta*

-corem-   comb. form meaning a "broom", better transliterated -korem-

-cori-   comb. form meaning "skin"

Cori cycle   the cycle of blood glucose to muscle glycogen to blood lactic acid to liver glycogen

coria   the membranes spanning the joint between segments in arthropods (= conjunctiva). The root is combined with the names of joints. Thus coxacoria are the membranes of the coxal joint, mandacoria of the mandible, etc. These terms are not separately defined since the meaning is apparent from the roots

Corixa   [kə·rĭks"·ə]   a genus of aquatic hemipteran insects often called "water boatman". Unlike Notonecta, they swim right way up

cork   the spongy outer layers of the bark of woody stems (see also pore cork)

-corm-   comb. form meaning "tree trunk", or "stump". The meaning is extended to "stem" or even "plant"

corm 1   (see also corm 2)   a swollen underground stem designed for food storage and which serves as a resting phase of the plants involved. Corms, such as those of Crocus and Gladiolus, are often mistakenly called bulbs

corm 2   (see also corm 1)   a group of individuals, such as a colony of hydrozoan coelenterates, all the members of which are both morphologically and physiologically united

cormophyte   [kôr'·mō·fit]   a plant that produces a stem and a root, as distinct from a thallophyte

-corn-   comb. form meaning "horn"

cornea   [kôr·nē"·ə]   the outer coat of the front of the eye

corneoscute   [kôr'·nē·ō·skyōōt"]   epidermal scales of the type that cover reptiles

cornified layer   the outer layer of the epidermis in adult mammals

cornual cartilage   [kôrn·yōō"·əl]   one of a pair of cartilages, articulating at the rear with the styloid cartilage, and running forward for about one fifth of the length of the piston cartilage in the chondrocranial complex of Cyclostomes

corolla   [kə·rŏl"·ə]   diminutive of corona and specifically the ring of petals in a flower

-coron-   comb. form meaning "crown"

corona 1   [kə·rōn"·ə]   (see also corona 2, 3) in the sense of a crown of cilia, specifically the equatorial girdle of very long cilia round the early embryo of ectoprocts or the ciliarly loop of Chaetognaths or the anterior ciliated area of rotifers

corona 2   (see also corona 1, 3) in the sense of a crown of calcareous structures. Specifically the body, as distinct from the stalk, of crinoid echinoderms and all that remains of the test of an echinoid echinoderm after the removal of the apical system

corona 3   (see also corona 1, 2) any part of a flower lying between the corolla and the stamens

corona radiata   [rād'·ē·āt'·ə]   a layer of follicle cells adhering on the surface of the zona pellucida of a mammal egg

coronary artery   the artery supplying blood to the walls of the ventricle of the heart

coronoid bone   [kŏr'·ən·oid]   one of a pair of chondral bones, found in the lower jaws of many vertebrates other than mammals, dorsal to the angular, posterior to the dentary, and anterior to the prearticular

coronoid process   the posterior process that rises from the dentary bone inside the zygomatic arch in many mammals

corpora quadrigemina   [kôr"·pə·rə kwăd"·rē·jĕm"·ĭn·ə]   two pairs of lobes developed in the mammalian brain from the roof of the mesencephalon corresponding to the single pair of optic lobes of the lower forms

corpus   (see also corpora, corpuscle, body) literally "body". The use, in English, of corpus, body and corpuscle is dictated by custom

corpus allatum   [kôr"·pəs əl·āt'·əm]   (plural corpora alata) one of a pair of wing-like structures projecting laterally from the hypocerebral ganglion of insects; they are primarily hormonal rather than nervous structures

corpus callosum   [kăl·ōs"·əm]   the dorsal of the two connections between the cerebral hemispheres in higher vertebrates

corpus cardiacum   [kär·dĭ"·ək·əm]   one of a pair of endrocrine glands lying immediately behind the brain of insects. The secretion influences growth and metamorphosis

corpus luteum   [lōō·tē"·əm]   the yellow secretory cells formed in the cavity of a ruptured Gräffian follicle

corpus pedunculatum = mushroom body

corpus striatum   [strĭ·āt"·əm]   a large ganglionic mass in the lower lateral wall of each of the cerebral hemispheres, usually considered part of the brain stem

corpuscle   the diminutive of corpus and thus applied

to any small body. The Latin form corpusculum, unlike corpus, appears to have disappeared from English biological literature. The following derivative terms are defined in alphabetic position:

CHORDAL CORPUSCLE   MECONIUM CORPUSCLE
CYLINDRICAL C.   MEISSNER'S CORPUSCLE
GENITAL CORPUSCLE   MERKEL'S CORPUSCLE
HASSALL'S C.   PACINI'S CORPUSCLE
HERBST'S CORPUSCLE   RUFFINI'S CORPUSCLE
KRAUSE'S CORPUSCLE   TACTILE CORPUSCLE
MALPIGHIAN C.   VATER'S CORPUSCLE

-cort-, -cortic-   comb. form meaning "bark", or by extension, "periphery"

cortex 1   [kôr″·teks]   (see also cortex 2) literally, bark, but in zoology used for the outer portion of any organ, such as the cerebrum, kidney or adrenal gland, in contrast to the inner medulla. In this sense the following terms are defined in alphabetic position:

HIPPOCAMPAL CORTEX   OLFACTORY CORTEX
NEOCORTEX   PYRIFORM CORTEX

cortex 2   (see also cortex 1) that part of the fundamental tissue of a plant which lies between the epidermis and the vascular region

corticospinal tract   [kôrt′·ik·o·spī′·nəl]   the voluntary motor pathway passing in mammals from the neocortex to the spinal medulla

cortisone   [kôrt′·ə·sōn]   one of several glucocorticoids (cf. corticosterone). Active in glycogen deposition in liver, muscle-work performance, potassium-sodium ratio, growth and the production of secondary sex characters

Corti's organ   that structure in the chochlea of mammals which is the principal receiver of sound

corticosterone   [kôrt′·ik·ō·stə·ōn′]   one of several glucocorticoids (cf. cortisone) active in glycogen deposition in liver, muscle-work performance, potassium-sodium ratio, growth and the production of secondary sex characters (see also deoxycorticosterone)

-coryn-   comb. form meaning "club"

-coryph-   comb. form meaning "summit"

coryphodont   [angl. kŏr′·if″·ō·dŏnt, orig. kŏr′·if·ō″·dŏnt]   possessing a dentition in which there is a gradual diminution in tooth length from front to back of the jaws

cosmobiotic theory   [kŏz′·mō·bi·ŏt″·ik]   the theory that life evolved elsewhere in the cosmos and was transferred to the planet Earth

cosere   [kō″·sēr]   a series of unit succession, usually of vegetation, in the same place

costa   [kŏst″·ə]   literally a rib, as much that of a plant leaf as a chordate thorax. The term is also used for the comb-rows of ctenophorans, the veins on an insect wing, and an extension of the scleroseptum of a coral connecting the thecea to the pseudothecae

costal angle   [kŏst″·əl]   the angle at the tip of an insect wing

coterie   [kō″·tə·ē]   a closed society consisting of several species

-cotyl-   comb. form meaning the "cavity of a cup" and by extension to the cup-like appearance of the first sprout on a seed

cotyl   [kŏt′·əl]   pertaining to a developing plant. The following derivative terms are defined in alphabetic position:

EPICOTYL   HYPOCOTYL   MESOCOTYL

cotyle   [kŏt′·il]   the socket of a ball-and-socket joint in arthropods

cotyledon   [kŏt′·əl·ē″·don]   the seed leaf: that is the leaf or leaves which first appear when the seed germinates

coupling   the condition of two loci on a chromosome when each has a recessive or dominant allele (cf. repulsion)

coverslip   a disk, or square, of very thin (usually about 0.1 mm) glass used to cover objects mounted on a microscope slide

coverts   [kŏv″·ərtz]   feathers which cover or lie between the flight feathers or tail feathers of a bird or are closely applied to other areas

Cowper's gland   an accessory sex gland of unknown function in male mammals

coxa   [kŏks″·ə]   the segment of the insect leg that articulates with the body (see also precoxa)

coxopodite   [angl. kŏks·ŏp″·ə·dīt, orig. kŏks′·ō·pōd″·it]   the basal segment of an arthropod appendage

cranial   pertaining to the skull or, by improper extension, pertaining to the head or brain

cranial capsule   that part of the skull which encloses the brain

cranial nerve   one leaving the central nervous system, within the cranium. They are commonly indicated in serial order by Roman numerals

cranial sclerotomite   the anterior division of a sclerotomite and which, therefore, forms the posterior half of one vertebra and the prezygapophysis of the vertebra behind (see also caudal sclerotomite)

cranium   [krān″·ēəm]   properly that part of a bony skull which surrounds the brain, but by extension used for any part of the head and branchial skeleton. The following derivative forms are defined in alphabetic position:

CHONDROCRANIUM   NEUROCRANIUM
ENDOCRANIUM   SPLANCHNOCRANIUM
EXOCRANIUM

-craspedo-   comb. form meaning "border", "edge" or, in biology, "velum"

Craspedocusta   see Microhydra

-crass-   comb. form meaning "thick"

-crater-   comb. form meaning "cup"

crateriform   in the form of a volcanic crater and its tube; used specifically in bacteriology to describe the liquefied portion of a stab culture

creatininase   [krē·ăt′·ən·ən·āz″]   an enzyme catalyzing the hydrolysis of creatinine to sarcosine and urea

-crem-   comb. form meaning "to hang" or "to overhang". Frequently confused with -cremno-

cremaster   [angl. krəm′·ăst·ə, orig. krēm′·ăst·ə]   suspensor; the term is used alike for the muscles suspending the testes in the scrotum or the hook on the end of the case by which some insect pupae can be suspended

-cremno-   comb. form meaning "cliff"

-cren-   comb. form hopelessly confused between Greek -kren- meaning "notch" and Latin -cren- meaning "spring" in the sense of a source of water

crenic   pertaining to springs and their immediately derivative waters

Crenobia   [krēn·ō·bēə′]   a genus of tubellaria of which one, C. alpina [ăl·pin·ə], has been the subject of much research on rheotaxis

-creo-   comb. form meaning "meat" or "flesh"

-crepi-   comb. form meaning "shoe" better transliterated -krepi-

Crepidula   [krĕp′·id″·yōō·lə]   a genus of gasteropod

mollusks with internal reflexed torsion so that the shell does not appear coiled. Early studies on the eggs did much to establish understanding of determinate cleavage

crepuscular [krə·pŭs′·kyül·ə] pertaining to the dim light of dusk or dawn (cf. auroral, matutinal, vespertine)

-cresc- comb. form meaning "grow"

crest a ridge, or projection, rising from a surface. The following derivative terms are defined in alphabetic position:

NEURAL CREST	ORBITAL CREST
NUCHAL CREST	SAGITTAL CREST

Cretaceous period [krə·tās″·ēəs] the most recent of the Mesozoic geological periods extending from 125 million years ago to 60 million years ago. It was preceded by the Jurassic and followed by the Tertiary. The reptiles were still dominant but the earliest mammals are of this period and there were numerous insects and flowering plants

-cribr- comb. form meaning "sieve"

cribriform plate 1 [krĭb″·rə·fôrm′] (see also cribriform plate 2 the part of the ethmoid bone that separates the cranial from the nasal cavity and therefore contains several foramina

cribriform plate 2 (see also cribriform plate 1) a fenestrated cartilage closing the gap between the paranasal cartilages and the nasal septum in embryonic chondrocrania

cricoid cartilage [krī″·koid] the anterior of the three laryngeal cartilages

-crin- comb. form meaning "hair". The Greek derivatives "lily" and "separate" are better transliterated -krin-

Crinoidea [krĭn·oid″·ēə] the only living pelmatozoan echinodermata. This group, often called sea lilies, is either attached by an adoral stalk or moves freely by adoral "legs". Most known species are Devonian fossils. The genus Antedon is the subject of a separate entry

-crist- comb. form meaning "crest"

crista (plural **cristae**) [krĭst″·ə (krĭst″·ē)] literally, a crest but usually applied to the transverse infolding membranes in mitochondria. The term is also applied to ridge-like membranes on the surface of some bacteria

-crit- comb. form meaning "select", "chosen", "notable" and better transliterated -krit-

Crithidia [krĭth·ĭd′·eə] a genus of trypanosomid protozoans parasitic in insects. It is not clear whether all are developmental stages of Trypanosoma, from which they differ in that the flagellum arises from the center of the body, or whether they are a distinct genus

Crocodylus [krŏk′·ō·dil″·əs] the genus of diapsid reptiles containing the true crocodiles. The principal difference from the genus Alligator is that the upper and lower teeth are in line, the fourth lower tooth fitting into a notch in the upper jaw

crop a food reservoir, lying between the mouth and the stomach, usually produced as an enlargement of the esophagus, particularly in birds and insects

crossing-over a genetic term used to describe that which occurs when homologous regions are interchanged between two members of a pair of chromosomes (see also somatic crossing over)

Crossopterygii an order of osteichthyes containing many fossil forms and the extant Latimeria. The primitive forms differ from the Dipneusti in having a single proximal bone in the paired fins which led directly to the limb of land forms

crossover the result of crossing-over

Crotalus [angl. krŏt′·ə·ləs, orig. crō·tāl′·əs] a genus of snakes (Squamata) containing the Rattlesnakes (but see also Sistrurus). C. horridus [hŏr″·id·əs] is the Timber Rattlesnake and C. adamanteus [ăd′·ăm·ănt″·ēəs] the Diamond Back

crozier [krō″·zēə] literally, a "shepherd's crook", but in botany commonly applied to the unfolding frond of a fern

-cruc- comb. form meaning "cross" (but see also chiasma)

Cruciferae [krōō·sif·″ər·rē] the family of dicotyledons that contains, the cabbages, the mustards, the radishes, the wall-flowers, and woad. The four sepals and four petals arranged in the form of a cross are typical of the family. The genus Brassica is the subject of a separate entry

-crumen- comb. form meaning "pouch" or "purse"

-cruor- comb. form meaning "blood"

cruorin [krōō′·ôr″·in] a general term for "blood pigment", at one time synonymous with hemoglobin (see also chlorocruorin)

crura the plural of crus

crus (plural **crura**) [krŭs, krōōr·ə] a leg

-crus- comb. form meaning "shell" in the sense of toughened integument

Crustacea [krŭs·tās·eə] a large class of the phylum Arthropoda containing, the crabs, lobsters, shrimps, beach hoppers, sow bugs, barnacles, water fleas and allied forms. They may be distinguished from other arthropod classes by the possession of two pairs of preoral antennae and at least three pairs of postoral appendages functioning as jaws. Almost all are aquatic. The Amphipoda, Anostraca, Branchiopoda, Cladocera, Conchostraca, Copepoda, Cirripedia, Decapoda, Isopoda and Ostracoda are the subjects of separate entries

crustose [krŭst″·ōz] having the appearance of a crust. Said particularly of lichens closely adpressed to a rocky surface in distinction from foliose

-cry- comb. form meaning "cold"

cryophilic [angl. kri·ŏf″·əl·ĭk, orig. kri′·ō·fĭl″·ĭk] thriving at low temperatures, particularly used of microorganisms

cryoplankton that of perpetually cold or icy waters, or organisms suspended in snow

-crypt- comb. form meaning "hidden" and thus, by extension, a cavity or vault. Better, but rarely, transliterated -krypt-

crypt of Lieberkuhn a tubular, epithelial, gland in the jejunum and ileum

cryptic color sematic color designed to blend an animal with its background (see also anticryptic color, procryptic color)

cryptobiosis the condition of an organism which must in theory have at one time existed, but which has left no fossil traces

Cryptobranchus [krĭp′·tō·brănk″·əs] a genus of primitive urodelan Amphibia containing the american hellbender (C. alleganiensis) [ăl′·ə·găn″·ē·ĕn·sĭs]. The adults lack external gills but retain one pair of open gill slits (see also Megalobatrachus)

Cryptococcus [krĭpt′·ō·kŏk″·əs] a genus of Endomycetales that do not form spores and are therefore sometimes referred to as the Deuteromycetes. C. hominis [hŏm″·ĭn·ĭs] (="*Torula histolytica*") may attack the brain, bones, lungs and skin of humans but the etiology is obscure

cryptogam [krĭpt″·ō·găm] a plant which reproduces by means other than flowers (cf. phanerogam)

cryptogenetics [krĭpt′·ō·jĕn·ĕt″·ĭks] the study of genotypic transfer as distinct from phenotypic

cryptomere [krĭpt″·ō·mēr′] a gene the phenotype of which is not known

cryptomitosis [krĭpt′·ō·mĭt″·ōs·ĭs] a type of mitosis found in some protozoans in which the acromatic figure is evident, but chromosomes are not apparent

cryptorchid [krĭpt·ôr′·kĭd] said of a mammal with concealed (i.e. abdominal) testes

crystalline style a transparent body, supposed to liberate enzymes, found in the alimentary canal of some pelecypod mollusks

CSF = cerebrospinal fluid

CSH = cysteine

-cten- *comb. form* meaning "comb"

ctene [tēn] the plate of fused cilia in the form of a comb, from which the Ctenophora derive their name

ctenoid scale [tēn″·oid] a teleost scale one side of which has finger-like or comb-like processes

Ctenophora [tēn·ŏf′·ər·ə] a phylum of animals containing "sea gooseberries". They were once united with the coelenterates but differ from them in the absence of nematocysts and the presence of mesenchymal muscles. The eight rows of ciliary plates (ctenes) from which their name derives is typical. The genera *Beroë* and *Cestus* are the subject of separate entries

CTP = cytidine triphosphate

cuboid bone a tarsal bone lying at the base of the fourth and fifth metatarsal

cuboidal epithelium an epithelium of which each cell is more or less cuboid

Cucumaria [kyōō′·kyōō′·mâr″·ēə] a genus of typical holothurians, cucumber-shaped and with ambulacral feet confined to the radii. The reddish-brown, foot long, *C. frondosa* [frŏn·dōz′·ə] is common on both sides of the Atlantic

Culex [kyōō″·lĕks] a genus of culicid dipteran insects. *C. pipiens* [pīp″·ē·ĕnz] is the commonest N. American mosquito. *C. fatigans* [făt″·ē·gănz′] is the vector of *Wucheria bancrofli*

Culicidae [kyōō′·lĭs″·ĭd·ē] a large family of dipteran insects called mosquitoes when they are blood sucking and phantom midges when they are not. The genera *Anopheles*, *Aedes* and *Culex* are the subject of separate entries

culm [kŭlm] a stem, such as that of a grass, which consists of hollow sections interrupted by solid nodes

cultigen [kŭlt′·ē·jĕn] an organism known only in cultivation or domestic association, not yet recognized anywhere as native or indigenous

cultivar [kŭlt″·ē·vär′] a plant cultigen, a variety of plant found only under cultivation

culture a laboratory induced association of organisms under controlled conditions (*see also* stab culture, streak culture)

-cuma- *comb. form* meaning "a wave" better transliterated -kuma-

-cumben- *comb. form* meaning "to lie down"

cumulative factor one of several that are not allelomorphic but each of which enhances the effect of the other

cumulative genes genes which, acting together, accentuate a character such as skin color

-cun- *comb. form* confused from two roots meaning respectively "wedge" and "cradle"

cuneiform bone [kyōōn′·ēə·fôrm] one of those tarsal bones that lie immediately at the base of the metatarsals. The medial cuneiform lies at the base of the first metatarsal, the intermediate at the base of the second, and the lateral at the base of the third, adjacent to the cuboid

Cuon [kyōō·ŏn] the genus of canid carnivores containing the Dholes, often referred to as "false dogs". The difference from the true dogs is in dentition, in having more nipples, and possessing long tufts of hair between their toes

Cupressus [kyōō·prĕs″·əs] the genus of coniferales containing the cypresses. They are distinguished by their small, scaly leaves and small globose cones

-cupul- *comb. form* meaning "cask"

cupula 1 [kyōō′·pyū·lə] (*see also* cupula 2) a domed mass of material, but applied particularly to the dome-shaped mucus protuberances from the lateral line sense organs of fish

cupula 2 (*see also* cupula 1) small, cup shaped, dense cytoplasmic inclusions disclosed by electron microscopy in many protozoans

curare [kyūr·är′·ē] an alkaloid extracted from the bark of *Strychnos toxifera*. It is one of the most powerful muscle relaxants known and is used by South American Indians as an arrow poison

cusp [kŭsp] a point or pointed structure. In zoology usually applied to substructures on teeth and in botany to fronds

cuspid tooth = canine tooth

cutaneous vein [kyōō·tān″·ēəs] a large respiratory vein running along the lateral skin of Amphibia to the brachial vein

cuticle the outer, dead, layer of an integument (*see also* endocuticle, epicuticle)

Cuvierian sinus [kyū·vēr′·ēən] a blood collecting vessel of vertebrate embryos running from the junction of the cardinal sini to the sinus venosus

Cuvierian tubule [kyū·vēr·ēən] one of a mass of tubules found at the base of the respiratory tree in holothurian echinoderms. These tubules emit a sticky, thread-like secretion, in which predators, or prey, become entangled

Cuvier's canal [kyūv·ēāz] = sinus venosus

-cyan- *comb. form* meaning "dark blue", better transliterated -kyan-

Cyanea [sī′·ān″·ēə] a genus of scyphozoan coelenterates. *C. capillata* [kăp′·ĭl·āt″·ə] is one of the largest known jelly fish sometimes reaching a diameter of six feet with forty foot tentacles. Contact with these has often proved fatal to human swimmers

cyanocobalamin [sī′·ən·ō·kō′·băl″·ə·mĭn] *see* vitamin B₁₂

cyanogenic [sī′·ən·ō·jĕn″·ĭk] gas-making, particularly of insects which will repel their enemies by this means

Cyanophyta a phylum of Algae containing the blue-green algae. They are distinguished by the lack of definite chromatophores and by the fact that the nuclear material is not organized into a nucleus. By no means all are blue-green (*see* for example, *Trichodesmium*). The genera *Anaboena*, *Eucapsis*, *Gleocapsa*, *Meris-*

mopeida, Nostoc, Synechococcus, Tolypothrix and *Trichodesmium* are the subject of separate entries

cyanopsin [sī′·ăn·ŏps″·ĭn] an analogue of iodopsin found in the cone cells of the retina

-cyath- *comb. form* meaning a deep cup or goblet (cf. -calyc-)

Cycadales [sī′·kə·dāl″·ēz] an order of gymnosperm plants that flourished in the Mesozoic epoch and of which many genera remain extant. Most were, and are, palm-like plants with the trunk armored with leaf bases. In all extant genera except *Cycas* the reproductive organs are strobili. The genera *Cycas* and *Zamia* are the subject of separate entries

-cycl- *comb. form* meaning "a circle" and every possible derivative of this word. Better transliterated -kykl-

cycle 1 *(see also* cycle 2) any rhythmic phenomenon of more or less uniform periodicity. In this sense the following derivative compounds are defined in alphabetic order:

ANOESTROUS CYCLE	LIFE CYCLE
ANOVULATORY CYCLE	SEASONAL CYCLE
DIESTROUS CYCLE	SEX CYCLE
ESTROUS CYCLE	

cycle 2 *(see also* cycle 1) a repetitive metabolic process. In this sense the following derivative compounds are defined in alphabetic position:

ARGININE CYCLE	KREBS CYCLE
CITRIC CYCLE	NITROGEN CYCLE
CORI CYCLE	ORNITHINE CYCLE
ENERGY CYCLE	TRICARBOXYLIC CYCLE
GLYOXALATE CYCLE	UREA CYCLE

cycloid scale [sī″·kloid] a large, smooth-edged, teleost scale showing concentric growth rings

cyclomeric theory [sī′·klō·měr″·ĭk] metameric segmentation arose in consequence of the rearrangement of radially symmetrical parts

cyclomorphism [sī′·klō·môrf″·ĭzm, sĭk·lō·môrf·ĭzm] a cyclic, usually seasonal, change in body form such as that found in some Cladocera and rotifers

Cyclops [sī″·klŏps″] a huge genus of fresh water copepods. Some species are vectors in *Dracunculus* infections

cyclosis [sī′·klōs″·ĭs, sĭk′·lōs″·ĭs] the regular movement along apparently predetermined paths of food vacuoles in some ciliate Protozoa and the streaming of protoplasm in plant cells

Cyclostomata = Marsipobranchia

cyesis [sī·ēs″·əs] the period, or the length of the period, between fertilization and birth in a viparous animal

cylindrical corpuscle an end organ consisting of a club shaped nerve termination in a connective tissue sheath

-cym- *comb. form* meaning "wave" better, and not infrequently, transliterated -kym-

-cymbo- *comb. form* meaning "cup" better transliterated -kymbo-

-cyn- *comb. form* meaning "dog", shortened from -cyon- and, in any case, better transliterated -kyon-

Cynocephalus [sīn′·ō·sěf″·əl·əs] the only genus of the mammalian order Dermoptera. There are two extant species called Flying Lemurs or Cobegos

Cynognathus [*angl.* sin·ŏg″·nə·thəs, *orig.* sīn″·ŏg·nāth″·əs] a typical genus of extinct theriodont synapsid reptiles from the lower Jurassic. The animal was

the size of a large wolf but with an elongate narrow skull. The teeth were clearly differentiated into incisors, canines and post-canines

Cynomys [sī·nōm″·ĭs] the genus of sciuromorph rodents containing the large ground squirrels known as "prairie dogs"

-cyon- *comb. form* meaning "dog" better transliterated -kyon- (cf. cyn)

cyphonautes larva [*angl.* sī·fŏn″·ôt·ēz, *orig.* sī′·fŏn·ôt″·ēz] a tent-shaped, free-swimming larva of some Ectoprocta

Cypraea [sĭp·rē″·ə] a genus of mostly Indian and Pacific gastropod mollusks of the kind called cowries. Most are bean-shaped with a toothed groove in the center of one flat side. *C. moneta* [mŏn·āt″·ə] was used as currency until quite recent times. The very beautiful shells were common ornaments in European parlors

Cyprinus [sĭp·rĭn″·əs] a genus of teleost fishes called Carp. *C. cyprinus* is the common carp. The widespread stories of its extraordinary longevity are untrue. They interbreed freely with *Carassius*

cypris larva [sĭp″·rĭs] the free-swimming, ostracod-like, larva of cirripedia

-cyst- *comb. form* meaning "bladder" better transliterated -kyst-

cyst [sĭst] in biology any cavity, more or less spherical structure, with a distinct wall, particularly one which encloses the resting stage of an organism. The following terms using this suffix are defined in alphabetic position:

BLASTOCYST	NEMATOCYST
CHONDROCYST	OOCYST
CNIDOCYST	OTOCYST
DAUGHTER CYST	SPIROCYST
HETEROCYST	SPOROCYST
HYDATID CYST	STATOCYST
MICROCYST	TRICHOCYSTS

cystidean larva that stage in the development of a crinoid in which the stalk first appears, but in which the arms are not developed

cysteic acid [sĭs·tē″·ĭk] α-amino-β-sulfopropionic acid. HOOCCH(NH₂)CH₂SO₃H. An amino acid found principally in hair and scales

1-cysteine [sĭs·tē″·ēn, sĭst′·ə·ēn] 2-amino-3-mercaptopropanoic acid. HSCH₂CH(NH₂)COOH an amino acid not essential in the nutrition of rats. Associated principally with hair and keratin (*see also* 1-homocysteine)

cysticercus larva [sĭst′·ē·sěrk″·əs] a larval stage in the development of a cestode platyhelminth consisting of a scolex attached to a bladder. There may be more than one scolex and one or more may be invaginated (cf. plerocercoid)

l-cystine [ěl·sĭst·ēn] 3, 3′-dithio-bis (2-aminopropanoic acid) that is dicysteine (*see* 1-cysteine). An amino acid not essential to rat growth formed predominantly from cystine and forming as much as 10% by weight of some hair structures

cystogenous gland [*angl.* sĭst·ŏj″·ən·əs, *orig.* sĭst′·ō·jěn″·əs] a gland for secreting a cyst. Best known in the cercariae larva of trematode platyhelminths

-cyte- *comb. form* meaning "a hollow vessel" or "gourd" but taken by most biologists to mean "cell". The following terms using this suffix are defined in alphabetic position:

AMOEBOCYTE
ARCHAEOCYTES
ASTROCYTE
CHOANOCYTE
CHONDROCYTE
COELOMOCYTE
COENOCYTE
COLLENCYTE
ERYTHROCYTE
GAMETOCYTE
GRANULOCYTE
HEMOCYTE
HISTIOCYTE
LEUCOCYTES
LYMPHOCYTE
MEGAKARYOCYTE
MEROCYTE

MONOCYTE
MYELOCYTE
MYOCYTE
NEMATOCYTE
OOCYTE
OSTEOCYTE
PHAGOCYTE
PINACOCYTES
PINOCYTE
PODOCYTE
POLYMORPHOCYTE
POROCYTE
RETICULOCYTE
SOLENOCYTE
SPOROCYTE
THIGMOCYTE
THROMBOCYTE

cytochrome [sit'·ō·krōm"] a hemoprotein utilized in biokinetic systems for electron transport in virtue of the reversible valency of hem -Fe

cytochrome A a cytochrome in which the Fe is in a formylporphyrin linkage

cytochrome B a cytochrome in which the Fe is in a protoporphyrin linkage

cytochrome C a cytochrome in which the Fe is in a mesoporphyrin linkage

cytochrome D a cytochrome in which the Fe is in a dehydroporphyrin linkage

cytogamy [*angl.* sit·ŏg"·a·mē, *orig.* sit'·ō·găm"·ē] the condition which results when both male nuclei remain in one of two conjugating animals and therefore both fuse with the female nucleus of the same animal; or the fusion of two cells into a haploid zygote

cytogenetics [sit'·ō·jĕn·ĕt"·iks] the study of the behavior of chromosomes in mitosis and meiosis and the correlation of this behavior with the transmission of heritable characters

cytokinins [sit'·ō·kin·ins] a group of compounds, mostly with a purine ring structure, that stimulate the division and enlargement of plant cells (*see also* plant hormones, auxin, giberellin)

cytomorphosis [si'·tō·môrf"·ōs·is] a change in the shape of a cell, particularly a bacterial cell, induced by environmental factors

cytopemphis [sit'·ō·pĕm"·fis] the passage, by a process analogous to pinocytosis of large molecules through the walls of a capillary. The passage of a leucocyte is diapedesis

cytopharynx [si'·tō·fâr"·inks] a permanently open food-entry passage in some protozoans

cytoplasm [sit'·ō·plazm'] that part of the cell substance which is not the nucleus

cytoplasmic inclusions [sit'·ō·plăz"·mik] any structure in the cytoplasm which has not so far been identified as an organelle

cytoplasmic sterility that which is due to discordance between the chromosomal gene complement and the cytoplasm

cytopyge [sit'·ō·pij"] the more or less definite point from which flagellates and ciliates eject the undigested remnants of their ingested food

cytostome [sit'·ō·stōm"] the external opening of the cytopharynx

cytotaxis [sit'·ō·tāks"·is] the movement of cells in relation to each other (*see also* negative cytotaxis, positive cytotaxis)

cytotaxonomy [sit"·ō·tāks'·ŏn"·əm·ē] that branch of taxonomy which is based on a study of the genotypic characters rather than the phenotypic

D horizon the bottom layer of soil lying below the C horizon and consisting of consolidated, unweathered, inorganic material

-dacry- *comb. form* meaning "tear" or "lachrymal" better transliterated -dakry-

-dactyl- *comb. form* meaning "finger" or "toe"

dactyl [dăk'·tĭl] in the sense of finger, or finger shaped object, or pertaining to such structures.

dactylopodite [*angl.* dăk'·təl·ŏp"·əd·it, *orig.* dăk'·tĭl·ō·pōd"·ĭt] the terminal joint of a crustacean appendage. Also a thin second tarsal segment rising from an enlarged first tarsal segment in some insects

dactylozooid [dăk'·tĭl·ō·zō"·id] a hydrozoan zooid, having tentacles furnished with nematocysts but no mouth

Daphnia [dăf"·nēə] a genus of fresh-water cladocerans so widely distributed as to have become, in English, almost synonymous with the order

daphnia an English word derived from the above often used for many other genera of cladocerans as in the phrase "the daphnias of Lake So-and-So"

dart a calcareous, hormone impregnated spike transferred between some land snails as a preliminary to copulation. Also sometimes used for the sting of an insect

Darwinian curvature bending induced at the apex of the root by mechanical stimulation

Darwinism properly the views expressed on evolution by Charles R. Darwin (1809–1882) who stressed the struggle for survival and the survival of the fittest. Among illiterates this term is sometimes thought to be synonymous with "evolution" (*see also* neodarwinism)

-dasy- *comb. form* meaning "thick"

Dasyatus [dăs·ē·ət·əs, dăs·ē·ā·təs] a genus of batoid elasmobranchs contain the stingrays. *D. sayi* [sā"·ĭ] is the Bluntnose Stingray common on the Atlantic Coast. The "sting" is a venom-bearing spine emerging from the dorsal surface just where the tail joins the body

Dasypus [dăs"·ē·pəs] the genus containing the mammals commonly called Armadillos. *D. novemcinctus* [nō'·vĕm·sĭnkt"·əs] is the common armadillo of the Southern States

Dasyurus [dăs'·ē·yūr"·əs] a genus of metatherian mammals closely paralleling the viverine cats in their general form and commonly called dasyures

daughter cell one derived from division of a mother cell or a cell produced in the interior of another cell

daughter chromosome those chromosomes developed through the separation of the chromatids in mitosis

day-eye an eye specifically adapted to diurnal habit, particularly among insects

de- *comb. prefix* used for "from" in the privative sense

deaminase [dē'·ăm·ĭn·āz"] a group of enzymes, more properly called amino hydrolases, which catalyze the liberation of ammonia by hydrolysis

Decapoda (Crustacea) [*angl.* dək·ăp'·əd·ə, *orig.* dĕk'·ā·pōd"·ə] a large order of malacostracan crustacea containing the shrimps, prawns, lobsters, crayfish, and crabs. They are distinguished from the Euphausiacea by the fact that some of the thoracic limbs are specialized as maxillipedes. The genera *Callinectes, Cambarus, Homarus, Mysis,* and *Palinurus* are the subject of separate entries

Decapoda (Mollusca) a suborder of cephalopod mollusks distinguished by the presence of ten arms, four pairs of which are shorter than the other pair

decarboxylase [dē'·kârb·ŏks"·əl·āz'] a large group of enzymes, more properly carboxylases, which catalyze the removal of CO_2

decomposers the group of organisms (bacteria and fungi) that degrade dead organic material in ecosystems

deconjugation precocious separation of chromosome pairs in meiosis

dedifferentiation the reversion of cells, particularly in the life history of many invertebrates, from a specific to a generalized form. This process differs from degeneration in conveying the implication that redifferentiation will take place (*see also* differentiation, redifferentiation)

deferent to convey downward (*see also* afferent, efferent)

deficiency the loss or inactivation of a section of a chromosome. In *Drosophila melanogaster* a general term for all mutants associated with such a chromosomal anomaly

definitive nucleus that which is formed in the plant embryo sac by the fusion of the two polar nuclei

degeneration the breakdown of specific cells or organs to an indeterminate form. It differs from dedifferentiation in carrying the connotation that the process will not reverse (*see also* differentiation, redifferentiation)

-dehisc- *comb. form* meaning "to yawn" or "gape"

dehiscence the opening of a dry fruit for the liberatiom of seed

dehydroepiandrosterone [dē·hi′·drō·ĕp′·ē·ăn′·drō·stĕr″·ōn] a hormone secreted by the testes having the same activity as testosterone

dehydrogenase a class of enzymes which catalyzes the removal of hydrogen. The reaction usually requires NAD or, more rarely, NADP, which becomes NADH₂ or NADPH₂, as the hydrogen receptor

Deiter's cells columnar cells supporting the hair cells in the organ of Corti

delayed dominant an allele the phenotypic expression of which does not appear until quite late in development

deletion the remainder of a chromosome after the occurrence of a non-terminal deficiency

-delph- *comb. form* meaning "womb" but extended in biology to the whole female reproductive system in both plants and animals. Frequently confused with -adelph-

-deme- *comb. form* hopelessly confused between "demos" (people) and "demas" (body or structure)

deme 1 [dēm] in the sense of "people" or "population," usually meaning, without qualification, an aggregate of similar cells or similar species. Also used for an isolated population within a species (*see* gamodeme, topodeme)

deme 2 in the sense of "structure" (*see* apodeme)

demersal [dē′·mĕrs″·əl] inhabiting the lowest layer of a lake

-demic *adj. term.* from deme in the sense of "people" or "population". In this sense the following terms using this suffix are defined in alphabetic position:

ENDEMIC   PANDEMIC
EPIDEMIC   POLYDEMIC

Demodex [dēm′·ō·dĕks″] a genus of minute, worm-like, Acarina living in the sebaceous glands of mammals. *D. folliculorum* [fŏl·ĭk′·yōō·lôr″·əm] is common in man where it is apparently a commensal but can cause mange if transferred to cats

demography [dē′·mŏg″·răf·ē] the field of human population analysis and also applied to the study of other animal populations

-dendr- *comb. form* meaning "tree"

dendrite [dĕn″·drĭt] those nerve fibers that extend from the nerve cell body to the source of stimulus (cf. axon)

dendritic [dĕn·drĭt″·ĭk] branched or tree-like

Dendrocoelum see *Procotyla*

dendrogaea [*angl.* dĕn·drŏj″·ēə, *orig.* dĕn′·drō·jē″·ə] the neotropical region

Dendronotus [*angl.* dĕn·drŏn″·ət·əs, *orig.* dĕn′·drō·nŏt″·əs] a cosmopolitan genus of nudibranch mollusks with the tentacles matted into a frontal appendage. *D. frondosus* [frŏn·dōs″·əs], a two inch, red and white spotted species, is common on all northern shores

Dendrostomum [*angl.* dĕn·drŏst″·əm·əm, *orig.* dĕn′·drō·stōm″·əm] a genus of large pear-shaped Sipunculida distinguished by having from four to eight branching tentacles. *D. zostericolum* [zŏs·tər·ik·ō·ləm] is found on the Pacific Coast

-dente- *comb. form* meaning "tooth"

dental formula a method of expressing mammalian dentition as a series of fractions, with maxillary teeth as the numerator and mandibular teeth as the denominator, representing successively the incisors, canines, premolars and molars

Dentalium [dĕnt·āl″·ēəm] a genus of scaphopod mollusks with a trilobed foot and a tusk shaped shell. There are numerous species, some of which are deepwater forms

dentary bone [dĕnt″·ə·rē] that bone in the lower jaw, or mandible, which carries teeth

dentate *see* -dont

dentine [dĕnt′·ēn] the layer between the enamel and the pulp cavity in teeth and placoid scales

dentition the sum of, or the arrangement of, the teeth in a given organ (*see also* dental formula)

dentosplenial bone [dĕnt′·ō·splĕn″·ēəl] a compound bone replacing the dentary in the lower jaw of actinopterygian fish (*see also* splenial bone)

deoxycorticosterone [dē′·ŏks·ē·kôrt″·ik·ō·stĕr·ōn′] a hormone secreted in trace quantities by the adrenal cortex and in general similar in its effects to aldosterone

deoxyribonuclease [dē′·ŏks·ē·ri′·bō·nyuk″·lē·āz′] an enzyme that catalyzes the hydrolysis of DNA to yield oligodeoxyribonucleotides

deoxyribonucleic acid a long chain polymer, the monomer of which is oxyribose plus a phosphate pentose plus a nucleotide. The nucleotides may be thyamine, adrenine, cytocine or guanine. DNA occurs in chromosomes as a double helix in which adenine pairs with thymine and guanine pairs with cytosine. Genetic information is conveyed by the sequence and number of such pairs (*see also* ribonucleic acid)

derivative cell a cell of potential specialized function derived from an initiating cell in the meristem

-derm- *comb. form* meaning "skin"

derm 1 (*see also* derm 2) in the sense of the surface layers of animals and their embryos. The term -dermis is identical in meaning but is rarely used except for the words epidermis and hypodermis. In this sense, the following terms using this suffix are defined in alphabetic order:

ECTODERM   MESENTODERM
ECTOMESODERM   MESODERM
ENDODERM   PARADERM
ENDOMESODERM   PERIDERM
ENTERODERM   PLASMOTROPHODERM
GASTRODERM   TROPHODERM
MESECTODERM

derm 2 (*see also* derm 1) in the sense of the surface layers of plants. In this sense the following terms using this suffix are defined in alphabetic position:

ENDODERM   PROTODERM

Dermacentor [dûrm′·ə·sĕnt″·ôr] a genus of ixodid acarines (ticks) containing many disease vectors dangerous to man. The worst is *D. andersoni* [ăn′·də·sōn″·ī], now of very wide distribution, that can transmit spotted fever, anaplasmosis, tularemia, brucellosis, *Salmonella*, Q fever, Colorado tick fever and virus encephalomyelitis. *D. variabilis* [văr′·ē·ăb·əl·əs] also transmits spotted fever and *D. parumapterus* [pär·ōōm·ăp·ər·əs] is suspected of doing so. Automobile traffic is the principal cause of the rapidly expanding distribution of dangerous ticks

dermal bone [dûr″·məl] bones that ossify directly in connective tissue masses

dermal denticles the tooth-like "scales" found in the skin of elasmobranch fish

dermal plate a dermal bone forming much of the side of the cranium of monotremes, and which is either a development of, or at least fused with, the petrosal

dermal scale any scale such, as those of chondrichthine and teleost fish, that is derived from dermal structures (cf. epidermal scale)

Dermaptera [dûrm·ăp″·tə·rə] the order of insects that contains the earwigs. They are distinguished by having wings which are folded in a complex manner under anterior elytra and in almost all species by the presence of a pair of forceps at the posterior end of the abdomen. The genus *Forficula* is the subject of a separate entry

dermis skin. The term is usually replaced by derm (q.v.) except for epidermis and hypodermis

Dermochelys [*angl.* də·mŏk″·əl·ĭs, *orig.* dûr′·mō·kēl″·ĭs] the monotypic genus of chelonian reptiles that contains *D. coriacea* [kŏr′·ē·ās″·ēd] the huge Leatherback Turtle of tropical seas, reaching a weight of 1200 lbs. The name is descriptive

Dermoptera [dûrm′·ŏp″·tə·rə] the group of placental mammals that contains the "flying" lemurs. They are distinguished by a double furred skin "parachute" which extends between the neck and front paws and between the front and hind paws as well as between the hind feet and tail. The unique feature is that the lower incisors are very wide and pectinate. The genus *Cynocephalus* is the subject of a separate entry

Descemet's membrane a membrane immediately above the inner endothelium of the cornea

descending colon the part of the colon in some mammals that runs from the transverse colon directly posterior

descending metamorphosis the metamorphosis of reproductive endosomatic parts in plants (e.g. the substitution of petals for stamens, etc.)

descending tract the ventral funiculus of the spinal cord which carries impulses in both directions

desert from the ecological point of view any area in which less than 20% of the ground surface is covered with permanent vegetation. Deserts are commonly classified according to the reasons that have caused them to exist such as low rainfall, cold, low nutrients in the soil, etc.

-desm- *comb. form* meaning "bundle" or "chain" and, by extension, "tie" or "ligament"

desma [dĕz″·mə] a bond, chain, link or group. The following terms using this suffix are defined in alphabetic position:
KINETODESMA   PLASMODESMA

Desmana [dĕz·män·ə] a genus of insectivores with typically mole-like front legs but with huge webbed hind feet

desmosome [dĕz″·mə·sōm′] thickened areas of closely apposed, but not fused, plasma membranes (cf. nexus)

desynapsis [dē′·sĭn·ăp″·sĭs] the precocious separation of synapsing chromosomes in meiosis

determinate cleavage holoblastic cleavage in which each cell is destined to form a specific part of the embryo

detritus [də·trĭt″·əs] dead plant and animal tissue in an ecosystem, usually including the live microorganisms engaged in the decomposition of the material

detrorse [dē·trôs″] directed downwards (*see also* antrose)

-deum suffix indicative of a "way", or "route" or "cavity through which things pass". The following words with this suffix are defined in alphabetic position:
PROCTODEUM   STOMODEUM   URODEUM

Deuter cell one of a row of large cells in the central strand of moss

-deutero- *comb. form* meaning "second" or "secondary" but often misused for "double"

deuterotoky [*angl.* dyōō·tə·ŏt″·ək·ē, *orig.* dyōō′·tĕr·ō·tōk″·ē] the parthenogenetic production of both male and female offspring

Deuteromycetes [dyōō′·tə·ō′·mi·sēt″·ēz] a botanical taxon erected to contain those fungi in which the formation of zygotes or spores has not been observed. The group is also referred to as fungi imperfectae. The genera *Diplodia, Monilia, Phymacotrichum* and *Sporotrichum* are the subjects of separate entries

Deuterostomia [dyōō′·tə·ō·stōm″·ēə] a term coined to describe those bilaterally symmetrical animals in which the embryonic blastopore becomes the anus. This includes the Chaetognatha, Echinodermata, Hemichordata, and Chordata (cf. Protostomia)

Devonian epoch one of the Paleozoic epochs extending from about 300 million years ago to 250 million years ago. It was preceded by the Silurian and followed by the Permian. Land forests of spore-bearing plants, mostly giant ferns, appeared and ostracoderms were plentiful

-dextr- *comb. form* meaning "right hand"

dextrorse [dĕk′·strôrs] towards the right hand

DFP di-isopropyl phosphorofluoridate

DFPase an enzyme that catalyzes the hydrolysis of di-isopropyl-phosphofluoridate to di-isopropylphosphate and hydrofluoric acid

-di- *comb. form* meaning "two" (cf. -dis-)

-dia- *comb. form* meaning "through"

diacmic plankton [dī′·ak″·mĭk] one showing two algal blooms per year (*see also* monacmic plankton)

diad [dī″·ăd″] one of a pair of chromosomes resulting from the separation of two homologous members of a tetrad in meiosis

diageotropism [*angl.* dī·ăj″·ēō·trōp′·ĭzm, *orig.* dī′·ə.jē′·ō·trōp″·ĭzm] a response which results in growth parallel to, and usually at or immediately under, the surface of the soil

diakinesis [dīə·kĭn′·ēs·ĭs] a terminal stage of the prophase in meiosis immediately following diplonema; it is characterized by the disappearance of the nucleolus and the even distribution of the bivalents throughout the nucleus

-dialy- *comb. form* meaning "to disband"

dialysis [dī·ăl′·əs·ĭs] the separation of large from small molecules by their passage through a membrane of suitable pore size

diapause [dī′·ə·pôrz″] a condition of larval resting development shown by many insect larvae overwintering in temperate climates

diapedesis [*angl.* dīə·pĕd″·ə·sĭs, *orig.* dī′·ə·pĕd·ēs″·ĭs] the process by which a leucocyte squeezes through the wall of a capillary

diaphototropism [dīə·fō′·tō·trōp″·ĭzm] a movement to orient an organism or part of an organism at right angles to the incident light

diaphragm [dīə·frăm″] a muscular partition separating the cavity of the thorax from that of the abdomen in mammals

diaphyseal center [dī·ə·fĭz″·ēəl] a center of osteogenesis in the middle of a shaft bone

diapsid [dī·ăp″·sĭd] pertaining to a reptilian skull in which there are two temporal openings

Diapsida a subclass of reptiles with a diapsid skull containing the extant Crocodilia but best known from the giant dinosaurs of the Triassic and Jurassic epochs.

The genera *Brontosaurus, Crocodilus, Diplodicus* and *Stegosaurus* are the subject of separate entries

diarthrosis [di'·är·þrō"·sĭs] a moveable joint. In human anatomy a synovial joint

diaschistic bivalent [di'·ə·shĭst"·ĭk] one which appears to divide transversely owing to the fact that the only chiasma is at the end (*see* bivalent)

-diastem- *comb. form* meaning "gap" or "interval"

diastole [di·ăst"·ə·li] the dilation of a contractile organ or vesicle (*see also* systole)

diatom common name of Bacillariophyceae

diatropism the tendency to become placed at right angles to an operative force

Diatrymiiformes [di·ăt·trĭm'·iə·fôrm"·ēz] an order of extinct giant terrestrial birds. The type genus *Diatryma* was represented in the U.S. by *D. steini* [sti"·ni"], about 7 ft tall, from the Lower Eocene of Wyoming

diaxenic *see under* monaxenic

Diceros [di"·sə·rŏs] a genus of Rhinocerotidae containing the two-horned *D. bicornis* [bi·kôrn"·ĭs], the African "black rhinoceros"

-dich- *comb. form* meaning "to disunite"

dichasial [di'·kāz"·ē·əl] said of a method of branching that results in a fork shape

dichogamy [*angl.* di'·kŏg"·ə·mē, *orig.* dik'·ō·gǎm"·ē] the condition of a hermaphrodite which cannot fertilize itself, because the two sexes mature at different times

Dicotyledoneae [di·kŏt'·ĭl·ēd"·ŏn·ā·ē] those gymnosperms in which the germinating seed produces a pair of cotyledons (cf. Monocotyledoneae)

-dictyo- *comb. form* meaning "net" better transliterated -diktyo-

dictyosome [dĭkt"·ē·ō·sōm'] any portion of the Golgi apparatus visible by light microscopy

dictyotic stage [dĭkt"·ē·ŏt'·ĭk] the terminal stage in nuclear division in which chromosomes are no longer apparent by normal techniques of light microscopy

Didelphis [di·dĕl'·fĭs] the genus of metatherian mammals containing the only North American metatherian *D. marsupialis* [mär'·syū·pē·āl"·ĭs], the opossum

Didinium [di·din"·ē·əm] a genus of predatory holotrich protozoans. They have a cone-shaped proboscis which is inserted into the prey which are then absorbed. *D. nasutum* [năs·yōō·təm] is predatory on *Paramoecium*

didiploid [di'·dip"·loid] the condition of a nucleus formed by the fusion of diploid nuclei

diel [dē·əl] a chronological day (24 hours) as distinct from the daylight portion of varying duration (cf. circadian, diurnal)

diencephalon [di'·ĕn·sĕf'·ə·lŏn] the portion of the brain that lies between the telencephalon and the mesencephalon

-diere- *comb. form* meaning "division"

diestrous cycle [di·ēs'·trəs] the period between estrous cycles (*see also* estrous cycle, anestrous cycle)

differential fertility in population genetics a term used to describe the result of genotypic selection

differential segment that portion of a chromosome which has no corresponding part in its bivalent

differentially permeable membrane one which permits the passage of some materials but not others (*see also* permeable membrane and semipermeable membrane)

differentiation the process by which cells, usually those of an embryo, develop specific characters or patterns (*see also* dedifferentiation, redifferentiation)

Difflugia [dĭf·lōō"·jēə] a genus of sarcodine protozoans distinguished by a flask-shaped test covered with particles derived from the substrate. The very common *D. pyriformis* [pĭr'·ē·fôrm"·ĭs] may reach a length of half a millimeter

diffuse muscle one in which the fibers are widely separated as in some Zoantharia

diffuse nervous system one in which nerve cells are spread over the surface of the body, and not concentrated in the ganglia

diffuse parenchyma apotracheal parenchyma evenly dispersed throughout the growth ring

diffuse placenta one having villi over the whole surface of the extraembryonic membrane

diffuse porous wood dicotyledenous wood in which the vessels are of equal diameter and universally distributed through a growth ring (*see also* ring porous wood)

digestive system the sum total of those hollow structures in which food is digested and the associated glands that assist in this process

-digit- *comb. form* meaning "finger"

digit the jointed terminations, or termination, of a limb; i.e. fingers, thumbs and toes

Digitalis [dĭj'·ĭt·āl"·ĭs] a genus of scrophulariacean flowering plants. *D. purpurea* [pyōōr·pyōōr"·ēə] (Foxglove) is well known both as an ornamental and as the source of a mixture of alkaloids (collectively called digitalis) which have been used in medicine since earliest times

digitigrade [dĭj·ĭt"·ē·grād'] an animal walks on the toes with the heel clear of the ground

dihybrid an individual which is heterozygous for two characters

1-3, 4-dihydroxyphenylalanine [di'·hi·drŏks'·ē·fĕn'·əl·ăl'·ən·ēn] an amino acid, 2-amino-3-(3,4-dihydroxyphenyl) propanoic acid, usually derived from leguminous seeds and widely known as dopa

dikaryon [di'·kâr"·ēən] a cell, or hypha, containing two haploid nuclei one of which is derived from another cell or hypha. Used principally of basidiomycete fungi

-diktyo- *comb. form* meaning "a net" frequently transliterated -dictyo-

Dimetrodon [di'·mĕt"·rō·dŏn] a genus of extinct synapsid reptiles of the Lower Permian. The neural spines of the vertebrae from the neck to the sacrum were enormously elongated apparently to support a huge semicircular sail

dimorphism [di·môrf"·ĭzm] having two forms. Sexual dimorphism and seasonal dimorphism are well known examples

Dinoflagellida [di'·nō flăj·ĕl"·ĭd·ə] an order of phytomastigophorous Protozoa the non-parasitic forms of which are distinguished by the presence of two flagella, one trailing in the body axis, the other transverse to the axis. They may also be considered as algae, in which case they form the class Dinophyceae of the phylum Pyrrophyta. The genera *Ceratium, Goniaulax* and *Noctiluca* are the subjects of separate entries

Dinosaur [dĭn"·ō·sôr'] a term loosely applied to many large extinct reptiles of the subclass Archosauria

Dioctophyme [*angl.* di·ŏkt·ŏf'·ə·mē, *orig.* di'·ŏkt·ō·fi"·mē] a genus of remarkable nematodes with a complex life cycle, strongly resembling that of the Nematomorpha, with a first stage in a leech parasitic on crayfish and a second stage in a fish. *D. renale* [rĕn·äl'·ē], the "Giant Kidney Worm", a 3 ft parasite, usually occurs in the mink but has been found in man

-diod- *comb. form* meaning "a passage"

dioecious [di·ēs'·ēəs] having a male and female sex organs in different individuals

Diphyllobothrium [di'·fil·ō·bŏth"·rē·əm] a genus of cestodes. *D. latus* [lā'·təs], the largest tapeworm found in man sometimes reaches a length of 30 feet. The coracidium larva infests *Cyclops* which is eaten by small fish, in which the procercoid larvas develop. These fish are then eaten by predators, in the U.S. particularly *Stizostedion vitreum* and *Esox lucius*, in which the pleurocercoids develop. The tapeworm is then acquired by man through eating undercooked walleyes and great northern pike

diphyodont [di'·fi·ə·dŏnt] said of a form with two generations of teeth

dipleurula larva [di·plûr"·yū·lə] that stage in the development of an echinoderm that lies between the gastrula and the pluteus or brachiolaria. It is often considered to represent the ancestral echinoderm type and by some to represent a common ancestor of echinoderms and chordates

diploblastic [dip"·lō·blăst'·ik] consisting of, or possessing, two embryonic layers

Diplocardia [dip'·lō·kärd"·ēə] a genus of oligochaete annelids. *D. communis* [kŏm'·yōōn·is] is probably the commonest large earthworm of the prairie states

Diplococcus [dip'·lō·kŏk"·əs] a genus of lactobacciliale schyzomycetes. *D. pneumoniae* [nyū·mōn"·ī·ē] is a dangerous human pathogen occurring not only in the respiratory tract but also in the meninges

Diplodia [dip·lŏd·ēə, dip·lŏd·ēə] a genus of deuteromycete fungi pathogenic to plants. *D. zeae* [zā'·ē] is the causative agent of dry rot in corn ears

Diplodicus [*angl.* dip·lŏd"·ə·kəs, *orig.* dip'·lō·dik"·əs] a genus of extinct diapsid reptiles of the Jurassic. They were elongate, long-necked, herbivores forms weighing up to fifty tons. The dorsal nostrils indicate that, like the modern Hippopotamos, they probably spent much of their time in water

diploid [dip"·loid] having twice the number (2N) of chromosomes found in a gamete, or in the other of two alternate generations. The diploid number is usually given as the chromosome number of an organism (*see also* -ploid, didiploid, hyperdiploid, syndiploid)

diplonema [*angl.* dip·lŏn"·əm·ə, *orig.* dip'·lō·nēm"·ə] that stage in the prophase of meiosis immediately following pachynema and in which the separation of the paired chromosomes commences

Diplopoda [*angl.* dip·lŏp"·ə·də, *orig.* dip'·lō·pōd"·ə] a class of progoneate arthropods containing the millipedes. The cylindrical, or hemicylindrical body, bearing from forty to a hundred pairs of legs is typical of the class. The genus *Spirobulus* is the subject of a separate entry

diplosome [dip·lō"·sōm] one of a pair of centrosomes

diplotene stage [dip"·lō·tēn'] the stage in meiotic division that immediately follows the pachytene stage and which is characterized by the first appearance of the chromatids

diplotonic [dip'·lō·tŏn"·ik] said of a haplobiontic life cycle in which the adult is diploid (*see also* haplotonic)

Dipneusti [dip·nyōōs"·ti] an order of Choanichthyes, mostly extinct, popularly referred to as lung fishes by virtue of the presence of one or more sac-like lungs communicating with the mouth. They differ from the Crossopterygii in the structure of the fins and in a general tendency to a reduction of the internal skeleton. The genera *Ceratodus*, *Lepidosiren*, *Neoceratodus* and *Protopterus* are the subject of separate entries

Dipnoi= Dipneusti

Diptera [dipt"·ər·ə] the order of insects containing the forms most commonly called flies. They are distinguished by possessing a single pair of wings, the place of the second pair being taken by halteres which function as gyroscopic stabilizers. The families Culicidae and Muscidae and the genera *Chrysops* and *Phlebotomus* are the subject of separate entries

direct metamorphosis the type of insect life history in which there is no pupa (= incomplete metamorphosis)

directive species one which attracts a predator, of which it is not the customary prey, to an area rich in the customary prey

-dis- *comb. form* meaning apart and also a frequent form of -di- meaning "two"

disc any thin circular object. The following compound terms are defined in alphabetic order:

A-DISCS	IMAGINAL DISC
EPIPHYSEAL DISC	INTERCALATED DISC
FIXATION DISC	ORAL DISC
GERMINAL DISC	PEDAL DISC
I-DISC	

discinid larva [disk·in'·id] a planktonic bivalve larva of brachiopods

disclimax *either* a climax which has been disturbed through the introduction, by man or other agency, of new organisms that destroy stability *or* a climax caused and maintained by human interference

discoblastula [disk'·ō·blăst"·yū·lə] a blastula consisting of a disc of cells resting on yolk

discoidal cleavage the type of cleavage of a telolecithal egg in which the cleavage is restricted to a germinal disc

discoidal placenta one in which the villi are confined to one or two discs on the surface of the extraembryonic membrane

disharmony law the logarithm of a dimension of any part of an animal is proportional to the logarithm of the dimension of the whole animal

dislocated segment a segment of a chromosome homologous with, but in a different position from, a similar segment in another chromosome

disomic [di·sōm'·ik] having an additional chromosome, usually in virtue of nondisjunction (*see also* monosomic, nullisomic)

dispersal the manner in which organisms are dispersed (cf. dissemination)

dissemination the methods by which individual offspring or seeds are liberated from the parent (cf. dispersal)

dissoconch [dis"·ō·kŏnk'] the shell of a veliger larva

dissogony [*angl.* dis·ŏg"·ə·nē, *orig.* dis'·ō·gŏn"·ē] the condition of an organism which reproduces sexually both as a larva and as an adult

distal [dist"·əl] that which lies further from. In general biological usage the anterior end and the main axis are used as the points of departure so that the stomach is distal to the esophagus and the arm is distal to the shoulder (*see also* proximal)

distal chiasma   one distal to an inversion loop and the centromere

-disto-   *comb. form* meaning "to stand apart", or by extension "to be remote from"

diurnal   [dī·ûrn″·əl] pertaining to daylight hours either as a rhythmic cycle or as an activity, in contrast to nocturnal or crepuscular (*see also* circadian, diel)

DNA = deoxyribonucleic acid

DNP = dinitrophenol

-dodeca-   *comb. form* meaning "twelve"

-dolabr-   *comb. form* meaning a "hatchet" or "mattock"

-doli-   *comb. form* meaning a "cask"

-dolicho-   *comb. form* meaning "along"

Dolichohippus   [dŏl′·ĭk·ō·hĭp″·əs]   a monotypic genus of the Equidae containing *D. grevii* (Grevy's Zebra), the only extant equid not in the genus Equus

doliolaria larva   [dōl′·ē·ō·lâr″·ēə]   a free-swimming ciliated postgastrular stage in the development of crinoids and a post auricularia holothurian larva. It possesses a large apical tuft and either four or five ciliated bands

Doliolum   [dōl·ē″·ōl·əm, dōl′·ĭ·ōl″·əm]   a genus of thaliacean urochordates distinguished for its polymorphic life history. The sexual generation produces the first asexual generation which buds the second asexual generation in a long chain. The central individuals of this chain break off into separate individuals resembling the sexual generation in everything except the possession of gonads. These bud off individuals of the sexual generation from a ventral stolon

Dollo's Law   structures lost in the course of evolution can never be regained (cf. Arber's Law), but structures gained may none the less be lost

-dome-   *comb. form* meaning "house" frequently confused with the totally unrelated root -deme- (*see* apodome, myodome)

domin   [dŏm″·in″]   an organism showing weak dominance in an association

dominance 1   (*see also* dominance 2) in the genetic sense, the condition of a character which appears both in the heterozygote and the homozygote; the character is said to be "dominant" and the effect to result from a dominant allele. In this sense the following compounds are defined in alphabetic position:

CONDITIONED D.   PARTIAL DOMINANCE
INCOMPLETE D.   PSEUDODOMINANCE
OVERDOMINANCE

dominance 2   (*see also* dominance 1) in the ecological sense, the condition of an organism which dominates a community either in virtue of its habits or of the shear weight of its numbers

dominant 1   (*see also* dominant 2) an allele which dominates another allele; for variation of this last condition see: dominance 1 (*see also* delayed dominant, double dominant)

dominant 2   (*see also* dominant 1) in the ecological sense, an organism or several organisms in a community which so behave, either passively or actively, as to dominate or control the whole habitat (*see also* subdominant)

dominant allele   one which determines the phenotypic expression in a heterozygous form

dominant character   the expression of a dominant allele

-domit-   *comb. form* meaning "tamed"

-domous-   "dwelling in the house of", or "providing a house for". The numerous compounds of this form are not separately defined since the meaning is obvious from the roots (e.g. polydomous, having many houses [or nests]; mymecodomous, housing, or being housed by, ants; etc.)

donor 1   (*see also* donor 2) the individual from which a tissue transplant is removed to a receptor

donor 2   (*see also* donor 1) a molecule from which atoms are removed in the course of an enzyme catalyzed reaction

-dont-   *comb. form* meaning "tooth" (cf. -dent-). The following forms using this suffix are defined in alphabetic position:

ACRODONT	LOPHODONT
BUNODONT	MEGADONT
CORYPHODONT	OLIGODONT
DIPHYODONT	OLIGOPHYODONT
EUPLEURODONT	PLEURODONT
HAPLODONT	POLYPHYODONT
HETERODONT	SUBPLEURODONT
HOMODONT	THECODONT

-dorm-   *comb. form* meaning "sleep" but used, by extension, for words descriptive of any structure or organism which remains inactive

dormancy   a condition in an organism in which the life processes are slowed down, usually for the purpose of surviving a temporarily inclement environment

dormant egg   an egg, in invertebrates usually thick-shelled, which is destined to remain for some considerable time before developing (*see also* subitaneous egg)

dorsal   pertaining to the back. In animals the back is usually taken to be the surface furthest from the ground when the animal is in a resting position. In erect animals (e.g. man) the term dorsal becomes synonymous with posterior. In plants the dorsal surface of a leaf is the lower face which was originally the outer surface in the bud

dorsal aorta   the blood vessel formed by the union of the two aortic roots and which carries blood back along the middorsal line of vertebrates (*see also* aorta, ventral aorta)

dorsal bronchus   a series of dorsal connections with the mesobronchus

dorsal fin   one or more median-dorsal unpaired fins of the fish

dorsal root   The metameric sensory ganglia, external to the cord, forming the root of spinal nerves

dorsalia = arcualia

double dominant   the condition in which a phenotypic expression is dependent on the presence of both of two dominant alleles

double fertilization   when nuclei in the pollen tube fuse both with the egg nucleus and with the polar nucleus

double recessive   having homozygous recessive alleles at different loci

double haploid   a haploid having a complete genome from each of two species (*see also* synhaploid)

double tetraploid   one which carries four genomes from each of two distinct species

Doyères eminence   a bump formed at the point where a motor nerve ending penetrates the sarcolemma

Draco   [drā″·kō′]   a genus of Sumatran lizards with a broad, scaley membrane between front and hind legs. This "wing" is additionally supported by outgrown ribs. The animals can glide 30 or 40 ft and are therefore called "Flying Lizards"

Dracunculus [drā·kŭn"·kyōō·ləs] a genus of filarioid nematode worms, the only important member of which is *D. medinensis* [mĕd'·ēn·ĕn"·sis] (the "Guinea Worm"). The adult, which may reach a length of four feet, lives in the subcutaneous tissues of man. The larvae, liberated by protruding the uterus through the skin, enter *Cyclops* and infest men who drink water containing this copepod. The range extends from central India to central Africa. There is a strong supposition that this worm was the "fiery serpent" (Numbers 21 :6) that inflicted the Israelites

dreen the shrubby, or sometimes woody, swamp which occupies the "channel" between a partially detached island and the mainland

-drom- *comb. form* meaning a "course" or "direction" usually in the form -dromo-

dromotropism the spiral growth of plants occasioned by a response to touch stimulus

dromous pertaining to a direction (*see* anadromous, catadromous)

Drosera [drŏs·ə·rə] a dicotyledenous plant usually called the sundew. The leaves bear mucilage-tipped hairs to which insects become stuck. The insects are then digested

-droso- *comb. form* meaning "dew"

Drosophila [*angl.* drəs·ŏf"·əl·ə, *orig.* drŏs'·ō·fil"·ə] a genus of dipteran insects properly called pomace flies or dew flies, though the misnomer fruit fly is common. *D. melanogaster* [mĕl·ăn·ō·găst·ə] has been the subject of intensive genetic studies and the science of genetics is still largely based on it. Several hundred phenotypically distinct mutants are known and the presence of giant chromosomes in the salivary glands make it easy to plot gene loci. Flies of this genus are also responsible for distributing yeasts among grapes in vineyards and thus making possible the production of wines

-drup- *comb. form* meaning "an olive" in the sense of the fruit (cf. -elaeo-)

drupe [drōōp] a fruit, such as a plum, cherry or peach in which there is a thick stony endocarp and a fleshy mesocarp

Dryopithecus [dri'·ō·pĭth"·ə·kəs, dri'·ō·pĭth·ē"·kəs] one of the earliest extinct pongid apes appearing as fossils in the Miocene

duct any tubular vessel in an organism carrying either fluids or gasses. The following derivative terms are defined in alphabetic position :

ALVEOLAR DUCT	HERMAPHRODITE DUCT
ANTERIOR CEPHALIC D.	LACTEAL DUCTS
ARANZIO'S DUCT	MÜLLER'S DUCT
BILE DUCT	OVIDUCT
BOTALLO'S DUCT	OVOVITELLINE DUCT
COELOMODUCT	PNEUMATIC DUCT
COMMON OVIDUCT	RESPIRATORY DUCT
EJACULATORY DUCT	SANTORINI'S DUCT
ENDOLYMPH DUCT	STENO'S DUCT
GÄRTNER'S DUCT	WHARTON'S DUCT
GONODUCT	WIRSUNG'S DUCT

ductless gland = endocrine gland

ductule [dŭk'·tyōōl"] diminutive of duct

ductus arteriosus the rudimentary remnants of the sixth aortic arch when the latter is primarily, but not entirely, converted to a pulmonary artery

duff the top layer of soil, consisting of partially decomposed vegetable matter and lacking any admixture of mineral soil from the lower layers

Dufour's gland a pheromone producing gland in the abdomen of ants : in other Hymenoptera a similar gland contributes to the venom

Dugesia [dyū·gĕs'·eə] a genus of turbellarian platyhelminths widely used in laboratories, particular in experiments on regeneration *D. tigrina* [ti·gri·nə] (often called *Planaria maculata*) is the very common small (10–15 mms), dark brown or black planarian found in all types of freshwater. *D. dorotocephala* [dŏr'·ō·tō·sĕf"·ə·lə] is much larger (30 mms) and frequently has a pale stripe down the back

Dugong [dyū"·gŏng"] one of the two extant genera of sirenian mammals. They inhabit the coasts of the Indian Ocean and are distinguished from the Manatees (*Trichecus*) by the larger head (with tusks in the male) and sickle-shaped tail

dun [dŭn] the sub-imago of an ephemopteran insect

duodenal bulb [dyū'·ə·dēn"·əl, dōō·ə·dēn"·əl] a thickened portion of the duodenum in some mammals, containing Brunner's glands

duodenum [dyū·ŏd"·ən·əm, dōō·ŏd"·ən·əm] that part of the small intestine which is immediately adjacent to the stomach (*see also* mesoduodenum)

duplex uterus a condition in which paired, tubular uteri open into the vagina

duplicate gene one of several genes at different loci all effecting the same phenotype in the same manner

dura spinalis [dyūr'·ə spin·āl'·ĭs] the outer of the two meninges of higher forms

Dusicyon [dyūs'·ē·si"·ŏn] the genus of canid carnivores to which belong the South American jackals and "Wild Dogs". They were all once called "dogs" and, indeed, they more closely resemble foxes than jackals in their habits

dymantic color [dĭ·măn"·tĭk] that which is easily remarked but the purpose of which is not immediately apparent

-dynam- *comb. form* meaning "power"

dynamic pertaining to power, energy or importance (*see* oligodynamic, trophodynamic)

dynamic synecology the study of the interaction of organisms within a community (*see also* synecology, geographic synecology, morphological synecology)

-dys- *comb. form* meaning "bad"

-dysis- *comb. form* meaning "to clothe". The following terms using this suffix are defined in alphabetic position :
ECDYSIS
METECDYSIS

dysphotic zone [dĭs·fŏt"·ĭk] that layer of water with enough light for animal response, but not for photosynthesis

dysploid [dĭs"·ploid] having a number of chromosomes which is not a multiple of the haploid number (= aneuploid)

dystrophic [dĭs·trŏf"·ĭk] causing, or resulting from, inadequate nutrition

dystrophic lake an old, heavily sedimented, vegetation-filled lake, often the first stage in the production of a peat bog

dyttja [dūt'·yə] the fine mud on the bottom of productive lakes with brown, humic-colored, water (*see also* hyttja)

E

-e- privative meaning "without" or "destitute of", but not in the sense of having had something removed (which is better expressed by de-)

ear a complex vertebrate phonoreceptor. The term properly applies to the whole apparatus but in mammals is often loosely applied to the pinna

ecad [ē·kăd′] an organism, frequently but not of necessity sessile, which is specifically, and sometimes uniquely, adapted to the environment in which it is found

Ecardines [ē′·kär·dĭn″·ēz] a class of Brachiopoda distinguished by the fact that the shells are not hinged together. The genus *Lingula* is the subject of a separate entry

eccrine [ek″·rĭn″, ek″·krĭn′] secretory. Used specifically to distinguish a gland that produces what its name indicates (e.g. an eccrine sweat gland produces sweat) as distinct from a gland which does not (e.g. an apocrine sweat gland that produces a milky fluid)

ecdysis [ĕk·dĭs″·ĭs] the shedding of the outer layer or molting

ecdyson [ĕk·dĭ″·sŏn] a hormone secreted in the prothoracic gland of insects which controls molting

-echin- comb. form meaning "spine"

Echinarachnis [ē′·kĭn·ə·răk″·nəs] a genus of echinoid echinoderms, distinguished by its very flattened shape. *E. parma* [pär′·mə] is the common "sand dollar"

Echiniscus [angl. ə·kĭn″·əs·kəs, orig. ē′·kĭn·ĭs″·kəs] a genus of Tardigrada with several single claws on each leg and 2 red eyes. There are more than 100 species, both marine and freshwater. *E. testudo* [tĕs·tyū″·dō] is a moss-dwelling form often called the "roof water bear" since it was readily found in moss on roofs when this habitat existed

Echinococcus [ē′·kĭn·ō·kŏk″·əs] a genus of cestode platyhelminthes (tapeworms) the minute adult of which is an intestinal parasite of cats and dogs. The coenurus larva of *E. granulosus* [grăn′·yōō·lōs″·əs] grows into a large hydatid cyst which, in man, usually occurs in the lung or brain

Echinodera [angl. ək·ĭn·ŏd″·ər·ə, orig. ē·kĭn′·ó·dēr″·ə] a small phylum of microscopic marine pseudocoelomate bilateral animals distinguished by superficial spines and a body superficially segmented into thirteen divisions.

This phylum may also be regarded as a class of the Aschelminthes. *Pycnophyes* is the subject of a separate entry

Echinodermata [ək′·ĭn·ō·dĕrm″·ə·tə, ē′·kĭn·ō·dĕrm″·ətə] the phylum of the animal kingdom that contains the sea urchins, the sea stars, the sea cucumbers, and the sea lilies. They are distinguished from other enterocoelous coelomate coelomate animals by a radial symmetry of five and the presence of a calcareous skeleton often consisting of external plates bearing spines. The classes Asteroidea, Crinoidea, Echinoidea, Holothurea, and Ophiuroidea are the subjects of separate entries

Echinoidea [ək·ĭn′·oid·eə, ē·kĭn′·oid·eə] the class of echinoderms that contains the sea urchins. They may be spheroidal, disk-shaped, or heart-shaped, but are distinguished from other echinoderms by the large number of moveable spines on the surface which is covered with a test of contiguous calcareous plates. The mouth is on the ventral surface, and is furnished with protrusible jaws activated by a system of articulated levers known as "Aristotle's lantern". An adoral filter disc (the madrepore) leads to the stone canal which takes water to the polian vesicle which is the pressure valve that controls the extrusion of tube feet. The genera *Arbacea*, *Echinarachnius* and *Echinus* are the subjects of separate entries

Echinus [ək·ĭn″·əs] a genus of echinoid echinoderms from which both groups were named. The apple-sized, obovate *E. esculentus* [ĕs′·kyū·lĕnt″·əs] is the commonest European sea urchin. The name comes from the custom of eating the ripe gonads

Echiurida [ē′·kĭ·yūr″·id·ə] a small phylum of coelomate bilaterally symmetrical animals closely allied to the Annelida and by some placed as a class in that phylum. Most possess a pair of setae but they lack meraneric segmentation. An outstanding characteristic is the very large spatulate prostomium. The genus *Bonellia* is the subject of a separate entry

ecobiotic isolation [ē′·kō·bi·ŏt″·ĭk] segregation in consequence of habitat, as the head louse and the body louse

ecoclimatic isolation segregation by virtue of the climate

ecocline [ē″·kō·klĭn′] a cline reflecting ecological conditions in general

ecogeographic isolation [ē'·kō·jē'·ō·grăf"·ĭk] isolation in virtue of geographic location

ecological age age in relation to the ability to breed and therefore divided into pre-, post- and actual

ecological divergence the production of races and subspecies adapted to varying ecological conditions

ecological efficiency an expression of the energy required by a given unit of protoplasm

ecological equivalent an animal occupying an ecological niche, and having in the broadest sense the same general appearance, as another animal to which it is not closely related

ecological isolation isolation of populations in virtue of living conditions

ecological niche either a microhabitat or a position occupied by an individual in an assemblage

ecological race variously used to mean geographic race or subspecies

ecological succession the gradual change of uninhabited land to a climax community

ecology a study of organisms in relation to each other and to their environment. The following derivative terms are defined in alphabetical position :

AUTECOLOGY
DYNAMIC SYNECOLOGY
GEOGRAPHIC SYNECOLOGY
MORPHOLOGICAL SYNECOLOGY
SYNECOLOGY

ecophene [ĕk"·ō·fēn'] a modification of a genotype through selective breeding, the selection being occasioned by environmental or ecological factors

ecophenotype [ĕk'·ō·fēn"·ō·tīp'] a phenotype showing non-heritable variations from the norm in consequence of an adaptation to an environment

ecospecies [ĕk'·ō·spēsh'·ēz] a species in a group of populations associated with a specific environmental niche but capable of interbreeding with other neighboring ecospecies; the term is sometimes used as a synonym of ecotype

ecosystem an ecological system comprised of both the biotic and abiotic components; the basic ecological unit of structure and function which is considered to be at least partially self sustaining and self regulating, usually requiring only a continuing supply of solar radiation

ecotone [ĕk'·ō·tōn"] a transitional area between two associations, climaxes or seres

ecotype [ĕk'·ō·tīp"] a phyletic unit adapted to a particular environment but capable of producing fertile hybrids with other ecotypes of the same ecospecies

-ecto- comb. form meaning "outside"

Ectocarpus [ĕk'·tō·cärp"·əs] a genus of fucale Phaeophyta occurring as brownish tufts rising from creeping filaments

ectochordal centrum [ĕk'·tō·kôrd"·əl] one in which the center of the notochord is replaced with cartilage surrounded by a cylinder of bone

ectoderm [ĕk'·tō·derm"] the outer layer of an animal, particularly an embryo (see also mesectoderm)

ectomesoderm [ĕk'·tō·měs'·o·darm"] mesodermal, usually invertebrate, mesenchyme cells budded off from the ectoderm (see also entomesoderm)

ectoparasite [ĕk'·tō·per"·ə·sīt] one the body of which is not lodged within the tissues of the host

ectoplasm [ĕk'·tō·plazm"] the outer, jell, layer of the protoplasm of rhizopod Protozoa. The term is also, rarely, applied to the capsule of bacteria

Ectoprocta [ĕk'·tō.prŏkt"·ə] a phylum of lophophorian coelomate animals at one time combined with the phylum Entoprocta into the no-longer valid taxon Bryozoa. They are distinguished from other lophophorians by their exoskeleton of small horny cases or gelatinous masses. They have often been referred to as moss-animalcules from their habit of growth. The genera *Bugula* and *Pectinatella* are the subject of separate entries

ectopterygoid bone [ĕk'·tō·tĕr'·ə·goid"] one of a pair of dermal bones in the palatoquadrate complex of actinopterygian fish lying immediately outside and below the metapterygoid (see also pterygoid bone, metapterygoid bone)

edaphic [ē·dăf'·ĭk] pertaining to the influence of soil upon organisms growing in or on it

edaphic climax one which has reached its condition in consequence of the nature of the soil

edaphic ecotype an ecotype effected by soil as distinct from climatic conditions (see also climatic ecotype)

edaphic factor the contribution of the inhabitants of the soil to an ecosystem

edaphonekton [ē·dăf'·ō·nĕkt"·ən] the total of all the organisms that live in the free water in soil

Edentata [ē'·dĕnt·ät"·ə] an assemblage, once regarded as an order, of mammals lacking true teeth. Nowadays this taxon is usually divided into the orders Pholidota, Xenarthra, and Loricata

edge effect the tendency for variety and density to increase at community junctions

effect (see also law) the result of an action, or interaction, so constant and so frequently observed that it has been recorded in the literature under a specific name. The terms "effect" and "law" (q.v.) are, in biological literature used almost interchangeably. The following derivative terms are entered in alphabetic position :

CIS-TRANS EFFECT   POSITION EFFECT
EDGE EFFECT   SEWALL WRIGHT EFFECT

efferent [əf"·ər·ənt] to convey away from (see also afferent, deferent)

efferent branchial artery an artery carrying blood away from the gill

efferent nerve one which carries impulses from the brain

egg properly the female gamete but generally applied to the entire female reproductive unit including extracellular food reserves and the case or shell. By some the term ovum is used for the gamete and egg for the unit (see also cleavage, lecithal). The following derivative terms are defined in alphabetic position :

AMICTIC EGG   MICTIC EGG
CLEIDOIC EGG   MOSAIC EGG
DORMANT EGG   REGULATIVE EGG
EPHIPPIAL EGG   TACHYBLASTIC EGG
ISOTROPIC EGG   WINTER EGG

egg pod a pouch full of eggs; this is typical of orthopteran insects in the original wide sense of this term

egg sac the term is particularly applied to the egg masses of copepod crustacea. It is also used, in botany, to describe membranes surrounding eggs, ovules, and oospheres

-eidos- comb. form meaning "resemblance". Frequently abbreviated in compounds to the termination -id

Eisenia = *Allolobophora*

ejaculatory bulb any structure which by compression forces spermatozoa through a penis or aedeagus

ejaculatory duct that portion of the male gonoduct which by contraction forces out the sperm

-elaeo- *comb. form* meaning "olive" either in the sense of its color, or of its oily nature

elaeodochon [ĕl'·ē·ŏd·ə·kŏn] the oil gland of birds at the base of the tail used in preening and waterproofing the feathers

elaioplasts [ĕl·ē"·ə·plast'] oil droplets in plant cells

elephantiasis [ĕl'·ə·fănt·ī"·əs·ĭs] a disease involving monstrous swelling of the lower limbs owing to the blocking of lymph vessels by *Wucheria bancrofti*

Elaphas [ĕl"·ə·făs] the genus of proboscidean mammals containing the Indian Elephant (*E. indicus*) [ĭn"·dĭk·əs] distinguished from the African Elephant (*Loxodonta*) by the smaller ears

Elasmobranchii [əl·ăs'·mō·brănk"·ēī] a subclass of Chondrichthyes distinguished from the Holocephali by possessing placoid scales, by lacking an operculum and in not having the upper jaw fused to the cranium. The orders Selachii (sharks) and Batoidea (rays) are subject of separate entries

elastic cartilage cartilage with elastic fibers in its intercellular substance

elastic connective tissue *see* yellow elastic connective tissue

elastic fibers fibers, possessing considerable elasticity and mechanical strength and which are bound with other connective tissues as branched filaments without organized pattern

-elat- *comb. form* meaning "drive" in the sense of causing movement

elater [əl·ā'·tər] that which causes an organism to jump as the furcula of Collembola, or the prosternal process of some beetles. In botany, any of numerous structures, which by their sudden expansion serve to disseminate seeds or spores

elbow in primates, particularly anthropoids, the joint between the humerus and the radio-ulna (cf. knee)

electric organ modified muscular tissue, found in some fish, capable of generating high potentials and storing considerable charges

electron microscope a device strictly analogous to an optical microscope but using a beam of electrons focused by magnetic lenses in place of a photon beam focused by glass lenses (*see also* microscope, resolution)

electroplaque [əl·ĕk"·trō·plăk', əl·ĕk'·trō·pläk"] a modified myoblast forming a unit of an electric organ in electric fish

-eleuthero- *comb. form* meaning "free"

elittoral [*angl.* ə·lĭt'·ər·əl, *orig.* ē·lĭt'·ôr·əl] the extension of the shoreline to the point under water where photosynthesis is no longer possible

Elodea [ĕl'·ō·dē"·ə] a genus of hydrocharitaceous water plant. *E. canadensis* is the common waterweed of the North Eastern U.S.

Eltonian pyramids the pyramidal arrangement of numbers, biomass or calories of the trophic levels in an ecosystem

-elytra- *comb. form* meaning sheath

elytron [əl·ī"·trŏn] a "sheath" or "shield", particularly the modified horny anterior wings of beetles and some other insects or the modified chaetae in the form of scales or plates found in some polychaete worms (*see also* hemielytron)

embolomerous [*angl.* ĕm·bŏl·ŏm"·ər·əs, *orig.* ĕm'·, bōl·ō·mĕr"·əs] having a persistent notochord running through the vertebra

emboly [ĕm'·bə·lĭ] the formation of the endoderm of a gastrula by invagination

embryo [ĕm"·brē·ō] in animal development, a stage incapable of supporting a separate existence in contrast to a larva or nymph. In plant development, a stage in which specific organs or organ systems are not visibly differentiated, particularly the partially developed sporophyte in a seed

embryo sac specifically the female gametophyte which develops from the megaspore within the nucellus

embryogenesis [ĕm'·brē·ō·jĕn"·əs·ĭs] reproduction through the medium of embryos

embryonic gradient any embryonic phenomenon which differs in intensity along any axis

embryonic membrane a name generally applied to the amnion, allantois, and yolk sac taken together

embryotroph [ĕm'·brē·ō"·trŏf] either embryonic nutritional material derived by mammalian embryos from the uterine wall in the course of implantation or the sum total of the nutritional materials secured by an embryo from any source

emeiocytosis [ĕm·mi'·ō·sit·ōs"·ĭs] the reverse of pinocytosis; that is, the expulsion of minute particles by a cell

-emphys- *comb. form* meaning "breathe on"

-en- *comb. form* meaning "in"

enamel the outer layer of a tooth consisting of overlapping prismatic scales of apatite cemented in an organic matrix

-enant- *comb. form* meaning "opposite"

-enchym- *comb. form* meaning "poured, or molded"

-enchyma termination used to indicate plant tissues. The termination -chyme is preferred for animals. The following terms using this suffix are defined in alphabetic position:

BOTHRENCHYMA	PLECTENCHYMA
COLLENCHYMA	PORENCHYMA
INENCHYMA	PROSENCHYMA
MESENCHYMA	SCLERENCHYMA
PARENCHYMA	

Enchytraeus [ĕn·kĭt"·rē·əs] a very large genus of slender oligochaete annelids mostly from one to three centimeters long and usually less than a millimeter thick; many inhabit freshwaters, though some are found on ocean shores and in decaying vegetation. *E. albidus* [ăl·bĭd'·əs] is commonly cultured by aquarists

encystment [ən·sĭst'·mənt] the process of forming, or becoming enclosed in, a cyst (*see also* excystment)

-end-, -endo- *comb. form* meaning inner

end bulb the swollen termination of an axon

endarch xylem [ĕnd'·ärk] that in which the most mature elements are located nearest to the center of the axis

-endeca- *comb. form* meaning "eleven"

endemic [ĕn·dĕm"·ĭk] confined to, but normally inhabiting, a limited area

endobiosis [ĕnd'·ō·bi·ōs"·ĭs] *either* the condition of an organism which lives within another (usually with the connotation of parasitism) *or* the condition of organisms that live in the surface of mud

endoblast [ĕn′·dō·blăst″]   the lower germ layer in early development of the telolecithal egg

endocardium [ĕn′·dō·kärd″·ē·əm]   the inner epithelial layer of the heart

endocarp [ĕn″·dō·kärp′]   the inner of the three layers into which a pericarp may differentiate

endochondral bone [ĕn′·dō·kŏn″·drəl]   a bone in which ossification commences in the center of cartilage

endocranium [ĕn′·dō·krān″·eəm]   the part of the bony skull that is derived by ossification of parts of the embryonic chondrocranium

endocrine gland [ĕn″·dō·krĭn,ĕn″·dō·krīn′]   a gland without ducts, the secretory products of which are accordingly distributed by the blood and known as hormones (*see also* exocrine gland)

endoderm [ĕn′·dō·dûrm″]   the innermost layer of cells, derived from the walls of the archenteron, in animals. Also the sheath of cells surrounding the vascular center of a root, or a conifer leaf. In this last meaning, endodermis is sometimes preferred

endogenous rhythm [ĕn·dŏj″·ən·əs]   a cyclic change in an organism induced by causes presumed to be within the organism itself

endogenous spore   one formed within the cell

endognath [ĕnd′·ŏg·nāth]   the modified endopodite of a crustacean mouth part

Endolimax [ĕn′·dō·lĭm·ăks]   a genus of amoebid protozoans symbiotic in the human intestine

endolymph [ĕn″·dō·limf′]   any contained fluid not otherwise specifically identifiable but particularly the fluid within the semicircular canals

endolymph duct   the primitive connection between the auditory vesicle and the exterior

endomeristem [*angl.* ĕnd·ŏm″·ar.ĭst′·ĕm, *orig.* ĕnd′·ō·mĕr″·ē·stem]   the initial of the central strand in mosses

endomitosis [*angl.* ĕnd′·ə·mət·ōs″·ĭs, *orig.* ĕn′·dō·mĭt″·ōs·ĭs]   the division of the nucleus without division of the cell, thus resulting in polyploidy

endomixis [ĕn′·dō·mĭks″·ĭs]   the reorganization of nuclear material within a protozoan, through the replacement of an existing macronucleus with the products of division of one or more micronuclei (*see also* hemimixis)

Endomycetales [ĕn′·dō·mi·sət·āl″·ēz]   a group of ascomycete fungi most of which are commonly called yeasts. They normally occur as single or conjugated cells though a few may have a transitory hyphal stage. Those that do not form spores, such as *Cryptococcus*, are sometimes placed in the "Cryptococcaceae" as a division of Deuteromycetes. The genera *Cryptococcus* and *Saccharomyces* are the subjects of separate entries

endomysium [ĕn′·dō·mĭs]·ēən]   a connective tissue sheet, containing capillaries, surrounding an individual muscle fiber

endoneurium [ĕn′·də·nyūr·ēəm, ĕn′·dō·nyūr″·ēəm]   connective tissue fibers surrounding nerve fibers

endoplasm [ĕn′·dō·plazm″]   the inner (sol) layer of the protoplasm of rhizopod protozoans

endoplasmic reticulum   a network or organization of membrane "tubules" spreading through the cytoplasm. It has at various times been reported to open both to the exterior of the cell and from the nucleus. The outer surface of the membrane bears ribosomes

endopodite [*angl.* ĕnd·ŏp″·ə·dĭt, *orig.* ĕn′·dō·pōd·ĭt]   the adaxial branch of a biramous crustacean appendage

endorhachis [*angl.* ĕnd·ər·ək·ĭs, *orig.* ĕnd·ō·răk″·ĭs]   the connective tissue lining of the vertebral column and skull

endoskeleton [ĕnd′·ō·skĕl″·ət·ən]   a skeleton under the surface, as that of vertebrates, in contrast to a surface skeleton as that of arthropods. Also any part of the skeleton which lies within another part, as the endoskeleton of the vertebrate skull

endosmosis [ĕnd′·oz·mōs″·is]   passage of solvent from a region of low solute concentration to one of high solute concentration

endosperm [ĕnd′·ō·spûrm″]   the food reserve of a seed

endospore [ĕnd′·ō·spôr″]   one that results from internal fragmentation of the cell and which is liberated without the forming a wall

endosteum [ĕnd·ŏst′·ēəm]   the membrane lining the marrow cavity of bones

endostyle [ĕnd·ō′·stil]   a flattened or trough-like structure, grooved on its dorsal surface by the hypobranchial groove, which runs along the ventral margin of the pharynx in prochordates

endosymbiosis [ĕnd′·ō·sĭm.bi·ōs″·ĭs]   a form of symbiosis in which one symbiont lives within another as do green algae in *Hydra* or *Convoluta*

endotheliochorial placenta   one in which both the epithelium and the connective tissue of the uterus are eroded away, so that the embryonic tissues lie in contact with the maternal blood vessels

endothelium [ĕn′·dō·thĕl″·ēəm]   an epithelium of mesodermal origin that lines a cavity as the coelom (*see also* epithelium)

endozoic [ĕn′·dō·zō·″·ĭk]   said of an animal living within another animal (*see also* holendozoic)

energid [ĕn″·ər·jĭd]   a unit consisting of a nucleus and the cytoplasm that it directly controls. Also used for "cell" as a unit of living matter

enhancer   a gene which intensifies the effect of a mutant, making it more extreme in its departure from wild type

-ennea-   *comb. form* meaning "nine"

-ennial   pertaining to a period of a year. The following terms using this suffix are defined in alphabetic position:

BIENNIAL   PERENNIAL
BIPERENNIAL   TRIENNIAL

-ensi-   *comb. form* meaning a "sword"

ensiform cartilage = xiphoid cartilage

Entamoeba [ent′·ə·mē″·bə]   a genus of amoebid protozoans all of which live in the alimentary canal of vertebrates and some invertebrates. *E. histolytica* [hĭst′·ō·lĭt″·ĭk′·ə] is responsible for amebic dysentery and liver abscesses in man. *E. coli* [kō″·li″] is a symbiote in the large intestine and there is no evidence that it is pathogenic. *E. gingivalis* [jĭn′·jĭv·ăl″·ĭs] is a symbiote in the mouth but may take part in secondary infections

-entell-   *comb. form* meaning to "order" in the sense of command

entellechy [ən·tel′·ək·ē]   the postulated condition that vital functions can be suspended by a hypothetical agent which cannot be perceived or measured

enteric fever *see Salmonella*

enteric nervous system   that part of the nervous system of invertebrates concentrated in ganglia supplying the visceral organs. It is thought by some to be ancestral to the sympathetic nervous system

Enterobius [ĕnt'·ər·ōb·ēəs] a genus of nematodes parasitic in both mammals and insects. *E. vermicularis* [vûrm'·ik·yōōl·âr"·ĭs] is the common human pinworm. Infection is by direct ingestion of eggs

enteroceptor [ĕnt'·ər·ō·sĕpt"·ər] a sensory structure or organ located in the visceral mass of an organism, particularly one that provides an organism with information about its own body

enterocoel [en'·tûr·ō"·sēl] a coelom derived as the cavity of an evaginated pouch

Enterocoelia = Deuterostomia

enterocrinin [*angl.* ən·tûr·ŏk"·rən·ĭn, *orig.* ĕnt'·ûr·ō·krĭn"·ĭn] a hormone secreted in the intestinal mucosa. It is active in controlling the rate of production and concentration of digestive enzymes in the alimentary canal

enteroderm [ĕnt'·ər·ō·dûrm'] that part of the endoderm which gives rise to the gut itself

enterogastrone [*angl.* ĕnt·ə·rŏg"·əs·tron', *orig.* ĕn'·tûr·ō·gǎst"·rōn] a hormone secreted by the duodenal mucosa. It is active in controlling motor activity and acid secretion of the stomach (*see also* urogastrone)

-enteron- *comb. form* meaning "gut." The terms archenteron and coelenteron are defined in alphabetic position

Enteropneusta [ent'·ər.ŏp·nyūs·tə] a class of the phylum Hemichordata containing the acornworms. They are distinguished by the acorn-shaped proboscis. The genus *Saccoglossus* is the subject of a separate entry

entoglossum [ĕnt'·ō·glŏs"·əm] the hyoid-derived skeleton of the tongue

entoglossal bone [ĕnt'·ō·glŏs"·əl] a tooth-bearing prolongation of the basihyal in some fish

-entomo- *comb. form* meaning "insect"

Entomostraca [ĕnt·ə·mŏs"·trə·kə] a no-longer acceptable zoological taxon at one time embracing all of the crustacea except the malacostraca. As so defined it contained the Branchiopoda, Ostracoda, Copepoda, and Cirripedia each now regarded as a separate subclass

Entoprocta [ĕn'·tō·prŏk'·tə] a phylum of pseudocoelomate bilateral animals distinguished by a distal circlet of ciliated tentacles. These sessile forms were once combined with the Ectoprocta into the no-longer valid phylum Bryozoa. The genus *Pedicellina* is the subject of a separate entry

envelope apparatus the sporocarp of ascomycetes

enzyme [ĕn"·zĭm] a protein or conjugated protein which, due to its configuration, both lowers the energy of activation of, and directs the stepwise pathway taken by, chemical reactions in a living organism. It has been estimated that a cell may require a thousand or more enzymes to maintain the vast number of cyclic reactions which characterize "living" systems. The following derivatives are defined in alphabetic position:

ANTIENZYME HOLOENZYME
APOENZYME ISOENZYME
COENZYME

-eo- *comb. form* meaning "dawn" and used by biologists extension to mean "early" or "earliest" in contrast to "·meso·" and "·neo·"

Eoanthropus [ē·ō·ǎn·þrō"·pəs] the notorious fake "Piltdown Man" ("*E. dawsoni*") carried out by some rogue planting a human skull and an ape jaw in a fossiliferous bed

Eocene epoch [ē"·ō·sēn, ě·əsēn'] the oldest of the tertiary geologic epochs extending from about 60 million years ago to about 35 million years ago. The hoofed mammals and carnivores became common at this time

Eohippus [ē'·ō·hĭp"·əs] the earliest known ancestor of the horse. It was abundant in N. America in the Eocene epoch. It was the size of a large dog and had four hoofed toes on each front foot, and four on each hind foot

Eolis [ē"·ōl·ĭs] a genus of nudibranch mollusks in which nematocysts, derived from the coelenterates on which it feeds, accumulate in the tips of the cerata (*see* ceras)

eosere [ē"·ō·sēr] a climax occurring in a specific era

eosinophil [ē'·ō·sĭn'·ə·fĭl] a myelocyte with a bi- or tri-lobed nucleus, and containing eosinophilic granules

eosinophilic granule [ē'·ō·sĭn'·ō·fĭl'·ĭk] one that selectively stains with eosin, particularly in eosin-azure, or eosin-blue combinations (cf. azurophilic granule)

eosinophilic metamyelocyte a descendant of an eosinophilic myelocyte in which the nucleus is "u" shaped

eosinophilic myelocyte a descendant of a promyelocyte, in which the eosinophilic granules are apparent, but the nucleus is still ovoid

eosins [ē"·ō·sĭnz'] a group of fluorescein halide dyes widely used as a cytoplasmic stain in histology and in compounding dyes for blood smears

epacme [ěp·ǎk'·mē] that phase in the development of a population of organisms when the vigor of the population is steadily increasing

epaxial muscle [ěp·aks"·eal] that part of the muscular system of a vertebrate which lies, or originates, dorsal to the vertebral column

ependymal cell [*angl.* ěp·ěnd'·əm·əl, *orig.* ěp'·ěn·dim"·əl] one in the coat lining the cavity of the central nervous system

Ephelota [ěf'·ə·lōt"·ə] a common genus of marine suctorian protozoans epizoic on hydroids

-ephem- *comb. form* meaning "short lived"

Ephemeroptera [ěf'·ěm·ŏp"·tə·rə] the order of insects containing the mayflies. They are distinguished by four pairs of membranous wings of which the anterior pair are usually triangular in shape

Ephestia [ěf"·ěst"·ēə] a genus of plant product pests. *E. kuhniella* [kōōn'·ě·ěl"·ə] (properly *Anagasta kuhniella*) is widely used for genetic research

ephippial egg [ěf·hĭp"·ēə] the winter egg of Cladocera

ephyra larva [ěf·ĭr"·ə] a primitive medusoid stage budded off from the scyphistoma of syphozoans. It has four lappets, each with a short bifurcated tip

-epi- *comb. form* properly meaning "upon" but often used in the sense of "uppermost", or "outermost"

epibiosis [ěp'·ē·bi·ōs·ĭs] *either* the condition of those organisms which live on the surface of another *or* the condition of those benthic organisms which live on the surface of bottom mud (cf. endobiosis)

-epiblem- *comb. form* meaning "a cloak"

epiboly [*angl.* ěp·ĭb"·əl·ē, *orig.* ěp'·ē·bō"·li] the process, as in the frog, by which the cells of the animal pole of the blastula grow down over the cells of the vegetal pole which become enclosed as the endoderm

epibranchial groove [ěp'·ē·brǎnk"·ēə] a groove

running along the dorsal edge of the pharynx in pro-chordates (*see also* hypobranchial groove, peribranchial groove)

epicardium [ĕp'·ē·kärd"·ēəm]   the outer covering of the embryonic heart

epichordal fold [ĕp'·ē·kôrd"·əl]   the dorsal of the two fin folds, from which the caudal fin of fish is derived

epicoel   metacoel

epicotyl [ĕp"·ē·kŏtəl']   the part of the primordial plant stem that lies above the cotyledons

epidemic [ĕp'·ē·dĕm"·ĭk]   pertaining to a rapid, great, temporary increase in the population of a given species

epidermis [ĕp'·ē·dûrm"·ĭs]   the cells covering the surface of an organism

epididymis [ĕp'·ē·did"·əm·ĭs]   that portion of the sperm duct of the male vertebrate which is derived from remnants of the mesenephros

epifauna [ĕp·ē·fôrn·əl]   that which lives on the surface of the bottom of the ocean or, less specifically, any encrusting fauna

epigamic character   tertiary sexual character

epigamic color   [*angl.* ĕp·ĭg"·əm·ĭk, *orig.* ĕp'·ē·gắm"·ĭk]   colors used in or developed for mating displays

epigamic display   mating display

epigamy [*angl.* ĕp·ĭg"·ə·mē, *orig.* ĕp'·ē·gām"·ē]   the type of behavior, such as the mating dance, which precedes or hastens copulation

epigastric artery [ĕp'·ē·gắst"·rĭk]   an artery which extends from the aorta to the ventral side of the body; supplies the muscles of the body wall and gives rise to the femoral artery

epigenesis [ĕp'·ē·jĕn"·ə·sĭs]   the theory that the egg is structureless and that the adult develops from it by a process of structural elaboration. The antithesis is pre-formation. Also loosely used to describe all pro-cesses of developmental elaboration above or after the level of generation

epiglottis [ĕp'·ē·glŏt"·ĭs]   a raised flap which folds back to assist in closing the glottis

epilemmal [*angl.* ĕp·ĭl"·ə·məl, *orig.* ĕp'·ē·lem"·əl]   literally "above the sheath," specifically used of that part of a motor nerve ending which lies outside the sarcolemma

epilimnion [ĕp'·ē·lim"·nē·ən]   the upper region of a well oxygenated, warmed, and thermally stratified lake

epimeningeal space [ĕpi'·mĕn·ĭn"·jē·əl]   the space between the single meninx and the endorhachis in primitive vertebrates

epimer [ĕp·ĭm"·ər]   a stereoisomer which differs from the other isomer in possessing more than one asym-metrical carbon atom only one of which, with its attached groups, contributes to the stereoisomerism

epimeral bone [*angl.* ĕp·ĭm"·ər·əl, *orig.* ĕp'·ē·mēr"·ăl]   one or a series of dorso-lateral, frequently "Y"-shaped, bones, lying between the somites of a fish

epimerase [ĕp·ĭm'·ər·āz"]   a group name for en-zymes that catalyze the production of one epimer from another

epimere [ep"·ē·mēr']   *either* a dorsal projection from the phallobase *or* the dorsal mesoderm of a ver-tebrate embryo before the differentiation of myotome and sclerotome

epimorphic regeneration [ĕp'·ē·môrf"·ĭk]   regen-eration which takes place from a blastema of undif-ferentiated cells, in contrast to morphallaxis

epimorphosis [ĕp'·ē·môrf·ōs"·ĭs]   the replacement, or partial replacement of tissue removed from an organ-ism (cf. morphallaxis)

epimysium [ĕp'·ē·mis"·ēəm, ĕp'·ē·mĭs"·eəm]   a con-nective tissue sheath surrounding an entire muscle

epinasty [*angl.* əp·ĭn"·əs·tē, *orig.* ep'·ē·năst"·ē]   an outward and downward bending movement of a flat surface of a plant

epinephrine [ĕp'·ē·nĕf"·rĭn]   a hormone secreted by the adrenal medulla and active in glucose metabolism and vaso-constriction (*see also* norepinephrine)

epineural [ĕp'·ē·nyŏŏr"·əl]   said of those animals (i.e. Chordata) in which the central nervous system is dorsal to the alimentary canal

epiostracum [ĕp'·ē·ŏst"·trə·kəm]   the horny, pig-mented, outer layer of a molluscan shell

epipelagic [ĕp'·ē·pĕl·ăj"·ĭk]   pertaining to organ-isms living in about the top forty fathoms of ocean water

epiphyseal center   [*angl.* ə·pĭf'·ə·sē"·əl, *orig.* ĕp'·ē·fĭs"·ē·əl]   a center of ossification at the end of a shaft bone

epiphyseal disk   a transverse disk of cartilage which separates epiphyseal bone from diaphyseal bone

epiphysial organ   one or more evaginations from the dorsal anterior end of the forebrain from which the pinneal subsequently develops

epiphysis [*angl.* ə·pĭf"·ə·sĭs, *orig.* ĕp'·ē·fĭs"·ĭs]   literally "an outgrowth above" specially growths above a bone which subsequently become joined to the bone. Occasionally used as a synonym for pineal body

epiphyte [ĕp'·ē·fĭt"]   a plant which grows on other plants; there is no connotation of parasitism

epiplankton   the upper or surface layer of plankton. There is no agreement as to the depth to which the epiplankton extends. A fairly common measure is the limit of light penetration in the waters in question. The term is also used for floating organisms attached to pelagic organisms

epipleuron [ĕp'·ē·plyŏŏr"·ŏn]   a lateral projection from the rib of teleost fish

epipodite [*angl.* ə·pĭp"·ə·dit, *orig.* ĕp'·ē·pŏd"·it]   a structure, usually a gill, but sometimes a gill separator, arising from the base of a crustacean appendage

epipterygoid bone [*angl.* ə·pĭp"·tə·rə·goid, *orig.* ep'·ē·ter"·ē·goid]   one of a pair of membrane bones extending upwards from the dorsal side of the pterygoid to the wall of the cranium in vertebrates other than mammals (*see also* pterygoid bone, ectopterygoid bone, metapterygoid bone)

episematic color [ĕp'·ē·sĕm"·ət·ĭk]   one thought to serve as a recognition signal (*see also* antiaposematic color, sematic color, pseudosematic color)

episome [ĕp'·ē·sōm"]   a genetic factor that can exist either in the chromosome or in the cytoplasm and thus applied to elements in the genetic makeup of bacteria which engender sensitivity of the bacteria to specific, usually phage-borne, lethal materials (*see also* plasmagene)

epistasis [*angl.* ə·pĭs"·tə·sĭs, *orig.* ĕp'·ē·stă"·sĭs]   the condition of having one gene which masks the effect of another to which it is non-allelic (*see also* genepi-stasis, hypostasis)

epithalamus [ĕp'·ē·þăl"·əm·əs] the dorsal portion of the thalamencephalon

epitheliochorial placenta [ĕp'·ē·thēl·ē·ō·kôr"·ē·əl] one in which the embryo and the maternal tissue lie in close interdigitating contact

epitheliomuscular cell [ĕp'·ē·thēl"·ē·ō·mŭs"·kyōō·lər] a cell containing one or more myofibrils in the epidermis of coelenterates (see also glandulomuscular cell)

epithelium [ĕp'·ē·thē"·lē·əm] any tissue that covers, or lines, an organ or organism. The curious use of a word which literally means "on the nipple" for this purpose derives from the fact that the word was originally used only for the covering of the nipples, transferred from this to lips and nipple, thence to the whole female skin and finally to any soft skin (see also endothelium). The following derivative terms are defined in alphabetic position:

CILIATED E.	SIMPLE EPITHELIUM
COLUMNAR E.	SQUAMOUS EPITHELIUM
CUBOIDAL E.	STRATIFIED EPITHELIUM
MYOEPITHELIUM	TRANSITIONAL EPITHELIUM
SENSORY EPITHELIUM	

-epithem- comb. form meaning "covering" (see also remarks under epithelium)

epitoke [ĕp'·ē·tōk"] the posterior sexual portion of the body of some polychaete worms

epitoky [angl. ə·pĭt"·ə·kē, orig. ĕp'·ē·tōk"·ē] the production of sexual organs by an apparently asexual form. Particularly the production of such organs at the posterior end of polychaete worms

epizoic [ĕp'·ē·zō"·ik] said of an animal living as an ectocommensal on another animal

epoch a tertiary division of geological time. The primary divisions are eras and secondary divisions are periods. Epochs are defined in alphabetic position and listed under the periods of which they are divisions

epoophoron [ĕp·ō·ə·fôr·ən] an anterior remnant of the mesonephros in adult amniotes

eposematic [ĕp'·ō·səm·ăt"·ĭk] said of coloration, such as the white tuft at the base of a deer tail, which is a recognition signal

epural bone [ĕp·yōōr'·əl] expanded, flattened neural spines at the posterior end of the vertebral column in fish

equatorial plate the wide central portion of the mitotic figure on which the chromosomes lie at metaphase

-equi- comb. form meaning "equal"

Equidae the family of perissodactyl mammals containing the horses and their immediate allies. All extant species except Grevy's Zebra are of the genus Equus. The genera Anchetherium, Dolichohippus, Equus, Merychippus, Mesohippus, Miohippus, Pliohippus are the subjects of separate entries

equilin [ĕk"·wə·lĭn] a hormone having properties similar to estradiol excreted in the urine of pregnant mares

Equus [ĕk"·wəs] the genus of equid mammals containing the horses and donkeys. The domestic horse is E. caballus [kă·băl"·əs] and the Siberian wild horse is E. przewalskii [prəzh·văl"·skē·i]. The donkey is a domesticated animal of mixed origin, though probably derived from E. onager [ŏn·ə·gər], the Nubian wild ass. The zebra is E. burchelli [bûr·chĕl"·i] but Grevy's zebra is in a distinct genus Dolichohippus

equilibrium receptor an organ (e.g. a statoblast) which orients an organism in relation to gravity

Equisetinae [ĕk'·wə·sē·tin·ē] the class of the Pterydophyta that contains the horsetails. They are distinguished by the erect axis with joints from which rise scale-like leaves

Equisetum the type genus containing the only extant representatives of the Equisetinae, having the characteristics of the class. E. hiemale [hi·ə·māl·ē] is the common scouring rush

era a primary division of geological time. The Azoic, Archeozoic, Paleozoic, Mesozoic and Cenozoic eras are the subjects of separate entries

-erem- comb. form confused from three roots and therefore variously meaning "desert", "solitary" and "gentle"

-eremu- comb. form meaning "hermit"

Eretmochelys [angl. ər·ĕt·mŏk"·əl·ĭs, orig. ĕr·et'·mō·kēl"·ĭs] the genus of chelonian reptiles containing E. imbricata [ĭm'·brĭk·āt·ə] (the hawksbill turtle) once widely hunted for its translucent carapace plates of "tortoise shell"

-ergas- comb. form meaning "labor"

ergot [ər'·gət, ûr"·gət] a dark, spongy, parasitic mass found on the ovaries of various grasses. It is the sclerotium of Claviceps

-erio- comb. form meaning "wool"

-erko- comb. form meaning "fence" usually transliterated -herco-

erotogenic [ər·ŏt'·ō·jĕn"·ĭk, ĕr·ŏt'·ō·jĕn"·ĭk] producing sexual desire; usually said of areas causing this effect under tactile stimulation

-erp- comb. form meaning "creep" commonly transliterated -herp-

Errantia [ĕr·ănt"·ēə] an assemblage, sometimes regarded as an order, of those polychaete worms which are free living and predatory (cf. Sedentaria)

-eruc- comb. form meaning "caterpillar" (cf. -camp-)

Eryops [ĕr'·ē·ŏps"] a genus of Permian Labyrinthodontia. They were about five feet long and not unlike a modern crocodile in appearance and probably also in habit

-erythro- comb. form meaning red

erythroblast [ə·rĭþ"·rō·blăst'] the precursor of an erythrocyte. The following derivative terms are defined in alphabetic position:

BASOPHILIC E.	PROERYTHROBLAST
POLYCHROMOPHILIC E.	

erythrocyte [ə·rĭþ"·rō·sit"] a mature red blood cell (see also polychromatophilic erythrocyte)

erythropoiesis [ə·rĭþ'·rō·pō·ēs"·ĭs] the formation of red blood cells

Erythroxylon [ə·rĭþ'·rō·zĭl"·ŏn] a large genus of tropical and subtropical shrubs. E. coca [kō"·kō"] yields cocaine and was used by the Aztecs to relieve fatigue

Escherichia [ĕsh"·ər·ĭk"·ēə, ĕsh"·ər·ich"·ēə] a genus of Enterobacteriaceae. E. coli [kō"·li], common in the alimentary canal of most mammals, has been the subject of intensive genetic and biochemical studies

esophagus [ē·sŏf·ə·gəs] in vertebrates that region of the alimentary canal which lies between the mouth and the stomach and, in invertebrates, that portion of the alimentary canal which lies between the pharynx and the stomach

Esox a genus of freshwater teleost fish. E. lucius

[lōōs'·ēəs] is the great northern pike of U.S. and Canada and the only pike of Europe. *E. masquinonge* [măs·kĭn·ŏnj·ē] (the muskellonge), *E. niger* [nī'·jer] (the chain pickerel) and *E. americanus* [ăm·ĕr·ĭk·än·əs] (the grass pickerel) are other American species

esterase [ĕs'·tĕr·āz"] a hydrolase which acts on ester bonds (*see also* cholinesterase, acetylcholinesterase)

estradiol [ĕs·trə·dī'·ŏl] a hormone secreted by the ovary necessary for the maintenance of pregnancy and to the development of female secondary sex-characters

estrone [ĕs·trōn', ĕs·trōn'] a hormone secreted by the ovary having properties similar to estradiol

estrous [ĕs'·trəs, ĕs'·trəs] the hormone induced changes which prepare the uterine epithelium for the implantation of a developing mammal (*see also* anestrous cycle, diestrous cycle)

-etes- *comb. form* meaning "annual"

ethmoid bone [ĕþ'·moid] a chondral bone of the skull lying between the anterior region of the orbit and the roof of the nasal chamber (*see also* pre-ethmoid bone)

ethmoid tooth a single posteriorally directed bony projection from the base of the palatine commissure in Myxine

etiolated [ēt"·ēō·lāt'·əd] said of a seedling which is drawn out and bleached through the absence of light

etionastic [ēt'·ēō·năst"·ĭk] descriptive of a change in direction, or bending, of a flattened plant organ caused by external forces

-eu- *comb. form* meaning "true" or "real" or "genuine"

Eubacteriales [yōō'·băk·tēr'·ē·ăl"·ēz] an order of schizomycetes; if motile, they have peritrichous flagella, and most are either spherical or rod shaped. Both gram-positive and gram-negative forms occur. This is the largest order of schizomycetes and contains most of those forms commonly thought of as "bacteria". There are saprophytic, symbiotic and pathogenic forms. The genera *Aerobacter, Azobacter, Bacillus, Clostridium, Escherichia, Klebsiella, Proteus, Rhizobium, Salmonella* and *Shigella* and also the family Lactobacillaceae are the subjects of separate entries

Eubranchipus [yōō'·brănk"·ĭp·əs] a genus of freshwater anostracan Crustacea. *E. vernalis* [vĕrn·āl'·ĭs], about an inch long, is the "fairy shrimp" that sporadically occurs in immense numbers in pools in the N.E.U.S.

Eucalyptus [yōō'·kə·lĭpt"·əs] a genus of Australian Myrtaceae containing some of the tallest trees in the world. The pungent oil produced by many was once thought to be a cure for the common cold

Eucapsis [yōō·kăps"·ĭs] a genus of colonial cyanophyte Algae that divide regularly in two planes, thus producing a cubical colony

eucephalous [yōō·sĕf"·əl·əs] having a true head, applied particularly to insect larvae in which all of the head appendages are present

Euciliata [yōō'·sĭl·ē·a"·tə] a subclass of ciliate Protozoa distinguished from the Protociliata by the presence of a cytostome

eucone eye [yōō"·kōn"] a compound eye in which the cones are quadriangular in section

Eudorina [*angl.* yōō·dor"·ən·ə, *orig.* yōō'·dôr·ĭn"·ə] a genus of chlorophyte algae in the form of a spherical or ovoid colony on the periphery of a gelatinous mass

Euglena [yōō·glēn"·ə] a genus of euglenid mastigophoran protozoans with a single flagellum and a flexible body. Actually electron microscopy has disclosed a vestigeal second flagellum in some species

Euglenida [yōō·glēn"·ĭd·ə] an order of Protozoa distinguished by their amoeboid like movements and their ability to exist either holophytically or saprozoically. As holophytes they are placed in the algal phylum Euglenophyta. *Euglena* and *Peranema* are the subject of separate entries

Euglenophyta [yōō'·glēn·ō·fīt"·ə] a phylum of algae containing both motile unicellular forms exhibiting euglenoid movement and a few sessile forms. The former are regarded as animals by zoologists who class them in the protozoan order Euglenida

euglobulin [yōō·glŏb"·yū·lĭn] a simple protein insoluble either in water or in a solution of ammonium sulfate (*see also* globulin, pseudoglobulin)

euhaline [yōō·hāl"·ĭn] said of waters containing between 30 and 40 parts per thousand of dissolved salts; that is, in most cases, normal sea water

eukaryotic [yōō'·kär·ē·ŏt"·ĭk] the condition of having a well defined, membrane limited, nucleus (cf. prokaryotic)

eulittoral zone [yōō·lĭt"·ər·əl] the true littoral zone which is variously defined in oceanography as the zone never covered by more than ten fathoms of water and in limnology as the zone of rooted aquatic plants or the zone which is subject to submerged wave action

Eumeces [yōō·mē"·sēz] a genus of skinks (Squamata) containing many U.S. forms. They are slender, long tailed lizards of small size

eumedusoid gonophore [yōō'·mĕd·yūz"·oid] a gonophore which resembles a medusa in every way except the production of tentacles

Eunectes [yōō·nĕk"·tēz] a monotypic genus of snakes (Squamata) containing the giant anaconda (*E. murinus* [myūr·in"·əs]). There is no authenticated record, though numerous legends, of an Anaconda longer than 25 feet

Eunice [yōō·nis"·ē] a genus of thin, elongate Polychaetae errantia. Several species (called Palolo worms) of the genus store their reproductive cells in the posterior end of the body. At a specific lunar period these portions break off and swarm to the surface, where, in the South Pacific, they are collected for food. This species, *E. viridis* [vĭr'·ĭd·ĭs] swarms during the last quarter of the October-November moon while the West Indian species swarms at the full of the July moon

eupelagic plankton that which is found exclusively in oceans

Euphausiacea [yōō·fôrz"·ē·ās·ē] an order of malacostracan Crustacea distinguished by the fact that none of the thoracic limbs are specialized as maxillipedes

euphotic [yōō·fŏt"·ĭk] literally, "well-lighted"; applied particularly to plants, such as submerged water plants, that are relatively poorly lighted but still capable of photosynthesis

euphotic zone that zone of water within which photosynthesis is possible

Euplectella [yōō'·plĕkt·ĕl"·ə] a genus of deep water hexactinellid sponges with a cylindrical body, some more than a foot long and two inches in diameter.

The skeleton of this is called "Venus' Flower Basket" and was once an object of keen competition among wealthy collectors

eupleurodont [yōō·plūr"·ō·dŏnt'] a pleurodont in which the teeth are replaced from beneath (*see also* subpleurodont)

euploid [yōō"·ploid] a polyploid, the chromosome number of which is an exact multiple of the chromosome number of its ancestral species

Euplotes [yōō·plŏt"·ēz] a large genus of spirotrichid ciliate Protozoa. Very common in "mixed cultures" of protozoans where they may be distinguished by the scuttling motion produced by their ventral compound cilia

-euri- *comb. form* meaning "broad", better written **-eury-**

-eury- *comb. form* meaning "broad or wide"

euryapsid [yūr'·ē·ăp"·sĭd] a reptile skull in which there is only a single temporal opening

Euryapsida [yūr'·ē·ăp"·sĭd·ə] a subclass of extinct reptiles distinguished by having the temporal openings high on the roof of the skull. Most were Mesozoic and adapted to aquatic life but were far less fish-like than the Ichthyopterygia. The genus *Ichthyosaurus* is the subject of a separate entry

eurybath [yūr'·ē·băth] an organism found in a wide range of depths of water

eurybenthic [yūr'·ē·běnth"·ĭk] pertaining to benthic organisms that occur at widely varying depths

Eurycea [yōōr·ēs"·ea] a genus of urodele Amphibia commonly called the brook salamanders. *E. bilineata* [bī'·lĭn·ē·ā"·tə], the two-lined salamander, is distributed all over the U.S. *E. lucifuga* [lōō'·sē·fyōō"·gə] is the cave salamander and is peculiar not only in its habitat but in having a prehensile tail

Eurypelma a genus of large (two inch) burrowing Araneae (spiders) commonly called tarantula. The genus *Tarantula*, however, belongs in the Uropygi

Eurypterida [yūr'·ĭp·tĕr"·ĭd·ē] a group of fossil chelicerate arthropods often united with the Xiphisura in a group Merostomata. They could well be described as giant carnivorous water scorpions and were the largest (up to 10 feet long) arthropods ever to have existed

eurytherm [yūr"·ē·thûrm'] an organism capable of maintaining itself over a wide temperature range

euryzonal [yūr'·ē·zōn"·əl] pertaining to an organism occurring in many zones either aquatic or aerial

Euspongia [yōō·spŏnj"·ea] a genus of more or less spherical keratose sponges. The collection of *E. officinalis* [ŏf'·ĭs·ĭn·āl"·ĭs] (the "Bath Sponge") was once an important industry in many parts of the world. A variety of this (*S. officinalis lamella*), well described as the "Elephant Ear Sponge", is still in use in the ceramic industry

eustachian tube [yōō·stăch"·əan] a duct connecting the tympanic cavity with the pharynx

eustachian valve one of the valves in the base of the postcava vein

eutele [yōō'·tĕl·ē] the condition of an animal which retains the same number of nuclei in the adult throughout its life, except in the gonad, and which therefore "grows" only through increase in cytoplasmic mass

Eutheria [yōō·ber"·ēə] the sub-class of mammalian chordates which contains all the orders except Marsupialia and Monotremata. This group is more familiar

as the placental, or true, mammals. The contained orders are recorded under the entry Mammalia

-euthy- *comb. form* meaning "immediately"

eutrophic [yōō·trŏf"·ĭk] rich in nutrients, said particularly of bodies of water and swamplands

eutrophic lake a highly productive lake characterized by abundant plankton and high turbidity

eversible [ē·vûrs·əb·əl] said of a portion of an organism, particularly the pharynx of an invertebrate, which is protruded and turned inside out at the same time

evocator [ē'·vō·kăt·ər] a chemical substance involved in induction

evolution [ĕv'·ŏl·yōō"·shŭn] a process of gradual change by which one form of something slowly changes into a similar but significantly different form of the same thing. The following derivative terms are defined in alphabetical position:

ACCIDENTAL E.	MACROEVOLUTION
ALLOMORPHOTIC E.	REGRESSIVE EVOLUTION
CLANDESTINE E.	SALTATORY EVOLUTION

-ex- *comb. form* variously meaning "out", "without", or "out of"

exalbuminous seed [ĕks'·ăl·byōō"·mĭn·əs] one in which the endosperm is completely absorbed by the time the seed is ready for dispersion

exarch xylem [ĕks'·ärk'] that in which the most mature elements are located furthest from the center of the axis

exclusive species a species of animal, the distribution of which is rigidly limited to environments showing specific types of plants or vice versa

exconjugant [ĕks·kŏn"·jū·gănt'] a freshly separated conjugant protozoan

excretion the process of removing unwanted materials from a cell, organ, or organism

excystment [ĕks·sĭst"·mənt] emergence from a cyst (*see also* encystment)

exite [ĕks'·ĭt] an outer lobe on the appendages of eubranchipod crustacea

-exo- *comb. form* meaning "out of"

exocarp [ĕks'·ō·kärp"] the outer of the three layers into which the pericarp may differentiate

exoccipital bone [ĕks'·ŏk·sĭp"·ĭt·əl] one of a pair of chrondral bones at the posterior region of the skull, lying on each side of the foramen magnum (*see also* occipital bone, basioccipital bone, supraoccipital bone)

exocoel [ĕks'·ō·sēl'] that portion of the coelenteron of an anthozoan which lies between widely separated pairs of septa which themselves include an endocoel between them

exocone eye [ĕks'·ō·kōn] a compound eye in which the crystalline cone is replaced by an ingrowth from the corneal facet

exocranium [ĕks'·o·krān"·ēəm] those bones of the skull which are of dermal origin

exocrine gland [ĕks'·ō·krĭn"] a gland the secretion of which is collected into a duct for transportation to the site of action (*see also* endocrine gland)

exogamy [*angl.* ĕks·ŏg"·gəm·ē, *orig.* ĕks'·ō·găm"·ē] reproduction between groups or organisms not usually interbreeding

exogenous rhythm a rhythmic behavior pattern in an organism induced by a cause external to the organisms such as seasonal changes or those produced by tides

exoisogamy [*angl.* eks'·ĭs·ŏg"·əm·ē, *orig.* ĕks'·ĭs·ō·

gam"·ē] the condition in which an isogamete will only fuse with another isogamete from a different brood

exopodite 1 [*angl.* ĕks·ŏp"·ō·dĭt, *orig.* ĕks'·ō·pōd"·ĭt] (*see also* exopodite 2) the third segment of the insect maxillary palp

exopodite 2 (*see also* exopodite 1) the abaxial branch of a biramous crustacean appendage

exoskeleton skeletal elements forming the surface of an animal as in arthropods

exosmosis [ĕks'·ŏz·mōs"·ĭs] the reverse of endosmosis

expressivity the extent to which a gene exercises a phenotypic effect (*see also* reduced expressivity and penetrance)

extension that movement of a jointed appendage which results in opening the angle subtended by the two segments

external auditory meatus the tube, at the base of which lies the tympanic membrane, in those organisms in which this last is sunk beneath the surface

external fertilization the production of a zygote outside the body

external gill a protuberance, finger-like, or filamentous process or series of processes arising from the body wall, and functioning as a gill

external iliac artery = femoral artery

exteroceptive system [ĕks'·tər·ō·rĕs·ĕpt"·ĭv] the sum total of those sensory receptors which perceive external stimuli (*see also* interoceptive system, proprioceptive system)

exteroceptor [ĕks·tə·rō·sĕp·tôr] a receptor which receives external stimuli other than by contact

extinction threshold the number below which a given population cannot fall without becoming extinct

-extra- *comb. form* meaning "outside"

extra-embryonic any part of a developing animal which is not structurally part of the embryo itself

extra-embryonic membranes the chorion, amnion, allantois and yolk sac

extrahyoid arch [ĕks'·trə·hī"·oid] one of a pair of thin cartilages lying backwards from the base of the styloid cartilage under the optic capsule and articulating with the first branchial arch in the cyclostome visceral skeleton

extrascapular bones [ĕks'·trə·skăp"·yōōl·ə] the antero-dorsal elements of the pectoral girdle of crossopterygian fish; there is a medial and two lateral extrascapulars (*see also* scapular bone)

exumbrella [ĕks'·ŭm·brĕl"·ə] the convex oral surface of the umbrella of a medusa

eye a photoreceptor organ particularly one in which there is a lens, or retina, or their equivalents. The following derivative terms are defined in alphabetic position:

CEREBRAL EYE   EXOCONE EYE
COMPOUND EYE   PARIETAL EYE
DAY EYE   PINEAL EYE
EUCONE EYE

eyelid a retractile flap of skin covering the eyes of many animals

eye spot a small area of photoreceptive pigment in some protozoans

F abbreviation for filial, itself meaning "pertaining to children"; the characters F_1, F_2, etc. are used to denote the first, second, etc. generations of offspring from a given mating

-fab- *comb. form* meaning "bean" particularly the "broad bean" (*Vicia faba*)

facial nerve Cranial VII. Arising from the medulla oblongata. It adheres to the face, the roof of the mouth and the hyoid nerve

faciation [fā″·shē·ā′·shŭn] an association within which one or more dominants has been replaced, thus differing from a consociation in which a dominant has dropped out but not been replaced

facies [fā′·shē·ēz] used in the sense of "over-all shape" of a group of organisms

factor in biology usually applies to an unidentified substance involved in, or the causative agent of, a specific reaction or process. The following terms are defined in alphabetic position:
CUMULATIVE FACTOR   GROWTH FACTOR
EDAPHIC FACTOR

FAD = flavin-adenine dinucleotide

faeces *see* feces

Fagaceae [făg·ās′·i·ē] the family of dicotyledons that contains the birches, beeches and oaks. Distinctive characteristics are that only the staminate flowers are born in catkins and that there are three carpals and a one-seeded fruit. The genus *Quercus* is the subject of a separate entry

-falc- *comb. form* meaning "sickle"

falces *plural of* falx

falciform ligament the term used to describe the ventral area of fusion of two liver lobes

Fallopian tube the female oviduct of the human

false hybrid a hybrid which exhibits the phenotypic characters of only one parent

false rib a pleural rib which reaches the sternum indirectly through a crest or cartilage

false suture one in which the bones do not interlock

false tissue the mycelially produced "tissues" of fungi

falx literally, a "sickle" and applied to almost any biological object of that shape from a piece of the genitalia of Lepidoptera to the septum extending vertically between the cerebral hemispheres of vertebrates

family a biological taxon ranking immediately above genus, and therefore composed of a number of related genera. In zoology all family names end in -idae and in botany (except for the Compositae and Umbelliferae) all end in -aceae. Most families take their name from a type genus. Thus *Amoeba* is the type genus of the *Amoebidae* and *Fagus* of the Fagaceae (*see also* subfamily and super-family)

-farious *comb. form* meaning "placed in rows". Compounds using this suffix are not separately defined. The meaning is obvious from the roots (e.g. trifarious, multifarious)

fascicle [făs″·ik·əl] a bundle, particularly of vessels in a plant stem (*see* vascular bundle). The term is also used for bundles of fused zoecia of some Ectoprocta

Fasciola [făs′·ē·ol″·ə, făs·i″·ō·lə] a genus of trematodes. The species *F. hepatica* [hĕp·ăt″·ik·ə], or sheep liverfluke, occurs also in many other herbivorous animals. Numerous water snails act as the intermediate host. Human infection is rare

Fascioloides [*angl.* făs′·ē·ol″·oid·ēz, *o.ig.* făs·ē·ō·loid″·ēz] a genus of trematodes lacking the anterior cone of *Fasciola* and *Fasciolopsis*. *Fascioloides magna* [măg′·nə] is a common parasite of N. American deer

Fasciolopsis [*angl.* făs·ē·ôl″·əp·sĭs, *orig.* făs′·ē·ō·lôp″·sĭs] a genus of asiatic trematodes. *F. buski* [bŭsk″·i], parasitic in pigs and men, may reach a length of 3 inches. The intermediate hosts are planorbid snails and human infection frequently occurs through ingestion of raw water chestnuts

fat cell a cell which stores fat

fate in embryology, the anticipated end of a developing part (e.g. the production of one lateral half of the whole from one blastomere after the first division of an echinoderm egg) (cf. potency)

-fauc- *comb. term* meaning "throat"

fauna [fôrn″·ə] the animal population of a given region. The undernoted derivative terms are defined in alphabetic position:
EPIFAUNA   INTERSTITIAL FAUNA
INFAUNA   MEIOFAUNA

feather a unit of the external covering of birds. The following derivative terms are defined in alphabetic position:
CONTOUR FEATHER   PIN FEATHER
FLIGHT FEATHER

feces [fē"·sez"]   the solid, or semi-solid, excreta of higher animals

fecundity [fĕk·ŭnd"·ət·ē]   the quality of being able to produce many offspring

Felidae [fēl"·ĭd·ē]   the family of fissipede carnivorous mammals that contains the true cats. They are distinguished by having a greatly inflated auditory bulla and in possessing a carnassial tooth of the upper jaw with three lobes to the blade. The commonly given characteristics of retractile claws applies to all save the cheetah. The genera *Felis, Panthera* and *Profelis* are the subjects of separate entries

Felis [fē"·lĭs]   the genus of fissipede carnivores that at one time contained all cats but is now restricted to the small cats. *F. domesticus* [dō·mest"·ĭk·əs], the domestic cat, is of polyphyletic origin

fell field   a stony area with dwarf, scattered plants

female   that form of a dioecious organism which produces eggs (*see also* neofemale, metafemale)

female parthenogenesis   that form of parthenogenesis in which the female gamete produces a new individual (cf. etheogenesis)

femoral [fĕm"·ər·əl]   adjective from femur

femoral artery   an artery running down the anterior side of the leg. It derives from the external iliac artery.

femoral gland   one of the glands constituting the femoral organ of lizards

femur [fē"·mə]   in vertebrates the proximal bone of the leg skeleton or thigh bone. In arthropods, that joint of the leg which is third from the articulation of the body, and which therefore, in insects, articulates with the trochanter at its proximal end and with the tibia at its distal end

fen   a peaty, moist tract, usually derived from the ageing or draining of a swamp

fenestrate [fĕn"·ĕs·trāt']   windowed

fenestrated membrane   the elastic structural unit in the wall of an artery

-fer- *comb. form* meaning "to bear" (cf. -phor-). The following terms with this suffix are defined in alphabetic position:

ALIFER	LATICIFER
LACTICIFER	PALPIFER

feral [fĕr"·əl]   a domestic animal, such as cat, which has adopted a "wild" existence

fermentation   anaerobic metabolism in which both the electron donor and acceptor are organic compounds

-ferous *comb. term.* meaning "bearing" (cf. -fer, -phor, -phorous). The following terms with this suffix are defined in alphabetic position:

CONIFEROUS	SEMINIFEROUS
LACTIFEROUS	STROBILIFEROUS
LATICIFEROUS	

fertilization   the union of a male and female gamete. The following derivative terms are defined in alphabetic position:

CLOSED F.	EXTERNAL F.
DOUBLE F.	PREFERTILIZATION

fertilization cone   a prominence extending from the surface of some eggs at the moment of, or in some cases allegedly shortly before, contact with the sperm

fertilization membrane   the vitelline and perivitelline membranes taken together

fertilizin [fĕrt"·il·iz"·ĭn]   a substance produced by an egg and causing agglutination of spermatozoa

fertility   the possibility, or a measure of the possibility, of a female becoming fertilized (*see also* differential fertility)

fetalization [fēt'·əl·īz·ā"·shŭn]   the persistence of certain foetal or immature characters of an ancestor in the adult stages of a descendant

fetus [fē"·təs]   commonly applied to the human embryo after two months of gestation

fiber 1 (*see also* fiber 2)   in plants an elongated sclerenchyma cell (cf. sclereid) and usually distinguished as an easily separable thread of high tensile strength. The following derivative compounds are defined in alphabetic position:

LABRIFORM FIBER	XYLEM FIBER

fiber 2 (*see also* fiber 1)   in animals any fine threadlike structure. The following compounds using this term are defined in alphabetic position:

CHOLINERGIC FIBER	RETICULAR FIBER
COLLAGENIC FIBERS	SEPTATE FIBER
CYTOSTOMAL FIBER	SHARPEY'S FIBERS
ELASTIC FIBERS	SPINDLE FIBER
PURKINJE FIBER	

fibril   diminutive of fiber and, in biology, used almost interchangeably with fiber. The following derivative terms are defined in alphabetic position:

ARGYROPHILIC F.	MYOFIBRILS
INTERCILIARY F.	NEUROFIBRILS

fibroblast [fi"·brō·blăst', fîb"·rō·blăst]   a cell found in areolar connective tissue arising directly from the primitive reticular cell and thought to give rise to fibers

fibrocartilage [fi·brō·kärt"·əl·əj, fîb·rō·kärt·əl·əj]   cartilage with collagen fibers in its intercellular substance

fibrocartilagenous joints   those joints in which the skeletal elements are united by fibrocartilage during some stage of their existence

fibrous connective tissue   see white fibrous connective tissue

fibrous joint   a joint between two bones united by fibrous tissues

fibrous protein   one in which the molecules are arranged in compound spirals with hydrogen bonds in many planes

fibula bone [fib"·yōō·lə]   the smaller of the two bones (the other is the tibia) which lie between the femur and the ankle

fibulare bone= calcaneus

-fid- *comb. form* meaning "cleft"

fidelity   in biology the degree to which a species is restricted to a given set of conditions

field   an area of ground in which crops are grown or, by extension, any restricted area of land (*see also* fell field)

field theory   an organism, more particularly an embryo, is divided into or surrounded by, areas which may mutually interact

filament   a thread-like structure

-fili- *comb. form* meaning "thread" frequently confounded with -filia- and -filic-

-filia- *comb. form* meaning "daughter" frequently confounded with -fili- and -filic-

filial generation   offspring of a cross, the first being indicated by F_1 the second, or grandchild, by F_2 etc.

-filic- *comb. form* meaning "fern" frequently confounded with -fili- and -filia-

filiform papilla [fil"·ē·fôrm]   one of the sensory papillae of the tongue in the form of numerous short, thread-like bodies arising from a thickened base

filopod [fil″·ō·pŏd′, fĭl″·ō·pŏd′] a pseudopodium composed entirely of ectoplasm, usually thin and with a pointed tip

filter-feeding a mechanism common in invertebrates but found also in the baleen whales by which planktonic food or food particles are filtered from large volumes of water

-fim- *comb. form* meaning "dung" (cf. -copr-, -kopr-)

-fin- *comb. form* meaning "limit"

fin an appendage, used for swimming, balancing, or steering, found in aquatic animals particularly fish

fingerling a post-larval teleost less than one year old, differing from a fry in that it can readily be identified through its resemblance to the adult form

finiform a taxon, usually plant, all of the known species of which are fossils

fin ray any fine bony structure supporting the fin of a fish

fission used in biology in the sense of division or splitting. The following derivative terms are defined in alphabetic position:

BINARY FISSION   PROGAMOUS FISSION
MULTIPLE FISSION

Fissipedia [fĭs′·ē·pēd″·ēə] the suborder of Carnivora which contains the land-dwelling forms distinguished from the aquatic Pinnepedia by having legs instead of flippers. The families Canidae, Felidae and Mustelidae, and the genus *Procyon* are the subjects of separate entries

fissure [fĭsh″·ər, fĭs″·yōōr] a space or split. The following derivative terms are defined in alphabetic position:

ORBITAL FISSURE   SCLEROTOMIC FISSURE
RHINAL FISSURE   SYLVIUS' FISSURE

Fissurella [*angl.* fish′·ə·el″·ə, *orig.* fĭs′·yōōr·el″·ə] a genus of gastropod mollusks often called volcano shells or keyhole limpets. They have uncoiled conical shells like a limpet but with a perforation at the tip

fixation disc the disc by which the free-swimming larvae of sessile invertebrates form their first attachment or by which the larvae of some motile invertebrates, particularly echinoderms, attach themselves before undergoing metamorphosis

fixative a mixture of reagents used to preserve organisms and parts of organisms intended for the preparation of wholemounts or sections for microscopical examination

flagellar cord a flagellum which is attached to the outside of the body wall of some mastigophora and which keeps the body in a condition of constant undulation

Flagellata = Mastigophora

flagellated band the two outer bands, parallel to the cnidoglandular band on the septum of Anthozoa

flagellum 1 [flə·jĕl″·əm] (*see also* flagellum 2) a motor organelle found on many free-living cells in both the plant and animal kingdom. A flagellum differs from a cilium only in being larger and occurring singly or in small groups (cf. tractellum)

flagellum 2 (*see also* flagellum 1) any long tactile arthropod appendage or part of an appendage

flame bulb the excretory organs of Entoprocta which are large compound flame cells

flame cell a primitive excretory cell. The cell is in the shape of an elongate flask to the inside of the base of

the bulbous end of which are attached elongate flagella which by lashing force the contained fluid along the neck of the flask either directly to the exterior or to an excretory canal

-flect- *comb. term.* meaning "bent"

flexion [flĕk′·shŭn] that movement of a jointed appendage which results in closing the angle about the joint

flexure [flĕk″·shyūr] a bend, particularly one in the central nervous system. The following derivative terms are defined in alphabetic position:

NUCHAL FLEXURE   PRIMARY FLEXURE
PONTAL FLEXURE

flight feather a stiff feather in a bird's wing

flipper a mammalian limb lacking digits and therefore adapted to swimming

floating rib a pleural rib that does not reach the sternum

flocculus [flŏk″·yū·ləs] a small lateral outgrowth from the base of the vermis

floor plate a narrow plate along the ventral surface of the brain stem or spinal cord separating the basal plates

-flor- *comb. form* meaning "flower"

flora [flôr″·ə] the sum total of the plants in any given area or environment (*see also* microflora)

floral diagram a diagrammatic section across a flower, on which the position of the various parts is indicated as though they lie in one plane

floral series the successive whorls or spirals of members such as petals, sepals, etc., which make up a flower

floret diminutive of flower; usually used for the components of composite flowers

floridean starch a polysaccharide accumulated by rhodophyte algae. It more nearly resembles glycogen than starch

florigen [flər·ē·gen″] the hormone which initiates the production of flowers

-florous *comb. form* meaning "flowered" in the sense of uniflorous, etc. The substantive termination, not recorded in this work, is -flory

-flos- *comb. form* meaning "flower"

flower a short stem bearing appendage specialized for sexual reproduction in an angiospermous plant (*see also* perfect flower, imperfect flower)

flower receptacle the swollen end of the axis on which the components of the flower are placed

fluid in biology any liquid. The following derivative terms are defined in alphabetic position:

AMNIOTIC FLUID   SYNOVIAL FLUID
CEREBROSPINAL FLUID   TISSUE FLUID
STIGMATIC FLUID

fluke popular name of trematode platyhelminths

-flumen- *comb. form* meaning "a river"

fluvial [flū″·vē·əl] pertaining to streams

flyway the established route of migratory birds

FMN = flavin mononucleotide

foetus *see* fetus

-folia- *comb. form* meaning "leaf"

folic acid [fŏl″·ĭk] N-[f-{[(2-amino-4-hydroxy-6-pteridyl)-methyl]-amino}-benzoyl]-glutamic acid. A water soluble nutrient frequently regarded as a vitamin and apparently having a hemopoietic function in some animals

foliose [fō·li·ōz, fōl″·ē·ōz] having the appearance of

a leaf, said particularly of lichens in contradistinction to crustose

follicle [fŏl"·ĭk·əl] literally, a small container, but used in histology for a more or less spherical aggregate of cells surrounding a single cell in fluid, particularly the follicle containing the egg in the ovary. The following derivative terms are defined in alphabetic position:

ATRETIC FOLLICLE   LYMPH FOLLICLE
GRAFFIAN FOLLICLE   PRIMORDIAL FOLLICLE
HAIR FOLLICLE

follicle cell one of the cells immediately surrounding the developing ovum in the follicle of the mammalian ovary; the term is applied to analogous cells in some invertebrates

follicle-stimulating hormone one produced by the adenohypophysis which stimulates growth of ovarian follicles (= FSH)

follicular gonad [fŏl·ĭk"·yōō·lər] one which consists of numerous small parts scattered through the tissues of the body, particularly in platyhelminths

Fomes [fō"·mēz"] a genus of aphyllophorale Basidiomycetes and one of the very few perennial fungi. They occur as plate-like woody outgrowths from tree trunks

fontanel [fŏn'·tə·nĕl] literally a "little fountain", but in biology applied to shallow depressions such as that which surrounds the frontal and the opening of the frontal gland on the head of termites, or the gap between the bones in the posterior region of the skull of a human infant

food chain a series of organisms that feed one on the other. It may start with a holophyte or saprophyte

food web an interconnected series of food chains

foot 1 (*see also* foot 2) the terminal structure of all four limbs of terrestrial vertebrates except birds, in which only the hind limb is furnished with a foot and primates in which the front foot is called a hand. The following derivative terms are defined in alphabetic position:

MESAXONIC FOOT   PARAXONIC FOOT

foot 2 (*see also* foot 1) the locomotory appendage of some invertebrates (*see* tube foot)

foot cell a cell in the mycelium of aspergillale fungi from which are derived those hyphi which later bear conidiophores

foramen [fôr·ā'·mĕn] (*plural* foramina) a hole, particularly one in a bone through which a nerve or blood vessel passes (*see* Monro's foramen, Panizza's foramen)

foramen magnum (vertebrates) [măg"·nəm] the posterior opening of the skull through which the spinal cord passes out

foramen ovale [ōv·äl'·ē] the posterior of the two foramina which pierce the alisphenoid bone

foramen rotundum [rō·tŭn'·dəm] the anterior of the two foramina which pierce the alisphenoid bone

Foraminifera [fôr·ăm'·ĭn·ĭf"·ərə] a large order of sarcodinous Protozoa characterized by a calcareous shell through the pores in which anastomizing pseudopodia protrude. The genera *Camarina* and *Globigerina* are the subjects of separate entries

forb any plant in a meadow or prairie which is not a grass

forebody an anterior body region, sharply distinguished from a hindbody, but not distinguished as a head or other morphologically acceptable division. The term is usually applied to trematode worms

forebrain = prosencephalon

foregut an ill-defined term. In vertebrate embryos, applied principally and usually to that portion of the alimentary canal, which runs from the mouth to the pyloric stomach. In arthropods, it is delimited to that anterior region which is lined with chitin

forest any considerable area of land covered with a heavy growth of trees. The following derivative terms are defined in alphabetic position:

HIGH RAIN FOREST   RAIN FOREST
MONSOON FOREST   TROPICAL RAIN FOREST
PURE FOREST

forest-tundra intermittent timbered tracts on the northern verge of the limit of growth of trees

Forficula [fôr·fĭk"·yū·lə] a genus of dermapteran insects containing the common european earwig (*F. auricularia* [ôr·ĭk'·yū·lär"·eə]). This form is occasionally found on the East Coast of the U.S.

-form 1 (*see also* -form 2) *comb. term* meaning "in the shape of". The several hundred derivatives with this suffix are not defined in this dictionary, since the meaning is obvious from the root prefix (e.g. cordiform, heart-shaped; cyathiform, flask-shaped; scalariform, ladder-like, etc.)

-form 2 (*see also* -form 1) in the sense of "being a type of". The following derivatives using this suffix are defined in alphabetic position:

FINIFORM   RAMIFORM
GROWTH FORM   VERSIFORM
NOVIFORM

forma the smallest category commonly used in botanical taxonomy and applying to trivial variations occurring among a population of any one species

formation a word which is defined, in ecology, as, or even more, variously than association (q.v.); it cannot, at present, be usefully defined and the sense must be determined from the context. It is sometimes used in botany with the meaning "assemblage" in the sense that a meadow is an assemblage of grasses. The following derivative terms are defined in alphabetic position:

CLOSED FORMATION   OPEN FORMATION
COMPLEX FORMATION   PANFORMATION
MIXED FORMATION

Formicoidea [fôrm'·ĭk·oid"·eə] a very large superfamily of aprocritan hymenopteran insects commonly called ants. They are social insects with three or more castes in the colony with short-lived males, sterile females which lose their wings after a mating flight, and usually one functional queen per colony (*see also* aner, ergate). The genus *Polygergus* is the subject of a separate entry

-foss- *comb. form* meaning "ditch"

fossa [fŏs"·ə] literally a ditch but also used in biology for a groove or trough (*see* cerebral fossa, Hatschek's fossa)

fourth ventricle the cavity of the medulla oblonga

fractional mutation a mutation in one part of an organism occurring through the production of dominant mutation in an early division

-frag- *comb. form* meaning "to break"

Fraser-Darling Law the relative number of breeding individuals and young in a population of birds increases as the size of the population increases; there is a corresponding shortening of the breeding season

freemartin an intersex in cattle induced in a female twin by the sex hormones of a male twin

frenulum [frĕn´·yōō·ləm]   diminutive of frenum, save in the case of bristles interlocking the fore and hind wing in lepidopterous insects for which the term frenulum is commonly used to distinguish it from frenum in the same animal

frenum [frē´·nəm]   literally, "reins" or a "strap", particularly any fold of tissue supporting an organ, such as that under the tongue, or a ridge on insects extending from the scutellum to the base of the anterior wing. The word is used almost interchangeably with frenulum

freshwater   from the biological point of view, water containing less than 0.5 parts per thousand of dissolved salts (*see also* haline, brakish water, sea water)

Fringillidae [frin·jil´·id·ē]   a large family of passeriform birds containing the finches. They are mostly small birds with short and pointed bills usually thick and very rarely hooked. Many are brightly colored. The genus *Geospiza* is the subject of a separate entry

fringing reef   a coral reef parallel to the shore and separated from it by a relatively shallow, usually narrow, lagoon

Fritschiella [fritch´·ē·ĕl´·ə]   a genus of chlorophyte Algae in which there is an erect stem branched at the end and with rhizoids at the base

-frond-   *comb. form* meaning "leaf" but usually used in the sense of a fern frond

frond   (*see also* acrorhagus) the aerial branch of a fern and thus corresponding to stem and leaf in higher plants

frons [frŏnz]   literally "forehead" but also used for the anterior region of an insect head

frontal [frŭn´·təl]   pertaining to the forehead of mammals or to the anterior end of the insect head

frontal bone   a membrane bone forming the anterior region of the roof of the skull, and therefore lying between the parietal and the nasal (*see also* postfrontal bone, prefrontal bone)

frontal gland   any gland imbedded in the anterodorsal region of an invertebrate though usually only specifically so designated in Nemertea, Platyhelminthes and termites; in this last case the gland produces a milky secretion through the frontal pore

frontal lobe   the most anterior region of the cerebral hemisphere

frontal plane   the longitudinal plane at right angles to the sagittal

froth glands   glands which produce the "spittle" of some hemopteran insects (= Batelli's glands)

-fruct-   *comb. form* meaning "fruit"

-frug-   *comb. form* meaning fruit

fruit   the reproductive body of seed plants in general. The following derivative terms are defined in alphabetic position:
AGGREGATE FRUIT   SIMPLE FRUIT
MULTIPLE FRUIT

frustule [frŭs´·tyōōl]   one of the two silicious plates that enclose a diatom

fry   a young fish, particularly those in a stage too immature to permit ready identification (*see also* fingerling)

FSH = follicle-stimulating hormone

Fucales [fyōō·kāl´·ēz]   a division of phaeophyte algae distinguished from the Laminales by the presence of pneumatophores. The genera *Ectocarpus*, *Fucus* and *Sargossum* are the subjects of separate entries

Fucus [fyōō´·kəs]   a genus of fucale Phaeophyta of world wide distribution. They are dichotomously branched, each branch having bladder-like pneumatophores. The ends of the branches swell into receptacles in which the sporangia develop

-fugal   pertaining to a movement away from. The adjectival termination -fugous is also frequent

fulcra   spines heavier than fin rays, supporting portions of the fin of fish

-fund-   *comb. form* meaning ",depth"

Fundulus [fŭn´·dyōō·ləs]   a genus of teleost fishes. *F. heteroclitus* [hĕt´·ər·ō·kli´´·təs] (the killifish) has been extensively used in experimental studies

Fungi [fŭn´´·jē, fŭng·gē]   a very large division of the plant kingdom, lacking chlorophyll and composed of a thallus of tangled, sometimes interconnecting, hyphae. Reproduction is by spore. The classes Chytriomycetes, Oomycetes, Zygomycetes, Ascomycetes, Basidiomycetes and Deuteromycetes are the subjects of separate entries

Fungi imperfectae = Deuteromycetae

Fungia [fŭn´´·jēə, fŭng·gēə]   a genus of Madreporia well described by their common name of "mushroom corals". They have a remarkable life history in that the initial cup-shaped colony, formed from the planula larva, casts off the cup which then grows a stalk, ever increasing in girth, grows another cup and so on until, ultimately, the cup is replaced by a large mushroom-shaped cap

fungiform papilla   sensory papilla of the vertebrate tongue, in the form of a pileate lobe, interspersed among the filiform papillae

-funic-   *comb. form* meaning "rope"

funicle [fyōō·nĭk´´·əl]   any cord, rope, or stalk-like body particularly those composed of a bundle of fibers or vessels.

funnel   a structure in the form of an open cone attached to a tube. Specifically, the lower expanded chamber of a pneumatophore. The following derivative terms are defined in alphabetic position:
COELOMIC FUNNEL   PERITONEAL FUNNEL
ORAL FUNNEL

fructofuranosidase [frŭk´·tō·fyūr·ăn´·ō·sid´´·āz]   an enzyme catalyzing the hydrolysis of fructofuranoside into alcohol and fructose

-furca-   *comb. form* meaning "fork"

-fus-   *comb. form* meaning "spindle" in the sense of the cigar-shaped spindle used in hand spinning and weaving

fused plasmodium   the stage of a myxomycophyte in which the free cells fuse and subsequently come to fruit

fusiform initial [fyūz´·ē·fôrm]   those cells on a vascular cambium which will give rise to all those cells of the xylem and phloem which have their axes parallel to the direction of growth

G

Gadus [gā″·dəs] a genus of teleost fish of great economic importance. *G. morrhua* [mŏr″·yōō·ə] (the cod), *G. pollachius* [pŏl·ăk″·ēəs] (the pollack) and *G. merlangus* [mĕr·lăng·gəs] (the whiting) are all extensively fished

-gae- *comb. form* meaning "earth", and therefore identical with -ge- and -geo-, but preferred in compounds denoting geographical zones. The *adj. term.* -gaen is similarly used in contrast to -gean. The following terms using this suffix are defined in alphabetical position :

AMPHIGAE	NEOGAEA
ARCTOGAEA	NOTOGAEA
DENDROGAEA	ORNITHOGAEA
GERONTOGAEA	PALAEOGAEA

galactopoiesis [găl′·ăkt·ō·pō″·ēs·is] the maintenance of lactation in a mammal

α- and β-galactosidases [găl·ăk′·tō·sid″·āz·əz] enzymes catalyzing respectively the hydrolysis of α and β-d-galactosides into alcohols and d-galactose

galea [gā″·lēə] literally a "helmet", but mostly usually applied to the outer of the two lobes which terminate an insect maxilla (cf. lacinia)

gall 1 [gôrl] (*see also* gall 2) = bile

gall 2 (*see also* gall 1) an abnormal growth induced in a plant. Those induced on oaks by cynipid hymenoptera were at one time the principal commercial source of tannin

gallbladder a sac-like structure found in the liver of many vertebrates which serves to accumulate bile

-galo- *comb. form* meaning "milk"

galvanotaxis [găl′·văn·ō·tăks″·ĭs] orientation or movement in relation to direct current

-gam- *comb. form* meaning "marriage" and thus, by extension, anything connected with sexual reproduction

-gam termination used in classifying plants according to their method of reproduction or to distinguish parts of plant reproductive systems (*see* cryptogam, phanerogam)

-gamae termination interchangeable with -gam

Gambusia [găm·byōō″·zé·ə] a genus of American freshwater fish known as mosquito fish. *S. affinis* [əf·in′·ĭs], from the Southeast U.S. is the actual species distributed all over the world for the control of mosquito larvas

-gamet- *comb. form* meaning "a spouse"

gametangial copulation [găm′·ət·ănj″·ēəl] the fusion of nuclei within a syncytium that subsequently divides into distinct gametes

gametangium [găm′·ət·ănj″·ē·əm] that organ of lower plants in which the gametes are developed

gamete [găm″·ēt] one of two cells that fuse and develop into an individual. Where the sexes can be distinguished the male gamete is a sperm and the female an egg. The following derivative terms are defined in alphabetic position :

AGAMETE	MEGAGAMETE
ANISOPLANOGAMETE	MEROGAMETE
APLANOGAMETE	MICROGAMETE
APOGAMETE	OBLIGATE GAMETE
HETEROGAMETE	PARTHENOGAMETE
HOMOGAMETE	PLANOGAMETE
ISOGAMETE	PROGAMETE
ISOPLANOGAMETE	ZOOGAMETE

-gametism termination indicating the condition of having special kinds of gametes. For example, heterogametism is the condition of an organism which produces heterogametes

gametocyte [gə·mēt″·ō·sit′] a gamete mother cell

gametogenesis [gə·mēt″·ō·jĕn″·ə·sĭs] the cytoplasmic and nuclear processes involved in the production of gametes

gametogonium [gə·mēt′·ō·gōn″·ēəm] a common term embracing both oogonium and spermatogonium

gametophore [gə·mēt″·ō·fôr′] that portion of a lower plant, particularly an alga, which produces gametes

gametophyte [gə·mēt″·ō·fit′] in those plants showing alternation of generations, that generation which reproduces sexually through the production of gametes

-gamety termination indicating the condition of producing special gametes or producing gametes by special means. For example, apogamety (*see* apogamete) is the formation of gametes by apomixis

-gamia a termination synonymous with -gamy, the form preferred in this work

-gamic an *adj. term.* derived from the substantive *term.* -gamy, under which definitions are given in this work. The alternative form -gamous is usually interchangeable though in a few cases (e.g. cryptogamic botany) the -ic form appears fixed

Gammarus a genus of amphipod Crustacea with

numerous species in both fresh and salt waters. *G. fasciatus* [făs'·ē·ăt"·əs, făs·ĭ·ăt"·əs] is the commonest freshwater amphipod in the U.S.

gamont [găm"·ŏnt"] that form of an organism which produces gametes (*see also* agamont)

-gamous an *adj. term.* derived from the substantive form -gamy under which definitions are given in this work. The alternative form -gamic is usually interchangeable though in a few cases (e.g. polygamous humans) the -ous form appears fixed

-gamy *comb. suffix* indicating the type or method of reproduction. The following words using this suffix are defined in alphabetic position:

ADELPHOGAMY	MEIOTIC EUAPOGAMY
AMPHIGAMY	MEROGAMY
APOGAMY	MONOGAMY
AUTOGAMY	OOGAMY
CYTOGAMY	PARTHENOGAMY
DICHOGAMY	PLASMOGAMY
EPIGAMY	POLYGAMY
EXOGAMY	PSEUDOGAMY
EXOISOGAMY	SPOROGAMY
HOLOGAMY	SYNCHRONOGAMY
HOMIOGAMY	SYNGAMY
ISOGAMY	ZOOGAMY

ganglion literally a swelling but most commonly in biology applied to a group or cluster of associated nerve cells. The following derivative terms are defined in alphabetic position:

AUTONOMIC GANGLION	NODOSE GANGLION
CARDIAC GANGLION	PARASYMPATHETIC G.
CEREBRAL GANGLION	PETROSAL GANGLION
CEREBROSPINAL G.	PREVERTEBRAL GANGLION
CHAIN GANGLION	SEMILUNAR GANGLION
COELIAC GANGLION	SENSORY GANGLION
GASSERIAN GANGLION	SYMPATHETIC GANGLION
MECKEL'S GANGLION	WRISBERG'S GANGLION

gangrene [găng'·grēn"] the destruction of part of an organism. Gas gangrene is caused by *Clostridium.* Dry gangrene, in which the part mummifies, is caused by ergot (*see* Claviceps) and was known in the Middle Ages as St. Anthony's fire

ganoid scale [găn'·oid] scales, found in many actinopterygian fish, in which the basal part is bony and the surface covered with a peculiar enamel called ganoin

Gärtner's duct the degenerate remnant in the female of such portions of the mesonephros duct as has not formed the oviduct

gas gland 1 (*see also* gas gland 2) the gland which secretes gas into the pneumatophore of a siphonophoran

gas gland 2 (*see also* gas gland 1) a mass of capillaries lying at the anterior end of the swim bladder of teleost fish and which, by the secretion or absorption of gas, varies the buoyancy of the fish

Gaskell's bridge = His's bundle

Gasserian ganglion = semilunar ganglion

-gaster- *comb. form* meaning "stomach" or "abdomen"

gaster [găst'·ər] an abdomen, and particularly one which is swollen, and specifically that portion of the hymenopteran insect abdomen which lies behind the petiole

gastralia 1 [găst·trāl"·ēə] (*see also* gastralia 2) sternal rib-like bones found in the ventral abdominal wall between the last true rib of the pelvis in *Crocodilia, Sphenodon,* and some fossil reptiles (= parasternalia)

gastralia 2 (*see also* gastralia 1) dermal ossifications which contribute to the plastron of turtles

gastric pit a pit in the wall of the stomach forming the common aperture of numerous glands

gastrin [găs"·trĭn] an enzyme secreted by the gastric mucosa. It is active in the control of hydrochloric acid production in the stomach

gastrodermis [găs·trō·dĕrm"·ĭs] the inner surface of the coelenteron

Gasteromycetales [găs'·tĕr·ō·mī'·sĕt·āl"·ēz] an order of basidiomycete Fungi in which the sporophore is enclosed in a ball-like peridium. Puff balls are typical

Gastropoda [găs'·trŏ·pō"·də] the class of Mollusca that contains the snails, slugs and their allies. Almost all (slugs and limpets are the exceptions) have spirally coiled shells. Land and freshwater forms have developed pulmonate respiration but most marine forms have gills. Many are vectors of trematode parasites. The orders Pulmonata (with some contained genera), Nudibranchiata (with some contained genera), and the separate genera *Arion, Busycon, Crepidula, Fissurella* and *Limax* are the subject of separate entries

gastrostege [găs"·trō·stĕj'] a ventral scale immediately anterior to the anus in snakes (*see also* urostege)

Gastrotricha a phylum of pseudocoelomatous bilateral animals. All are microscopic and have an unsegmented cuticle furnished with spines or scales. The group may also be regarded as a class of the phylum Aschelminthes. The genus *Chaetonotus* is the subject of a separate entry

gastrozooid [găst"·rō·zō'·ĭd] a zooid, particularly of a siphonophoran coelenterate, designed to digest, and frequently to capture, prey

gastrula [găst"·rōō·lə] the stage in the early development of an embryo in which the rudimentary enteron is established but the nervous system is not yet apparent. This stage therefore lies between the blastula and the neurula. The following derivative terms are defined in alphabetic position:

ARCHIGASTRULA	DISCOGASTRULA
COELOGASTRULA	

Gause's Law two species occupying the same ecological niche cannot survive in the same geographic location

GDP = guanosine diphosphate

-ge- *comb. form* meaning "earth"—both in the sense of the soil (for which -edapho- is a better combining form) and of the planet (*see also* -gae- and -geo-)

-gean *comb. term.* usually employed to indicate "earth" (soil) as distinct from "Earth" (planet) for which -gaean is preferred

-geito- *comb. form* meaning "neighbor"

Gelidium [jĕl·ĭd'·ēəm] one of the many genera of red algae from which agar is manufactured

-gemm- *comb. form* meaning bud

gemma [jĕm"·ə] literally a little bud particularly a globose bud becoming detached from the parent and thus being an initial stage of asexual development, particularly in the lower plants

gemmule 1 [jem"·yōōl] (*see also* gemmule 2) a bud-like body of any plant except an angiosperm; the term has also variously been applied to the plumule and the ovule of angiosperms

gemmule 2 (*see also* gemmule 1) an asexual reproduction body found in some aquatic invertebrates; it consists normally of a mass of undifferentiated tissue surrounded by a protective case

-gen- *comb. form* combined or abbreviated from numerous Greek roots, meaning "bring forth", "parent", "beginning or origin", "birth", "ancestor", "nation" or "pertaining to birth". The English derivatives are confused and frequently have arbitrary meanings. Entries in this dictionary are indicated under all the following combining forms: -gen, gene, gener, generation, generic (1, 2), -genesis, -geneous, -genetic, genetics, -genia, -genic, -genous, genus, -geny, Many adjectival forms are not separately indicated but will be found only with the substantives with which they combine

-gen 1 (*see also* -gen 2) *comb. suffix* in the sense of "that which produces". The following terms using this suffix in this sense are defined in alphabetic position:

AEROGEN	PHELLOGEN
ANDROGEN	SCLEROGEN
FLORIGEN	TRICHOGEN
MUCIGEN	UROBILINOGEN
PATHOGEN	ZYMOGEN

-gen 2 (*see also* -gen 1) *comb. suffix* in the sense of "ancestor" or "ancestry". The following terms using this suffix in this sense are defined in alphabetic position:

CULTIGEN	SYNGEN

gena [jē″·nə] literally cheek and applied to the lateral facial area of many animals, particularly arthropods; it is also used of the basal, feathered portion of a bird's bill or jaw

gender the specification of sex

gender name a name which of itself indicates sex (e.g. bull and cow)

gene [jēn] a functional unit of heritable information occupying a specific locus on a chromosome and is that section of the DNA which determines the amino acid sequence of a single peptide chain. The following derivative terms are defined in alphabetic position (*see also* allele):

BUFFERING GENE	MODIFIER GENE
COMPLEMENTARY GENE	MODIFYING GENE
CUMULATIVE GENE	OLIGOGENE
DUPLICATE GENE	PLASMAGENE
INDEPENDENT GENE	PLASTOGENE
INHIBITOR GENE	POLYGENE
MAJOR GENE	TRIPLICATE GENE
MIMIC GENE	WILD TYPE GENE

gene-flow the wide distribution of genes within a population by interbreeding

gene interaction the mutual effect of non-allelic genes

gene substitution the replacement of one allele by another

gene symbol arbitrary symbols, not separately listed in this dictionary, for specific genes the locus of which has been mapped

gene theory heritable characters are controlled by small units (genes) strung along the length of a chromosome

genepistasis [*angl.* jĕn·ə·pĭst″·ə·sĭs′, *orig.* jĕn′·ĕp·ē·stās″·ĭs] the condition of a form which has failed to show evolutionary changes over a long period of time, as for example, the brachiopod genus *Lingula* which has remained unchanged since the Ordovician epoch

generation 1 (*see also* generation 2) the act of generating or producing. In this sense the following derivative terms are defined in alphabetic position:

DEGENERATION
HOMOETIC REGENERATION
REGENERATION
SPONTANEOUS GENERATION

generation 2 (*see also* generation 1) a group of offspring descended from common parents, or a group of parents (*see* antithetic generation, filial generation)

generation time the period between the sexual maturity of one organism and the sexual maturity of its offspring; the term is also used in bacteriology to denote the length of time required to double the number of bacteria in a culture

generative cell = gamete

generative nucleus one of the two nuclei derived from the primary division within a pollen cell (*see also* tube nucleus)

generative parthenogenesis parthenogenesis by a haploid cell and particularly the production by asexual means of a sporophyte from a haploid germ cell of a gametophyte

generic pertaining to a genus (*see monogeneric*)

generitype [jĕn·ĕr″·ē·tīp′] the type species of a genus (*see under* genotype)

Gene's gland an accessory gland of the acarine female reproductive system, functioning only during the period of egg laying

-genesis 1 (*see also* -genesis 2) *comb. form* used in the sense of "a type of reproduction". The following derivative terms using this suffix in this sense are defined in alphabetic position:

ALLIOGENESIS	METAGENESIS
AMPHIGENESIS	ORGANOGENESIS
ANTHOGENESIS	PAEDOGENESIS
CYTOGENESIS	PARTHENOGENESIS
EMBRYOGENESIS	PATROGENESIS
GAMETOGENESIS	SPOROGENESIS
HETEROGENESIS	TACHYGENESIS
HISTOGENESIS	ZYGOGENESIS

-genesis 2 (*see also* -genesis 1) *comb. form* used in the sense of "origin" or "origination". The following terms using this suffix in this sense are defined in alphabetic position:

ABIOGENESIS	PALINGENESIS
BIOGENESIS	PANGENESIS
EPIGENESIS	PHYLOGENESIS
MORPHOGENESIS	POLYGENESIS
NEOBIOGENESIS	

-genetic 1 (*see also* -genetic 1, 2) *comb. form* used in the sense of "reproduction" and thus the adjectival form of -genesis 1 under which derivatives of this form are given

-genetic 2 (*see also* -genetic 1, 3) *comb. form* used in the sense of "origin" and thus the adjectival form of -genesis 2 under which derivative terms are indicated

-genetic 3 (*see also* -genetic 2, 3) adjective used in the sense of "pertaining" to heredity

genetic assimilation the fixation of a genetic character, not evident in the original phenotype, by artificial environmental changes

genetic drift the irregular change in gene frequencies in a population from generation to generation as a result of random processes

genetic dwarfism a condition produced in some plants, particularly legumes, caused by the mutation of a single gene. It is overcome by the application of gibberellin

genetic isolation isolation through mutual sterility

genetic marker a symbol used in microbial genetics,

corresponding to the gene symbols used in higher forms. Genetic markers cannot, for obvious reasons, be mapped as are genes on chromosomes but those of *Escherichia* can be arranged in terms of minutes of elapsed time before transfer is complete

genetic spiral an imaginary line drawn down a plant axis and which passes successively through the origins of lateral bodies in order of their age

genetics the study of the causes and effects of heritable characteristics. The following derivative terms are defined in alphabetic position :
CRYPTOGENETICS   CYTOGENETICS   SYNGENETICS
-genic 1 (*see also* -genic 2, 3) pertaining to genes (*see* isogenic)
-genic 2 (*see also* -genic 1, 3) *comb. form* used in the sense of "production" (*see* cyanogenic, erotogenic)
-genic 3 (*see also* -genic 1, 2) used in the sense of "causation" and therefore *comb. form* usually synonymous with -genous the form mostly preferred in this work

genic balance [jĕn″·ĭk] the balance of phenotypic genes of many chromosomes especially sex chromosomes, particularly those that control secondary and tertiary sexual characters

genic sterility a variety of hybrid sterility, due to failure to produce functional gametes

geniculate [jĕn·ĭk″·yū·lāt] being in the possession of or having the form of, a knee or knee joint

genital armature those portions of the reproductive system of an arthropod which are directly used in copulation

genital bursa in insects, the pouch in which the female receives and stores sperm

genital corpuscle a specialized type of Vater's corpuscle found in the erotogenic areas

genital ridge the ridge along the dorso-lateral portion of the embryonic coelom to which primary sex cells migrate. Also the analogous structure in some invertebrates

genitalia [jĕn′·ĭt·al″·ēə] secondary sex characters (e.g. those sexual organs directly used in fertilization)

Gennari's line = Baillarger's line

genocline [jĕn″·ō·klin] a gradual change of character across a geographical region due to gene flow

genome [jē″·nōm] one complete haploid set of chromosomes (*see also* phenome)

genophore [jĕn″·ō·fôr′] a term applied to the gene-carrying complex in those forms in which discrete chromosomes are not apparent

genotype [jē″·nō·tip′] a term used to designate the genetic composition as distinct from the appearance (cf. phenotype) ; frequently incorrectly used for generitype

-genous 1 (*see also* -genous 2) *comb. term.* used in the sense of "causation". The synonymous form -genic (*see* -genic 3) is often used. The following terms using this suffix in this sense are defined in alphabetic position :
AUTOGENOUS   MUTAGENOUS
HETEROGENOUS   SCHIZOGENOUS
-genous 2 (*see also* -genous 1) *comb. form* used in the sense of "sex" or "sex organs" (*see* cenogenous)

genus [jē″·nəs] an assemblage of species considered to be more closely related to each other than they are to members of another genus. The names of genera are invariably capitalized and usually, as in this volume, printed in italics or underlined in manuscript (*see also* subgenus, supergenus)

-geny *comb. form* used in the sense of a "condition or type of reproduction" and of a "condition or type of inheritance". Many more meanings are referred to under -genic and -genous. The suffix -gony is also used for a type of reproduction and there is no uniformity in the application of any of these suffixes. The following terms using this suffix are defined in alphabetic position :
ARRHENOGENY   LYSOGENY
CONTINUED EMBRYOGENY ONTOGENY
DISSOGENY   PHYLOGENY
-geo- see -ge- and -gae-

geocline [jē″·ō·klin′] a cline reflecting geographical rather than ecological conditions

geodiatropism [jē′·ō·dīə·trōp″·izm] the orientation of an organ at right angles to the force field of gravity

geographic synecology the relation of environmental factors to the distribution of communities (*see also* synecology, dynamic synecology)

geological epochs, eras and periods *see* epoch, era, period

Geometridae [jē′·ō·mĕt″·trĭd·ē] a very large family of moths commonly called measuring worms or geometers from the habit of their larvae which progress in a looping manner. The genus *Biston* is the subject of a separate entry

geonastic [jē′·ō·năst″·ĭk, jē′·ə·năst″·ĭk] said of curvature towards the ground of a flattened plant structure such as a leaf

Geophilus [*angl.* jē·ŏf″·əl·əs, *orig.* jē′·ō·fil″·əs] a cosmopolitan genus of Chilopoda. Some species have nearly two hundred segments, each divided into 2 sub-segments. All are long and very slender. There are several common U.S. species

Geospiza [*angl.* jē·ŏs″·pīz·ə, *orig.* jē′·ō·spiz·ə] a genus of fringillid birds of the Galapagos Islands commonly called "Darwin Finches", since it was their remarkable morphological adaptation to a wide variety of ecological niches that convinced him of the principle of natural selection. It now seems probable that the ancestor of these birds must have reached the islands long before any other small land bird. They were thus without competition and so able to fill every type of niche that in other places would have already been occupied by other birds

geotaxis [jē′·ō·tăks″·ĭs] movement in relation to gravity

geotropism [*angl.* jē·ŏt″·rə·pĭzm, *orig.* jē′·ō·trōp″·ĭzm] growth in response to gravity (*see also* diageotropism)

-geous *adj. term* synonymous with -gean, the form preferred in this work

Gephyrea [jĕf′·ə·rēə] at one time a phylum containing the Echiurida, Priapulida and Sipunculoidea. This taxon is no longer acceptable

gephyrocercal [jĕf′·ĭr·ō·sĕr″·kəl] said of a caudal fin in which the axial skeleton is truncated and without hypurals. The fin is usually vestigial and without lobes

germ ball reproductive cells in some larvas from which other larvas may be produced. Particularly a clump of cells, actually a rudimentary embryo, found in the rear of a miracidium larva

germ band the thickened area in an arthropod egg from which the embryo is produced

germ layer the three layers (ectoderm, mesoderm, endoderm) into which many embryos are clearly differentiated

germ nucleus a nucleus produced by the fusion of gametes

germ plasm the cytoplasm of germ cells constituting a germ line from generation to generation, opposite of *somaplasm*

germ tube the initial beginnings of a hypha coming from a spore

germinal area a morphologically undifferentiated area in a gastrula, or other early embryonic stage, which has developed the potency to form a specific organ

germinal disc that area of yolk-free protoplasm on the upper surface of a telolecithal egg, to which early development is confined

germinal localization the specific localization of parts of the embryo in the egg or in the cleaving egg

germinal vesicle the resting nucleus of an oocyte

germination growth resulting from the breaking of dormancy in a seed

germovitellarium [jûrm′·ō·vit′·ēl·âr″·ēəm] an ovary producing both eggs and yolk

-geron-, -geront- *comb. form* meaning "old man" but in biology extended to mean "old world"

gerontogaea [jĕr′·ŏnt·ō·jē″·ə] the "old world" as applied to the distribution of plants (cf. neogaea)

gerontomorphosis [jĕr·ŏnt′·ō·môrf·ōs″·ĭs] a supposed evolution of new groups from neotenic larvas of other groups

gestation [jĕst·ā″·shŭn] the period between fertilization and birth in oviparous animals

giant chromosomes unusually large chromosomes, particularly those found in the salivary glands of certain Diptera. Those in *Drosophila* have been the basis of much genetic investigation

Giardia [jē·ärd″·ēə] a genus of zoomastigophorous Protozoa. *G. muris* [myūr′·ĭs] is so common in the intestines of laboratory mice that it is often used as a teaching specimen

Gibberella [jĭb′·ə·rĕl″·ə] a genus of sphaeriale ascomycete fungi parasitic on grasses. Investigations on *G. fujikuroi* [fōō·jē·kyūr·oi], a parasite of rice, lead to the discovery of gibberellins

gibberellins [jĭb′·ər·ĕl″·ĭnz] a group of plant hormones causing elongate stem growth. They are also responsible in some cases for breaking seed dormancy and inducing flowering (*see also* auxins, cytokinins)

gill 1 (*see also* gill 2) a thin-walled process, or series of processes designed to promote osmotic exchange between the blood and the environment in water-dwelling animals (*see also* external gill)

gill 2 (*see also* gill 1) the spore-bearing plate of a basidiomycete fungus

gill arches those visceral arches that bear gills

gill book a series of leaf-like pads functioning as gills found in marine arachnids (*see also* book lung)

gill raker a series of protuberances varying in shape from nobs to filaments on the inner edge of the gills of fish

gill slit the aperture between the pharynx and the exterior which, in the adult animal, is kept open by gill arches bearing gills

-gingin- *comb. form* meaning "gum"

-ginglym- *comb. form* meaning "a hinge" usually transliterated -gingly-

ginglymoidy [gĭng″·glē·moid′·ē] the condition of a vertebra in which the articular surfaces are doubled or asymmetrical

Ginkoales [gĭnk′·ō·āl″·ēz] an order of vascular plants widespread in the Mesozoic epoch and now represented by the single extant species *Ginko biloba* [bi·lōb·ə]. All the group appear to have been dioecious with androsporangia on loose strobili and with single ovules born on stalks

girder sclerenchyma a sclerenchyma which in section has the shape of a "T" or "H" (*see also* sclerenchyma, protosclerenchyma)

-gito- *see* -geito-

gizzard [giz″·əd] in birds a muscular sac, with sclerotized internal teeth, immediately behind the proventriculus. In insects the proventriculus is sometimes called the gizzard

gland 1 (*see also* gland 2) a group of cells, frequently globular or flask-shaped, which secrete a specific substance or group of substances. In invertebrates the term is also extended to similar structures of excretory function. The names (e.g. parathyroid) and types (e.g. holocrine) of glands are defined in alphabetic position

gland 2 (*see also* gland 1) used as synonymous with glans (from which the word is indeed derived)

glandulomuscular cell [glănd′·yōō·lō·mŭsk″·yū·lə] a glandular cell with a contractile extension such as is found in the pedal disk of many coelenterates (cf. epitheliomuscular cell)

glans literally an acorn but also used for acorn-shaped structures such as the end of the human penis or the proboscis of Enteropneusta

-glea *see* -gloea

gleba [glē′·bə] the mass of interwoven mycelia which form the interior of the immature fruiting head of a puffball

-gli- *comb. form* meaning "glue" and thus applied to some forms of connective tissue

glia [glē′·ə] a general term applied to connective tissue of the vertebrate central nervous system. The following derivative terms are defined in alphabetic position:

ASTROGLIA   MICROGLIA
MACROGLIA   OLIGOGLIA

Glisson′s capsule the connective tissue sheathing of vessels, both bilary and hepatic, within the liver

Globigerina [glōb′·ē·jûr·in″·ə, glŏb″·ē·jûr″·in·ə] a genus of foraminiferan Protozoa consisting of a globular aggregate of spherical chambers. It is so common that 60% of the floor of the Atlantic is covered to a depth of many feet with globigerina ooze

-globin a general term for the protein portion of respiratory pigments

globulin [glŏb″·yōō·lĭn] general term for a group of water insoluble simple proteins (*see also* euglobulin, pseudoglobulin, thyroglobulin)

glochidium larva a bivalved larva of freshwater clams which attaches by its valves to fish for distribution

-gloe- *comb. form* meaning "glue", and applied generally to apparently structureless colloid layers in organisms

Gloecapsa a genus of cyanophyte algae forming small colonies in which a number of mucous-coated individuals are enclosed in a gelatinous sheath. *G. polydermatica* [pŏl′·ē·dûrm″·ăt″·ĭk·ə] is common in fresh water

Gloger′s law a cool, dry environment tends to produce races having paler hair or feathers than those found in other environments

-glom- *comb. form* meaning "ball" or "sphere"

glomerular capsule [glŏm'·ər·yū"·lə]   the swollen termination of a metanephric or mesonephric unit which contains the glomerulus

glomerulus [glŏm'·ər·yū"·ləs]   literally a "little ball" but applied in anatomy to a ball-shaped network of blood vessels projecting into a cavity. Without qualification in vertebrate anatomy usually refers to the glomerulus of a kidney unit

-gloss-   *comb. form* meaning "tongue"

glossa [glŏs'·ə]   literally a tongue, but specifically applied to the inner of the two terminal lobes of an insect labium (*see also* paraglossa)

Glossina [glŏs·in"·ə]   a genus of muscid dipteran insects many of which are disease vectors. *G. morsitans* [môrs·it·ănz], among others, transmits *Trypanosoma*

Glossiphonia [glŏs'·ē·fōn"·ēə]   a genus of freshwater hirudinean annelids (leeches) with the posterior sucker on a peduncle. They cannot swim and feed on worms and snails

glossopharyngeal nerve   Cranial IX. Arising from the medulla, close to the root of the tenth nerve

glossum   a word without classic justification but apparently an attempt to latinize the Greek glossa. The following derivatives using this term as a suffix are defined in alphabetic position:
ENTOGLOSSUM   HYPOGLOSSUM   PARAGLOSSUM

glottid   a variant form of glottis, presumably derived from the plural glottides but usually used as though singular (e.g. proglottid)

glottis [glŏt"·ĭs]   properly the back of the tongue but now used only for the opening of the pharynx to the trachea. The following derivatives using this word as a suffix are defined in alphabetic position:
EPIGLOTTIS   PROGLOTTIS

glucagon [glük"·ə·gŏn]   a hormone secreted in the α-cells of the islets of Langerhans in the pancreas. Its action is antagonistic to that of insulin

glucanase [glük"·ən·āz"]   general term for enzymes catalyzing the hydrolysis of α-1,4 glucan links

glucoamylase [glük'·ō·ăm'·il·āz"]   an enzyme that catalyzes the removal of successive glucose units from the non-reducing ends of polysaccharide chains (*see also* amylase)

glucosidase [glük'·ō·sid'·āz"]   general term for enzymes catalyzing the hydrolysis of glucosides

-glum-   *comb. form* meaning "husk"

glume [glōom]   a bract, particularly one of a grass

l-glutamic acid [ĕl·glyōō·tăm"·ĭk]   a widely distributed amino acid, *2-aminopentanedioic acid*. $H_2OCCH_2CH_2CH(NH_2)CO_2H$. Its sodium salt is derived in very large quantities from the soy bean and sold in crystalline form under various names as a food seasoning or in solution as soy sauce

gluteal artery = sciatic artery

Glycera [glĭs'·ər·ə, glĭs"·ĕr·ə]   a genus of Polychaetae errantia that live in sand burrows. There is a very long proboscis and retractile gills rise from the small parapodia. *G. dibranchiata* [dī'·brăk·ē·ā"·tə] of the Atlantic Coast is a common laboratory specimen

glycine [glī'·sēn]   aminoacetic acid. NH_2CH_2COOH. A widely dispersed amino acid not necessary for the growth of rats

glycoprotein [glī'·kō·prō·tēn]   a compound protein having a sugar-like prosthetic group

glyoxalate cycle [glī·ŏks'·ə·lāt]   a fat to carbohydrate mechanism involving the entry of glyoxalate

(from isocitrate) into the tricarbocylic acid cycle. Fatty acids are then utilized to yield malate as the precursor of oxaloacetic acid

-glyph-   *comb. form* properly meaning "carved" or "notched". The common usage as "fang" (of a snake) is difficult to justify. The following derivatives using this term as a suffix are defined in alphabetic position:
AGLYPH   SIPHONOGLYPH
OPISTHOGLYPH   SOLENOGLYPH

GMP = guanosine 5'phosphate

-gnath-   *comb. form* meaning "jaw" (*see* endognath, paragnath)

-gnathal   a synonym of -gnathous

Gnathostomata [năþ'·ō·stōm"·ə·tə]   a division of craniate chordate animals erected to contain those classes possessing a hinged lower jaw. Thus defined, this group contains all extant craniates except the Marsipobranchii

-gnathous   pertaining to a jaw

-gnot-   *comb. form* meaning "known"

gnotobiosis [nō'·tō·bi·ōs"·ĭs]   the condition of animals which are sterile both internally and externally, or of the procedures used to secure and maintain these conditions. The adjectival form gnotobiotic is sometimes misused as applying to animals or cultures of known bacterial content (*see* axenic)

goblet   a cavity or organism in the shape of a wine glass without the stem

goblet cell   an epithelial cell that secretes mucous

goggyl-   *comb. form* meaning "round" usually transliterated gongylo-

Golgi apparatus   a mass of lipoidal reticulating fibers, granules and vesicles, present in all animal and some plant cells. The function is doubtful though the complex is apparently continuous with the endoplasmic reticulum

Golgi body = Golgi apparatus

-gomph-   *comb. form* meaning "club" and since bolts are club-headed, also "bolt" or "fasten"

-gon-   *comb. form* used as a great number of compound words concerned in reproduction but also, from another root, in the meaning "angle". Entries, and in some cases lists of derivative forms defined in alphabetic position, will be found under gonad, -gonalis, -gonangium, -gone, -gonia, -gonic, -gonidium, -gonimium, -gonium, -gonous and -gony

gonad [gŏn'·ăd, gŏn'·ad]   an organ producing gametes. The male gonad is the testis, the female is the ovary (*see also* follicular gonad)

gonadotroph cell [gŏn·ăd'·ō·trôf"]   a basophilic cell in the pars intermedia of the pituitary, the glandular content of which varies with the production of gonadotrophic hormone

gonangium [gŏn·ănj"·ē·əm]   a hydrozoan coelenterate gonotheca with its enclosed blastostyle

gonapophysis [*angl.* gŏn·a·pŏf"·əs·ĭs, *orig.* gŏn'·ăp·ō·fĭs"·ĭs]   the sum total of the insect genitalia (*see also* apophysis, anterior apophysis, parapophysis, zygapophysis)

gone [gŏn]   the asexual equivalent of gamete

-gone   *comb. term.* usually used in the sense of "a reproductive structure" (*see* hormogone)

goneoclinic [gŏn'·ē·ō·klĭn"·ĭk]   said of a hybrid that shows the phenotypic characters of only one parent

-goneutic   *comb. term.* used in the sense of the "frequency of reproduction" (*see* digoneutic)

-gonia   *comb. form* frequently used as synonymous

with -gon- (q.v.) but better confined to the meaning "angle" (see protogonia)

goniangium [gŏn′·ē·ănj″·ēəm]   a common term for cystocarp and scyphi

Goniaster [*angl.* gŏn·ē″·əs·tər, *orig.* gŏn′·ē·ăst″·ər] a large, cosmopolitan genus of asteroid echinoderms, often called "cushion stars"

-gonic *comb. term.* used both in the sense of pertaining to reproduction and pertaining to reproductive structures. Most references to derivative terms are given under the substantive suffixes -gony and -gonium. The alternative adjectival form -gonous appears to be used mostly in a rather different sense. The following terms using the suffix gonic are defined in alphabetic position :

HETEROGONIC   HOMOGONIC
HOLOGONIC   TELOGONIC

gonidangium [gŏn′·id·ănj″·ē·əm]   an organ producing a sexual spore

gonidial layer [gŏn·id″·ē·əl]   properly applied to a layer of gonidiophores, but widely used for the algal layer in a lichen thallus (cf. gonohyphema)

gonidium [gŏn′·id·ēəm]   the diminutive of gonad generally used for a reproductive cell, or structure of, a plant but also applied to the algal component of a lichen

Gonionemus [*angl.* gŏn·ē·ōn″·əm·əs, *orig.* gŏn′·ē·ō·nēm″·əs]   a genus of hydrozoan coelenterates, the medusoid form of *G. murbachi* [mŭr·băk′·ī] being widely used in teaching

Gonium [gŏn″·ē·əm, gŏn·ē·əm]   a genus of colonial chlorophyte algae with the form of a flat plate of from two to 16 cells, all of which are reproductive

-gonium *comb. form* used in the widest sense for "reproductive structure". The adjectival form -gonic has a few specialized meanings. The following derivative terms are defined in alphabetic position :

ARCHEGONIUM   OVOGONIUM
CARPOGONIUM   SPERMATOGONIUM
GAMETOGONIUM   SPOROGONIUM
OOGONIUM

gonocoel [gŏn′·ō·sēl]   the cavity of a gonad

gonocoel theory   the coelom is the expanded cavity of a gonad. Sometimes called Bergh's theory

gonoduct [gŏn″·ō·dŭkt′]   any duct connected to the reproductive system, but frequently used as synonymous with coelomoduct

gonohyphema [gŏn′·ō·hi′·fēm″·ə]   the fungal layer of lichens (cf. gonidial layer)

gonophore [gŏn′·ō·fôr″]   any structure bearing reproductive cells but particularly the reproductive zooid of a hydrozoan coelenterate, or the reproductive polyp of a siphonophore

gonopod [gŏn′·ō·pŏd″]   an arthropod appendage modified for copulation

gonopore [gŏn′·ō·pôr]   the opening of a gonoduct

gonotheca [gŏn′·ō·thēk″·ə]   that portion of the perisarc of a hydrozoan coelenterate which encloses the gonozooid

-gonous *comb. term.* logically synonymous with -gonic but rarely used except in the sense of "breeding" as distinct from reproduction. Occasionally found in the sense of "angled" (e.g. "digonous")

gonozooid 1 [gŏn′·ō·zō″·id] (*see also* gonozooid 2) an ectoproct heterozooid with a bulbous brood chamber

gonozooid 2 (*see also* gonozooid 1) a polyp of a hydrozoan coelenterate colony that is modified to produce medusae

-gony *comb. form* used in the sense of a method of reproduction. The adjectival forms -gonic and -gonous can be derived from any of these words but both have other specialized uses. The termination -geny is used in the same sense and there is no uniformity in the application of these suffixes. The following derivatives using the suffix -gony are defined in alphabetic position :

DISSOGONY   SCHIZOGONY
GAMOGONY   SPOROGONY
HETEROGONY   TELEGONY
MEROGONY

Gonyaulax [gŏn′·ē·ôr″·lăks]   a genus of dinoflagellid algae. *G. polyedra* [pol′·ē·ē″·drə], a phosphorescent marine planktonic form, normally occurs at the rate of about 20 per liter. It may, however, suddenly bloom to a rate of about ten million per liter. The water becomes discolored (*see* red tide) and, as the protozoan secretes a toxin intensely poisonous to vertebrates, there is a massive fish kill. Man is only affected through eating pelecypod mollusks which, though immune themselves, accumulate the toxin

Gordiacea = Nematomorpha

Gordius [gôrd″·ēəs]   a genus of Nematomorpha. As with other gordioids the adults are long and hair like ; for example, *G. robustus* [rō·bŭst″·əs] may be 20–30 inches long and 1 mm thick. The young are parasitic in insects. They were once called horsehair worms and were once believed, both from their shape and from the frequency of their occurrence in horse's drinking troughs, to be horsehairs come to life

Gorgonacea [gôr′·gŏn·ās″·ea, gôr′·gŏn·ās″·ea]   the order of alcyonarian Anthozoa that contains the horny corals, the gorgonians, the sea fans, and the sea feathers. The axial skeleton, made either of horny materials or calcareous Spicules, is typical

Gorgonia [gŏr·gŏn″·ea, gôr·gōn″·ea]   a genus of gorgonacean coelenterates with anastomosing branches. *G. flabellium* is flattened and often called the "sea fan"

Gorgonocephalus [gôr′·gŏn·ō·sěf″·əl·əs]   a genus of ophiurid echinoderms, the arms of which are so numerously branched as to give the appearance of a Gorgon's head. *G. arcticus* [ärk·tĭk·əs], sometimes called the "Basket Fish", is common on both sides of the Atlantic

Gorilla [gə·ril″·ə]   the pongid primate containing the single species *G. gorilla*. It is distinguished from the chimpanzee and orang-utan by the pigmented skin and naked face

Graafian   *see* Graffian

-grad- *comb. form* meaning "to walk". The following derivative terms are defined in alphabetic position :

DIGITIGRADE   PLANTIGRADE
ORTHOGRADE   PRONOGRADE

Graffian follicle [grăf′·ē·ən]   the functional follicle in which the mammalian ovum develops in the ovary

graft   the action of inserting one part of an organism into another, or the part so inserted. The following derivative terms are defined in alphabetic position :

AUTOLOGOUS GRAFT   HOMOGRAFT
HETEROGRAFT

graft hybrid   one resulting from the interaction of scion and stock

Graminales = Glumiflorae

Gramineae [grăm′·ĭn·ē]   the grass family of monocotyledons. The grasses are easily distinguished from the sedges by the hollow stem. The genera *Avena* and *Zea* are the subjects of separate entries

gram-negative said of bacteria that do not retain dyes after washing with an iodine solution

gram-positive said of bacteria that do retain dyes after washing with an iodine solution

gram-variable an organism which is gram positive in one stage of its existence and gram negative in another

grana [grăn·ə, grän·ə] membranous portions of the chloroplasts of higher plants which contain all of the photosynthetic pigments

Grantia [gränt'·ēə, grănt'·ēə] a genus of calcareous sponges with radial tubes diverging from a central cavity into flagellated chambers which divide the tubes into inhalent and exhalent portions. Closely allied to *Sycon*

granule any small particle. The following derivative terms are defined in alphabetic position:

AZUROPHILIC G.	NISSL GRANULE
BASAL GRANULE	PALADE'S GRANULE
EOSINOPHILIC G.	"SAND" GRANULE
KÜHNE'S GRANULE	SECRETORY GRANULE
METACHROMATIC G.	

granulocyte [grăn'·yōōl·ō·sīt'] a granular leucocyte

granum *singular* of grana

gray crescent an area on the fertilized egg of anuran amphibians, exactly opposite the point of entry of the sperm

grassland biomes dominated by various species of grasses

gravel rock particles between 1 mm and 1 inch in size (*see also* clay, sand, silt)

gravity receptor = equilibrium receptor

gray matter that part of the central nervous system which consists principally of cell bodies (*see also* white matter)

great ape a term properly applied to pongine primates. Extant genera include *Gorilla, Pan, Simia* and, in the opinion of most biologists, *Homo*

greater omentum a membranous sack in the omentum, frequently containing fat bodies

-gregar- *comb. form* meaning flock or herd

grid a 4 mm disk of very fine copper mesh used to hold objects to be examined in an electron microscope

groove any fold or channel. The following derivative terms are defined in alphabetic position:

ATRIAL GROOVE	ORAL GROOVE
EPIBRANCHIAL G.	ORONASAL GROOVE
HYPOBRANCHIAL G.	PERIBRANCHIAL GROOVE
LARYNGOTRACHEAL G.	

ground meristem partially differentiated meristem which will later give rise to the fundamental system of a plant

group name a term used to designate a group of animals such as a "flock" of birds or a "school" of fish

growth an increase in size. The following derivative terms are defined in alphabetic position:

AMPHITROPHIC GROWTH	PRIMARY GROWTH
INSTERSTITIAL G.	SECONDARY GROWTH

growth factor plant or animal hormones involved in the initiation, or maintenance, of growth of higher forms or any substance required for the growth of a microorganism

growth hormone one produced by the adenohypophysis which stimulates growth particularly of

bones. In adults, its action is antagonistic to insulin. The term is sometimes also applied to auxins

growth ring an annular growth layer such as that seen on the scale of a fish or on the stem of a plant in transverse section

Grylloblattodea [grĭl'·ō·blăt·ō"·dēə] an order of insects at one time included, together with the Dictyoptera and Phasmida, in the Orthoptera, a group restricted by this definition to grasshoppers, locusts and crickets. The Grylloblattoidea have ten evident abdominal segments but are distinguished from the Dictyoptera by lacking ocelli

GSH = reduced glutathione

GSSG = oxidized glutathione

guanine a break-down product of nucleic acids present in fish scales and the excrement of fish-eating birds

guard cell one of the two bean-shaped epidermal cells which together form a stoma on the surface of a leaf or other plant structure. The expansion and contraction of the guard cells control the passage of gases through the stomata

gubernaculum [gōō'·bə·nak"·yōōl·əm] the ligament anchoring the testes to the scrotum

-gula- *comb. form* "throat"

gular bones [gōōl"·ər] one of a series of chondral bones, derived from the hyoid arch and which form the bony support of the floor of the mouth in fish. The central pair are the lateral gulars, with the marginal gulars outside them. They come together in front in the unpaired anterior medial gular

gum 1 (*see also* gum 2) those portions of the mucous membrane of the mouth that cover the base of the teeth

gum 2 (*see also* gum 1) any water soluble, or water miscible, exudate of trees. Water insoluble exudates are properly resins

-gust- *comb. form* meaning "taste" and applied by extension to anterior regions of the alimentary canal (*see* epigusta)

-gutt- *comb. term.* meaning a "drop"

guttation [gə·tā"·shŭn] the voiding of drops of excess contained water onto the surface or edge of a leaf

gymnoblastic [jĭm'·nō·blăst"·ĭk] having medusa buds not enclosed in a sheath

gymnosere [jĭm"·nō·sēr'] a sere in which gymnosperms predominate

Gymnospermae [jĭm'·nō·spûrm"·ē] that division of the Spermatophyta in which the seeds are not contained in an ovary. The group contains, in addition to the Coniferales, the Cycadales, Ginkgoales and Gnetales (*see also* Angiospermae)

-gyn- *comb. form* meaning "female"

gynandromorph [jĭn'·ăn·drə·môrf'] an organism, of a type not normally hermaphroditic, which is partly perfect male and partly perfect female; most commonly (and particularly in birds), the two sides are affected

gynecomorph [jĭn'·ə·kō·môrf'] an organism with primary sexual characters of the male and tertiary sexual characters of the female

gyne [jĭn"·ĕ, jĭn"·ə] any female organism or part (*see* pseudogyne, trichogyne)

gynecophoric canal [*angl.* jĭn·ə·kŏf"·ər·ĭk, *orig.* jĭn'·ə·kō·fôr"·ĭk] that in-curved portion of the ventral surface of male schistosomatid flukes in which the female is carried

-gynic *see* -gynous

-gynism *comb. term.* meaning "a female characteristic" (*see* heterogynism)

gynoecious [jĭn·ēs″·ē·əs]   pertaining to the housing of female organs

gynoecism [jĭn·ēs″·ĭzm]   the condition that only female forms of an organism are known, as in many species of Rotifera

gynoecium [jĭn·ēs″·ēəm]   the female portion of a flower

gynopedium [jĭn′·ō·pēd″·ē·əm]   an assemblage consisting of a female and her offspring, or a few females and their direct offspring

gynosynhesmia [jĭn′·ō·sĭn′·hĕz″·mēə]   a group of females gathered together during the breeding season (*see also* androsynhesmia, synhesmia)

-gynous *comb. term.* pertaining to "females", "female functions" and "female structures". -gynic and -gynicous are rare variants. The substantive suffix is "-gyny". The following terms using -gynous as a suffix are defined in alphabetic position:

METAGYNOUS	PROTOGYNOUS
MONOGYNOUS	PROGYNOUS
PERIGYNOUS	PROTEROGYNOUS

-gyny *comb. form* pertaining to "a female condition". Reference to compounds is made under the adjectival form -gynous

-gypsoph- *comb. form* meaning "chalk"

-gyrate *adj. term* synonymous with -gyrous the form preferred in this work

gyre [jĭr]   a spiral form, or part of a spiral form. Thus it can be said of a spiral that the gyres are numerous

-gyro- *comb. form* meaning "round"

-gyrous pertaining to a turn

gyrus [jĭr″·əs]   a fold, or convolution, in the cerebral hemisphere

gyttja [gŭt′·yə]   the fine mud on the bottom of colorless but productive lakes (*see also* dyttja)

H

habenular body [hə·běn'·yū·lər]   the nerve center in the epithalamus

habenular nucleus   a group of nerve cells in the epithalamus

habitat   [hăb"·ĭt·ăt']   the sum total of the environment of a particular species (*see also* microhabitat)

Habrobracon   [hăb'·rō·brā"·kən]   a genus of braconid wasps containing *H. juglandis* [jyōō·glănd"·ĭs] (properly *Bracon hebetor*) a parasite of *Ephestia* and widely used in genetic research

Haeckel's law = biogenetic law

haemal, etc.   *see* hemal, etc.

Haematoxylon   [hē'·măt·ō·zi"·lŏn]   a genus of leguminous trees. *H. campechiana* [kăm·pēch"·ē·än·ə] yields the dye hematoxylin

Haemosporidia   [hē'·mō·spôr·ĭd"·ēə]   an order of Sporozoa distinguished by the motile zygotes producing naked sporozoites. They are, as the name indicates, blood parasites. The genera *Plasmodium* and *Babesia* are the subject of separate entries

hemocoel   [hē'·mō·sēl"]   a blood-filled body cavity, particularly that of the mollusca and the arthropods; in such a system, the organs are bathed directly in blood and the true coelom occurs only as scattered remnants

hair 1   (*see also* hair 2) a thin strand of keratin secreted by follicles in the mammalian skin and typical of that order

hair 2   (*see also* hair 1) hair-like structures other than those forming the pelt of mammals. In this sense the following derivative terms are defined in alphabetic position (*see also* seta):

COLLECTING HAIR   ROOT HAIR
PLANT HAIR

hair follicle   the pit in mammalian skin at the base of which lie the glands secreting the hair

-hal-   *comb. form* meaning "salt or marine"

halarch succession   [hăl'·ärk]   one occurring under saline conditions

Haldane's law   in those F_1 hybrids in which one sex does not occur, or at least is sterile, that sex is the heterogametic one

Haliclystus   [hăl'·ē·clĭst"·əs]   a genus of stauro-medusan Scyphozoa with the characteristics of the order. There are eight lobes, each with a cluster of tentacles rising from a cup-shaped body, itself attached by a quadrate stalk to a rock. *H. auricula* [ôr'·ik"·yōōl·ə], about 1 inch tall, is common on all shores bordering the Atlantic

-haline-   *comb. form* meaning "salt" or "salty". The following terms using this suffix are defined in alphabetic position:

EUHALINE   MIXOMESOHALINE
EURYHALINE   MIXOOLIGOHALINE
HYPERHALINE   MIXOPOLYHALINE
MIXOEUHALINE   STENOHALINE
MIXOHALINE

hallucinogen   [hăl'·ōō·sĭn"·ō·jĕn]   a substance that causes hallucinations. Many are produced by fungi (*Claviceps, Panaeolus*), and some (peyote, mescalin) by cacti

hallux   [hăl'·əks"]   the first digit on the hind limb which is the hind toe of birds or the great toe of mammals

halolimnetic   [hăl'·ō·lim·nĕt"·ik]   pertaining to salt lakes, though the term is sometimes incorrectly used for marine

halolimnic   [hăl'·ō·lĭm"·nĭk]   said of typically marine organisms capable of surviving in freshwaters

haltere   [hălt'·âr]   a peg-like organ replacing the hind wings of dipterous insects; it vibrates at the same speed as the wings and thus serves as a gyroscopic stabilizer

-ham-   *comb. form* meaning "hook"

hamate bone   [hăm'·āt]   the carpal that lies at the base of the fourth metacarpal alongside the capitate bone

Hanström's organ   a neurosecretory gland in the pedicel of many stalk-eyed Crustacea

Hapaloidia   [hăp'·əl·oid"·ēə]   a sub-order of primate mammals containing the South American marmosets and their immediate allies; they differ from the monkeys and apes in possessing claws instead of nails

haplodont   [hăp'·lō·dŏnt"]   said of a form with simple conical teeth

haploid   [hăp"·loid]   having half the number of chromosomes found in a somatic cell or half the number found in the other of two alternating generations. The haploid is usually said to have N chromosomes in contrast to the diploid 2N. This term differs from monoploid in that the chromosomes need not of necessity comprise a complete genome (*see also* double haploid, synhaploid, diploid, triploid, polyploid)

haplomict [hăp′·lō·mĭkt″] a hybrid of which the genome is derived from chromosomes and portions of chromosomes from various sources

haplosis [hăp·lōs″·ĭs] the reduction of the diploid number to the haploid number at meiosis

haplostele [hăp′·lō·stēl″] one which consists of xylem surrounded by phloem

haplotonic [hăp·lō·tŏn″·ĭk] said of a haplobiontic life cycle in which the dominant is haploid (*see also* diplotonic)

-hapto- *comb. form* meaning "contact" or, by extension, "togetherness" or even "binding"

-haptor- *comb. term.* meaning sucker

Hardy-Weinberg law the genetic constitution of a population tends to remain stable, even in the absence of selection, since the frequency of the pairs of allelic genes is an expansion of a binomial equation

Hassall′s corpuscles concentrically layered globose corpuscles found in the developing thymus

hashish *see Cannabis*

-hasta- *comb. form* meaning "halberd"

Hatschek′s nephridium a tube carrying numerous solenocytes which project forward along the right vento-lateral side of the notochord from just above the mouth in *Branchiostoma*

Hatschek′s theory all invertebrates having a trochophore larvae are descendants of a postulated form, called a trochozoon, which was, in effect, an adult trochophore

Hatschek′s fossa an asymmetrical, hollow, ciliated, blindly ending tube, rising from one of the languets of the wheel organ in *Branchiostoma* to a position just left of the notochord

haustoria [hŏrs·tôr′·ēə] hyphi of parasitic fungi which penetrate tissues and absorb nourishment

Haversian canal the longitudinal blood vascular channel in bone

head fold that area at the anterior end of the embryo which rises from the blastoderm in the early development of a telolecithal egg

heart any organ which circulates a fluid within an animal. The following derivative terms are defined in alphabetic position:
ACCESSORY HEART   LYMPH HEART

heart node a modified lump of muscular tissue exercising nervous functions

heathland an area covered with low growing shrubs interspersed with occasional areas of grass

-hebe- *comb. form* meaning "puberty"

hekistotherm [hĕk″·ĭst·ō·thĕrm′, hĕk·ĭst″·ō·thĕrm′] an organism living above or beyond the tree line and frequently in areas of heavy snow

-helic- *comb. form* meaning "twisted" but often used in biology in the sense of "snail"

helicine arteries [hĕl′·ĭs·ĭn″] the coiled arteries in the flaccid penis

heliotropism [hēl′·ē·ō·trōp″·ĭzm] the condition of growing towards light

Heliozoa an order of rhizopod Protozoa distinguished by their spherical shape and stiff radiating pseudopodia. The genera *Actinophyris* and *Actinosphaerium* are the subjects of separate entries

Helix [hē″·lĭks] a genus of terrestrial pulmonate molluscs. The large edible snail *H. pomatia* [pōm·āsh″·ēə], of European origin, is now widely distributed in California

-helo- *comb. form* meaning "a marsh"

Heloderma [hĕl′·ō·dûrm·ə] a genus of stout, blunt-nosed, thick-tailed lizards (Squamata). Both extant species are venomous. *H. suspectum* [sŭs·pĕkt″·əm] is the Gila monster of the South Western States and *H. horridum* [hŏr″·ĭd·əm] is the beaded lizard of Mexico

hemal arch [hē′·məl] the arch that descends ventrally from the centrum of a vertebra and through which run the caudal vein and artery

hemal rib a rib which lies inside the muscles forming the body wall

hematoblast [hē·măt″·ō·blăst′] the first stage in the development of an erythrocyte

hematoxylin [hĕm′·ə·tŏks″·ə·lĭn] a dye, usually compounded in solution with various alums, extracted from the heartwood of the leguminous tree *Haematoxylon campechiana*. Widely used as a nuclear stain in sections

heme [hēm] a general term for the iron-porphyrin group in hemoglobin

hemerythrin [hĕm″·ə·rĭth″·rĭn] an iron-protein respiratory pigment, found in brachiopods, sipunculids and some annelids

-hemero- *comb. form* meaning "cultivated"

-hemi- *comb. form* properly meaning "half", but widely misused in biology to mean "partly"

hemibranch [hĕm″·ē·brănk′] a branchial arch having the gill only along one side (cf. holobranch)

Hemichordata [hĕm′·ē·kôrd″·ā·tə] a phylum of enterocoelous coelomate animals. At one time they were thought to be allied to the chordates but are now not so regarded. The phylum contains the classes Enteropneusta (acorn worms), Pterobranchia and Planctosphaeroidea

hemielytron [hĕm′·ē·ĕl·ĭt″·rŏn] the anterior wing of hemipterous insects; one half of this wing is thickened

hemimetabolous [hĕm′·ē·mĕt·ăb″·əl·əs] in its broad sense is said of any insect which is not holo-metabolous. In its restricted sense (e.g. contrasted with ametabolous and paurometabolous metamorphoses) it is said of those insects in which the egg hatches into a naiad, which, though differing markedly from the adult, does not pass through a pupal stage

hemimixis [hĕm′·ē·mĭks″·ĭs] the break up and refusion of the parts of the macronucleus of a protozoan cell without contributions from the micro-nuclei (*see also* endomixis)

Hemiptera [hĕm·ĭp″·tĕr·ə] an order of insects distinguished by the fact that the basal portion of the front wing is thickened while the apical portion is membranous. The term "bug", used without qualification, usually refers to a hemipteran. The Hemiptera are closely allied to, and were at one time fused with, the order Homoptera

hemizoic [hĕm′·ē·zō″·ĭk] an organism capable of existing either by the ingestion of particulate matter or by photosynthesis. *Euglena* is an example (*see also* holozoic)

hemizygoid parthenogenesis [hĕm′·ē·zĭg″·oid] asexual reproduction from haploid eggs

hemochorial placenta one in which the walls of the maternal blood vessels are penetrated by the embryonic blood vessels, which are therefore directly bathed in maternal blood

hemocyanin [hĕm′·ō·sī″·ən·ĭn] a blue, copper containing, respiratory pigment found in crustacean blood.

The copper is directly attached to an amino-acid and not, as the iron in hemoglobin, to a porphyrin

hemocyte [hē'·mō·sit″] a cell found in blood fluid. The term is usually restricted to invertebrates

hemocytoplast [hēm'·ō·sit'·ō·plăst″] the first stage in the development of a polymorph leucocyte

hemoglobin [hēm'·ō·glōb″·ĭn] a general name for a group of respiratory pigments in which ferrous iron, linked to a protoporphyrin, is conjugated to a protein. All hemoglobins are about 94% protein but the molecular weight varies from about 17,000 to as high as 2,750,000 in some invertebrates (see also carbohemoglobin)

hemolymph gland [hēm″·ō·lĭmf'] a lymph gland distinguished by a presence of erythrocytes

hemopoiesis [hē'·mō·pō·ēs″·ĭs] blood formation (see also prehepatic hemopoiesis)

hemotrophy [hēm'·ō·trŏf″·ē] the nourishment of an embryo through a connection with the maternal blood supply

hemovanadin [hēm'·ō·văn″·ə·dĭn] a green, vanadium containing, blood pigment found in urochordates

Henle's layer the outer of the three layers of which the internal hair sheath is composed

Henle's loop the portion of a metanephron that runs from Bowman's capsule to the convoluted tubule

Hensen's cells columnar cells surrounding the hair cells in the organ of Corti

Hensen's node the anterior end of the primitive streak corresponding to the dorsal lip of the blastopore

heparinase [hĕp'·ə·ĭn·āz″] an enzyme catalyzing the break down of heparin through the hydrolysis of α-1,4-links between 2-amino-2-deoxy-D-glucose and D-glucuronic acid

hepatic caecum [hə·păt″·ĭk] the rudimentary pouch at the anterior end of the embryonic alimentary canal from which the liver is subsequently developed

hepatic diverticulum either a diverticulum running anteriorly from the junction of the esophagus and gut in Amphioxus or the outgrowth from the primitive gut, precursor to the liver

-hepato- comb. form meaning liver

Hepaticae [hə·păt″·ĭk·ē] the phylum of bryophyte plants that contains the liverworts. The group is of very diverse form and habitat and differs from other bryophytes principally in lacking the leaf-like structures of the mosses and the large pyrenoid containing chloroplasts of the hornworts. The genus Marchantia is the subject of a separate entry

hepatopancreas [hə·păt'·ō·păn″·krē·əs] a gland found in many invertebrates which combines the food storage functions of the liver of higher forms with the secretion of digestive enzymes

-hepta- comb. form meaning "seven"

herbaceous [hĕrb·ās″·ē·əs] said of a plant that lacks a woody stem and is therefore neither a shrub nor a tree

herbarium [hĕrb·âr'·ēəm] a collection of dried plants arranged taxonomically for reference purposes

Herbst's corpuscle a type of Pacinian corpuscle found in birds

-herco- see -erko-

heredity the transmission of characters from parent to offspring

Hermaea [hûrm″·eə] a genus of nudibranch mollusks distinguished by the fact that the brilliantly colored cerata do not open to the exterior

hermaphrodite an organism which has both male and female primary sex characters (see also pseudohermaphrodite, somatic hermaphrodite)

hermaphrodite duct a duct which carries, usually in alternate seasons, either sperm or eggs

heroin [hĕr″·ō·ĭn] a derivative (diacetyl morphine) of morphine. The manufacture of this dangerously addictive narcotic is now prohibited in the U.S.

herpon [hûr″·pən] the sum total of those microorganisms that live on the surface of slimy bottoms of freshwater lakes (cf. gyttja)

-hesia comb. term presumably derived by extension from adhaesere, used by ecologists to indicate a "swarm" (see synhesia)

-hesmia comb. form presumably related to -hesia, used by ecologists for "reproductive swarm". The following terms using this suffix are defined in alphabetic position: ANDROSYNHESMIA GYNOSYNHESMIA SYNHESMIA

Hesperornis [hĕs'·pûr·ôrn″·ĭs] a genus of extinct birds, with vestigeal wings apparently adapted to swimming. Sharp conical teeth were present in both jaws

heterobrachial chromosome [hĕt·ər·ō·brăk″·ēəl] one in which the centromere is not in the middle

heterocercal [hĕt·ər·ō·sĕrk·əl] said of a caudal fin in which the dorsal fold is lacking and in which the ventral fold has a much longer posterio-dorsal than antero-ventral lobe. The tail of sharks is typical

heterochelate [hĕt'·ər·ō·kēl″·āt] said of an arthropod in which the right and left chelae are different, either in structure (as in the lobster) or size (as in the fiddler crab)

heterochromosome [hĕt·'ər·o·krŏm″·ō·sōm'] one of two chromosomes, that are morphologically distinguishable but pair in meiosis, as, for example, the sex chromosomes

heterocyst [hĕt″·ər·ō·sĭst'] an enlarged cell occurring at intervals along the length of some filamentous cyanophyte Algae. The sections between the heterocysts are called hormogonia

heterodont [hĕt″·ər·ō·dŏnt'] a term used of a form possessing many different kinds of teeth

heteroecious parasite [hĕt'·ər·ēs″·ē·əs] either one which requires more than one host for its life cycle or which is not host specific

heterogamete 1 [hĕt'·ər·ō·găm″·ēt] (see also heterogamete 2) one of two gametes which can be distinguished from each other

heterogamete 2 (see also heterogamete 1) that produced by the sex with the potential of producing either sex

heterogametic sex [hĕt'·ər·ō·găm·ēt″·ĭk] that sex which is determined by the XY arrangement of chromosomes

heterogenesis [hĕt'·ər·ō·jĕn″·əs·ĭs] alternation of generation; the term was at one time also applied to the appearance of mutants

heterogenetic association the pairing of chromosomes derived from different ancestors in an allotetraploid

heterogony [angl. hĕt·ər·ŏg″·ən·ē, orig. hĕt'·ər·ō·gŏn″·ē] the alternation of parthenogenetic and zygogenetic generations

heterograft [hĕt·ər·ō·grăft] one made between two organisms of different species

heterogynism [hĕt'·ər·ō·jĭn"·ĭzm] the condition in which the female of the species is more variable than the male in its phenotypic expression

heterohemolysin [hĕt'·ər·ō·hēm'·ō·lĭs"·ĭn] an antibody that will lyse a blood cell from another individual

heterokaryon [hĕt'·ər·ō·kâr"·ē·ŏn] a cell, or an individual containing cells, formed by the coalescence of the cytoplasm, but not the nuclei, of hyphal cells

heterokaryosis [hĕt'·ər·ō·kâr'·ē·ōs"·ĭs] the association of a reproductive structure of nuclei of varied genetic content, a circumstance common in some fungi

heteromastigote [hĕt'·ər·ō·măst"·ĭg·ōt'] having an anterior functional tractellum and posterior functionless flagellum

heteromerous [angl. hĕt·ə·rŏm"·ər·əs, orig. hĕt'·ər·ō·mēr"·əs] when there is no uniformity in the number of parts between organisms of the same species or organs on the same individual. For example, in many beetles the number of tarsi differ on the first and third legs

heteromorphic [hĕt'·ər·ō·môrf"·ĭk] having different forms, but pertaining particularly to cases where the difference is a function of the life history, as distinct from polymorphy where the different forms occur in the same instar

heteromorphosis [hĕt'·ər·ō·môrf·ōs"·ĭs] the regeneration of an organ different from the one that has been removed or lost

heteronereis [hĕt'·ər·ō·nēr"·ē·ĭs] a term applied to nereidiform worms, such as *Nereis* and *Neanthes* when the sexually mature posterior end (epitoke) differs markedly from the sexless anterior end (atoke)

heteroparthenogenesis [hĕt'·ər·ō·pär'·thĕn·ō·jĕn"·əs·ĭs] the condition of producing parthenogenetically, either offspring which themselves reproduce parthenogenetically, or alternatively, offspring which reproduce sexually

heteroploid [hĕt'·ər·ō·ploid"] any polyploid condition other than a diploid

heteropolar [hĕt'·ər·ō·pōl"·ər] a term applied to a radially symmetrical object, the two ends of which are different

Heteroptera [hĕt·ə·rŏp"·tər·ə] at one time a suborder (=present Hemiptera) of Hemiptera when this order included the Homoptera

heteropterous [angl. hĕt'·ə·rop"·tər·əs, orig. hĕt'·ər·ō·tĕr"·əs] either having different kinds of wings or, as in hemipterous insects, having two textures on the same wing

heteropycnotic chromosome [hĕt'·ər·ō·pĭk·nŏt"·ĭk] one which is shortened and thickened more than the other chomosomes of the set

heterosis [hĕt'·ər·ōs·ĭs] a group of economically desirable characters associated with inter-variety crosses in domestic plants and animals, leading to increased vigor of hybrid populations

heterotaxia [hĕt'·ər·ō·tăks"·ēə] movement of organs to an unusual position or twisted in an unusual direction, as when the anterior end of the chicken embyro twists to the left instead of the right

heterotopic bone [angl. hĕt·ə·rŏt"·ə·pĭk, orig. hĕt'·ər·ō·tŏp"·ĭk] a bone which is not a part of either the axial or the appendicular skeletons

Heterotricha = Holotrichida

heterotrophic [angl. hĕt·ə·rŏt"·rə·fik, orig. hĕt'·ər·ō·trŏf"·ĭk] literally, feeding on mixtures; used specifically for Protozoa that ingest a variety of food materials or for symbiotic forms such as lichens which may themselves become dependent upon a symbiotic relationship with another form

heterozonic [hĕt'·ər·ō·zōn·ĭk] pertaining to organisms that can, but need not, occupy different environments in the course of their life history

heterozooid [hĕt·ər·ō·zō·ĭd] all ectoproct zooids other than autozooids are grouped under this term

heterozygote 1 [hĕt'·ər·ō·zi"·gōt] (see also heterozygote 2) one having different alleles at the same locus and therefore one derived from the union of two genetically dissimilar chromosomes (see also inversion heterozygote)

heterozygote 2 (see also heterozygote 1) the cell which results from the fusion of two morphologically distinguishable gametes

heterozygous [hĕt'·ər·ō·zig"·əs] the condition of an individual having two different alleles at the same locus on a pair of homologous chromosomes

-hexa- comb. form meaning "six"

hexacanth larva [hĕks'·ə·kănþ"] a spherical or ovoid six-hooked larva hatching directly from the egg of many cestode platyhelminthes

Hexactinellida [hĕks'·ăkt·ĭn·ĕl"·ĭd·ə] an order of non-calcareous sponges with triaxial silicious spicules. They are often called "glass sponges". The genus *Euplectella* is the subject of a separate entry

-hibern- comb. form usually thought to mean "winter" (cf. -hiemal-) but actually meaning "winter quarters"

hibernating gland [hī'·bə·nāt"·ĭng] a gland found in hibernating mammals possessing some of the characteristics both of fat glands and hemolymph glands

hibernation [hī'·bə·nā"·shun] the condition of physical inactivity and decreased metabolic activity in which some organisms pass the winter months (see also aestivation, artificial hibernation)

-hidro- comb. form meaning "sweat" and by extension excretion (cf. hydro-)

-hiemal- comb. form meaning "winter"

hierarchy [hī'·ə·rär"·kē] the ranks established within a group of organisms by means of the peck order or some such method; the term caste is more usually confined to those forms, such as the social insects, in which there are morphological differences between the social ranks

high-rain forest a rain forest having more than seventy-two inches of rainfall annually (see also tropical rain forest)

hilum 1 [hī"·ləm] (see also hilum 2) literally, a trifle, but used by both plant and animal anatomists for a reentrant position on an organ (such as a kidney) through which blood vessels or other structures penetrate to the interior

hilum 2 (see also hilum 1) a central point, around which rings, not of necessity either circular or symmetrical, radiate; a hilum is to be observed on a starch grain, a teleost scale, the valves of pelecypods, and on the scar left on a seed by the detachment of the funicle

high moor an area rich in peat but which no longer has bog water available to it

hindbrain = rhombencephalon

hindgut that posterior region of the arthropod alimentary canal which is lined with chitin

hinge cell a compressed cell in the upper face of the leaf of a grass, which therefore renders folding easy

hippocampal commissure [hĭp′·ō·kămp″·əl] the ventral of the two connections between the two cerebral hemispheres in the higher vertebrates

hippocampal cortex that portion of the cortex of the brain which runs back from the olfactory bulb to meet at the posterior pole of the hemisphere

Hippoglossus [hĭp′·ō·glŏs″·əs] a genus of enormous teleost fish called halibut. *H. hippoglossus*, the Atlantic Halibut reaches a weight of 700 lbs but the Pacific Halibut *H. stenolepsis* [stē′·nō·lĕp″·sĭs] rarely exceeds 400 lbs (*see also Paralichthys*)

Hippopotamus [hĭp′·ō·pŏt″·əm·əs] a genus of artiodactyl mammals distinguished, apart from its well known shape, by possessing four toes and lacking a caecum. The African *H. amphibius* [ăm″·fĭb″·ēəs] is the only extant species. The Pigmy Hippopotamus is *Choeropsis liberiensis*

-hippos- *comb. form* meaning "horse"

Hirudinea [hĭr′·ōō·dĭn″·eə] the class of annelid worms containing the leeches. They are distinguished from the Oligochaetae and Polychaetae by the absence of bristles and the presence of suckers. The genus *Hirudo* is the subject of a separate entry

Hirudo [hĭr·ōōd″·ō] a genus of Hirudinea containing the once widely used "medicinal leech" *H. medicinalis* [mĕd′·ĭs′·ĭn·ăl″·ĭs]. This European animal is now widely distributed in the U.S.

His's bundle a mass of nervous impulse-conducting fibers connecting the auricles and ventricles of the heart

His's cells those extraembryonic mesoderm cells which give rise to blood vessels

l-histidine [ĕl·hĭst′·id·ēn″] α-amino-4-imidazole-propionic acid. A common amino acid known to be essential in the nutrition of rats

histioblast [hĭst″·ē·ō·blăst′] a stage intermediate between a hemocytoblast and a histiocyte *or* a small, active cell involved in germination of the gemmule of a sponge [= tissue-producing cell]

histiocyte [hĭst″·ē·ō·sĭt′] a loose, large, wandering cell in areolar connective tissue directly descended from the primitive reticular cell, and distinguished from a fibroblast by its large size and ovoid shape (= macrophage)

-histo- *comb. form* meaning "web" and, like -istem of the same meaning, has come to be extended to cellular structures particularly when seen in section

histogenesis [hĭst″·ō·jen″·əs·ĭs] the development and differentiation of tissues, particularly reorganization after histolysis

histology [hĭs·tŏl″·ō·jē] the study of tissues

histolysis [*angl.* hĭs·tŏl″·əs·ĭs, *orig.* hĭst′·ō·lĭs″·ĭs] literally, the breakdown of cells or cellular structure, but specifically the initial phase of the reorganization of material in, for example, the pupa of an insect or the brown body of an ectoproct

histone [hĭst″·ōn″] any of several proteins complexed, at one time or another, with DNA

holarctic [hŏl·ärk″·tĭk] pertaining to those areas of the planet Earth which are not tropical and therefore comprising both the nearctic and palaearctic regions

holendozoic [hŏl′·ĕnd·ō·zō″·ĭk] said of a parasite which passes its entire life within an animal host. Used almost exclusively of fungi (*see also* endozoic)

holdfast any large adhesive organ or structure, particularly those of some trematode platyhelminthes,

the terminal discs on the rhizoid of an Ectoproct or the thickenings of the wall of those sipunculids which inhabit abandoned gastropod shells. In botany used specifically for the rhizoids of large marine algae, that portion of a thallophyte plant body which attaches the organism to a base. Used, less specifically, for many attaching structures

-holo- *comb. form* meaning whole; the correct transliteration, sometimes still used, is -olo-

holoblastic cleavage [hŏl′·ō·blăst″·ĭk] the type of cleavage shown by isolecithical eggs in which the entire egg segments into separate blastomeres of approximately equal size

holobranch [hŏl′·ō·brank″] a gill bar with a row of filaments on each side (cf. hemibranch)

Holocephali [hŏl·ō·sĕf″·əl·ī] a group of gnathostomatous craniate chordates containing the rabbitfishes and ratfishes. They are distinguished by the presence of a cartilaginous skeleton but differ from the Elasmobranchs in that the upper jaw is fused to the cranium. The genus *Chimaera* is the subject of a separate entry

holochordal centrum [hŏl′·ō·kôrd″·əl] one in which the segment of notochord from which the centrum is derived becomes calcified

holocrine gland [hŏl′·ō·krin″] a gland in which the secretion is produced by the complete breakdown of some of the glandular cells (*see also* merocrine gland)

holoenzyme [hŏl′·ō·ĕn″·zim] a complete functional enzyme, that is the apoenzyme and the coenzyme taken together

hologamy [*angl.* hō·lŏg″·ə·mē, *orig.* hŏl′·ō·găm″·ē] reproduction through the fusion of two entire protozoans

hologonic [*angl.* hō·lŏg″·ən·ĭk, *orig.* hŏl′·ō·gōn″·ĭk] said of sac-like ovaries in which germ cells are proliferated only from the blind proximal end (cf. telogonic)

holometabolous [hŏl′·ō·mĕt·ăb″·əl·əs] said of insects, though applicable to other forms, in which the eggs, larva, pupa, and imago are morphologically distinct each from the other

holonephros [hŏl′·ō·nĕf″·rŏs] a postulated organ ancestral to all vertebrate kidneys

holophyte [hŏl′·ō·fit″] a plant which derives its nourishment solely from inorganic materials, in contrast to a saprophyte or a parasite. It is inherent in this definition that photosynthesis is necessary

holosaprophyte [hŏl′·ō·săp″·rō·fit″] a plant which is solely dependent on decomposing organic matter for its nutrition

holospondylous vertebra [hŏl·ō·spŏnd·əl·əs] a vertebra consisting of a single piece

Holostei [hŏl·ŏst″·ē·ī] a superorder of actinopterygian bony fish best characterized as being intermediate between the Chondrostyei and the Teleostei. The typical genus *Lepisosteus* is the subject of a separate entry

Holothuroidea [hŏl′·ō·thûr·oid″·ēə] a class of Echinodermata commonly called sea-cucumbers. They are distinguished by the elongate shape and the reduction of the skeleton to spicules or plates embedded in the leathery body wall. They are mostly cucumber-shaped and usually have tentacles, tube feet and respiratory trees. The Apoda are worm-like sand burrowing forms. The genera *Cucumaria*, *Leptosynapta* and *Thyone* are the subjects of separate entries

Holotricha [hŏl′·ō·trĭk″·ə] a subclass of ciliate protozoans differing from the Spirotricha in not having the cilia arranged in spiral ranks. The genera *Balantidium*, *Didinium* and *Tetrahymena* are the subjects of separate entries

holotrophic [hŏl·'ō·trŏf"·ĭk] said of a plant the nutrition of which is based on the synthesis of carbohydrates (= holophytic nutrition) or of a predator that preys on a single species

holotype [hŏl'·ō·tīp"] the one individual, recorded and preserved specimen from which a species is named

holozoic [hŏl'·ō·zō"·ĭk] an organism only capable of existing by ingesting solid or particulate matter. Usually used of Protozoa such as *Paramoecium* (*see also* hemizoic)

-homal- *comb. form* meaning "equal" (cf. homo- and homoe-)

Homarus [hō·mär"·əs] a genus of marine decapod crustaceans. *H. americanus* [ə·měr"·ĭk·ăn·əs] is the American lobster and *H. vulgaris* [vŭl·gär·ĭs] the European lobster. "Rock", or "spiny" lobsters are of the genus *Palinurus*

homeosis [*angl.* hŏm·ē"·əs·ĭs, *orig.* hŏm'·ē·ōs"·ĭs] the alteration, by mutation, of one organ or appendage into another or the transfer, also by mutation, of an appendage to a segment on which it does not usually appear

homeostasis [hŏm'·ē·ō·stās"·ĭs] organic stabilization either of parts or of conditions within an organism or of the relations between an organism and its environment. Broadly the maintenance of a stable internal environment in harmony with the external

Hominoidea [hŏm"·ĭn·oid'·eə] a sub-order of primate mammals erected to contain man and the great apes, now usually regarded as the families Pongidae and Hominidae

Hominidae [hō·mĭn"·ĭd·ē] a family now considered to contain the single genus *Homo*. It differs from the Pongidae in the size of the cranium, the less protrusive jaw, the more erect carriage and the relative lack of hair

-homo- *comb. form* meaning "identical" (cf. -homal-, -homoe-). The h is sometimes omitted from all these forms

Homo [hō"·mō"] the genus of primates containing man. There is no doubt that all extant humans are of the single species *H. sapiens* [săp"·ē·ənz] as are many prehistoric men (Neanderthal, Heidelberg, etc.) previously accorded specific rank. The extinct species *H. erectus* [ə·rěk"·təs] contained many races previously accorded generic rank, such as *Pithecanthropus* [pĭth·ə·'·kan"·thrə·pəs], *Australopithecus* [ô·străl"·ō·pĭth·əkəs] and *Sinanthropus* [sĭn·ăn·thrō"·pəs]. *H. habilus* [hăb'·ĭl·əs], dating back nearly two million years, is thought by some to have been a primitive race of *H. erectus*

homocercal [hō'·mō·sûrk"·əl] said of the tail fin of a fish that has equal ventral and dorsal lobes. Both lobes are derived from the hypochordal fold

l-homocysteine [ĕl·hō'·mō·sĭs·tē"·ēn] essentially cysteine with the addition of one HCH group. Apparently formed in tissues by the demethylation of methionine (*see also* l-cysteine)

l-homocystine [el·hō'·mō·sĭst·ēn'] essentially l-cystine (q.v.) with the addition of 2 HCH groups apparently directly derived from homocysteine (*see also* cystine)

homodont [hō'·mō·dŏnt"] said of a form in which all of the teeth are of a similar size and shape

-homoe- *comb. form* meaning "similar to" (cf. -homal-, -homo-)

homoeomerous [hŏm'·ē·ə·měr"·əs] the condition of a lichen in which the algal and fungal components are uniformly mixed and not layered

homoeotic regeneration [hŏm'·ē·ə·ŏt"·ĭk] an abnormal form of regeneration in which a serial homologue replaces the lost structure

homogamete [hō'·mō·găm"·ēt] that produced by the sex with only one sex-producing potential

homogametic sex [hō'·mō·găm·ĕt"·ĭk] that sex which is determined by the XX arrangement of chromosomes

homogonic [hō'·mō·gon"·ĭk] said of the life cycle of a parasite, in which the parasitic form develops from the egg

homograft [hō'·mō·grăft"] one made between two organisms of the same species

homoimerous [hō'·moi·měr"·əs] the condition of a lichen in which the algal and fungal components have about the same volume, but are not of necessity uniformly distributed

homoiogamy [*angl.* hō'·moi·ŏg"·əm·ē, *orig.* hō'·moi·ō·găm"·ē] the fusion of two gametes of the same sex

homoiosmotic [hō'·moi·ŏz·mŏt"·ĭk] the condition of an organism that possesses an osmotic regulating system and the body fluids of which need not, therefore, be in equilibrium with the environment

homoiotherm [hō·moi·ō·bûrm"] an animal that maintains a constant body temperature in spite of variations in the temperature of the environment

homologous [hō·mŏl"·əg·əs] used in biology to describe structures of similar phylogeny but different in appearance (e.g. the flipper of a whale and the wing of a bird) (cf. analogous)

homologous chromosome one which ‚is identical to another with respect to the position of the gene loci

homologous theory alternate generations of plants are not distinct entities but have arisen one from the other (cf. antithetic theory)

homonomous [*angl.* hō·mŏn"·əm·əs, *orig.* hō'·mō·nŏm"·əs] similar or specialized along similar lines as successive identical segments of an annelid (cf. heteronomous)

homonym [hō'·mō·nĭm"] a name rejected on the ground of prior occupancy. Also said of two different names for the same organism, usually of a species variously assigned to either of two genera

homonym law a rule of nomenclature which requires that two different taxa cannot have the same name

Homoptera [hŏm·ŏpt"·ər·ə] the order of insects that contains the cicadas, spittle bugs, tree hoppers, scale insects, and white flies. They are distinguished, except for the scale insects, by the presence of four similar membranous wings typically roofed over the body when at rest

homosynapsis [hō'·mō'sĭn·ăps"·ĭs] that which occurs between homologous chromosomes

homotypical association [hō'·mō·tĭp"·ĭk·əl] an assemblage of organisms of the same species occurring together because they are descendants of the same parent (primary homotypical association) or parents (secondary homotypical association)

homozygote 1 [hō'·mō·zīg"·ōt] (*see also* zygote 2)

the cell which results from the fusion of two morphologically indistinguishable gametes

homozygote 2 (*see also* zygote 1) one having identical alleles at the same locus and therefore derived from the union of gametes of identical genetic composition

homozygous [hō'·mō·zīg"·əs] pertaining to an individual having identical alleles at the same locus on a pair of homologous chromosomes

Hopkin's Law spring, and its accompanying biological phenomena, in the eastern U.S. is four days later for each degree of northward latitude, five degrees of eastward longitude and four hundred feet of elevation

-hor- *comb. form* meaning "hour" or "time of day". In compounds it is frequently confused with -horamo- **-horam-** *comb. form* meaning "what is perceived" or "noticed"

horizon a horizontal level or layer particularly, in biology, of soil (*see* A-, B-, C-, and D-horizons in alphabetic position)

horizontal plane a plane parallel to the longitudinal axis and at right angles to the sagittal axis. Sections through this plane are often called "frontal" sections

-hormo- *comb. form* confused from three Greek roots and variously meaning "necklace", "base" and "excitation"

hormocyst [hôr'·mō·sīst"] a hormogonium in a thick sheath

hormogone [hôr'·mō·gōn"] a filamentous algae which reproduces by breaking off terminal portions

hormogonium [hôr·mō·gōn·ē·əm] the sections of a filamentous cyanophyte Alga that lie between the heterocysts

hormone [hôr"·mōn] a substance which exercises an effect on organs that do not secrete it. The following derivative terms are defined in alphabetic position:
CHROMATOPHOROTROPIC H.
FOLLICLE STIMULATING H.
GROWTH HORMONE
LACTOGENIC HORMONE
LUTEINIZING HORMONE
MELANOCYTE STIMULATING H.
MELANOPHORE HORMONE
MOLTING HORMONE
NEUROHORMONE
OXYTOCIC HORMONE
PARAHORMONE
PARATHORMONE
PHEROMONE
PLANT HORMONE
PUPATION HORMONE
THYROTROPIC HORMONE

hormozonic [hôr'·mō·zōn"·īk] said of an organism in which the larval stages are restricted to the same environment occupied by the adult

Horneophyton [hôrn'·ē·ō·fīt"·ŏn] a genus of extinct plants found in the Devonian period. They are now placed in the Rhyniophytina with the characteristics of that order

horotelic [hôr'·ō·tēl"·īk] evolving at what appears to be a normal rate

Howship's lacuna a pit on the surface of bone in which osteoclasts are clustered

humeral [hyū"·mə·əl] pertaining to the humerus bone but specifically used for a flight feather attached to the humerus of a bird

humerus bone [hyū"·mər·əs] the bone of that part of the forelimb which lies between the shoulder and the elbow

humoral transmission [hyū"·mə·əl] communication between cells by the diffusion of chemicals

humus [hyū"·məs] decomposing organic matter in the soil (*see also* necron)

humus-necron partially decomposed vegetable matter not yet humus, but in a condition in which the origin from leaves or stems, etc. can still be determined

Huxley's layer the middle of the three layers of which the internal hair sheath is composed

hyal bone = hyoid bone

hyaline [hī"·ə·līn] translucent

hyaline cartilage cartilage consisting of chondrocytes and intercellular substances only

-hyalo- *comb. form* meaning crystal or crystalline

hyaloid artery [hī"·ə·loid'] an embryonic artery supplying blood to the developing vitreous body of the eye

hybrid [hī"·brĭd] an offspring resulting from the union of two different forms. The differences may be varietal or racial but inter-specific and inter-generic hybrids are known. The following derivative forms are defined in alphabetic position:
DIHYBRID
RECIPROCAL HYBRID
SESQUIRECIPROCAL H.

hybrid cline one composed of interspecific hybrids

hybrid sterility the inability of an F_1 generation to reproduce

hybrid swarm a large number of interspecific hybrids occurring at the boundaries of two species populations

hybrid vigor = heterosis

hydathode [hĭd'·ə·þōd"] the structure through which water is passed in the process of guttation

hydatid [hī·dăt"·ĭd] any fluid-filled cyst or cyst-like structure

hydatid appendix the rudimentary remains of the oviduct in the male mammals

hydatid cyst the coernurus of the tapeworm *Echinococcus granularis*

hydatid sand small secondary cysticerci found in hydatid cysts

-hydr-, -hydro- *comb. form* meaning "water" or "hydrogen"

Hydra [hī"·drə] a genus of freshwater hydroid coelenterates. There is much confusion as to the taxonomy of this genus but the cosmopolitan green *H. viridissima* [vĭr'·ĭd·ĭs"·ĭm·ə] is generally accepted. *H. oligactis* [ŏl'·ĭg·ăkt"·ĭs] (= *H. fusca* [fŭsk'·ə]) is now placed in the genus *Pelmatohydra*

Hydractinia [hī'·drăk·tĭn"·ēə] a genus of hydroid coelenterates in the form of a colony of individual zooids of three kinds, for feeding, for reproduction and for defense. *H. echinata* [ĕk'·ĭn·ăt"·ə] is sometimes commensal with hermit crabs (*see Pagurus*)

Hydramoeba [hĭd'·rə·mēb"·ə] a monotypic genus of amoebine protozoa containing the only ectoparasitic ameba. It lives on the surface of *Hydra* the cells of which it ingests

hydratase [hī'·drə·tāz"] a group of enzymes that catalyse the addition of water to a double bond—hence the opposite of hydrolases

-hydro- *comb. form* meaning "water"

hydrogenase [hĭ·drŏj″·ĕn·āz′] a little known bacterial enzyme, or group of enzymes, which utilizes molecular hydrogen in the reduction of a variety of materials

hydrolase [hĭ′·drə·lāz″] a group term for those enzymes which catalyze a hydrolytic reaction. Among common trivial names for hydrolases are lipase, esterase, lactonase, phosphatase, deaminase and many others. Most of the enzymes with other than standard names (e.g. pepsin, trypsin, papain, etc) are hydrolases

hydrophyll [hĭ′·drō·fĭl″] the submerged leaf of a hydrophyte

hydrosere [hĭ″·drō·sēr′] an association in a moist environment = hydrarch

hydrosphere [hĭ′·drō·sfēr″] the sum total of all the waters on the planet Earth

hydrostatic [hĭ′·drō·stăt″·ĭk] used by biologists for an assemblage of organisms which is not attracted to a region of higher moisture. The term is also, of course, used in its normal meaning of equilibrium pressure

hydrostatic succession one which does not change with variations of moisture

hydrotheca [hĭ′·drō·thĕk″·ə] that part of the periderm of a hydrozoan coelenterate which surrounds the hydranth

hydrotropic succession [hĭ′·drō·trŏp″·ĭk] one which changes in virtue of an increased humidity or water volume

l-γ-hydroxyproline [el·găm′·ə·hĭ·drŏks′·ē·prō″·lēn] an amino acid, 4-hydroxy-2-pyrrolidine-carboxylic acid. Found, together with choline, in many collagens

Hydrozoa [hĭ′·drō·zō″·ə] a class of the phylum Coelenterata containing the hydroids, millepores, and Portuguese man-of-war. The group is distinguished in that both the hydroid and the medusoid forms are known for almost all. The order Siphonophora and the genera *Goneonemus, Hydra, Hydractinia, Microhydra, Obelia* and *Pelmatohydra* are the subjects of separate entries

-hyema- comb. form meaning "winter"

Hyenia [hĭ·ĕn″·ēə] a genus of fossil plants of the group Sphenophytina. It possessed forked, leaf-like structures. The jointed stem was similar to the extant *Equisetum*, also a sphenophytine plant

-hygro- comb. form meaning "moisture", as distinct from -hydro- meaning "water"

hygroscopic cell a cell which, through its hygroscopic properties, causes a change of shape in a plant structure under conditions of varying atmospheric humidity

-hyl- comb. form meaning "forest" precisely the same combining form is used for "material"

Hyla [hĭ″·lə] a very large genus of Salientia known as tree-frogs. They are distinguished by the adhesive pads on the toes and notorious for the noise they make. *H. crucifer* [krōō′·sĭf·ə] is the common American spring peeper

-hyli- comb. form meaning "forest"

Hylobatinae that subfamily of anthropoid primates which contains the gibbons and the siamang; the family differs from the Ponginae (great apes) on the very doubtful qualification of having enlarged canines

-hymen- comb. form meaning "membrane"

hymen [hĭ″·men] a membrane, usually, without qualification, the first of the membranes partially closing the vagina of some virgin mammals

Hymenolepis [hĭ′·měn·ō·lĕp″·ĭs] a genus of cestodes. *H. nana* [nä′·nä′] is the commonest human tapeworm. It is almost unique among tapeworms in requiring only a single host

Hymenoptera [hĭ′·měn·ŏp″·tə·rə] the order of insects that contains the ants, bees, and wasps. They are distinguished by possessing at some stage of their life history two pairs of membranous wings. Many form social colonies. The suborders Aprocrita and Symphyta are the subjects of separate entries

hyoid [hĭ″·oid] in the form of a "Y" with recurved ends to the fork as in the capital Greek letter upsilon

hyoid arch the branchial arch immediately posterior to the mandibular arch

hyoid bone any bone derived from the hyoid arch (*see also* stylohyal bone, tymphanohyal bone, urohyal bone)

hyoid rays a series of cartilaginous rays, rising from the ceratohyal in some chondrichthian fishes, and serving the same function as the branchiostegal of bony fishes

hyomandibular bone [hĭ′·ō·măn·dib″·yū·lər] one pair of membrane bones derived from the hyoid arch of Actinopterygian fish, which lies immediately behind the orbital complex and which, together with the sympletic and quadrate, forms the attachment for the lower jaw (*see also* mandibular bone)

hyostyly [angl. hĭ·ŏst″·əl·ē, orig. hĭ′·ō·stīl″·ē] the condition of having the lower jaw articulated to the skull through the medium of the hyoid

hypantra [hĭ′·păn″·trə] vertical articulations below the level of the zygapophyses

-hyper- (sometimes, in error, -hyp-) comb. form meaning "upper", "above", "more than"

hyperdiploid [hĭ′·pər·dĭp″·loid] the condition in which the normal complement of chromosomes is augmented by a translocated portion of another chromosome

hyperhaline [hĭ′·pər·hāl″·ĭn] said of water containing more than 40 parts per thousand of dissolved salts

hypermetamorphosis [hĭ′·pər·mĕt′·ə·môrf·ōs″·ĭs] a form of life history, such as that found in the blister beetles, in which there are numerous active larval instars but with very different forms (cf. triungulin larva) or in which a sub-imago is interpolated (cf. dun)

hypermorph [hĭ′·pər·môrf″] a mutant allele which has an overriding effect over the ancestral allele

hyperstomatous [hĭ′·pər·stōm″·ə·təs] said of a leaf with stomata only on the upper surface (*see also* hypostomatous)

hypertonic [hĭ′·pər·tŏn″·ik] having a lower osmotic concentration (*see also* hypotonic)

hypertrophy [angl. hĭ·pûr″·trŏf·ē, orig. hĭ′·pər·trŏf″·ē] overgrowth. The term is not confined to overgrowth from excessive nutrition (*see also* compensatory hypotrophy)

hypha [hĭ″·fə] a fungus thread. Numerous matted hyphae form a mycelium

hypnotoxin [hĭp′·nō·tŏks″·ĭn] the material obtained from an aqueous extract of the nematocyst bearing tentacles of coelenterates (cf. congestin and thalassin)

-hypo- (sometimes, in error, -hyp-) comb. form meaning "lower", "under", "less than" (*cf.* -hyper-)

hypoblast [hĭp′·ō·blăst″] the lower, loose, involuted endodermal cells in the early development of the telolecithal egg. By some used as synonymous with endoderm

hypobranchial groove [hī'·pō·brănk"·ēəl] the groove along the ventral surface of the atrium, particularly in *Branchiostoma* where its ciliated floor and glandular walls constitute the endostyle (*see also* epibranchial groove, peribranchial groove)

hypobranchial ridge a strip dividing the pharyngeal and digestive portions of the alimentary canal in some Hemichordata (*see also* parabranchial ridge)

hypocentrum [hī'·pō·sĕnt"·rəm] the anterior equivalent of the pleurocentrum

hypocercal [hī'·pō·sûrk"·əl] said of a caudal fin in which the tail of the fish is directed downward, and there is only a single dorsal lobe. It is, in effect, the reverse of a heterocercal type

hypochord [hī'·pō·kôrd"] a thin rod immediately below the notochord in some embryos

hypochordal fold the ventral of the two fin folds from which the caudal fin of the fish may be derived

hypocleideal bone [hī'·pō·klīd"·ē·əl] a projecting median process at the symphysis of the clavicle in birds

hypocotyl [hī'·pō·kŏt"·əl] that part of the axis of a plant embryo which lies below the cotyledons

hypodermal cell [hī'·pō·dûrm"·əl] in botany the apical cell of the nucellus from which the embryo sac is derived. In zoology the term simply refers to any cell in the hypodermis

hypodermic impregnation [hī'·pō·dûrm"·ĭk] a term used for the fertilization process in those invertebrates (e.g. many leeches) in which the sperm are injected through the epidermis of the female

hypodermis [hī'·pō·dûrm·"ĭs] the innermost layer of the skin of a higher animal

hypogastric artery [hī'·pō·găst"·rĭk] an artery arising from the dorsal aorta at or near the origin of the iliac artery, and which supplies blood to the posterior viscera

hypoglossum [hī'·pō·glŏs"·əm] a cartilage found in the floor of the mouth of turtles and which is generally thought not to be part of the hyoid apparatus

hypolemmal [hī·pō·lĕm"·əl] literally "under the sheath", specifically used of part of a motor nerve ending which lies immediately under the sarcolemma

hypolimnion [hī'·pō·lĭm"·nē·ən] lower zone of a thermally stratified lake, removed from circulation and usually depleted in oxygen during summer stagnation

hypomeral bones [*angl.* hī·pŏm"·ər·əl, *orig.* hī'·pō·mēr"·əl] one of a ventro-lateral series of bones, usually confined to the tail region, corresponding to the dorso-lateral epimeral bones

hypomere [hī'·pō·mēr"] the lower segment of mesoderm which divides into somatopleure and splanchnopleure in a vertebrate embryo

hypomorph [hī'·pō·môrf] a mutant allele the effect of which is overridden by the normal ancestral allele

hyponastic [hī'·pō·năst·ĭk] said of a flattened plant structure that curves upwards because the ventral surface grows more rapidly

hyponeural [hī'·pō·nûr·əl] said of those animals in which all save one pair of ganglia of the central nervous system is ventral to the alimentary canal

hyponym [hī'·pō·nĭm"] a species name not supported by an actual specimen

hypopharynx [hĭp'·pō·fâr"·ĭnks] a chemo-receptor organ on the upper surface of the insect labium frequently referred to as the "tongue"

hypophyseal cartilage [*angl.* hī·pŏf"·ə·sē·al, *orig.* hī'·pō·fĭs"·ē·əl] one of a pair of cartilages in the embryonic chondrocranium lying on each side of Rathke's pouch

hypophyseal plate a cartilage formed in the embryonic chondrocranium from the fusion of the hypophyseal cartilages

hypophysis [*angl.* hī·pŏf"·əs·ĭs, *orig.* hī'·pō·fĭs"·ĭs] literally "an outgrowth beneath" but almost invariably referring to an endocrine gland lying between the floor of the brain and the roof of the mouth in vertebrates. In botany sometimes used for the initial of primary root and rootcap (*see also* adenohypophysis, neurohypophysis)

hypoploid [hī'·pō·ploid"] having less than the normal number of chromosomes

hypopus [hī·pōp'·əs] the second nymphal stage in the life history of some acarines. Most hypopi have suckers or claspers for grasping other animals for dispersion

hypostasis [*angl.* hĭp·ŏst"·ə·sĭs, *orig.* hī'·pō·stās·ĭs] the condition of having one gene which is masked by another to which it is non-allelic (*see also* epistasis)

hypostomatous [hī'·pō·stŏm"·ət·əs] said of a leaf with stomata only on the under surface (*see also* hyperstomatous)

hypostome 1 [hī'·pō·stōm"] (*see also* hypostome 2, 3) in insects, the lower anterior part of the face above the mouth

hypostome 2 (*see also* hypostome 1, 3) a raised area bearing the mouth in coelenterates

hypostome 3 (*see also* hypostome 1, 2) the piercing organ of ixode acarines (ticks)

hypothalamus [hī'·pō·thăl"·ə·məs] the ventral portion of the thalamencephalon

hypothermia [hī'·pō·thûrm"·ē·ə] the condition of a homoiothermous animal which has been artificially cooled

hypothesis [hī·pŏth"·ə·sĭs] a hypothesis is properly a postulate which may, bolstered by facts, turn into a theory and later a law. In biology hypothesis, theory and law are so confused in the literature that only the last two terms are used in this dictionary. Lists of those laws and theories that are defined will be found under "law" and "theory"

hypotonic [hī'·pō·tŏn"·ĭk] having a greater osmotic concentration (*see also* hypertonic)

hypotrophic [*angl.* hī·pŏt"·rə·fik, *orig.* hī'·pō·trôf"·ĭk] said of an organism, such as a virus, that utilizes the substances of the host cell

hypoxanthine [hī'·pō·zăn"·thēn] a purine (6-hydroxypurine) found in living tissue and usually produced by the hydrolysis of adenine

-hyps- *comb. form* meaning "high" or "lofty"

Hyracoidea [hīr'·ə·koid"·ē·ə] a small order of placental mammals containing the African animals known as dassies. They are distinguished by the possession of one upper and two lower pairs of enlarged incisors, which, unlike those of the rodents, are prismatic in section and have no enamel on the hind surface. The genus *Procavia* is the subject of a separate entry

-hystero- *comb. form* meaning "following"

I

I-band *see* I-disc
I-disc the lighter of the alternate light and dark discs of which striated muscle appears to be composed. They represent the portion of the small actin fibrils which are singly refractive or isotropic (*see also* A-disc, Z-band)
Ichthyopterigia [ĭk′·thē·ŏp′·tĕr·ĭj″·ē·ə] a subclass of reptiles adapted to aquatic life and closely resembling fish in their external form. All are extinct but occurred from the late Carboniferous through the Jurassic
Ichthyornis [ĭk′·thē·ŏrn″·ĭs] a genus of extinct birds found in the Upper Cretaceous. They resembled modern birds in having a keeled sternum ánd well developed wings but it is not known whether teeth were present in the jaws
Ichthyornithiformes [ĭk′·thē·ôr·nith·ē·fôrm″·ēz] an order of extinct birds of which *Ichthyornis* is typical
Ichthyosaurus [ĭk′·thē·ō·sôr″·əs] a genus of extinct eurapsid reptiles from the Jurassic. They were streamlined, fish-like, forms with elongate sharp-toothed jaws and limbs modified as paddles. They were also ovoviviparous
-icosa- *comb. form* meaning "twenty"
id a term used for the heredity unit before gene or allele
-idae [ĭd′·ē] suffix indicating familial rank in animal taxonomy. The botanical equivalent is -aceae
-idio- *comb. form* meaning "peculiar"
idiosome [ĭd′·ē·ō·sōm″] the immediate forerunner of the acroblast in spermatogenesis
ileocolon [ĭl′·ē·ō·kō″·lŏn] the undifferentiated hindgut of arthropods
ileum [ĭl″·ē·əm] that portion of the small intestine which lies between the jejunum and the large intestine
iliac artery [ĭl″·ē·ăk] the most posterior of the major arteries leaving the dorsal aorta and which supplies blood to the hind-limb
iliac process a pair of lateral projections from the cartilaginous pelvic girdle of primitive fish
iliac vein the vein drawing blood from the hindlimb; it continues in the lateral body wall as the lateral abdominal vein
illegitimate name the name of a taxon which does not follow the International Rules of Nomenclature
imaginal [ĭm·āj·ən·əl] pertaining to the imago

imaginal bud = imaginal disc
imaginal disc those histoblasts, or new formations, in the pupa of holometabolous insects from which the parts of the imago are formed; also called imaginal cells and imaginal buds
imagines plural of imago
imago [ĭm·ā′·gō] the terminal, or sexually mature, stage in the life history of an arthropod (*seé also* subimago)
imbibition theory water ascends plants by a chemical process in the cell walls, rather than by transport through vessels
-imgioc- *comb. form* meaning "claw", or "nail"
immersion objective a microscope objective designed to be immersed in oil to permit an increase in the numerical aperture and hence the resolution
immunity the condition of being resistant to an infection. The following derivative terms are defined in alphabetic position:

ACQUIRED IMMUNITY	PASSIVE IMMUNITY
ACTIVE IMMUNITY	RACIAL IMMUNITY
CONGENITAL IMMUNITY	SPECIES IMMUNITY
INNATE IMMUNITY	

imperfect flower one which does not contain both male and female sex organs (*see also* perfect flower)
implantation the attachment of the blastocyst (early embryo) of a mammal to the uterine wall
inactive phloem that part of the phloem in which the sieve elements no longer function
-inae [ĭn′·ē] suffix indicating subfamilial rank in animal taxonomy or subtribal rank in botanical taxonomy. The names of plant subfamilies terminate in -oideae
incipient species a variation which, if stabilized, would become a sub-species
incisor tooth [ĭn·sĭz″·ər] literally a "cutting tooth" but specifically applied to the anterior teeth in a mammalian jaw
incomplete dominance the incomplete masking of a recessive character by a dominant character in the F_1 generation
incomplete metamorphosis an insect life history having instars fewer than, or different from, those found in complete metamorphosis
incomplete penetrance = partial penetrance

incus [ĭnk″·əs]   literally, an anvil.   Particularly the combination of ramus and fulcrum in the mastax of a rotiferan but usually the incus bone

incus bone   the center of the three bones which conduct vibration from the tympanum to the inner ear (cf. malleus, stapes)

independant genes   non-allelic genes on different chromosomes

independent assortment   the condition when, in meiosis, genes segregate independently of one another

indeterminate cleavage   holoblastic cleavage in which the cleavage pattern bears no definite relation to the embryo

index species   one which is typical of, and unique to, a particular habitat or a geological stratum

indigoid biochrome [ĭn·dĭg″·oid]   a group of blues and purples derived as the end product of the metabolism of tryptophane. They are best known in the plant kingdom but also account for the mollusk-derived Tyrean purple of the ancients (see also quinone biochrome)

individualism   in biology applied to a type of symbiosis, in which the aggregate differs from any of its components; a lichen is a case in point

induction   the production of a part under the influence of another, as the induction of the embryonic lens by the embryonic eyecup

-indus-   comb. form meaning "clothing" or "covering"

indusium [ĭn·dyū″·zē·əm]   literally, a petticoat, but used for any structure of much that shape, such as the epidermal outgrowth round the sorus of a fern, or the ring of collecting hairs below a stigma. It is also used for the annulus of some basidiomycetes, the case of a caddis fly larva and the inner of the membranes which surround an arthropod larva

-ineae [ĭn·ē·ē]   suffix indicating subordinal rank in plant taxonomy

inenchyma [ĭn·ĕn′·kə·mə]   tissue, such as the spiral cells in sphagnum, which have the appearance of being conducting vessels, but which are not

infauna [ĭn′·fôrn·ə]   those bottom-dwelling forms that burrow

inferior ovary   one which appears to be below the calyx

inflorescence [ĭn′·flôr·ĕs″·ĕns]   the grouping of flowers set apart from the foliage into cymes, racemes, umbels, etc.

influent   a motile organism which moves into a stable community which it disturbs by the destruction of organisms or parts of organisms (see also sub-influent)

infundibulum [ĭn′·fŭn·dĭb″·yōō·ləm]   literally, a "funnel" and used in biology for any organ of this shape including the calyx of a kidney, the ostium of an oviduct, the pouch which grows down from the diencephalon, and joining with Rathke's pouch forms the pituitary gland, and the outer chamber of the coronal funnel in Rotifera

ingression [ĭn·grĕsh″·ĭn]   in biology the movement of cells from an outer layer towards the interior (see also multipolar ingression, unipolar ingression)

inheritance   the sum total of genetically transmitted characters

inhibitor gene   one which masks or inherits the effect of a gene at another locus

initial   a word used in botany to designate a cell or group of cells destined to produce a specific type of cell, tissue or organ. The following derivative terms are defined in alphabetical position:

APICAL INITIAL	RAY INITIAL
FUSIFORM INITIAL	SUB-MARGINAL INITIAL
MARGINAL INITIAL	SUBAPICAL INITIAL

initial meiosis   meiosis immediately following fertilization

initiating cell   a cell in the meristem which remains unchanged while budding off derivative cells

ink sac   a gland found in some cephalopod mullusks that produces black ink-like fluid used to confuse predators

innate immunity   that which results from the presence of antibodies not acquired by exposure to the infectious agent

inner plexiform layer   that layer of the retina in which the axons of the bipolar cells synapse with the dendrites of the ganglion cells (see also outer plexiform layer)

innominate artery = brachiocephalic artery

innominate bone [ĭn·nŏm′·ĭn·āt]   that part of the pelvic girdle which is formed by the fusion of the illium, ischium, and pubis

inosinase [ĭn·ōs′·ĭn·āz″]   an enzyme catalyzing the hydrolysis of inosine into hypoxanthine and ribose

inositol [ĭn·ōs′·ĭt·ŏl]   hexahydroxycyclohexane. A water soluble nutrient frequently classed as a vitamin and acting as a growth factor in some animals and microorganisms. No human need has been demonstrated

inquiline [ĭn″·kwĭl·ĭn′, ĭn′·kwə·lĭn″]   one who dwells in the house of another; specifically one organism which shares the quarters of another or lives on another without damage or help to either, and particularly an insect developing in a gall produced by another species (see also symphile)

Insecta [ĭn·sĕkt″·ə]   a class of arthropoda very variously defined. If classed as possessing antennae and six legs it must include the Collembola and Protura which many omit. To exclude these groups it is necessary to add to the definition of Insecta that they possess fourteen post cephalic segments which develop without adding body segments in postembryonic stages. The Insecta is the largest class, both in numbers and kinds, in the animal kingdom. The orders Coleoptera, Corrodentia, Collembola, Dermaptera, Diptera, Hemiptera, Homoptera, Hymenoptera, Isoptera, Lepidoptera, Mallophaga, Mecoptera, Neuroptera, Odonata, Orthoptera, Plecoptera, Protura, Psocoidea, Siphonaptera, Strepsiptera, Thasmida, Thysanoptera and Trichoptera are the subjects of separate entries

Insectivora [ĭn′·sĕk·tiv″·ōr·ə]   a heterogenous order of mammals containing the solenodonts, the tenrecs, the moles, the hedgehogs, and the shrews. They may be defined as plantigrade placental mammals with primitive sharp-cusped teeth, a long snout, and a primitive skull. Most are omnivorous, rather than insect-eaters. The genera Condylura, Desmana, and Scalopus, are the subjects of separate entries

insertional translocation   that which involves the transfer of materials to a body, as distinct from the end, of a nonhomologous chromosome

instar [ĭn·stär″]   any of the several stages through which a metamorphosing arthropod passes before reaching the adult or imago

insulin [ĭn·syū″·lĭn] a hormone secreted by the β-cells of the islets of Langerhans in the pancreas. It is active in the control of glucose metabolism (cf. glucagon)

integument the outer skin of an organism or organ. In botany specifically that portion of the ovule which surrounds the nucellus

intercalary [ĭn′·tĕr·kāl″·ər·ē] literally "inserted" but referring usually to growth which takes place neither at the base nor at the tip of an elongate structure

intercalary appendages any appendage, usually rudimentary, interposed between those of the regular series

intercalary cell a cell which, in the formation of an ascidiospore mother-cell, is equivalent to a polar cell

intercalary segment one which is inserted between two regular segments, though the term is also applied to the premandibular segment of the head of insects

intercalated disc any disc interposed between two other objects, particularly vertebrae

intercaloid bone [in′·tər·kāl″·oid, in·tûr″·kə·loid] one of a pair of dermal bones in the region of the opisthotic bone, which serves for the ligament of attachment of the pectoral girdle in Actinopterygian fish

intercellular bridges supposed connections between the cytoplasm of neighboring cells particularly in epithelia. Those investigated under the electron microscope have proved to be opposed prominences not actual connections

intercellular substance the material between cells (often called ground substance)

interchange translocation between non-homologous chromosomes (see also segmental interchange)

interciliary fibril [ĭn′·tər·sĭl″·yûr·ē] the conduction fibrils connecting the basal bodies of cilia in Protozoa

interclavicle bone [ĭn′·tər·klăv″·ĭk·əl] a bone jointed to and between the two clavicles in some reptiles and birds (see also clavicle bone)

interferon [ĭn′·tər·fēr″·ŏn] a postulated substance apparently elaborated by proteins under the influence of a virus, and which is antagonistic to another virus

intermediate association one in which a little room for growth remains

intermitosis [ĭn′·tər·mĭt·ōs″·ĭs] the period between two cell divisions

internasal bone [ĭn′·tər·nāz″·əl] one of one or more membrane bones in the dermocranium of crossopterygian fish, lying between and partly anterior to the nasals (see also nasal bone)

internuncial neuron [ĭn′·tə·nŭn″·sē·əl] one which connects sensory and motor pathways

interoceptive system [ĭn′·tər·ō·sĕpt″·ĭv] the sum total of those receptors which perceive internal stimuli (see also exteroceptive system, proprioceptive system)

interopercular bone [ĭn′·tə·ō·pûrk″·yū·lə] one of a pair of chondral bones derived from the hyoid arch in Actinopterygian fish, lying between the preopercular and the subopercular (see also opercular bone, preopercular bone, subopercular bone)

interorbital cartilage [ĭn′·tər·ôrb″·ət·əl] a cartilage rising vertically from the fused trabecula in the embryonic chondrocranium (see also orbital cartilage)

interoreceptor [ĭn′·tər·ō·rē·sept″·ər] one which responds to internal stimuli

interpapillary peg [ĭn′·tər·păp′·il″·ər·ē] a mass of cells which support the neck of a sweat gland as it passes through the papillary layer

interparietal bone = postparietal bone

interphase = intermitosis

interphase nucleus [ĭn′·tər·fāz″] a nucleus not actively dividing (= "resting" nucleus)

interradius [ĭn′·tər·rād″·ē·əs] a radius of a radially symmetrical organism which lies between the perradii

interrenal body [ĭn·tər·rēn·əl] cells, or a discrete body, found between the kidneys of lower vertebrates and corresponding to the medulla of the adrenal gland of higher forms

intersex [ĭn′·tər·seks″] an organism, not normally hermaphroditic, which shows characters intermediate between those of the two sexes

interstitial-cell-stimulating hormone = luteinizing hormone

interstitial fauna [ĭn′·tər·stĭsh″·əl] animals living within crevices, particularly those between sand grains

interstitial growth growth within the substance of a tissue

interstrial connectives [ĭn′·tər·strē″·əl] conducting fibrils running parallel to and between the interciliary fibrils in protozoans

intertemporal bone [ĭn′·tər·tĕm″·pər·əl] one of a pair of membrane bones in the dermocranium of fish, lying on each side of the posterior of the parietal, and immediately anterior to the supratemporal (see also temporal bone, post-temporal bone, supratemporal bone)

intervertebral body [ĭn·tə·vûr″·tə·brəl] a vertebra lacking spinous processes, lying between the vertebrae, in diplospondylous vertebral columns

intestine [ĭn·tĕst″·ĭn] a term loosely applied to any portion of the alimentary canal which lies posterior to the stomach. The following derivative terms are defined in alphabetic position:

LARGE INTESTINE   TERMINAL INTESTINE
SMALL INTESTINE

-intra- comb. form meaning "within"

intralemmal [ĭn′·trə·lĕm″·əl] literally "that which lies inside the sheath", specifically one of the extreme tips of a motor nerve ending which lies within the sarcolemma

intromittent organ any male organ, in many forms known as a penis, adapted for insertion into the female for the transfer of sperm (see aedeagus, penis, telopod)

intussusception [ĭn′·təs·sŭs″·sĕp″·shŭn] the increase in thickness of a plant cell wall, or other similar structure, through the deposition of materials in its interior (cf. apposition)

inulase [ĭn·yōōl′·āz] an enzyme catalyzing the break down of inulin through the hydrolysis of β-1,2-fructan links

invagination [ĭn·văj′·ĭn·āsh″·ən] the pushing of a layer of cells into a cavity as in the formation of a gastrula from a simple blastula (cf. involution)

inverse ocellus one in which the free ends of the retinal cells face the light

inversion the condition of a portion of a chromosome in which the gene loci run in the reverse direction to their normal position. The following derivative terms are defined in alphabetic position:

ACENTRIC INVERSION   PARACENTRIC INVERSION
OVERLAPPING I.   PERICENTRIC INVERSION

inversion heterozygote   a heterozygote in which the homologous chromosomes have a different linear order as the result of inversions

invertebrates [ĭn·vêrt"·ə·brāts] a loose term, not an acceptable taxon, embracing all animals except the chordates

involuntary muscle = smooth muscle

involution   the ingrowth of one or more layers of cells as in the later stages of gastrula production in the frog

iodinase [ĭ·ō·dĭn'·āz"] an enzyme that catalyzes the production of iodine and water from iodide and hydrogen peroxide. This is part of a complex system once thought to be a single enzyme ("tyrosine iodinase") which produces monoiodotyrosine from tyrosine

iodopsin [ĭ·ō·dŏp"·sĭn] an analog of rhodopsin (q.v.) found in the cones of the retina

Iridaceae [ĭr'·ĭd'·ās"·ē] that family of liliflorous monocotyledons which contains not only *Iris* but also the gladiolus, the crocus and the shell flowers. Distinguishing characteristics are the three extrose stamens, the inferior 3-celled ovary and the typical, showy, iris type of flower

iris [ir"·ĭs] an anterior projection of the ciliary process, lying immediately in front of the lens of the eye and which may be opened or closed

irritability   in biology the ability of organisms or parts of organisms to respond to stimuli

ischiadic artery = sciatic artery

ischial process [ĭs"·kē·əl] an unpaired, posterior projection from the cartilaginous pectoral girdle of primitive fishes

ischiopodite 1 [*angl.* ĭs·kē·ŏp"·ə·dĭt, *orig.* ĭs'·kē·ō·pŏd"·ĭt] (*see also* ischiopodite 2) in insects, variously the second segment of the telopodite, the second trochanter or the prefemur

ischiopodite 2 (*see also* ischiopodite 1) the joint of a crustacean appendage that lies between the meriopodite and the basipodite

Ischnochiton [ĭsh'·nō·kĭt"·ən] a genus of long, narrow, polyplacophoran mollusks with large valves and girdle covered with overlapping scales. Common on the California coast where the three inch *I. magdalenensis* [măg'·dăl·ēn·ĕn"·sĭs] is one of the commonest chitons

islets of Langerhans   groups of cells in the pancreas which are responsible for the secretion of insulin

-iso-   *comb. form* meaning "equal"

isoallele [ĭs'·ō·əl'·ēl"] one allele which so closely resembles another that the two can only be distinguished by special techniques

isocercal [ĭs'·ō·sŭrk"·əl] said of the caudal fin of a fish which lacks any lobes, and therefore appears as a smooth prolongation of the tail end of the body

isochromosome [ĭs'·ō·krôm'·ō·sōm"] a metacentric chromosome with two identical arms

isocies [ĭ·sō'·shēz] a group of organisms associated together but of different taxonomic affinities sometimes used merely in the sense of habitat groups (*see also* socies, associes, consocies, subsocies)

isoenzyme [ĭs"·ō·ĕn'·zim'] one of two or more molecular forms of the same enzyme

Isoetes [ĭs'·ō·ĕt"·ēz] a genus of lycopid plants commonly called quillworts. They are mostly aquatic and have quill-like fronds, either microsporophylls or megasporophylls, arising from a corm-like expanded base. Fossil forms are known from the late Palaeozoic on

isogamete [*angl.* ĭs·ōg"·ə·mēt, *orig.* ĭs'·ō·gām"·ēt] one of a group of gametes which cannot be distinguished from each other

isogamy [*angl.* ĭs·ŏg"·ə·mē, *orig.* ĭs'·ō·găm"·ē] the fusion of apparently identical gametes (*see also* exisogamy)

isoleucine [ĭs'·ō·lyū·sēn] α-*amino-*β-*methylvaleric acid* $CH_3CH_2\text{-}CH(CH_3)CH(NH_2)COOH$. An amino acid, necessary to the nutrition of rats and an isomer of l-leucine

isoholotype [ĭs'·ō·hō'·lō·tĭp"] a plant specimen taken at a later period from the same bush or tree from which the holotype was originally taken

isokont [ĭs'·ō·kont"] having two flagella of equal length (*see also* anisokont)

isolation   the segregation of organisms. The following derivative terms are defined in alphabetic position:

ECOBIOTIC I.   ECOLOGICAL ISOLATION
ECOCLIMATIC I.   GENETIC ISOLATION
ECOGEOGRAPHIC I.   REPRODUCTIVE ISOLATION

isomastigote [ĭs'·ō·măst'·ĭg·ōt"] possessing several flagella of equal length (*see also* isokont)

isomerase [ĭs·ŏm"·ə·āz'] a group of enzymes that synthesize reactions leading to the internal reorganization of molecules. Among trivial names in this group are isomerase, mutase and racemase

isoplanogamete [ĭs'·ō·plăn'·ō·găm"·ēt] a motile gamete the sex of which cannot be distinguished, particularly in fungi

Isopoda [ĭs·ŏp"·əd·ə, ĭs·ō·pōd"·ə] an order of malacostracan Crustacea. The order is distinguished by the fact that there is no distinct carapace but the first thoracic somite is coalesced with the head. The order includes many more or less aberrant parasitic forms, mostly marine, and numerous free-living marine and freshwater forms and the terrestrial "woodlice" or "sow bugs". The genera *Armadillum* and *Asellus* are the subject of separate entries

Isoptera [ĭs·op"·tər·ə] the order of insects which contains the termites. The wingless forms are easily distinguished by their white larva-like bodies. Winged forms have equal-sized wings with indistinct veins and a prothorax smaller than the head. They are wood borers, the injested particles being digested by commensal mastigophoran Protozoa

isoschist [ĭs'·ō·shĭst"] one of a brood of identical cells developed from one parent cell

isospore [ĭs'·ō·spôr"] a spore produced in only one kind, in contrast to anisospore

isotonic [ĭs'·ō·tŏn"·ik] having the same osmotic concentration

isotropic egg [ĭs'·ō·trŏp"·ik] one which lacks a predetermined axis

isozygous [ĭs'·ō·zĭg"·əs] having identical genetic composition in two zygotes

-istem-   *comb. form* meaning a "web". The root derives from -isto-, spelt by zoologists -histo-. Both usages refer to the web-like appearance of cells in crude sections studied with primitive microscopes (*see* meristem)

isthmus [ĭs″·məs] a sharp constriction in the brain separating the mesencephalon from the rhombencephalon

-ite *comb. term.* abbreviated from -podite (e.g. "exite" is sometimes used in place of "exopodite")

iter [ĭ″·tûr] the connection between the third and fourth ventricles in higher vertebrates (cf. mesocoele)

-iul- *see* -jul-

Ixodidea a division of acarine arachnida containing the ticks. They are distinguished by having the hypostome modified as a piercing organ and provided with recurved teeth. Many are vectors of blood parasites. The genera *Boophilus* and *Dermacentor* are the subject of separate entries

J

Jacobson's organ an accessory, olfactory organ, in many vertebrates, supplied by the first and fifth nerves, usually situated toward the rear of the mouth

Jamin's chain alternate bubbles of air and drops of water running through the length of the stem of a water plant

Java man the first race of *Homo erectus* (*see* Homo) to be discovered and originally placed in the separate genus *Pithecanthropus*

jejunum [jĕ·jōō·nəm] that part of the small intestine which lies between the duodenum and the ileum

jetsam [jĕt'·səm] considered by some ecologists to be that area of the beach on which floating material is washed up and moreover to be synonymous with flotsam ; the correct meaning of these words is given in standard English dictionaries

joint a movable, but inseparable, junction between two parts of anything (*see also* arthrosis, which is synonymous). The term has also become confused with segment. The following derivative terms are defined in alphabetic position:
FIBROCARTILAGENOUS J.   SYNOVIAL JOINT
FIBROUS JOINTS

Jordan's law 1 the nearest relatives of the species are found immediately adjacent to it, but isolated from it by a barrier

Jordan's law 2 fish of a given species develop more vertebrae in a cold environment than in a warm one

jugal bone [jōō"·gəl] one of a pair of membrane bones of the skull which, together with the zygomatic arch, encloses, or partly encloses, the orbit. In many forms, the anterior end is fused to the nasal bone (*see also* quadratojugal bone)

jugular vein [jōō'·gyū·lə, jŭg'·yōō·lə] one of a pair of major veins draining blood from the head region to the heart. In fishes, there is an inferior jugular vein serving the lower jaws and the lower side of the gill arches

jugum [jōō"·gəm] literally a yoke and used in biology for many structures having this function

-jul- *comb. form* meaning catkin. Properly, but practically never, spelled -iul-

jungle an association of trees interspersed with brush, particularly in India

Jurassic period [jōōr·ăs"·ĭk] the middle of the Mesozoic era extending from about 150 million years ago to 125 million years ago. It was preceded by the Triassic and followed by the Cretaceous. It was the age of the great reptiles and saw the beginnings of flowering plants and hardwood forests

juvenile name names, such as puppy or kitten, that in themselves are indicative of youth

juxtaglomerular cells [jŭks'·tə·glŏm"·ər·yū·lə] a cuff of cells round the neck of the arterioles entering the glomerulus of a kidney

K

-kaino- *comb. form* meaning "recent", usually transliterated -caeno-

-kako- *comb. form* meaning "bad" usually transliterated -caco-

Kallima [kə·lim"·ə] a genus of butterflies that, with the wings folded, strongly resemble leaves. Often used as an example of protective mimicry

-kalyb- *comb. form* meaning "cottage" usually transliterated -calyb-

-kalyptr- *comb. form* meaning "a veil" usually transliterated -calyptr-

-kampt- *comb. form* meaning "bend" frequently transliterated -campt-

-kapnod- *comb. form* meaning "smoke" usually transliterated -capnod-

kappa particle [kăp"·ə] a particle, self-perpetuating in Paramecium, containing a K gene and capable if this gene is present of developing a substance poisonous to other Paramecium

-kaps- *comb. form* meaning "box" frequently transliterated -caps-

-karpho- *comb. form* meaning a "splinter" or "twig"

-karp- *comb. form* meaning "fruit". Frequently transliterated -carp- and thus confused with "wrist"

-karyo- *comb. form* meaning "nut" or by extension "nucleus" sometimes transliterated -caryo-

karyokinesis [kâr'·ē·ō·kin·ēs"·is] an obsolete term still occasionally used to denote mitosis, and later the role played by the chromosomes in mitosis

karyon [kâr'·ē·ən] a term for nucleus now rarely used except in compound words. The following terms using this suffix are defined in alphabetic position:

HETEROKARYON   SYNKARYON

karyotic [kâr·ē·ŏt"·ik] pertaining to nuclei. The following terms using this suffix are defined in alphabetic position:

AKARYOTIC   POLYKARYOTIC
EUKARYOTIC   PROKARYOTIC

karyotype [kâr"·ē·ō·tip'] two individuals are said to be karyotypic when their chromosomal constitution appears to be identical

-kata- *comb. form* meaning "down" usually transliterated -cata-

-kaus- *comb. form* meaning "burn" usually transliterated -caus-

-kentro- *comb. form* meaning "a sharp point" frequently transliterated -centro- and thus confused with -centr- meaning "middle"

-kera- *comb. form* meaning "horn" frequently transliterated -cera-

Kerckring's valves circular folds on the inner surface of the large intestine of mammals

-kerkid- *comb. form* meaning "a small comb" frequently transliterated -cercid-

key in biology a table listing the characteristics of a group of organisms arranged in such a manner as to facilitate the identification of its members

kinase [kin'·āz"] a large group of kinases, more properly called ATP:phosphotransferases, catalyse phosphorylation with energy derived from the reduction of ATP to ADP. These are all named with simple compound words descriptive of their function (e.g. xylulokinase yields xylulose 5-phosphate, protein kinase yields a phosphoprotein, etc)

-kine- *comb. form* meaning "move" or "movement", frequently transliterated -cine-

kinesis 1 [kin'·ēs"·is] (*see also* kinesis 2) movements of cells, or parts of cells. The following terms using this suffix are defined in alphabetic position:

BLASTOKINESIS   DIAKINESIS   KARYOKINESIS

kinesis 2 (*see also* kinesis 1) movement between parts of a skull

kinetodesma [kin'·ĕt·ō·dez"·mə] one of the numerous fine strands forming a network connecting the kinetosomes of ciliate protozoa

kinetoplast [kin'·ĕt·ō·plăst"] the blepharoplast and parabasal body combined. The term is also applied to a body associated with, and larger than, the blepharoplast of hemoflagellate protozoans

kinetosome [kin'·ĕt·ō·sōm"] the basal body of a cilium or flagellum

-kino- *comb. form* meaning "movement", frequently transliterated -cino- (cf. -kine-)

Kinorhyncha = Echinodera

-klado- *comb. form* meaning "branch" almost invariably transliterated -clado-

-klas- *comb. form* meaning "fracture" usually transliterated -clas-

Klebsiella [klĕbs"·ē·ĕl·ə] a genus of eubacteriale Schizomycetes most species of which live symbiotically in the mouth and respiratory tract. *K. pneumoniae*

[nyū·mōn"·i·ē] ("Friedlander's bacillus") is sometimes the cause of lobar pneumonia

-kleid- *comb. form* meaning "key", frequently transliterated -cleid-. This form is used both in the sense of shape (e.g. hypocleideal bone) and function (e.g. cleidoic egg)

Kleinfelter's syndrome an XXY (sex chromosome) male with substandard secondary sex development

-klept- *comb. form* meaning "thief"

-klesi- *comb. form* meaning "closing" frequently transliterated -clesi-

-klima- *comb. form* meaning "climate"

-klimac- *comb. form* meaning "ladder". Frequently transliterated -clima-

-klin- *comb. form* hopelessly confused between two roots meaning "bed" and "bend". Frequently transliterated -clin-

-klino- *comb. form* meaning a bed, almost invariably transliterated -cline- or -klin-

klinotaxis [klin'·ō·taks"·is] the response of an organism that moves its head from side to side symmetrically in moving towards the stimulus (cf. tropotaxis)

-klito- *comb. form* meaning a "hillside" or "slope" frequently transliterated -clito-

-kloster- *comb. form* meaning spindle almost invariably transliterated -closter-

knephopelagile [něf'·ō·pěl·ăg"·il] that zone of the ocean which extends from about five fathoms to the depth, in that particular area, at which photosynthesis is no longer possible

knephoplankton that which occurs between 15 and 250 fathoms

-kodo- *comb. form* meaning "fleece" frequently transliterated -codio-

-koelo- *comb. form* meaning "hollow" usually transliterated -coel-

-koino- *comb. form* meaning "sharing" or "togetherness", usually transliterated -coen- and therefore confused with a Latin derivative properly so spelled

-koito- *comb. form* meaning "a bed chamber" usually transliterated -coeto-

-kokko- *comb. form* meaning "berry" usually transliterated -cocco-

-kole- *comb. form* meaning "sheath" frequently transliterated -cole-

-koll- *comb. form* meaning "glue" frequently transliterated -collo-

Kollicker's pit a ciliated invagination on the dorsal surface of *Branchiostoma* just in front of the forward end of the neural tube

-kolp- *comb. form* meaning "bosom" frequently transliterated -colp-

-komo- *comb. form* meaning "hair" frequently transliterated -como-

-kon- *comb. form* meaning "cone" frequently transliterated -con-

-kondyl- *comb. form* meaning "a knuckle", but extended to any knobbly joint. Frequently transliterated -condyl-

-koni- *comb. form* meaning "dust", frequently used in the sense of "spore" and also transliterated -conid-

-konop- *comb. form* meaning "gnat", frequently transliterated -conop-

-kont- *comb. form* meaning "a rod" or "pole", sometimes, but rarely, transliterated -cont-. Also used for "flagellum". The following terms using this suffix are defined in alphabetic position:

ISOKONT   MONOKONT   POLYKONT

-kopr- *comb. form* meaning "dung" frequently transliterated -copr- (cf. -fim-)

-korac- *comb. form* meaning "raven" frequently transliterated -corac-

-korem- *comb. form* meaning "a broom" frequently transliterated -corem-

-korm- *comb. form* meaning "a tree trunk" frequently transliterated -corm-

-kotyl- *comb. form* meaning "tube", almost invariably transliterated -cotyl-

-krater- *comb. form* meaning "cup" usually transliterated -crater-

Krause's corpuscle an end organ, sensing cold, consisting of a spherical mass of connective tissue with a much branched central nerve ending

Krebs cycle *see* tricarboxylic acid cycle

-kremno- *comb. form* meaning "cliff", almost invariably transliterated -cremno-

-kren- *comb. form* meaning "notch" but see -cren-

-kreo- *comb. form* meaning "meat" or "flesh", usually transliterated -creo-

-krepi- *comb. form* meaning "shoe" often transliterated -crepi-

-krin- *comb. form* meaning "lily" or "separate" frequently transliterated -crin- and thus becoming confused with hair

-krit- *comb. form* meaning "chosen" frequently transliterated -crit-

Krogh's law the rate of a biological process is directly correlated to the temperature

-krypt- *comb. form* meaning "hidden" and thus, by extension, "cavity" or vault. Usually transliterated -crypt-

Kuhne's granule one located between the meshes of the nerve net which terminates a motor nerve ending

-kuma- *comb. form* meaning "a wave" frequently transliterated -cuma-

Kupffer's canal an outgrowth from the mesonephric duct to metanephric units

Kupffer's cells reticulo-endothelial cells of the liver

-kyan- *comb. form* meaning "cornflower" and giving directly -cyan- meaning "blue"

-kykl- *comb. form* meaning "circle", almost invariably transliterated -cycl-

-kym- *comb. form* meaning a "wave" frequently transliterated -cym-

-kymbo- *comb. form* meaning "cup", frequently transliterated -cymbo-

-kyo- *comb. form* meaning "contain" frequently transliterated -cyo-

-kyon- *comb. form* meaning "dog" frequently transliterated -cyn- or -cyon-

-kyst- *comb. form* meaning a "bladder" almost invariably transliterated -cyst-

L

labellum [lə·bel″·əm]   a term applied both to an extension of the labrum in insects, and also, in honeybees, to the spoonlike tip of the glossa

labia [lā″·bē·ə]   plural of labium but also used for that part of the exoskeleton of an insect which immediately surrounds the spiracle

labia majora   the outer lips of the mammalian vulva. Part of the female genitalia

labia minora   the inner lips of the vulva

labium 1   (*see also* labium 2) in botany the lip of a labiate flower

labium 2 [lā″·bē·əm]   a compound structure forming the "lower lip", or floor of the mouth, in insects

labrum [lā″·brəm]   literally upper lip, and specifically that portion of the insect head which covers the base of the mandible and forms the roof of the mouth. The term is also applied to the thickened edge round the mouth of a gastropod shell and to the lip-like projection of the posterior interambulacrum over the peristome in some echinoderms

labyrinthine sense [lăb′·ər·ĭn·thīn″]   that sense which derives from the activities of the inner ear

Labyrinthodonta [lăb′·ər·ĭn·thō·dŏnt″·ə]   a subclass of extinct Amphibia, deriving their name from the labyrinthine infolding of the enamel of the teeth. They developed in the Carboniferous, as the first land animals, and extended through the Permian. They had an enlarged flattened skull and heavily built pelvic and pectoral girdles. The genus *Eryops* is the subject of a separate entry

Lachesis [lăk·ēs′·ĭs]   a monotypic genus of pit vipers. *L. muta* [myū′·tə] is the notorious bushmaster of Central America, probably the most dangerous of all venomous snakes

-lacin-   *comb. form* meaning "torn"

lacinia [lăs·ĭn″·ēə]   literally a "lappet", but typically applied to the inner of the two lobes which terminate an insect maxilla (cf. galea)

-lacrim-   *comb. form* meaning "tear" or "tear drop"

lacrimal bone [lăk″·rĭm·əl]   one of a pair of membrane bones in the skull which lies in the antero-ventral part of the orbit, immediately anterior to the ethmoid bone

lacrimal gland   a lubricating (tear-producing) gland on the outer side of the vertebrate orbit

lactation   the production of milk

lacteal ducts [lăk″·tē·əl]   lymph ducts carrying chyle

lactic   pertaining to milk

lacticifer [lăk·tĭs′·ĭf·ər]   a cell that produces milk (cf. laticifer)

lactiferous   milk carrying

Lactobacillaceae [lăk′·tō·băs′·ĭl·ās″·ē]   a family of gram-positive eubacteriale schizomycetes having the form of rods which may divide into chains and even become filamentous. Usually divided into two tribes, the Streptococceae almost all of which are dangerous pathogens and the Lactobacilleae a few of which are pathogenic but many of which occur in the production of cheese and other milk products. The genera *Diplococcus* and *Streptococcus* are the subjects of separate entries

lactogenesis [lăk′·tō·jĕn″·ə·sĭs]   the initiation of the secretion of milk in the mammalian mammary gland

lactogenic hormone [lăk′·tō·jĕn″·ĭk]   one produced in the adenohypophysis that stimulates the production of milk

lactonase [lăk″·tən·āz′]   general name for a group of enzymes that catalyze the hydrolysis of lactones

lacuna [lăk′·yū·nə]   a hole, particularly, in botany, an air-space in tissue. In zoology lacuna is frequently used for invertebrate structures which, if vertebrate, would be called sinus (*see* Howship's lacuna)

lacustrine [lə·kŭs′·trĭn]   pertaining to lakes

-lagen-   *comb. form* meaning "flask"

lagena [lə·gēn·ə]   a side pouch from the sacculus which gives rise to the lower part of the cochlea

Lagomorpha [lā′·gō·môrf″·ə]   the order of placental mammals which contains the rabbits. They are distinguished from the rodents, with which they were once fused, by the possession of two pairs of upper incisors and a single, relatively short, pair of lower incisors. The genera *Lepus*, *Oryctolagus* and *Sylvagus* are the subjects of separate entries

lake   any large body of fresh water. The following derivative terms are defined in alphabetical position:

BOG LAKE   OLIGOTROPHIC LAKE
DYSTROPHIC LAKE   OXBOW LAKE
EUTROPHIC LAKE   RESERVOIR LAKE

Lama [lä′·mä, lä·mə]   a genus of camelid artiodactyls commonly known by their Spanish name *llama*. *L.*

huanicos [hwän′·ik·ôs] ("guanico") is the common Andean beast of burden

Lamarckism a view of evolution derived from the postulate that characters acquired through exposure to the environment can be inherited by the offspring of the individual acquiring these characters

lamella [lə·měl″·ə] a little plate, or structure resembling a little plate. Among biological structures so designated are the gill of an agaricale fungus, the gill of a pelecypod mollusk, the thin projections found on the pygidium of coccid insects, and the calcareous platelets surrounding the Haversian canals in bone

lamellar bone [lə·měl″·ər] a bone consisting of parallel platelets

Lamellibranchia = Pelecypoda

lamina [lăm″·in·ə] literally a "sheet" and used to describe any flattened organ or structure. The following derivative terms are defined in alphabetic position:
TERMINAL LAMINA   VERTEBRAL LAMINA

Laminaria [lăm′·in·är″·ēə] a cosmopolitan genus of laminariale Phaeophyta. They are usually unbranched and are known as kelp in the Atlantic. They were at one time a major industrial source of iodine and are still used in many coastal areas as a fertilizer

Laminariales [lăm′·in·är′·ē·āl″·ēz] an order of heterogenerate Phaeophyta commonly called the "kelps". They are clearly distinguished by the blade-like thallus of the sporophyte provided with a hold fast at its lower end. The genera *Laminaria*, *Macrocystis* and *Nereocystis* are the subjects of separate entries

lamp-brush chromosomes chromosomes from which, as the name indicates, lateral loops stick out so as to provide a lamp-brush or bottle-brush appearance. They are principally found in the midprophase of the first meiotic division particularly in the ooctye nuclei of those vertebrae having heavily yolked eggs

Lampropeltis [lăm′·prō·pěl″·tis] the genus of Squamata containing the king snakes. Some of the smaller ones are milk snakes (*L. doliata* [dŏl·ē″·ā·tə], for example) while some red and yellow ringed species are the false coral snakes

languet [lăng′·gwět] literally a "little tongue". Used of any soft projection, particularly the ciliated processes that project into the oral funnel in *Branchiostoma* or the pharynx of tunicates

lanthionine [lăn·ḇi·ō·nēn] bis(2-amino-2-carboxyethyl)sulfide. An amino acid not known to be necessary to the nutrition of rat and not definitely established as a protein precursor

lanugo [lə·nyū′·gō] the prenatal fur of some mammalian embryos and, by extension, any downy growth of hair. The term is also sometimes used for fine, "single" hairs on insects

-lapid- *comb. form* meaning "stone"

lappet [lăp′·ət] a flap of tissue, specifically the lateral wattle of a bird and the protuberant portion of the scalloped edge of a medusa

large intestine that portion of the alimentary canal which runs from the small intestine to the anus (= colon)

Larix [lăr′·iks] the genus of Coniferales containing the larches. They have deciduous acicular leaves in clusters

Larrea [lăr′·ē·ə] a genus of zygophyllaceous shrubs. *L. tridentata* [tri′·děnt·āt″·ə] (creosote bush) is a classic example of adaptation to dry conditions by reduction in size

larva [lär″·və] a developmental stage differing from an embryo in being able to secure its own nourishment. Almost all larvas differ markedly from the adult in appearance and attain the adult shape by metamorphosis. Larvas owe their fanciful names (e.g. miracidium, pluteus) to early misidentification as separate generic entities. In the United States entomologists confine the term to the prepupal stage of holometabolous insects, but European entomologists use the word for the prenymphal stage of any metabolizing insect. The various larvas are defined in alphabetic position

larviparous an animal from which larvae are born, as in some dipteran insects

Larvacea [lär·vās″·ēə] a class of tunicates distinguished by the retention throughout life of a "tail" with metameric muscle segments, and a notochord with a "principal nerve" dorsal to it. This tail is at right angles to the trunk which resembles a minute ascidian except that there is no peribranchial cavity, the pharynx opening directly to the exterior through two apertures ("stigmata")

laryngeal cartilage [lăr′·in·jē″·əl] a general name for the cricoid, aretinoid, and thyroid cartilages which form the skeletal support of the larynx

laryngotracheal groove [lă·rin′·jō·trāk″·ē·əl] a ventral, embryonic diverticulum from the posterior floor of the pharynx which gives rise to the lung

larynx [lär·ingks″, lə·ringks′] that proximal portion of the trachea which is modified in some animals to special functions

-lasi- *comb. form* meaning "shaggy"

latebra [lăt·ěb″·rə] a flask shaped mass of white yolk, extending from the center of the yolk mass to Pander's nucleus in telolecithal eggs

laterad [lăt′·ər·ăd] moving, or pointing in, a lateral direction

lateral abdominal vein *see* iliac vein

lateral band a band of cilia running along the edge of the hypobranchial groove in the endostyle

lateral geniculate body that part of the mammalian brain in which the optic tract terminates

lateral line in fish, a chain of lateral sense organs, almost certainly tonoreceptors, found in cyclostomes, fish and aquatic amphibians. In insects a line running along the side of a caterpillar and some other larvae

lateral meristem meristematic tissue running parallel to the stem or root as distinct from the apical meristem

lateral nucleolus the second, smaller of two, nuclei

lateral nucleus a second, usually smaller nucleus, as the micronucleus of a ciliate

lateral plate the nonsegmented, lateral part of the membranal mantle of an embryo

lateral ventricle the cavities of the cerebral hemisphere which develop from the common ventricle, and which are referred to as ventricles I and II of the brain

laterosphenoid bone [lăt′·ər·rō·sfēn″·oid] one of a pair of bones lying, in some vertebrates, immediately on each side of the basiophenoids (*see also* sphenoid bone, alisphenoid bone, basesphenoid bone, orbitosphenoid bone, parasphenoid bone)

latex [lā″·těks] a milky fluid, found in many plants. Rubber is present in the latex of many genera though it is highest in *Hevea* (the rubber plant)

laticifer [lăt·is″·if·ər] a cell that produces latex (cf. lacticifer)

Latimeria [lăt′·ē·mēr″·ēə] the single extant genus of Crossopterygian fishes. A single specimen was taken in

1952, since when numerous specimens have been taken from the vicinity of the Comoro Islands

latolent [lăt'·ō·lĕnt']   said of those forms which smell by moving their head from side to side (cf. dirolent)

Latrodectus [lā'·trō·dĕkt"·əs]   a genus of araneid Arachnida containing the black widow spider *L. mactans* [măk"·tănz]

Laurer's canal   a copulation canal in some trematode platyhelminths

law   a statement of a biological principle that appears to be without exception at the time that it is made. The terms "rule" and "effect" are, in contemporary literature, used almost interchangeably with "law". Many hypotheses have unfortunately become "laws" without even passing through the stage of being a theory. The following "laws" are defined in alphabetic position:

ALLEE'S LAW
BERGMAN'S LAW
BIOGENETIC LAW
DISHARMONY LAW
DOLLO'S LAW
FRASER DARLING LAW
GAUSE'S LAW
HAECKEL'S LAW
HALDANE'S LAW
HARDY-WEINBERG L.
HOMONYM LAW
HOPKIN'S LAW
JORDAN'S LAW
KROGH'S LAW
LEIBIG'S LAW
MENDEL'S LAW
MURPHY'S LAW
PRIORITY LAW
RECAPITULATION LAW
RENSCH'S LAWS
SEWALL WRIGHT'S LAW

layer   one, of several, sheet-like structures. The following derivative terms are defined in alphabetic position:

ABSCISSION LAYER
BASIDIAL LAYER
CORNIFIED LAYER
GERM LAYER
GONIDIAL LAYER
HENLE'S LAYER
HUXLEY'S LAYER
HYMENIAL LAYER
INNER PLEXIFORM L.
LIGNIFIED LAYER
MALPIGHIAN LAYER
MARGINAL LAYER
OUTER PLEXIFORM LAYER
PALISADE LAYER
PILIFEROUS LAYER
PLEXIFORM LAYER
PRICKLE CELL LAYER
PROTECTIVE LAYER
RETICULAR LAYER
SEPARATION LAYER
SPORE LAYER

leaf   a lateral appendage born on the stem of plants. The following derivative terms are defined in alphabetic position (*see also* cladode):

COMPOUND LEAF
PINNATE LEAF
SIMPLE LEAF
WATER LEAF

leaf scar   that portion of the abscission layer which remains attached to the branch

leaf stalk = petiole

leaflet   each separate blade of a compound leaf

-lecith-   *comb. form* meaning "yolk". The following terms using this suffix are defined in alphabetic position:

ALECITHAL
CENTROLECITHICAL
MEDIALECITHICAL
MEGALECITHICAL
MESOLECITHICAL
TELOLECITHICAL

lecithoprotein   [lĕs'·ith·ō·prō"·tēn]   a compound protein with lecithin as the prosthetic group

Leguminosae   [lə·gyūm"·in·ōs'·ē]   a very large family of dicotyledons clearly distinguished from all other families by the shell-like pods of seeds, large numbers of which are of economic importance. The genera *Haematoxylin, Mimosa, Pisum* and *Vicia* are the subjects of separate entries

Leibig's law   the minimal requirement, in respect to some specific factor, is the ultimate determinant in controlling the distribution or survival of a species

Leishmania   [lish·mān"·ēə]   a genus of trypanosomid protozoans. They have no visible flagellum but occur in a *Leptomonas* form with a terminal flagellum. *L. tropica* [trŏp'·ik·ə] causes boils. There is some question as to whether they are developmental stages of *Trypanosoma,* most of which pass through a leishmania phase, or a separate genus

-lemma-   *comb. form* meaning "bark" or "rind" and, by extension, "sheath"

lemma   [lĕm"·ə]   any sheath or coating but, without qualification specifically the lower bract of a grass floret (*see* neurolemma, sarcolemma)

lemmal   [lĕm'·əl]   pertaining to a sheath. The following terms using this suffix are defined in alphabetic position:

EPILEMMAL   HYPOLEMMAL   INTRALEMMAL

Lemna   [lĕm'·nə]   the type genus of Lemnacea, commonly called duckweeds, with the characteristics of the family

Lemnaceae   [lĕm·nās"·ē]   the family of monocotyledons that contains the duckweeds. The duckweeds are clearly distinguished by their small size and habit of growth as well as by the rarely occurring minute flowers born in pits on the edge or upper surface of the leaf. The genera *Lemna* and *Wolffia* are the subjects of separate entries

lemnoblast   [lĕm'·nō·blăst"]   a neuroglial cell, of the oligoglia variety, which accompanies peripheral nervous fibers and is differentiated into a Schwann cell

Lemuroidea   [lē·myōōr"·oid·ēə]   a suborder of primate mammals, confined to Madagascar, containing the lemurs and their allies. They are distinguished by their fox-like muzzle, large eyes and crepuscular habit. All other prosimians belong in the Lorisoidea

lenitic   [lə·nit'·ik]   pertaining to rapidly flowing waters (cf. lotic)

lens   a transparent lenticular mass in the eye, designed to produce images

-lent-   *comb. form* meaning a "lentil", by extension, first a lentil-shaped (plano-convex) lens and finally lenses in general

lentic   [lĕnt'·ik]   pertaining to standing water

lenticel   [lĕnt'·ē·sĕl"]   a lens-shaped area containing a pore on the stem of woody plants in which the cells are not suberized, and in which there are many intercellular spaces. They serve for gas exchange

lenticel phellogen   an area developed beneath a stoma which gives rise to a lenticel

Leodice = *Eunice*

Lepadomorpha   [lĕp'·ə·dō·môrf"·ə]   a division of cirripedes containing those forms which possess a stalk and are for the most part called goose barnacles. The genus *Lepas* is the subject of a separate entry

Lepas   [lĕp'·əs, lĕp·əs]   a genus of lepadomorph cirripedes widely distributed in the oceans of the world. The name "goose barnacle" derives from the legend that migratory geese, for the unexpected arrival of which some primitive people could not account, developed from them

-lepido-   *comb. form* meaning "scale"

Lepidodendron   [lĕp'·id·ō·dĕn"·drŏn]   a genus of giant fossil Lycopsida that were conspicuous features of Carboniferous forests, some reaching 100 feet in height

Lepidoptera   [lĕp'·id·ŏpt"·ər·ə]   the order of insects containing the butterflies and moths. They are distinguished from all insects except Trichoptera by their scaly wings and from this group by the absence of biting mouth parts. Many are brilliantly colored, mostly from Tyndall colors rather than pigment. The genera *Basilona, Bombyx, Ephestia, Kallima, Samia* and *Tegeticula* are the subjects of separate entries

Lepidosauria [lĕp'·ĭd·ō·sôr"·ēə]   a subclass of reptiles with two temporal openings in the skull. Of the three orders one is extinct, one (Rhyncocephalia) represented by a single extant species (*Sphenodon*) while the third (Squamata) contains all of the lizards and snakes

Lepidosirenidae [lĕp'·ĭd·ō·sir·ēn"·ĭd·ē]   a monotypic family of dipneustian fish containing the South American lung fish *Lepidosiren paradoxa* [lĕp'·ĭd·ō·sir"·ən pər·ə·dŏks"·ə]. It is distinguished from the Protopteridae by the possession of five gill arches and four gill clefts

Lepisosteida [lĕp'·ĭs·ŏst·ti"·də]   an order of Osteichthyes distinguished by having the anterior end of each vertebra convex and the posterior end concave. The single extant genus is *Lepisosteus* [lĕp'·ĭs·ŏst"·ē·əs], containing the gars or garfish. *L. osseus* [ŏs"·ē·əs], the longnose gar is widely distributed in the Mississippi basin

Lepisosteus   see Lepisosteida (above)

-lepo-   *comb. form* meaning "scale"

lepospondylous vertebra [lē·pō·spŏnd'·əl·əs]   one which is shell-like or husk-like

-lepto-   *comb. form* of hopelessly varied meaning resulting from the confusion of two real and one imaginary Greek words. *Leptos* means "small", or any of the attributes of smallness, such as "weakness" or "thinness". *Leptes* means an individual, or institution, which "receives" or "accumulates". The meaning "solid", as in the word leptom has no etymological justification

Leptocardii [lĕp'·tō·kärd·ē·ē]   a division, sometimes referred to as a subphylum of the acraniate chordates, containing the single class Amphioxi and the single genus *Branchiostoma*. The characteristics of the class are those of this group

leptocephalus larva [lĕp'·tō·sĕf"·əl·əs]   the marine larva of the freshwater eel *Anguilla*

leptom [lĕp'·tŏm]   the soft-walled parts of phloem

Leptomonas [lĕp'·tō·mōn"·əs]   a genus of trypanosomid protozoans with a terminal flagellum. Most *Crithidia, Leishmania,* and *Trypanosoma* pass through a *Leptomonas* stage but the genus is probably valid

leptonema [*angl.* lĕp·tŏn"·əm·ə, *orig.* lĕp'·tō·nē"·mə]   first stage of the prophase of meiosis in which the elongate thread-like chromosomes show chromosomeres

leptopel [lĕpt'·ə·pĕl]   finely divided dead matter dispersed in water

Leptostraca [*angl.* lĕpt·ŏst"·trə·kə, *orig.* lĕpt'·ō·sträk"·ə]   a subdivision of crustacea uniquely possessing eight abdominal segments. The genus *Nebalia* is the subject of a separate entry

Leptosynapta [lĕp'·tō·sĭn·ăp"·tə]   a genus of holothurians, differing from many genera in the order by having only one Polian vesicle and a sand-burrowing habit. *L. inhaerens* [ĭn·hēr"·ĕnz], about 5 inches long and half an inch thick, is common in sand on both sides of the Atlantic

leptoxylem [lĕp'·tō·zī"·lĕm]   the water-conducting tissue of mosses

Lepus [lē"·pəs]   a genus of lagomorphs of circumpolar distribution properly called "hares". The common European hare is *L. europaeus* [yōōr'·ō·pē"·əs] but the term jackrabbit is most unfortunately used for many American hares such as *L. townsendi* [touns·ĕnd"·ĭ] (the "whitetail jackrabbit")

Lernaea [lər·nē·ə]   a very aberrant genus of parasitic copepod crustacea. The adult female is reduced to a rhizoid anterior projection, buried in the tissues of the host, with an external unsegmented abdomen from which the egg masses protrude. The nauplius larva assumes, after many molts, a *Caligus*-like form. Mating occurs in this form and the male then dies. Many species on both fresh and salt water fishes

lethal mutation   one which causes death

lethical [lĕth·ĭk·əl]   pertaining to yolk

Leucanora [lyū'·kə·nôr"·ə]   a genus of lichens. The species *L. tinctoria* [tingkt·ôr"·ē·ə] yields the dye orcein

Leuchloridium [lyū'·klôr·ĭd"·ē·əm]   a genus of trematodes in which the sporocyst grows out as fungus-like bodies from the tentacles of land snails

l-leucine [ĕl·lyū'·sēn]   α-*aminoisocaproic acid.* (CH₃)₂CHCH₂CH(NH₂)COOH. An amino acid isomeric with isoleucine and necessary for the nutrition of rats (*see also* norleucine)

-leuco-   *comb. form* meaning "white", properly, but rarely, transliterated -leuko-

leucocyte [lyū'·kō·sĭt"]   a white blood corpuscle

leuconoid [lyū"·kŏn·oid']   a grade of sponge structure in which the choanocytes are in many small chambers which communicate with central chamber through canals (cf. asconoid, syconoid, sylleibid)

Leucosilenia [lyū'·kō·sil·ĕn"·ēə]   a genus of simple calcareous sponges consisting of thin walled tubes, sometimes colonial. There are no radial canals from the gastral cavity

levator muscle [lĕv·āt"·ə]   one that raises a part

libriform cell [lĭb·rē·fôrm, lib·rē·fôrm]   a thick-walled woody plant cell

libriform fiber   extremely long, woody fibers, occurring in phloem

lichen   *see* Lichenes

lichen tundra [li"·kən]   an arctic area in which the predominant, or only, vegetation is lichen

Lichenes [lik·ēn"·ēz]   a plant taxon containing the forms commonly called lichens which are a symbiotic association between a fungus and an alga. The genera *Cladonia, Leucanora, Rocella* and *Umbilicaria* are subjects of separate entries

Lieberkuhn's glands   simple tubular epithelial glands of the vertebrate intestine

lienal [li'·ən·əl]   pertaining to the spleen

life   an evanescent phenomenon dependent for its continued existence, and perpetuation, on cyclic enzymatic reactions in an environment consisting principally of protein and water

life cycle   the sum total of the changes through which an organism passes between the fertilization of an egg and the production of another mature egg; sometimes used as representing the sum total of the change between any two periods in the life history, such as that which intervenes between one type of larva and the subsequent production of a similar type

life zone   a geographic area having a characteristic fauna and flora

ligament [lĭg"·ə·ment]   a mass of yellow elastic connective tissues holding the articulations of bones together. Used also for any structure which supports an organ or part of an organ, including the springy chitinous tissue which serves to open the valves of pelecypod mollusca (*see also* falciform ligament)

ligase [li"·gāz]   ligases, also called synthetases, are enzymes which catalyze the combination of two molecules using energy derived from the breakdown of a phosphate bond, usually in ATP. Most of the trivial names are synthetases save for the carboxylases

lignified layer that part of the abscission layer which remains attached to the branch

-ligul- *comb. form* meaning a "little tongue" or "strap"

ligule [lĭg'·yūl] used of almost any ligulate structure including a fleshy prominence on the notopodium of polychaete Annelids, a small outgrowth round the adaxial surface of the leaf of lycopodiale Lycopsida, a small leaf-like structure at the junction of the blade and petiole of a leaf and the small tongue-shaped corolla of compositaceous flowers

Liliaceae [lĭl'·ē·ās"·ē] an enormous family of liliiflorous monocotyledonous angiosperms containing the lilies, tulips, trilliums, asparagus, the onions, and many other forms. They are distinguished from similar families, except the Juncacea, by the general habit of growth and large showy flowers. The genus *Yucca* is the subject of a separate entry

-limac- *comb. form* meaning a "slug"

" limax amebas " *see Naegleria*

limb bud the primordium of a limb

limbic lobe [lĭm"·bĭk] a cerebral lobe surrounding the corpus callosum and therefore forming an arch under the medial borders of the cerebral hemisphere in mammals

-limn- *comb. form* meaning "lake", frequently confused with -limno-

limnetic [lĭm'·nĕt"·ĭk] pertaining to lakes. The following derivatives using this suffix are defined in alphabetic position :

BATHYLIMNETIC   TYCHOLIMNETIC
HALOLIMNETIC

limnion [lĭm'·nē·ən] a horizontal division, or stratum, of the waters of a lake. The following derivatives using this suffix are defined in alphabetic position :

EPILIMNION   HYPOLIMNION   MESOLIMNION

-limno- *comb. form* meaning "marshy", frequently confused with -limn-

Limulus [lĭm"·yū·ləs] the only extant genus of Xiphisuran arthropods (*see* Xiphisura for description). *L. polyphemus* [pŏl'·ē·fēm"·əs] is the common east coast form

line 1 (*see also* line 2) a narrow area of demarcation, or the structure forming the demarcation or a linear group of structures. The following derivative terms are defined in alphabetic position :

BAILLARGER'S LINE   LATERAL LINE
CERVICAL LINE   WALLACE'S LINE
GENNARI'S LINE   Z-LINE

line 2 (*see also* line 1) in the sense of lineage (*see* pure line)

-ling *comb. term.* meaning "juvenile" (e.g. codling) and by extension appended to many names of small forms

-lingu- *comb. form* meaning "tongue" (cf. -ligul-)

lingual [lĭng"·gwəl] pertaining to the tongue

lingual cartilage one of a group of three cartilages (anterior, medial, and posterior) lying along the floor of the buccal cavity in Myxine

lingual shelf one on the inside of the dentary bone in some reptiles

Linguatulida [lĭng'·gwə·tyūl"·ĭd·ə] a class of arthropods, by some considered to be a separate phylum or an order of Arachnida, containing two genera of worm-like parasites found in the air passages, and occasionally other organs, of some vertebrates. The only appendages are a pair of retractile hooks on the very short cephalothorax. The eggs are voided in the feces of the host. They are picked up by an intermediate host, usually a small rodent, where they hatch into a mite-like larva with two pairs of appendages and boring mouth parts. This encysts in the liver of the intermediate host

Lingula [ling·gyōō·lə] a genus of brachiopods which first developed in the Ordovician and a few species of which still exist in the Indian and Pacific Oceans. The extant species, however, are all of relatively recent development

linin [lī'·nĭn] a term at one time applied to chromatin reduced to an apparently thread-like form by acid fixatives

linkage the type of inheritance exhibited by genes located in the same chromosome and therefore having a tendency to remain together in passing from one generation to the next (*see also* autosomal linkage)

-lip- *comb. form* meaning "fat"

lip the free edge of any opening, particularly the fleshy flap of skin lying anterior to the teeth in mammals, the area immediately dorsal to the yolk plug in the late amphibian gastrula and the lobes surrounding the mouth in nematode worms. In both zoology and botany the word is used interchangeably with labium

lipase [lī'·pāz'] an enzyme that catalyzes the hydrolysis of triglyceride into diglycerides and a fatty acid

-lipo- *comb. form* meaning "to depart from", frequently confused with -lip-

lipoblast [lĭp'·ō·blăst'] a fat-storing cell in an embryo

lipoid [lī'·poid] a fat, or fat-like, substance

lipoprotein [lĭp'·ō·prō'·tēn] a compound protein with fatty acid prosthetic group

lipoxenous [lĭp'·ŏks·ĕn"·əs] said of a parasite which leaves its host after it has secured adequate nourishment

liquefaction used specifically by microbiologists to indicate the hydrolysis of a jelled culture medium by a microorganism

-liss- *comb. form* meaning "smooth"

-lith- *comb. form* meaning "stone". The following terms using this suffix are defined in alphabetic position :

CYSTOLITH   OTOLITH   STATOLITH

Lithobius [lĭth·ō"·bē·əs] the commonest genus of Chilopoda with more than 200 species, less than 50 being American. The body has 15 leg-bearing segments of which nine are normal and six so reduced as to look like intersegmental rings. *S. forficatus* [fŏr'·fĭk·ā"·təs], a light brown or yellow form about an inch long is common all over the world

lithosere [lĭth"·ō·sēr] a stage of vegetation inhabiting relatively bare rocks

littoral pertaining to the shore and, in freshwaters, confined to those zones in which rooted vegetation occurs. The following derivative terms are defined in alphabetic position (*see also* riparian) :

ELITTORAL   SUBLITTORAL

Littorina [lĭt'·ər·ĭn"·ə] a genus of littoral marine pelecypod mollusks. *L. littorea* [lĭt'·ər·ē"·ə] is the edible periwinkle of Europe, introduced to but rarely eaten on, the Atlantic coast

liver an organ developed as an outgrowth of the alimentary canal primarily concerned with the storage and distribution of foodstuffs

llama [lä″·mə, lyä″·mǎ] a genus of South American camelid artiodactyles. Used as an English word it usually means *L. huanacos* [hwän″·ä·kŏs′] (= huanaco)

llano [län′·ō, lyän″·ô] South American prairie

Loa [lō″·ə] a monotypic genus of African filarioid nematode worms containing *L. loa*, the human "eye worm". The 2 to 3 inch adult lives in the subcutaneous tissues, often appearing under the cornea of the eye. Sheathed embryos enter the blood stream where they are picked up by any of several species diptera of the genus *Chrysops*. The larvas develop in the fly's alimentary canal and are transmitted directly

loa an abbreviation used for "length over-all" and frequently printed, as here, without separating periods

lobe any blunt prominence arising from a surface, or any of several parts into which an organ is divided by constrictions or indentations. The following derivative terms are defined in alphabetic position:

CEPHALIC LOBE	OPTIC LOBE
FRONTAL LOBE	PARIETAL LOBE
LIMBIC LOBE	PYRIFORM LOBE
OCCIPITAL LOBE	TEMPORAL LOBE

lobopod [lōb″·ō·pŏd′] a bluntly ending pseudopodium as that of most amoebas

-lochm- *comb. form* meaning a "thicket"

lociation [lōsh′·ē·ā″·shŭn] a modification of an association which, in contrast to a faciation, has been produced by the removal or addition of other than dominants

locies [lōsh′·i·ēz] the term is to associes as lociation is to association

lock and key theory 1 (*see also* lock and key theory 2) the specificity of an enzyme or a substrate is due to an interlocking molecular configuration

lock and key theory 2 (*see also* lock and key theory 1) the exact structural interlocking of insect genitalia causes those with even minute variations to be incapable of copulation

-loco- *comb. form* meaning "place"

locomotory metamerism [lō′·kō·mōt″·ər·ē] the theory that metameric segmentation arose in consequence of a sinuous method of progression (*see also* metamerism)

locular pertaining to, or having, cavities. Derivatives are not separately defined since the meanings are obvious from the roots (e.g. multilocular, trilocular, etc.)

locus (*plural* loci) [lō″·kəs] the place or site at which a gene is located on a chromosome

Loganiaceae [lō′·găn·ē·ās″·ē] a family of dicotyledons, mostly tropical shrubs, distinguished by the opposite stipulate leaves and by the bilocular superior ovary. The genus *Strychnos* is the subject of a separate entry

-logy termination derived for Greek logos meaning "word" or "speech" but which is now used in the sense of "the study of". The meanings of most derivative terms are too well known (e.g. biology, zoology) to warrant definition

Loligo [lŏl·i″·gō, lŏl″·əg·ō] a genus of cephalopod mollusks commonly called squid though the Italian *calamari* is not infrequently used. They are distinguished from *Sepia* by the terminal fins. *L. pealei* [pē·āl′·ē·i] is the common Atlantic and *L. opalescens* [ōp″·əl·ĕs″·ĕnz] the common Pacific squid

loma [lō·mǎ] a seasonal prairie in Latin America

lomma [lŏm″·ə] a plate-like lobe or flap

-longi- *comb. form* meaning "long" (cf. -macro-)

longitudinal axis the axis running the length of a biradially or bilaterally symmetrical object. It differs from the sagittal axis in that it does not divide the object into symmetrical parts

longitudinal plane a plane parallel to the longitudinal axis. It may be sagittal, horizontal, or at any position between these

-loph- *comb. form* meaning "crest", "mane" or "ridge"

Lophius [lōf″·ē·əs] a genus of marine teleost fish. *L. piscatorius* [pĭs′·kə·tôr″·ē·əs] is the angler fish that dangles a lure, attached to a projection from the top of its head, in front of its enormous mouth

lophodont [lōf″·ō·dŏnt′] having ridges on the crown of the teeth

Lophophora [*angl.* lō·fŏf″·ər·ə, *orig.* lōf′·ō·fōr″·ə] a genus of cactus with tuberculated stems. *L. williamsii* [wil·yǎmz″·ē·i] is the principal natural source of mescaline. The crude extract is peyote

lophophore a horseshoe-shaped crown of ciliated tentacles at the anterior end of many invertebrates (cf. Lophophoria)

Lophophoria [lōf′·ō·fōr″·ē·ə] a term coined to designate those animals possessing a lophophore. This taxon therefore comprises the phyla Phoronida, Ectoprocta and Brachiopoda

lorate [lōr″·āt] strap-shaped

loreal scale [lōr″·ē·əl] one or more scales lying between the nares and the eye in snakes

Lorenzini's ampulla blindly-ending, slime-filled pits, presumably of sensory function, found on the cheek and snouth of Chondrichthyean fish

lorica [lōr′·ĭk·ə] a hardened or thickened cuticle forming a shell or case. There is no clear distinction between cuticle, lorica and exoskeleton in invertebrates

Loricata [lōr′·ĭk·ä″·tə] an order of placental mammals, or a suborder of Xenarthra, which contains the armadillos. This group is distinguished by the transverse plates of fused hairs which cover the body. The genus *Dasypus* is the subject of a separate entry

Lorisoidia [lōr′·ĭs·oid″·ē·ə] a sub-order of primates erected to contain the families Lorisidae and Galagidae; they are principally distinguished from the Lemuroidia by the relatively large number (14 to 16) of dorsal vertebrate. The genera *Perodicticus* and *Tarsius* are the subjects of separate entries

lotic [lō″·tik] pertaining to rapidly flowing streams

low moor a peaty swamp

-lox- *comb. form* meaning "slanting"

Loxodonta [lŏks′·ō·dŏnt″·ə] the genus of proboscidean mammals containing the African elephant (*L. africanus* [ăf′·rĭk·än″·əs]). The Indian elephant is *Elaphas indicus*

LSD *see* lysergic acid diethylamide

luciferase [lōō·sif″·ə·āz′] an enzyme which catalyzes the oxidation of luciferin

luciferin [lōō·sif″·ər·ĭn] a flavin-related compound, found in many light-producing organisms, which emits light when oxidized

lumbar [lŭm″·bär, lŭm″·bər] literally, pertaining to the loins

lumbar vertebrae one in the region that lies between the thorax and the sacrum

Lumbricus [lŭm″·brĭk·əs] a genus of terrestrial oligochaetae. Though almost all earthworms are presented to classes as *L. terrestris* [tər·ĕs"·trĭs], many in point of fact belong to the allied genus *Allolobophora* or to the genus *Diplocardia*

-lumen- *comb. form* originally meaning "opening" but later "light" (coming through the opening). The adjective luminous, for example, can mean either light-producing or hole-producing

lumirhodopsin [lōōm'·ē·rōd·ŏps"·ĭn]   the first breakdown product of rhodopsin under the influence of light

-lun- *comb. form* meaning "moon"

lunate bone   the carpal that lies between the capitate bone and the junction of the radius and ulna

lung 1 (*see also* lung 2) heavily vascularized, usually loculated, sacs into which air for respiration is drawn by terrestrial vertebrates and a few terrestrial gastropod mollusks

lung 2 (*see also* lung 1) any device used for gas exchange by animals (*see* book-lung)

lung bud   one of a pair of embryonic diverticula from the laryngotracheal groove which subsequently form the lungs and adnexed ducts

luteal [lōōt"·ē·əl]   pertaining to the corpus luteum

luteinizing hormone [lōōt'·ē·ə·nīz"·ĭng]   one produced by the adenohypophysis that stimulates ovulation and the growth of the corpus luteum

lyase [lī"·āz]   a group of enzymes which catalyze the removal of groups, by other than hydrolysis, in such a manner as to leave a double bond or to add groups to double bonds. Among the better-known trivial names of lyases are decarboxylases, dehydratases and synthases (cf. synthetase)

Lycaenops [lī·kēn"·ŏps]   a genus of extinct Permian sphenapsid reptiles, probably ancestral to the theriodonts and thus a mammalian precursor. The dentary bone was large and the teeth were differentiated but not highly specialized

Lycaon [lī"·kā·ən, lī"·kā·ōn']   the genus of canid carnivores containing the cape hunting dogs. They differ from true dogs in having four toes on the front and back feet. Their vicious hunting packs are a byeword for ferocity

-lyco- *comb. form* meaning "wolf"

lycophore larva [lī"·cō·fôr']   a ciliated cestode larva with large frontal glands and ten hooks. It remains in the shell until eaten by a secondary host

Lycopodium [lī'·kō·pōd"·ē·əm]   one of the few extant genera of the Lycopsida. The sporophyte has a much branched stem

Lycopsida   a phylum of pteridophyte plants containing the club mosses. The name derives from the club shaped strobila which are typical. The Isoetales are the subject of a separate entry as are the genera *Asteroxylon, Lepidodendron* and *Lycopodium*

-lym-   *comb. form* meaning "ruinous"

Lymnaea [lĭm·nē"·ə]   a genus of freshwater pulmonate gastropod mollusks. Many species of this genus act as host to the miracidia of *Fasciola*, the liver fluke

lymph [lĭmf]   literally meaning "water" and so used as a *comb. form*. More usually refers, either alone or a *comb. form* to the fluid found in lymph vessels, which is a diluted blood plasma containing a few lymphocytes

lymph follicle   a patch of reticular tissue in the intestinal wall crowded with lymphocytes (Peyer's patches)

lymph gland   aggregates of connective tissue, leucocytes and lymph vessels

lymph heart   a pulsating portion of a lymph vessel that returns lymph to a blood vessel

lymph vessel   a vessel analogous to a blood vessel, but carrying lymph

lymphatic nodule [lĭm·făt"·ĭk]   an area tightly packed with lymphocytes

lymphatic tissue   a tissue composed of a stroma of reticular cells supported on fibers and with lymphocytes in the mesh of the stroma

lymphocyte [lĭm"·fō·sĭt']   a totipotent type of blood cell. The dominant component of lymphoid organs (*see also* large lymphocyte, small lymphocyte)

-lys-   *comb. form* meaning to "separate" or "loosen" and thus, by extension, to "liquify". The substantive form -lysis and the adjectival form -lytic do not, in compounds, always have identical meanings

lysergic acid diethylamide   [lī·sûrg"·ĭk ăs·ĭd di'·ep·əl·ăm"·ĭd]   a powerful hallucinogen (LSD). Since lysergic acid is the principal hydrolysis product of ergot (*see Claviceps*), there is probably some LSD in some samples of ergot. The syndrome "ergotism", well known in the early Middle Ages as the result of eating moldy rye, included not only death from spasms, cramps and dry gangrene but also some cases of hallucinations

-lysis   *subs. suffix* from -lys- (cf. -lytic). The following derivatives using this suffix are defined in alphabetic position:

AUTOLYSIS   PLASMOLYSIS
DIALYSIS   PHOTOLYSIS
HISTOLYSIS

lysogeny   [*angl.* lis·ōj"·ĕn·e, *orig.* lis'·ō·jen'·ē]   the condition of a microorganism that has a prophage in its genome

lysosome [lis"·ō·sōm']   a membrane bounded (circ. 0.1 microns in diameter) cytoplasm showing acid phosphatase activity (cf. ribosome)

lysozyme [lis"·ō·zim']   an enzyme catalyzing the hydrolysis of the wall substances of certain bacteria (= muramidase)

-lyte   *comb. term.* indicating a separated part

lytta [lĭt'·ə]   a longitudinal cartilaginous rod found in the tongue of many mammals

M

Macaca [măk·à″·kə] a genus of catarrhine primates commonly called "Macaques". *M. mulata* [myōōl·ät″·ə] is the "Rhesus Monkey" of research laboratories
Macracanthorhynchus [măk′·rə·kăn′·thŏ·rĭnk″·əs] a genus of acanthocephalans. *M. hirudinaceus* [hĭr′·ōō·dĭn·ās″·ē·əs], of which the female may reach a length of nearly 18 inches, is a common parasite of pigs, occasionally infesting man. The intermediate host is the larva of any of several scarabeid beetles
-macro- *comb. form* meaning "long", but almost universally used as a substitute for -mega- which properly means "big"
Macrochelys [mā′·krō·kĕl″·ĭs] a genus of chelonian reptiles containing the alligator snapping turtle (*M. temmincki* [tə·mĭnk″·ĭ]) of the Southeastern States. This, which may reach a weight of more than 200 lbs, is the largest known freshwater turtle
macroconjugant [măk′·rō·kŏn″·jōō·gănt] the larger of the two products of progamous fission
Macrocystis [măk′·rō·sĭst″·ĭs] a genus of laminareale Phaeophyta. There are several species of these giant kelps; *M. pyrifera* [pĭr·ĭf″·ĕr·ə] can reach a length of about 60 feet
macroevolution [măk′·rō·ĕv′·ŏl·yōō″·shŭn] the evolution of broad groups as distinct from individual organisms
macroglia [măk′·rō·glē″·ə] a general term for glial cells of ectodermal origin and therefore comprising both oligodendria and astrocytes
macronucleus [măk′·rō·nyūk″·lē·əs] a large nucleus, without recognizable chromosomal organization, found in ciliate protozoa (cf. micronucleus)
macrophage = histiocyte
macrophotograph in biological technique refers to a photograph, usually at magnifications from × 1 to × 20, taken of objects visible to the naked eye (cf. photomicrograph)
macroplankton those planktonic organisms that can be classified without the aid of a microscope (cf. megaplankton)
macroscopic [măk′·rō·skŏp″·ĭk] visible to the naked eye
madreporic plate [măd·rə·pŏr″·ĭk] the sieve plate which connects the internal water vascular system of echinoderms with the exterior

madreporic vesicle an aboral extension of the axial sinus in Echinoderms, placed close to the madreporic plate, but having no connection with it
madreporite = madreporic plate
major gene one of which the effects are readily identifiable
-malaco- *comb. form* meaning "soft", but by extension often used to mean "slug", "snail" or even "molluscan"
malaria [mə·lâr″·ē·ə] a disease caused by the protozoan blood parasite *Plasmodium*. Mosquitos of the genus *Anopheles* are the usual vector of human malaria
malarial crescents the gamonts of a malarial parasite
male that one, of two, heterosexual forms of the same species, which normally yields motile gametes called sperm (*see also* complemental male, supermale)
malleolus bone a projection from the lower end of the fibula
malleus bone the innermost of the three ossicles, which conduct vibration from the tympanum to the inner ear; the others are the incus and the stapes
Mallophaga [*angl.* məl·ŏf″·ə·gə, *orig.* măl′·ō·fäg″·ə] the order of insects containing the biting lice. They are distinguished as wingless parasitic insects with biting mouth parts, save in some bird lice in which the chewing mandibles are adapted to piercing. Most are parasitic on birds
Malpighian corpuscle [măl·pĭg″·ē·ən] the capsule and glomerulus, taken together, of a mesonephric or metanephric unit
Malpighian layer the basal layer of the epidermis
Malpighian tubule either blindly ending (presumable excretory) tubules opening into the midgut of many arthropods or a chordate nephron
mamillary recess [măm·ĭl″·ər·e] an evagination from the floor of the embryonic forebrain where the latter joins the midbrain
-mamm- *comb. form* meaning "breast", specifically a human breast and thus a mound bearing a prominence
Mammalia [măm·āl″·ē·ə] a class of craniate chordates distinguished by the presence of hair on the skin and by having milk secreting glands. The orders Artiodactyla, Carnivora, Cetacea, Chiroptera, Dermoptera,

Hyracoidea, Insectivora, Lagomorpha, Loricata,| Perissodactyla, Pholidota, Primates, Proboscidea, Rodentia, Sirenia, Tubulidentata, Xenarthra are the subjects of separate entries

mammary [măm″·ər·ē]   pertaining to the mammalian breast or to a structure of similar shape

mammary gland [măm″·ər·ē]   one of the prominences, known as breasts when pectoral, dugs when abdominal, and udders when lying between the hind legs of artiodactyls, that house the mammalian milk glands

mammillary body [măm·ĭl″·ər·ē]   a small dome-shaped prominence on the underside of the hypothalamus

-man-   *comb. form* meaning "hand"

man   *see* human

mandible [măn″·dĭb·əl]   in vertebrates, the lower jaw as a whole. Also the most anterior of three pairs of mouthparts in many arthropods

mandibular arch [măn·dĭb″·yū·lə]   the visceral arch that supports the lower jaw

mandibular bone   the tooth-bearing, and in mammals the only, bone of the lower jaw (= dentary bone) (*see also* hyomandibular bone)

Mandibulata [măn·dib″·yūl·ăt′·ə]   a term coined to describe those classes of the phylum Arthropoda that have a true mandible. As so defined this taxon contains the Crustacea, Pauropoda, Symphyla, Diplopoda and Insecta

Manis [mā″·nĭs]   the only extant genus of Pholidota (pangolins or scaley anteaters). The genus has the same characteristics as the order

α and β-mannosidase [mən·ō·sĭd″·āz]   enzymes respectively catalyzing the hydrolysis of α and β-D-mannosides into alcohols and mannose

mantle   an envelopment of the body usually meaning, without qualification, the outer soft coat of Mollusca and Branchiopods

Mantodea [măn·tōd″·ē·ə]   an order of insects containing the mantids now usually united with the cockroaches in the Dictyoptera

manubrium [mə·nūb″·rē·əm]   literally, a handle. Specifically, one of a pair of trophi in the mastax of a rotiferan, the handle-like extension on the end of which is the mouth of coelenterate medusae, the projection from the mesosternum of elaterid Coleoptera which fits into a corresponding cavity in the prothorax and enables them to "click" and the base of the furcula in Collembola

maquis [mä′·kē]   an association of hard-leaved shrubs, principally on the north shore of the Mediterranean

Marchantia [măr·kănt″·ē·ə]   a typical genus of Hepatica (liver-worts) with a flattened thallose vegetative structure from which arise stalked antheridia and archigonia

Margaropus   see Boophilus

marginal band   a band of cilia running along the outer edge of the endostyle of Urochordates

marginal canal   a canal formed by the fusion of mesonephric units growing towards the testes, where they will subsequently become vasa efferentia

marginal initial   one of the line of cells from which marginal meristem is developed, giving rise to the outer surface of the blade of a leaf (*see also* submarginal initial)

marginal layer   an envelope of longitudinal nerve fibers around the outside of the spinal cord

marginal meristem   meristem, lying in a band along each side of a developing leaf axis, and from which the blade is produced

marijuana   *see Cannabis*

marsh   an area of normally wet ground differing from a bog in being less soggy and in frequently possessing transitory dry areas (*see also* salt marsh)

marsipobranch [măr·sĭp″·ō·brănk′]   a branchial cleft in the form of a pouch. Found only in the group once called Cyclostomes, which have therefore been renamed Marsipobranchia

Marsipobranchia   an order of agnath Chordata with the characteristics of the class. They differ from the other order Ostracodermi in lacking anterior external armor. The extant forms are the lampreys and hagfishes. The genera *Myxine* and *Petromyzon* are the subject of separate entries

marsupial [măr·syū″·pē·əl]   literally "pouched" and therefore usually referring to the Metatheria

marsupial bone   a bone in the pelvic girdle of Metatheria which extends forward from the anterior margin of the pubis

Marsupialia = Metatheria

marrow   the vascularized tissue which fills the cavities of bones

masseter muscle [măs″·ə·ter]   the large adductor muscle of the jaw

-mast-   *comb. form* meaning "breast" or "breast shaped" (*see* gynecomast, neuromast)

mast cells [măst]   large mesenchymal cells containing numerous polychromatic granules

mastax [măs″·tăks″]   the complex pharynx of Rotifera. The term is sometimes used as synonymous with trophi. It consists of several chitinous rods and may be used to grind food or extrude jaws

-mastig-   *comb. form* meaning "whip" and thus "flagellum"

mastigoneme [măs′·tĭg″·ō·nēm″]   one of the lateral thread-like projections from a flagellum (*see* tractellum)

Mastigophora [*angl.* măs·tĭg″·ŏf·ər·ə, *orig.* măst′·ĭg·ō·fôr″·ə]   a class of protozoa distinguished by the presence of one or more flagella at all stages of the life history. At one time universally, and still occasionally, called the Flagellata. The subclasses Phytomastigophora and Zoomastigophora are the subjects of separate entries

mastigopod [măs′·tĭg·ō·pŏd″]   the swarm spores of a myxomycophyte

mastoid process   a bony extension just behind the tympanic bulla. In many mammals, it contains air spaces communicating with the inner ear

mating dance   the pre-copulatory activities of an animal, usually a male

maturation divisions = meiosis

matutinal [mə·tūt″·ĭn·əl]   referring to the morning (cf. crepuscular, auroral)

maxilla [măks″·ĭl·ə]   literally the upper jaw. In arthropods applied to the mouth parts immedi-ately posterior to the mandibles (*see also* second maxilla)

maxilla bone   a dermal bone of the splanchnocranium forming the main portion of the upper jaw, and the anterior end of the cranium. In mammals it bears all the upper teeth, except the incisors (*see also* premaxilla bone, septomaxillar bone)

maxillary gland [măks·il"·ər·ē] any arthropod gland in the same segment as or opening on the maxilla

Meandrina [mē'·ăn·drĭn"·ə] a common genus of Madreporia forming massive colonies. *M. sinuosa* [sĭn'·yōō·ōs"·ə] is the common "brain coral" of Caribbean waters

mechanism the doctrine that all living processes can be explained in terms of inorganic concepts. This doctrine is the antithesis of vitalism

mechanistic theory all animal activity is controlled objectively by stimuli received from the environment

mechanoreceptors [měk'·ən·ō·rē·sěp"·tôrz] a general term for receptors responding to physical forces

Meckelian ossicle a small ossification on Meckel's cartilage at the insertion of the adductor muscles of the jaw in crossopterygian fish

Meckel's cartilage the cartilaginous lower jaw of chondrocrania, and thus the cartilaginous precursor of the mandibular bone in embryos

Meckel's ganglion = sphenopalatine ganglion

Meckel's tract the portion of the small intestine that comprises the jejunum and ileum

meconium corpuscle [mək·ōn"·ē·əm] a granule in the intestinal epithelium of embryos of higher vertebrates and the excreta of a fetus in the uterus or of an insect in the pupal case

Mecoptera [mək·ŏpt"·ər·ə] the order of insects which contains the scorpion flies and their allies. Their distinguishing characteristic is the prolongation of the head into a trunklike beak and, in most groups, the possession of four equal-sized wings frequently mottled

medialecithal [mē'·dē·ə·les'·ĭb"·ĭk·əl] said of an egg with a considerable amount of yolk but still capable of holoblastic cleavage

mediastinum [mē'·dē·ă·stĭn"·əm] a cavity between the two pleural sacs of a mammal

mediterra- *comb. form* meaning "inland", but almost invariably used in the sense of the "inland sea", i.e. the Mediterranean

Mediterranean occurring in the area of the Mediterranean Sea

mediterranean dwelling far from the sea

medulla [mə·dŭl"·ə] literally "marrow" or "thick". Used generally to distinguish the central ("medullary") from the outer ("cortical") portions of an organ, cf. cortex. In botany frequently used as synonymous with pith

medulla oblonga the thickened floor of the myelencephalon

medullary ray plate of parenchyma tissue radiating from the center of a tree

medusa [mə·dūz"·ə] the free-swimming jellyfish-like form of a polymorphic coelenterate in contrast to the polyp

-mega- *comb. form* meaning "large"; the term -macro-, which properly means "long", is frequently substituted for -mega- in biological combinations

megadont [měg"·ə·dont'] possessing some teeth markedly larger than others

megagamete [měg·ə·găm"·ēt] the larger of a pair of heterogametes, and by convention, regarded as female

megakaryoblast [měg·ə·kâr"·ē·ō·blast'] the first

stage in the development of a thrombocyte from a hemocytoblast

megakaryocyte [měg·ə·kâr"·ē·ō·sit'] a giant blood cell with a giant polylobular nucleus directly descended from the hemocytoblast and thought to give rise to platelets

megalecithical = telolecithal

Megalobatrachus [měg'·ə·lō·bə·trăk"·əs] a genus of urodelan Amphibia, closely related to *Cryptobranchus*, containing the largest extant amphibian *M. japonicus* which reaches a little over 5 feet in length

megalopa larva [měg'·ə·lō"·pə] the pre-imago stage in the development of a crab in which the eyes and chelae are both very prominent

meganephridium [měg'·ə·něf·rĭd"·ē·əm] a large annelid nephridium furnished with both a peritoneal funnel and a duct leading to the exterior

megaplankton that which includes large floating organisms such as the water hyacinth, jelly fish, etc. (cf. macroplankton)

megasclerite [měg'·ə·slěr"·ĭt, měg'·ə·sklěr"·ĭt] a large sponge spicule, frequently uniting with others to form a definite skeleton (*see also* microsclerite)

megasporangium [měg'·ə·spôr·ănj"·ē·əm] that part of the base of the pistil which contains the ovule

megaspore [měg'·ə·spôr'] the larger of two spores or the ovule of spermatophytes

megasporocyte [měg'·ə·spŏr'·ō·sit"] a cell in the ovule of gymnospermous plants. It normally develops into four megaspores

megasporophyll [měg'·ə·spŏr"·ō·fĭl, měg'·ə·spôr"·ō·fĭl] the female cone of the cycad or, in angiosperms, the carpel

megastrobilus [měg'·ə·strō"·bĭl·əs] the female cone of a cycad

megazooid [měg'·ə·zō"·ĭd] literally a large zooid but usually used in the sense of a female algal gamete

Meibomian gland a gland in the eyelid secreting a fatty lubricant

-meio- *comb. form* meaning "less"

meiofauna [mi'·ō·fôrn·ə] the smaller, invertebrate, fauna of the sea bottoms

meiosis [mi·ōs"·ĭs] that form of cell division in gametogenesis in which the chromosomes are reduced from the diploid to the haploid number. The following derivative terms are defined in alphabetic position: BRACHYMEIOSIS   INITIAL MEIOSIS

meiotic apogamy [mi·ŏt"·ĭk] the condition in which a sporophyte plant is developed from the oospore

meiotic euapogamy the condition in which the mother cells of a sporophyte plant have the haploid chromosome number

meiotic parthenogenesis parthenogenesis from a gamete which has become haploid by a normal process of meiotic reduction divisions

Meissner's corpuscle a touch-sensing end organ in the form of ovoid structures with the central mass of irregular cells penetrated by irregularly curved nerve endings

Meissner's plexus a parasympathetic ganglion found in the intestinal submucosa

-melan- *comb. form* meaning "black"

melanin [měl·ə·nin] a general term for a group of indole biochromes best known for the black melanin but also occurring as yellow, orange and brown compounds

melanoblast [měl″·ə·nō·blast′] the precursor of a melanocyte which is itself an immature melanophore

melanocyte [měl″·ə·nō·sĭt′] an immature melanophore

melanocyte-stimulating hormone a hormone secreted in the pars intermedia of the pituitary of lower vertebrates. The name is descriptive of the action

melanophore [měl′·ăn·ō·fôr] a cell containing melanin. Melanophores expand and contract under the stimulus of melanocyte-stimulating hormones and melatonin

melanophore hormone = melanocyte-stimulating hormone

melatonin [měl′·ə·tōn″·ĭn] a hormone secreted by the pineal in lower vertebrates which is antagonistic to melanocyte-stimulating hormones

membrane any thin sheet, living or dead, of organic material. In cytology, membrane usually means unit membrane. The following derivative terms are defined in alphabetic position:

ARACHNOID MEMBRANE	NICTITATING M.
ARTICULAR MEMBRANE	NUCLEAR M.
BASEMENT MEMBRANE	PERIVITELLINE M.
BOWMAN'S MEMBRANE	PERMIABLE M.
CELL MEMBRANE	PLASMA M.
CLOACAL MEMBRANE	SEMIPERMIABLE M.
DESCEMET'S MEMBRANE	SYNOVIAL M.
DIFFERENTIALLY PERMIABLE M	TYMPANIC M.
EMBRYONIC MEMBRANE	UNDULATING M.
FENESTRATED MEMBRANE	UNIT MEMBRANE
FERTILIZATION MEMBRANE	VITELLINE M.

membrane bone a bone formed in a membranous area

membranelle [měm′·brə·nel″] a triangular plate, composed of two or more rows of fused cilia, in Protozoa

-men- comb. form hopelessly confused from many roots and variously meaning "moon", "crescent", "mouth", "courage", "vigor", "unchanging", "permanent" and "wrath"

Mendelian population a group of individuals who interbreed and form a community by themselves

Mendel's first law that there are pairs of factors in sexual organisms which are segregated in the parent but reunited in the offspring

menopause [měn′·ō·pôrz″] the time of cessation of the menstrual cycle in human females or the oestral cycle in other mammals

menotaxis [mě′·nō·tăks″·ĭs] movement at a fixed angle in relation to a source of stimulation

mental bone a small membrane bone occurring in some vertebrates at the anterior articulation or synthesis of the mandibles

mentum [měn′·təm] literally chin but in insects usually applied to the distal segment of the labrum

Mephitis [měf·ĭt″·ĭs] a genus of mustelid carnivorous mammals containing the striped skunks. *M. mephitis* is the common skunk. The spotted skunks are in the genus *Spilogale*

-mer- comb. form meaning "a part of". The substantive forms mer, mere, mera, merism, merile and the adjectival forms meric, meristic have all developed more or less specialized meanings

-mere- comb. form meaning "a part of" (see -mer-). The following derivative terms using this suffix are

defined in alphabetic position:

ACTINOMERE	EPIMERE
ANTIMERE	HYPOMERE
ARTHROMERE	METAMERE
BRANCHIOMERE	MYOMERE
CENTROMERE	PARAMERE
CEPHALOMERE	PROSTHOMERE
CHONDRIOMERE	RHABDOMERE
CHROMOMERE	SARCOMERE
CRYPTOMERE	TELOMERE

-meric pertaining to parts of, most frequently in the sense of more or less similar parts. The following derivative terms using this suffix are defined in alphabetic position:

DIMERIC	POLYMERIC
METAMERIC	

mericlone [měr′·ē·klōn″] a clone of flowering plants derived from a tissue culture of apical meristem. Used principally of orchids

meriopodite [*angl.* měr·ē·ōp″·ō·dit′, *orig.* měr′·ē·ō·pōd″·it] the joint of the crustacean appendage that lies between the carpopodite and ischiopodite

merism the division of something into parts. Botanical usage differs from zoological in that plant parts may be reproduced by increasing the number of radial axes as distinct from longitudinal divisions. The following derivative terms are defined in alphabetic position:

LOCOMOTORY M.	TRIMERISM
METAMERISM	

Merismopeida [měr′·ĭz·mō·pē″·də] a genus of colonial cyanophyte Algae that divide in one plane only to form a flattened sheet

meristem [měr′·ĭs·těm″] those plant tissues from which growth originates and from which various other tissues are formed. The following derivative terms are defined in alphabetic position:

APICAL MERISTEM	MARGINAL MERISTEM
ENDOMERISTEM	PROMERISTEM
GROUND MERISTEM	RESIDUAL MERISTEM
LATERAL MERISTEM	VASCULAR MERISTEM

-meristic pertaining to merism. That is to parts, and particularly to the number of parts (*see* allomeristic)

-merite generally used in biology in the sense of a specific segment. Derivative terms are not defined separately since the meaning is usually obvious from the roots (e.g. epimerite, the upper part, etc.)

Merkel's corpuscle an end organ consisting of two or more tactile cells in a connective tissue sheath

meroblastic cleavage [měr′·ō·blăs″·tĭk] the form of cleavage shown by large-yolked eggs in which only the animal pole divides into separate blastomeres

merocrine gland [měr′·ō·krĭn, měr′·ō·krĭn] a gland, the product of which is secreted without the loss of any part of a cell except the secretory granules (*see also* holocrine gland)

merocyte [měr′·ō·sĭt′] a nucleus in the non-segmenting portion of an egg showing meroblastic cleavage including nuclei formed from supplementary spermatozoa; the term is also sometimes used as synonymous with schizont

merogamete [*angl.* měr·ŏg″·ə·mēt, *orig.* měr′·ō·găm″·ēt] a gamete which is smaller than the organism, usually a protozoan, from which it is derived by fission

merogamy [*angl.* měr·ŏg″·ə·mē, *orig.* měr·ō·găm″·ē] fertilization through the union of merogametes

merogony [*angl.* měr′·ŏg″·ən·ē, *orig.* měr′·ō·gōn″·ē]

development of an enucleated egg initiated by a sperm

meroplankton that which is seasonal in its appearance or composed of temporary plankton, present in the plankton for only a portion of the life cycle

meropodite [*angl.* mĕr·ŏp"·ə·dĭt, *orig.* mĕr'·ō·pŏd"·ĭt] the equivalent of the femur in Chelicerata

merotype [mĕr"·ō·tīp'] a taxonomic type derived by vegetative reproduction from the original, or holotype

-merous a termination synonymous with -meric but rarely used interchangeably. The following derivatives using this suffix are defined in alphabetic position:

EMBOLOMEROUS   HOMOEOMEROUS
HETEROMEROUS   HOMOIMEROUS

merozoite [mĕr'·ō·zō"·ĭt] agametes produced by multiple fission of a trophozoite

Merychippus [*angl.* mĕr·ē·chĭp"·əs, *orig.* mĕr'·ĭk·hĭp"·əs] the first of the three-toed horses in which the two outside toes were significantly smaller than the central toe. *Merychippus,* which lived during the Pliocene, also had teeth adapted to grazing

-mes- *comb. form* meaning "middle". The pronunciations -mĕz- and -mēs- are almost invariably interchangeable

mesaxonic foot [mĕz"·ăks·ŏn'·ĭk, mēs"·ăks·ŏn·ĭk] one in which the axis of symmetry passes through the middle of the third digit. This situation pertains in the even-toed ungulates or Perissodactyla

mescaline [mĕs"·kə·lĭn] a hallucinogenic drug (3-4-5 trimethoxyphenetholamine) derived from the cactus *Lophophora williamsii* (*see also* peyote)

mesectoderm [mĕz'·ĕk·tō·dûrm", mēs"·ĕk·tō·dûrm'] mesenchyme cells supposedly derived from the ectoderm (q.v.) as those budded off from the primitive streak of an avian egg

mesencephalon [mĕz"·ĕn·sĕf'·əl·ŏn, mēs"·ĕn·sĕf'·əl·on] the midbrain; that is, the part which lies between the diencephalon and the metencephalon

mesenchyma [mĕz"·ĕn·kə·mə, mēs"·ĕn·kĭm"·ə] that tissue which lies between xylem and phloem in vascular bundles in roots

mesenchyme [mĕz"·ĕn·kĭm, mēs"·ĕn·kĭm] those portions of the embryonic mesoderm which are not segregated into layers or blocks

mesencoel [mĕz'·ĕn·sĕl, mēs'·ĕn·sĕl] a coelom produced through the rearrangement of mesodermal cells to enclose a space

mesentery [mĕz'·ĕn·tə·rē] the fold of mesoderm in which the viscera of coelomate animals is suspended

mesenteric artery [mĕz"·ĕn·tĕr'·ĭk, mēs"·ĕn·tĕr·ĭk] one or more derivatives of the omphalomesenteric artery

mesentoderm [mĕz"·ĕnt·ō·dûrm, mēs"·ĕnt·ō·dûrm] the layer of cells over the archenteron, produced by the involution of cells through the blastopore of the gastrula

mesoblast [mĕz"·ō·blăst, mēs"·ō·blast] the middle germ layer in early development of the telolecithal egg

mesobronchus [mĕz'·ō·brŏnk"·əs, mĕz·ō·brŏnk"·əs] a bronchus passing down the center of the lung in birds and reptiles

mesocardium [mĕz"·ō·kârd"·ē·əm, mēs·ō·kârd"·ē·əm] the vertical plate which primitively unites the ventral side of the heart in the embryo to the ventral surface of the body

mesocoel [mĕz'·ō·sĕl", mēs·ō·sĕl"] the cavity of the mesencephalon and thus the connection between the third and fourth ventricles in lower vertebrates (cf. iter)

mesocotyl [mĕz"·ō·kŏt'·əl, mēs"·ō·kŏt·əl] the node between the sheath and the cotyledon of seedling grasses

mesoderm [mĕz"·ō·dûrm', mēs"·ō·dûrm] the middle of the three layers of triploblastic animals. In coelomate forms it is divided into an outer and an inner layer (*see also* endomesoderm, ectomesoderm). It gives rise to the muscular, skeletal and blood vascular systems

mesodermic band [mĕz'·ō·dûrm"·ĭk, mēs'·ō·dûrm"·ĭk] a strip of mesoderm cells in the larva of annelids and some mollusks

mesogloea [mĕz'·ō·glē"·ə, mēs'·ō·glē"·ə] a transparent gelatinous matrix found in the walls of sponges and between the ectoderm and endoderm of coelenterates

Mesohippus [mēs'·ō·hĭp"·əs] the earliest three-toed horse, found in the early and middle Oligocene. It appears to have been the direct ancestor of *Miohippus*

mesolecithal = medialecithal

mesolimnion [mĕz'·ō·lĭm"·nē·ən, mēs'·ō·lĭm"·nē·ən] the middle layer of a thermally stratified lake

mesonephros [mĕz'·ō·nĕf"·rŏs, mēs'·ō·nĕf"·rŏs] a kidney composed of units each of which has both a Bowman's capsule containing a glomerulus and a coelomic funnel. It occurs as a developmental stage in all vertebrates and is the functional kidney of aquatic vertebrate larvas

mesomitosis [mĕz'·ō·mĭt·ōs"·ĭs, mēs'·ō·mĭt·ōs"·ĭs] *either* mitosis taking place within a complete nuclear membrane *or* that form of division in which the endosome furnishes the chromosomes, but not the centriole

mesopelagic [mĕz'·ō·pĕl·ăj"·ĭk, mēs'·ō·pĕl·ăj"·ĭk] referring to organisms living at moderate depths (sixty to one hundred fathoms) in the ocean

mesophyll [mĕz"·ō·fĭl", mēs"·ō·fĭl'] all that portion of an angiosperm leaf which is enclosed within the epidermis. In gymnosperm leaves, the tissue lying between the central bundle and the epidermis

mesopterygium [mĕz'·ō·tə·rĭj"·ē·əm, mēs'·ō·tə·rĭj"·ē·em] the cartilaginous rod which supports the base of the pectoral fin in elasmobranch fish

mesosaprobic [mĕz'·ō·săp·rŏb"·ĭk, mēs'·ō·săp·rŏb"·ĭk] descriptive of an aqueous habitat with a relatively low oxygen, and relatively high decomposing organic matter, content

mesosere [mēs"·ō·sēr'] an intermediate between two other stages

mesothorax [mĕz'·ō·thôr"·ăks, mēs'·ō·thôr"·ăks] the central of the three divisions of the insect thorax

mesotrophic [mĕz'·ō·trôf"·ĭk, mēs·ō·trôf"·ĭk] moderately nutrient, used particularly of water or wet lands

mesotrophic peat that which is found in moist, but not boggy, environments

Mesozoic era [mĕz'·ō·zō"·ĭk, mēs'·ō·zō"·ĭk] a geologic era extending from about 200 million years ago to 60 million years ago. It is divided into the Cretaceous, Jurassic and Triassic periods. It was preceded by the Paleozoic era and followed by the Cenozoic era

-mesto- *comb. form* meaning "filled"

-meta- *comb. form* which properly denotes "among".

Biologists have extended this meaning to "behind", "between", "adjacent", "later" and numerous other fanciful ideas, none of which, however, are as bad as the chemists "having least water". The use of meta- as the third of a series commencing with pro-, meso- is common, though without classical justification

metabolic [mĕt'·ə·bŏl"·ĭk] pertaining to the sum total of the physiological activity of living matter (see also ametabolic, photometabolic)

metabolous [mĕt·ăb"·ŏl·əs] pertaining to metamorphosis (see ametabolous, hemimetabolous, holometabolous, paurometabolous)

metacarpal bone [mĕt'·ə·kär"·pəl] one of those bones in the hand which lie between the phalanges and the carpals (see also carpal bone)

metacentric chromosome [mĕt'·ə·sĕn"·trĭk] one in which the centromere is medially located (see also acrocentric chromosome, telocentric chromosome)

metacercaria larva [mĕt'·ə·sûr·kår"·ē·ə] a cercaria which is encysted and has metamorphosed into a juvenile fluke

metachromatic granules [mĕt'·ə·krōm·ăt"·ĭk] the variously staining granules of some bacteria particularly those of diphtheria

metacoel [mĕt'·ə·sēl"] the cavity of the metencephalon and also an extension of the fourth ventricle into the cerebellum in lower vertebrates

metafemale [mĕt'·ē·fē"·māl] one having additional female sex determinants, such as an XXX chromosome

metagenesis [mĕt'·ə·jen"·ə·sĭs] a term preferred by some to "alternation of generations" for polymorphic coelenterates

metagynous [angl. mĕt·ăj"·ən·əs, orig. mĕt'·ə·jĭn"·əs] condition of an organism in which the male matures before the female

metamere [mĕt"·ə·mēr'] one division of a metamerically segmented organism

metameric [mĕt'·ə·mĕr"·ĭk] having parts in company with one another

metameric segmentation the form of segmentation involving the replication of similar parts throughout the length of one organism

metamerism [angl. mĕt·ăm"·er·ĭzm, orig. mĕt'·ə·mēr"·ĭzm] the condition of having many parts joined together as the segments of a polychaete annelid

metamorphosis 1 (see also metamorphosis 2) [angl. mĕt·ə·môrf"·ə·sĭs, orig. mĕt'·ə·môrf·ōs"·ĭs] transformation of one part into another, particularly in plants. The following derivative terms are defined in alphabetic position:
DESCENDING M.   REGRESSIVE M.
PROGRESSIVE M.

metamorphosis 2 (see also metamorphosis 1) change of shape, particularly applied to the changes of one larval form to another larval form or of a larval form to a nymph or adult. The following derivative terms are defined in alphabetic position:
COMPLETE METAMORPHOSIS
DIRECT METAMORPHOSIS
HYPERMETAMORPHOSIS
INCOMPLETE METAMORPHOSIS
SECONDARY METAMORPHOSIS

metamyelocyte [mĕt'·ă·mī"·ə·lō·sĭt'] the penultimate stage in the development of a polymorph leucocyte. The following derivative terms are defined in alphabetic position:

BASOPHILIC METAMYELOCYTE
EOSINOPHILIC METAMYELOCYTE
NEUTROPHILIC METAMYELOCYTE

metanauplius larva [mĕt'·ə·nôrp"·lē·əs] the stage immediately following the nauplius larva in which three further rudimentary pairs of appendages are apparent

metanephromixium [mĕt'·ə·nĕf'·rō·mĭks"·ē·əm] a nephridium which serves also as a genital duct

metanephros [mĕt'·ə·nĕf"·rōs] a kidney composed of units which possess a Bowman's capsule but lack a coelomic funnel. It is the functional adult kidney of reptiles, birds and mammals

metanym [mĕt'·ə·nĭm] a species name rejected on the basis that another organism of the same genus already bears this specific designation

metaphase [mĕt'·ə·fāz'] the second stage of mitotic division in which a spindle is formed and the chromosomes lie midway between the plates

metaphysis [angl. mĕt·ăf"·əs·əs, orig. mĕt'·ə·fĭs"·ĭs] a segment of a long bone occupied by an epiphyseal disk and by newly formed bone on the diaphyseal side of the disk

metaplasia [mĕt'·ə·plāz"·ē·ə] the production of one kind of tissue by cells belonging to another kind of tissue

metapleural fold = metapleure

metapleure [mĕt"·ə·plōōr'] one of a pair of integumentary folds on the latero-central surface of cephalochordates

metapodium [mĕt·ə·pōd"·ē·əm] the middle region of the foot, either in vertebrates or invertebrates

metapodosoma [mĕt'·ə·pōd'·ō·sōm"·ə] that division of the acarine body which carries the third and fourth pair of legs

metapterygium [mĕt'·ə·tĕr·ĭj"·ē·əm] cartilaginous rod supporting the pelvic girdle in elasmobranch fish

metapterygoid bone [mĕt'·ə·tĕr"·ĭg·oid] one of a pair of bones forming part of the palatoquadrate complex of actinopterygian fish, lying slightly posterior and dorsal to the pterygoid bone and thought by many to be homologous with the epipterygoid bone (see also pterygoid bone, ectopterygoid bone, epipterygoid bone)

metarhodopsin [mĕt'·ə·rōd·ŏp"·sĭn] the intermediate product between lumirhodopsin and scotopsin and retinene

Metasequoia [mĕt'·ə·sə·kwoi"·ə] a genus of coniferale trees having oppositely arranged flattened aciculate leaves. *M. glyptostroboides* [glĭp'·tō·strō·boid"·ēz], the sole extant species, was only discovered a few decades ago even though the genus had been named from Ordovician fossils

metastoma [mĕt'·ə·stōm"·ə] the chitinous lobe immediately below the mouth in Crustacea; if divided into two lobes, each is called a paragnath

metatarsal bone [mĕt'·ə·tär"·səl] one of those bones of the vertebrate foot which lie between the tarsals and the phalanges (see also tarsal bone)

metatarsus [mĕt'·ə·tär·səs] in insects the proximal tarsal joint, particularly when it is sufficiently conspicuous to be differentiated easily

Metatheria a subclass of mammals often called the marsupials. They do not develop a placenta, save in a few cases, so that the young are born in an embryonic condition and continue their development attached to a teat in a pouch. The skull is easily distinguished from the eutherian skull by the infolding of the lower angle of the dentary under the floor of the mouth. The genera *Didelphys, Notoryctis, Phascolarctus, Sarcophelus* and *Thalacinus* are the subjects of separate entries (*see also* Protheria, Eutheria)

metathorax [mĕt'·ə·thôr"·ăks]   the posterior of the three divisions of the insect thorax

metatrochophore larva [mĕt'·ə·trŏk'·ō·fôr] a trochophore larva possessing secondary ciliated bands in the posttrochal hemisphere

metatrophic [mĕt'·ə·trŏf"·ĭk] existing on organic nutrients only

metaxylem [mĕt'·ə·zĭl"·əm] the central, compacted wood of a tree or xylem maturing after the completion of elongation

Metazoa [mĕt·ə·zō·ə] a rarely used taxon of the animal kingdom into which are grouped all animals other than Protozoa

metecdysis [mĕt'·ĕk·dĭs"·ĭs] the period immediately following ecdysis in arthropods

metencephalon [mĕt'·ĕn·sĕf"·əl·ən] that portion of the brain which rises immediately behind the mesencephalon

metestrous [mĕt·ēs"·trəs] the period immediately following estrous

methionine [mə·þĭ·ŏn·ĭn] α-amino-γ-methylmer-captobutyric acid. $CH_3SCH_2CH_2CH(NH_2)COOH$. An amino acid essential to the nutrition of rats. It is important not only as a source of sulfur in the biosynthesis of other sulfur-containing amino acids but is also a common methyl donor in transmethylation reactions

metoecious parasite [mĕt·ēs"·ē·əs] one that is not host-specific

Metridium [mə·trĭd"·ē·əm] a genus of actiniarian coelenterates. The relatively large, brown or yellow *M. dianthus* [dĭ·ănth"·əs] is the commonest sea anemone on the coasts of North America

micelle [mĭ·sĕl"] a unit of a colloid built of polymeric molecules

micellar system [mĭ·sĕl"·ər] the portion of a plant cell wall that consists of interconnecting chain molecules of cellulose

-micr- *comb. form* meaning "minute"

microclimate [mĭ'·krō·klĭm"·ăt] a small restricted area, such as that under a decomposing log, which differs markedly from the general surrounding climate

micrococcal nuclease [mĭ'·krō·kŏk"·əl] an enzyme that hydrolyzes RNA and DNA, particularly at the adenine-thymine linkage

Micrococcus [mĭ'·krō·kŏk"·əs] a genus of eubacteriale schizomycetes usually occurring in irregular clusters or in tetrad. *M. glutamicus* [glōō·tăm"·ĭk·əs] is used in the commercial production of amino acids. *M. radiodurans* [rā'·dē·ō·dyōōr"·ănz] is the most resistant to radiation of any known organism

microconjugant [mĭ'·krō·kŏn"·jōō·gənt] the smaller of the two products of progamous fission

microcyst [mĭ'·krō·sĭst"] a small cyst, though used as a specific technical description of the cysts of Myxomycophyta

microenvironment = microclimate

microflora [mĭ'·krō·flôr"·ə] either the sum total of those plants in any locality that are too small to be distinguished by the naked eye or the dwarf plants of high mountains

microgamete [mĭ'·krō·găm"·ēt] the smaller, and by convention male, of a pair of heterogametes

microglia [mĭ'·krō·glē"·ə] a general term for glial cells of mesodermal origin

microhabitat [mĭ'·krō·hăb"·ĭt·ăt] the condition produced by the existence of a microclimate

Microhydra [mĭ'·krō·hĭd"·rə] a minute freshwater hydroid usually living in association with colonies of Ectoprocta. The medusoid stage was known long before the hydroid and placed in a separate "genus" *Craspedocusta*

micromutation [mĭ'·krō·myū·tā"·shŭn] one which does not markedly alter the phenotype though the ultimate result of many micromutations may be a new geographic race or subspecies

micron [mĭk"·rən] micromillimeter (0.001 mm) usually represented by the symbol μ (*see also* millimicron)

micronephridium [mĭ'·krō·nĕf·rĭd"·ē·əm] an annelid nephridium lacking a coelomic funnel

micronucleus [mĭ'·krō·nyūk"·lē·əs] the diploid nucleus of ciliate protozoa (cf. macronucleus)

microphage = neutrophil

microphotograph [mĭk'·rō·fōt'·ō·grăf"] a term often misused for photomicrograph. A microphotograph is actually a minute photograph and has no place in biological technique

Micropterus [mĭ·krŏpt"·ər·əs] the genus of teleost fish containing the freshwater basses. *M. salmoides* [săl·moid"·ēz] is the large-mouth and *M. dolomieui* [dŏl'·ĕm·myū"·i] the small mouth bass

micropyle [mĭ'·krō·pĭl"] a perforation in the envelope of an egg; in insects, the aperture in the egg membrane through which the sperm enters. In botany an opening from the integument from the nucellus to the exterior of an ovule

microsclerite [mĭ'·krō·sklēr"·ĭt] a small sponge spicule, that does not unite to form a definite skeleton (*see also* megasclerite)

microscope [mĭ'·krō·skōp"] a device for viewing small objects consisting essentially of a magnifying system called the objective and a viewing system called the ocular. Magnification is of less importance than resolution (q.v.) (*see also* electron microscope, phase contrast microscope)

microscopic [mĭ'·krō·skŏp·ĭk] invisible to the naked eye but visible under an optical microscope (*see also* ultramicroscopic)

microsome [mĭ'·krō·sōm"] originally any cell granule, not otherwise identified as to structure or function, but now usually confined to a small ribonucleic acid-containing granule within the endoplasmic reticulum

microspecies [mĭ'·krō·spēsh"·ēz] a genetically isolated species, usually one reproducing only parthenogenetically

microsporangium [mĭ'·krō·spôr·ănj"·ē·əm] a sporangium producing microspore and specifically that part of the stamen which contains the pollen

microspore [mĭ'·krō·spôr"] literally any small spore but usually confined to the pollen grain of spermatophytes

microsporocyte [mĭ'·krō·spôr'·ō·sīt"] pollen mother-cell

microsporophyll [mĭ'·krō·spôr'·ō·fĭl"] a leaf-like microsporangium or, in angiosperms, the stamen. The term is also used as synonymous with microstrobilus

microstrobilus [mĭ′·krō·strŏb″·əl·əs] the male cone of cycads

microtomy [mĭ·krŏt″·ō·mē] the art of cutting thin sections but sometimes extended to all phases of the preparation of materials for examination with the microscope

Microtus [mĭ·krŏt″·əs] a very large genus of myomorph rodents usually called "voles". They are capable of carrying, but do not die from, *Pasteurella pestis*, so that they act as a natural endemic reservoir of bubonic plague

microvillus a villus-like protuberance of a cell surface visible only by electron microscopy

microzooid [mĭ′·krō·zō″·ĭd] a male algal gamete

micrurgy [mĭ·krûr″·gē] surgical manipulations at the cellular level

Micrurus [mĭ·krûr″·əs] the genus of Squamata containing the venomous coral snakes. The false, or harmless, coral snakes are in the genus *Lampropeltis*

mict that which is produced by mixing. The following terms using this suffix are defined in alphabetic position:
AMPHIAPOMICT   APOMICT
AMPHIMICT   HAPLOMICT

mictic egg [mĭkt″·ĭk] an egg that is capable of developing either parthenogenetically or after fertilization (*see also* amictic egg)

midbrain = mesencephalon

migrant any organism that wanders, but particularly those that move seasonally between fixed places

migrarc [mig″·rärk′] the zone within which migration takes place

milk gland 1 (*see also* milk gland 2, 3) the milk-producing portions of a mammary gland

milk gland 2 (*see also* milk gland 1, 3) the enlarged sweat glands providing nourishment to the newly hatched protherians

milk gland 3 (*see also* milk gland 1, 2) any gland which produces a nutrient for the nourishment of young, specifically applied by entomologists to such a gland in some dipterans

milk tooth one of the first generation of deciduous teeth in a mammal

Milleporina [mĭl′·ē·pôr·ĭn″·ə] an order of hydrozoan coelenterates often called the false corals. The hydroid colony lies on the surface of a massive calcareous deposit in special cavities in which the medusae, subsequently becoming free, are formed

millimicron [mĭl″·ē·mĭ·krŏn′] 0.000001 mm usually represented by the symbol mμ (*see also* micron)

mimic gene one which produces a similar effect to another gene to which it is nonallelic

mimicry the assumption of the form of one animal by another (*see also* Batesian mimicry, Müllerian mimicry)

Mimosa [mĭm·ōz″·ə] a large genus of leguminous plants. *M. pudica* [pyōō″·dĭk·ə] (the sensitive plant) shows a remarkable thigmonastic reaction

mimotype [mĭm″·ō·tĭp′] a form which resembles another in general shape, but not in genetic makeup, and which occurs in a similar ecological niche in another part of the world

minimum lethal dose *see* MLD

Miocene epoch [mī″·ō·sēn′, mī″·ə·sēn′] one of the tertiary geologic epochs extending from about 25 million years ago to 10 million years ago. Anthropoids appeared in this epoch

Miohippus [mī′·ō·hĭp″·əs] a three-toed horse which appeared in the Oligocene and survived into the Miocene

miracidium larva [mĭr′·ə·sĭd″·ē·əm] the free-swimming, conical ciliated larva which hatches from the egg in trematodes

-mist- *comb. form* meaning "mingled"

mitochondrion [mĭt′·ə·kon″·drē·ən, mĭt′·ō·kon″·drē·ŏn] an organelle, usually sausage-shaped but varying from spherical to filamentous, found in the cytoplasm of all cells except bacteria and Cyanophyceae. The interior is imperfectly partitioned with usually transverse, but sometimes longitudinal or concentric, cristae which, or granules in which, are a primary source of ATP and hence control the oxidative metabolism of the cell

mitosis [mĭ·tōs″·ĭs] the sum total of the nuclear and chromosomal activities in those cell divisions in which the diploid chromosome number is retained. The following derivative terms are defined in alphabetic position:
AMITOSIS   MESOMITOSIS
C-MITOSIS   PROMITOSIS
CRYPTOMITOSIS   PSEUDOMITOSIS
ENDOMITOSIS
EUMITOSIS
INTERMITOSIS

mitotic figure [mĭ·tŏt″·ĭk] the general appearance of cell in mitosis at some particular phase (cf. -phase)

mitotic spindle the figure formed in a dividing cell by the actual rays spreading out from each centriole to the equatorial plate

mitr- *comb. form* originally meaning "headdress" but later confined to a bishop's mitre

mitral valve [mī″·trəl] one of the valves in the left-hand side of the atrio-ventricular canal of a four chambered heart

-mix- *comb. form* meaning "mixing" and, by extension, "breeding" and "fertilization"

mixed association one in which several species compete for dominance

mixed formation an assemblage of several kinds of vegetation, such as a veldt

mixed gland one which secretes both mucous and serous material

-mixis- = -mix- the following terms using this suffix are defined in alphabetic position:
AMIXIS   ENDOMIXIS
AMPHIMIXIS   HEMIMIXIS
APOMIXIS   PANMIXIS
AUTOMIXIS

mixoeuhaline [mĭks′·ō·yōō·hāl″·ĭn] said of estuarine waters that contain more than 30 parts per thousand of dissolved salts but less than the concentration of the adjacent seas

mixohaline [mĭks′·ō·hāl″·ĭn] said of all waters that contain anywhere from 0.5 to 30 parts per thousand dissolved salts

mixomesohaline [mĭks·ō·mĕz′·ō·hāl″·ĭn] said of brackish water containing from 5 to 18 parts per thousand dissolved salts

mixooligohaline [mĭks′·ō·ŏl′·ĭg·ō·hāl″·ĭn] said of brackish waters containing from 0.5 to 5 parts per thousand dissolved salts

mixopolyhaline [mĭks′·ō·pŏl′·ē·hāl″·ĭn] said of brackish water containing from 18 to 30 parts per million dissolved salts

mixotrophic [mǐks'·ō·trôf·ǐk] said of a plant capable of both holophytic and saprophytic nutrition

MLD minimum lethal dose. "50 MLD" is the minimum dosage that kills 50% of the test organisms

modifier complex an assemblage of modifier genes

modifier gene one which affects the expression of other loci

modifying gene one which has no other function than to modify the expression of another gene

Mola [mō'·lə] a genus of teleost fish containing the gigantic ocean sunfish *M. mola*. The vertically compressed body, with a relatively small crescentic tail, may weigh as much as a ton

molar tooth literally a grinding tooth but specifically applied to the posterior teeth in the jaws of mammals

Molgula [mŏl"·gyōō·lə] a large genus of ascidians with a much folded branchial sac. The test is often encrusted with sand. *M. manhattenensis* [măn'·hăt·ĕn·ĕn"·sǐs], an ovoid form about an inch in diameter by two inches long, is the commonest East Coast ascidian

Mollusca a phylum of bilateral coelomate animals containing such forms as the clams, snails, and squids. The soft body is sheathed in a mantle which, in most forms, secretes a calcareous shell. The strong, muscular "foot" is variously modified for locomotion or predation. The classes Amphineura, Cephalopoda, Pelecypoda and Scaphopoda are the subjects of separate entries

molting hormone an arthropod neurohormone inducing ecdysis

monacmic plankton one showing a single bloom per year (*see also* diacmic plankton)

monad [mŏn"·ăd] a general term applied to minute, colorless flagellates, frequently showing amoeboid motion, and to zoospores

monagamy [*angl.* mŏn·ăg"·ə·mē, *orig.* mŏn'·ō·găm"·ē] the condition of having only one sexual partner

monarch stele [mŏn'·ärk] a xylem bundle with a single thread

monaxenic [mŏn'·ăks·zĕn"·ǐk] said of an axenic culture to which one *other* kind of organism has been added. Similarly a diaxenic culture contains three types of organism and so on

monaxial symmetry [mŏn·ăks"·ē·əl] a type of symmetry, such as radial symmetry, in which only one axis can be cut, to produce identical halves

Monera [mŏn"·ər·ə] a taxon erected to contain those acellular organisms (i.e. Bacteria, Cyanophyta) which have a haploid nucleus lacking a nuclear membrane

-monili- *comb. form* properly meaning "necklace", but usually, by extension, "in the form of a string of beads"

Monilia [mŏn·ǐl"·ē·ə] a genus of deuteromycete fungi of which one (*M. albicans* [alb"·ē·kănz']) is pathogenic to man (*see also Monilinia*)

Monilinia [mŏn'·ǐl·ǐn"·ē·ə] a genus of pezizale ascomycete fungi often referred to the deuteromycete genus Monilia when only the asexual form is known. Thus the asexual form of *Monilinia fructigena* [frŭk·tǐj·ə·nə] (the cause of brown rot in stone fruit) is still often called *Monilia fructigena*

-mono- *see* -mon-

monoblast [mŏn"·ō·blăst"] a descendant of a hemocytoblast, and direct ancestor of a monocyte

monoclimax theory [mŏn'·ō·klim"·aks] the concept that a given region has only a single potential climax which is determined by the climate

Monocotyledoneae a class of Angiospermae named from the fact that the germinating seed produces a single cotyledon. They are also distinguished by the fact that vascular bundles are scattered through the stem but not arranged in a cylinder. The leaves are very commonly parallel veined. The genera *Allium, Avena, Tradescantia, Trillium, Valisneria,* and *Zea* are the subjects of separate entries

Monocystis [mŏn'·ō·sǐs"·tǐs] a genus of sporozoan protozoans parasitizing earthworms. Trophozoites and sporozoites, the latter called "pseudonavicellae" from their superficial resemblance to the diatom of that name, can almost always be found in smear preparations from the seminal vesicle

monocyte [mŏn'·ō·sǐt"] a large leucocyte with a slightly indented nucleus, and occasionally very fine neutrophil granules

Monodon [mŏn"·ō·don'] the genus of cetacean mammals (whales) containing the narwhal (*M. monoceros* [mŏn'·ō·sĕr"·ŏs]). The upper left canine tooth grows into an enormous anteriorly protruding "horn"

monoecious [mŏn·ēs"·ē·əs] said of organisms in which the male and female sex organs occur in the same individual, particularly plants bearing both male and female flowers. The following derivative terms are defined in alphabetic position:
AGAMOGYNOMONOECIOUS   COENOMONOECIOUS
ANDROMONOECIOUS   DIMONOECIOUS

monoestrous [mŏn·ēs"·trəs] said of mammals in which only a single estrous cycle occurs at one breeding season

monogeneric [mŏn'·ō·jĕn·ĕr"·ǐk] said of a taxonomic unit, such as a family, containing only one genus (*see also* monotypic)

monogynous [*angl.* mŏn'·ŏj"·ə·nəs, *orig.* mŏn'·ō·jin"·əs] the condition of having a single fertile female as in the colony of many social insects or, in higher forms, having only one female sexual partner

monokaric [mŏn'·ō·kâr"·ǐk] possessing a single nucleus

monokont [mŏn'·ō·kŏnt] having a single flagellum

monomorphic [mŏn'·ō·môrf"·ǐk] having only one form, but particularly applied to organisms, such as many cladoceran Crustacea, in which only one sex is known

monophyletic [mŏn'·ō·fil·ĕt"·ǐk] descending from a single stock or individual

monopodial [mŏn'·ō·pōd"·ē·əl] the type of growth in a hydrozoan in which the branch is topped by a terminal hydranth, and which elongates by reason of a growth zone below the hydranth

monosomic [mŏn'·ō·sōm"·ǐk] the condition of lacking one chromosome from the genome (*see also* disomic, nullisomic)

Monotremata [mŏn'·ō·trĕm"·ə·tə] a subclass of primitive mammals, containing the only extant Prototheria. They are egg-laying and with duck-like bills. The two families are the Tachyglossidae and the Ornithorhynchidae containing respectively the spiny anteaters and the duck-billed platypus

monotrophic [mŏn'·ō·trôf"·ǐk] said of a parasite that is host-specific, a predator that feeds only on one species, or a bee that visits only one species of flower

monotypic [mŏn'·ō·tĭp·ĭk] said of a taxon, usually a genus, represented by only one species

monoxenic [mŏn'·ō·zēn"·ĭk] said of a parasite which is host specific

Monro's foramen one of the two foramina which connect the third ventricle of the brain to the first two ventricles

Monro's sulcus the longitudinal groove that divides the embryonic central nervous system into dorsal and ventral regions

monsoon forest [mŏn·sōōn"] a tropical deciduous forest fluctuating seasonally between wet and dry

Montgomery's gland one of the areolar glands around the nipple in the mammalian breast

moor [môr] an area of predominantly peaty soil. The following derivative terms are defined in alphabetic position:

HIGH MOOR   LOW MOOR   MOSS MOOR

Moraceae [mŏr·ās"·ē] the family of dicotyledons that contains the mulberries, bread fruits, hemps and hops. The family is easily distinguished from its nearest neighbors, the nettles, by the milky juice and from the elms by its elastic stamens. The genus *Cannabis* is the subject of a separate entry

Morchella [môr·kĕl"·ə] a genus of pezizale ascomycete fungi known as morels. The pitted or wrinkled elongate-ovoid head is born on a short stalk. *M. edulis* [ĕd"·yōō·lĭs] is one of the gastronomically most esteemed mushrooms

mores [môr"·ez] plural of mos, but mostly used when referring to the habits or customs of the mos

morgan a unit of genetic recombination. For example a centimorgan refers to a one per cent recombination rate

Morgan's canon a biological restatement of the principle better known as Ockham's [Occam's] razor

Morgani's hydatid a remnant of the pronephric duct in an adult animal

-morph- *comb. form* meaning "shape", "form" or "type" of something. The following terms using this prefix are defined in alphabetic position:

ALLELOMORPH   HYPOMORPH
ANTIMORPH   NEOMORPH
GYNECOMORPH   POLYMORPH
GYNANDROMORPH
HYPERMORPH

morphallaxis [môrf'·əl·ăks"·ĭs] the reorganization of existing tissues in the course of regeneration (cf. epimorphosis)

-morphic *comb. form* meaning "having the shape of" or "like". Many of these terms duplicate in meaning words ending in -form. The following terms using this suffix are defined in alphabetic position:

CLINOMORPHIC   MONOMORPHIC
DIMORPHIC   POLYMORPHIC
HETEROMORPHIC

morphine the principal alkaloid derived from opium

-morphism the condition of having a shape. The following terms having this suffix are defined in alphabetic position:

CYCLOMORPHISM   PLEOMORPHISM
DIMORPHISM   POLYMORPHISM

morphogenesis [môrf'·ō·jen"·əs·ĭs] the development of shape and form

morphological synecology the relation of the morphology of an organism to its position in the community in which it lives (*see also* synecology, dynamic synecology, geographic synecology)

morphosis [môrf·ōs"·ĭs] the manner of development. The following terms using this suffix are defined in alphabetic position:

ALLOMORPHOSIS   HETEROMORPHOSIS
ANAMORPHOSIS   HYPERMORPHOSIS
CYTOMORPHOSIS   METAMORPHOSIS
EPIMORPHOSIS   NEOMORPHOSIS
GERONTOMORPHOSIS   THIGMOMORPHOSIS

-morphous *adj. term.* synonymous with -morphic, the form preferred in this work

morula [môr"·yū·lə] the solid ball of cells which results from cleavage of most mammalian eggs (*see also* amphimorula)

mos [mŏs] an assemblage of organisms "held together by the will and humor of each other" (Publius Terence, 185 B.C.). Used by contemporary ecologists in much the same sense to describe a community of one or more species living in amity but without mutual interdependence (cf. mores)

mosaic the occurrence of two types of genetically homozygous tissue in a heterozygote as the result of a mitotic chromosomal exchange

mosaic egg one in which developmental patterns are established before cleavage so that induced modification of cleavage patterns result in modification of development (cf. regulative egg)

moss moor a boggy moor, which is sufficiently dry to grow sphagnum

mother cell one which divides to form a specific group of cells

-motor- *comb. form* meaning "movement" (*see* pilomotor, vasomotor)

motor nerve= afferent nerve

motor neuron a neuron that originates stimuli

motor nucleus a group of nerve cells which originates nerve impulses for transmission to a muscle

mountant the material in which wholemounts and sections for microscopical examination are preserved between the coverslip and the slide

mouth any aperture leading to a cavity or tube, particularly the anterior aperture of the alimentary canal. The term is frequently improperly used for the buccal cavity

mucigen the precursor of mucous

mucopolysaccharide [myū'·kō·pŏl'·ē·săk"·ər·id] a class of polysaccharides widely distributed in the animal body and forming the principal ingredient of mucous secretions

Mucor [myū"·kôr] a genus of Zygomycetes commonly found on cereals. *R. javanicus* [jăv·ăn"·ĭk·əs] is used in the commercial reduction of starches to sugars

mucous gland one which secretes mucus

mucus [myū"·kəs] any slimy secretion of animal origin

Müllerian association an assemblage of different organisms in one geographical locality all showing similar aposematic colors

Müllerian mimicry *either* that shown by several species having the same aposomatic colors in common *or* that form which appears unnecessary since both the imitator and the imitated are inedible

Müller's duct the embryonic mesonephric duct which in the female adult becomes the oviduct

Müller's larva an elongate post-trochophore stage

in the development of Turbellaria distinguished by the presence of eight, backwardly directed, post-oral lobes

mulm amorphous organic detritus in freshwater

multangular bones [mŭlt·ăng"·gyōō·lə] those carpals which lie under the first and second metacarpals. The greater multangular lies under the first metacarpal between the radius and the navicular. The lesser multangular lies under the first metacarpal (see also angular bone, supraangular bone)

multi- comb. prefix meaning many

multiparous in zoology, bringing forth many young at one birth

multiple fission a division of a single cell into many cells, as in spore formation in Protozoa

multiple fruit an aggregate fruit in which the whole appears to grow as a single fruit. The pineapple is an example

multipolar ingression [mŭlt'·ē·pōl"·ər] the production of an endoderm by the budding off of cells from all parts of the blastula wall

multivoltine [mŭlt'·ē·vŏl"·tin] reproducing several times each season

-mur- comb. form meaning either "mouse" or "wall"

Murphy's law the bills of birds dwelling on islands are longer than those of the same species inhabiting the mainland

Mus a large genus of rodents containing more than a score of species of which one, M. musculus [mŭs"·kyū·ləs] (the house mouse), has become a commensal of man in every corner of the globe

-musc- comb. form meaning either "moss" or "dipteran fly"

Musci a phylum of bryophyte plants containing the mosses. The distinctive character, apart from the moss-like appearance, is the simple thallose protonema bearing upright gametophores

Muscidae [mŭs"·kĭd·ē] a very large family of dipteran insects containing the house flies, stable flies, tsetse flies, etc. The mouth parts vary greatly some, as in the house fly, being adapted to the absorption of fluid while others, such as those of the stable fly are piercing. The genus Glossina is the subject of a separate entry

muscle 1 (see also muscle 2) the contractile tissue of animals. The following derivative terms are defined in alphabetic position:

CARDIAC MUSCLE	SMOOTH MUSCLE
CIRCUMSCRIPT M.	STRIATED MUSCLE
DIFFUSE MUSCLE	VOLUNTARY MUSCLE
INVOLUNTARY M.	

muscle 2 (see also muscle 1) a mass of muscular tissue of specific function. The following derivative terms are defined in alphabetic position:

AMBIENS MUSCLE	PAPILLARY MUSCLES
APPENDICULAR M.	SKELETAL MUSCLE
CILIARY MUSCLE	SOMATIC MUSCLE
EPAXIAL MUSCLE	SPINDLE MUSCLE
LEVATOR MUSCLE	TRUNK MUSCLE
MASSETER MUSCLE	VISCERAL MUSCLE

muskeg [mŭs"·kĕg] moss bogs of the northern Canadian region

Mustela [mŭs"·tə·lə, mŭs·tēl"·ə] the type genus of the Mustelidae. They are small predators of very wide geographic distribution. Most are called weasels. M. rixosus [rĭks·ōs"·əs], for example, is the least weasel, and is circumpolar in distribution. Others, not called weasels, are M. erminea [ĕr·min"·ē·ə], the winter phase of which is the ermine, M. vison [vī"·sŏn] (the mink) and M. nigripes [nig"·rĭp·ēz] (the black-footed ferret). The european ferret, domesticated for rabbit hunting, is either a separate species (M. furo [fyōōr'·ō]) or a derivative of the European polecat (M. foetidus [fēt"·ĭd·əs])

Mustelidae a family of low slung carnivorous mammals containing the weasels, mink, badgers, skunks, otters and their immediate allies; they are distinguished by having only one molar in each jaw, or at the most two in the lower jaw, and in lacking an alisphenoid canal. The genera Mephitis, Mustela and Spilogale are the subject of separate entries

mutagenic [myōō'·tə·jĕn"·ĭk] that which produces mutations

mutant the result of a mutation

mutase [myōō·tāz'] a general term for those enzymes which catalyze intramolecular transfers

mutation [myōō'·shun] a heritable change in a genetic character resulting either from a change in the gene at a locus, or alterations in chromosomal structure. The following derivative terms are defined in alphabetic position:

BACK MUTATION	MICROMUTATION
BLOCK MUTATION	POINT MUTATION
FRACTIONAL M.	PREMUTATION
LETHAL MUTATION	REVERSE MUTATION

muton [myōō"·tŏn] the smallest unit of mutation (a single nucleotide pair) within a cistron (q.v. and c.f. recon)

mutual adaptation an adaptation resulting in a symbiosis of harm to neither partner

mutualism at one time a synonym of symbiosis, later extended to include both symbiosis and commensalism, and now frequently used for any association between two organisms including parasitism

mutualistic symbiosis symbiosis resulting in mutual advantage

mycelium [mi·sĕl"·ē·əm] a matted mass of fungus hyphae (see also promycelium, pseudomycelium)

myceloid [mi'·sə·loid, mi·sĕl"·oid] said of a bacterial culture having a filamentous appearance as though it were composed of mycelia

Mycetozoa see Myxomycophyta

Mycobacterium [mi'·kō·băk·tēr"·ē·əm] a genus of Actinomycetales. M. tuberculosis [tyōō'·bûrk·kyōō·lōs"·ĭs] is the causative agent of that disease in man and M. leprae [lĕp"·rē] of leprosy

Mycoplasmatales [mi'·kō·plăz"·mə·tāl"·ēz] an order of schizomycetes containing the single genus Mycoplasma. These are highly pleomorphic organisms most of which have a filterable stage in the life history. They were originally called pleuropneumonia like organisms

mycorhiza [mi'·kō·riz"·ə] fungus hyphae symbiotic with the roots of higher plants

mycotrophic [mi'·kō·trŏf"·ĭk] a plant having a mycorhizal association

-myel-, -myelo- comb. form meaning "marrow" or, by a curious extension, "spinal cord"

myelencephalon [mi'·ə·lĕn·sĕf"·ə·lŏn] the most posterior portion of the brain

myelin [mi'·ə·lĭn] a variable mixture of lipoid and proteinaceous materials investing nerve fibers

myeloblast [mī'·əl·ō·blăst"] a direct descendant of a hemocytoblast, destined to form myelocyte

myelocyte [mī'·əl·ō·sīt"] the antipenultimate stage in the development of a polymorph leucocyte. The following derivative terms are defined in alphabetic position:

BASOPHILIC M.   NEUROPHILIC M.
EOSINOPHILIC M.   PROMYELOCYTE
METAMYELOCYTE

myeloid tissue [mī"·əl·oid] bone marrow

-myi- comb. form meaning "fly" often improperly rendered -myo-

-myo- comb. form meaning "to contract" or to "close", but widely used for "muscle" which is properly -mys- and "fly" which is properly -myi-

myocardial [mī'·ō·kärd"·ē·əl] pertaining to cardiac muscle

myocardium [mī'·ō·kärd"·ē·əm] that part of a developing heart which gives rise to the muscular wall

myocoel [mī'·ō·sēl"] that portion of the coelom which lies within a myotome

myocomma [mī'·ō·kŏm"·ə] the septum separating two myotomes

myocyte [mī'·ō·sīt"] literally, a muscle cell; but usually used of the fusiform contractile cells in some sponges

myodome [mī"·ō·dōm] a cavity between the base of the cranium and the parasphenoid in actinopterygian fish

myoepicardial mantle [mī'·ō·ĕp'·ē·kärd"·ē·əl] the walls of the embryonic tube which give rise to the muscular and epicardial walls of the heart

myoepithelial cells [mī'·ō·ĕp'·ē·thēl"·ē·əl] stellate cells underlying the basement membrane of salivary, and some other, glands; their contraction is presumed to facilitate the movement of the secretion

myofibril [mī'·ō·fī"·brĭl] the unit fibril of muscle

myomere [mī"·ō·mēr] the dorsal, metamerically segmented, portion of the mesoderm from which the segmental muscles of chordates are derived

myosclerotome [mī'·ō·sklēr'·ō·tōm"] a term applied to the myotome before the sclerotomic cells have become differentiated (see also sclerotome)

myoseptum [mī'·ō·sĕp"·təm] the interval between, or the collagen filling the interval between, successive myotomes

myosin = ATPase

Myotis [mī·ōt"·ĭs] a genus of small chiropteran mammals called "brown bats". They are the most numerous, both in kind and number, of all North American bats

myotome [mī'·ō·tōm"] those dorsal mesodermal metameres which give rise to the dorsal muscles in chordates

Myriapoda a no longer acceptable taxon of the animal kingdom at one time containing the Diplopoda and their allies as well as the Chilopoda

Myrmecophaga [mûr'·měk·ō·fä·gə] the genus of xenarthran mammals containing the giant anteater

M. jubata [jōō·bāt'·ə]. The elongate snout and long protrusible tongue are distinctive

Myrtaceae [mûr·tās"·ē] a large family of dicotyledons containing the eucalyptus tree and myrtles, and a number of spice trees including the clove. The pungent oil glands are distinctive of the family. The genus *Eucalyptus* is the subject of a separate entry

-mys- comb. form meaning "muscle", often improperly rendered -myo-

Mysidacea [mī'·sīd·ās"·ē] an order of crustaceans in which the carapace extends over the greater part of the thoracic region but does not coalesce dorsally with more than three of the thoracic somites. The periopods are all biramous and function as swimming organs. They have the general appearance of small, more or less transparent, shrimps

Mysis [mī'·sĭs] a typical genus of mysidacean Crustacea, most species of which are completely transparent. *M. relicta* [rĕl·ĭkt"·ə] is one of the few invertebrates to inhabit both fresh and salt water being found both in the Great Lakes, in lakes in Ireland, Norway and Russia and on all the shores of the Atlantic Ocean. All other species of the genus are exclusively marine

mysium muscle. The following terms using this suffix are defined in alphabetic position:

ENDOMYSIUM   PERIMYSIUM
EPIMYSIUM

Mytilus [mĭt"·ĭl·əs] a genus of pelecypod mollusks with wedge-shaped shells narrow in front and wide behind, where they attach to rocks by a byssus. Many species, particularly the cosmopolitan *M. edulis* [ĕd"·yōō·lĭs], are eaten but this must be done with caution, particularly on the Pacific coast, since these filter feeding forms accumulate *Goniaulax* toxin during periods in which this alga blooms

Myxine [mĭks·īn"·ē] a genus of Marsippobranchia containing the Common Hagfish (*M. glutinosa* [glōō'·tĭn·ōs·ə]), a form differing from *Petromyzon* (the lampreys) in having anterior, instead of dorsal, nares and a poorly developed branchial basket

Myxobacterales [mĭks'·ō·băk'·tə·rāl"·ēz] an order of Schizomycetes commonly called slime bacteria. They occur either as unicellular rods, or in the form of a slime-like sheet, rapidly spreading over the substrate. Resting cells are formed either directly from vegetative cells or in the fruiting bodies formed from an aggregate of a large number of vegetable cells

Myxomycophyta [mĭks'·ō·mī'·cō·fī"·tə] a botanical taxon erected to contain the forms called slime-molds, regarded by botanists as related to the fungi, but differing from all other fungi in lacking a cell wall in all stages of vegetative development. These forms are regarded as animals by zoologists who place them among the sarcodine protozoa in the order Mycetozoa. The genus *Plasmodiophora* is the subject of a separate entry

myxopod [mĭks'·ō·pŏd'] the amoeboid stage of a myxomycophyte

myxotrophic [mĭks'·ō·trōf"·ĭk] being nourished through the ingestion of particles

N

N.A. *see* numerical aperture

NAD = nicotinamide-adenine dinucleotide

NADP = nicotinamide-adenine dinucleotide phosphate. A frequent receptor in oxidoreductase systems

Naegleria [nē·glēr″·ēə, nā·glēr″·ēə] a genus of amebas, often referred to as "limax amebas" because their method of locomotion resembles that of the slug *Limax*. These forms also have transitory flagellate stages

naiad [nā″·əd] the aquatic stage of any metabolous (in the restricted sense) insect

Nais a genus of transparent aquatic oligochaete worms. In spite of its frequent use to illustrate texts, the genus does not occur in North America where it is replaced by *Stylaria*

Naja [nā·jə] the genus of Squamata containing the African and Indian cobras. *N. naja* is the well known Indian cobra

Nakamura's gland the nuchal venom gland of snakes

-nan- *comb. form* meaning "dwarf"

nannander [năn′·ănd″·ər] a dwarf male

nannoplankton [năn′·ō·plănk″·tən] planktonic organisms so small that they pass through the meshes of the finest available nets

-nap- *comb. form* meaning turnip

narial bone = septomaxilla bone

nasal bone [nā″·zəl] one of a pair of membrane bones in the anterior region of the skull, lying above the nasal cavity, immediately in front of the frontal, and dorsal to the ethmoid (*see also* internasal bone)

nasal capsule that portion of the chondrocranium which encloses the nasal organs

nasal pit the precursor of the nasal cavity in embryos

nasal septum a vertical plate of the ethmoid bone dividing the right and left nares

nasal tube cartilage one of a series of angular cartilages supporting the nasal tube of Myxine

nasopharynx [nā′·zō·fâr″·ĭnks] that portion of the amniote pharynx which lies above the plane of the soft palate

-nast- *comb. form* meaning "pressed close"

nastic [năs″·tĭk] pertaining to curvature produced by the differential growth of the dorsal or ventral side of a flat plant structure such as a petal or leaf. The following terms using this suffix are defined in alphabetic position:

AUTONASTIC   HYPONASTIC

GEONASTIC   PHOTONASTIC

-nasty substantive suffix applying to responses of flattened plant organs (e.g. leaves, petals, etc.) to external stimuli. Some compounds are indicated under the adjectival suffix -nastic

natality [nā·tăl″·ĭt·ē] inherent ability of a population to increase

natriferine [năt′·rĭ·fə·ēn, nə·trĭf″·ər·ēn] a hormone controlling the transport of sodium across the skin of anuran amphibia

naturalization the enforced adaptation, usually with human aid, of an organism to a foreign environment (cf. acclimatization)

natural selection the selection, either of individuals or breeding rates by environmental conditions

nauplius larva [nôrp″·lē·əs] the one-eyed larva, with three anterior pairs of appendages, hatched from the egg of most crustacea

Nautilus [nôrt′·əl·əs] a genus of cephalopod mollusks with a coiled, multiloculate shell. A projection (siphuncle) from the base of the animal, which lives in the terminal chamber, extends through all of the chambers

navicular [năv·ĭk·yōō·lə] boat-shaped

navicular bone the tarsal bone that lies between the base of the medial and intermedial cuneiform, and the astragalus (= radiale)

neallotype [nē′·ăl″·lō·tip′] a type of the opposite sex from the holotype or any of the series of paratypes and of which a description is published after the description of the original type

Neanthes [nē′·an·þēz″] a genus of nereidiform polychaete annelids once included in the genus Nereis. *Neanthes virens* [vĭr′·enz] is the large (8″) green "Clam Worm" of the Atlantic Coast

nearctic [nē·ärk″·tĭk] pertaining to those areas of the New World which are not tropical, i.e. Greenland, Canada, Labrador together with the mountainous and Northern parts of the United States

neascus larva [nē·ăsk″·əs] a type of metacercaria with a cup-shaped forebody and a well developed hindbody

Nebalia [nĕb·ăl″·ē·ə] the commonest genus of leptostracan Crustacea, with the general appearance of a small (1/2 inch) shrimp. *N. bipes* [bī″·pēz] is very common in shallow water on both sides of the North Atlantic

Necator [něk·ā″·tôr] a genus of ancyclostomid nematodes containing *N. americanus* [ăm·ĕr′·ĭk·än″·əs] commonly called the American hookworm though it is African in origin. It is an intestinal parasite of man, infection occurring directly through skin in contact with feces contaminated soil or water (*see also Anchylostoma*)

necron [něk″·rən] living material which has died but which has not yet decomposed; applied particularly to plant forms (*see also* humus necron)

-nect- *comb. form* meaning "to join". Often confused with -nekt- (q.v.)

nectar a sugary, frequently aromatic, fluid secreted by many flowering plants

nectary a glandular trichome secreting a sweet substance. Usually, but not always confined to the flowers

nectochaeta larva [něk′·tō·kĕt″·ə] a posttrochophore larva of polychaete annelids in which one or more pairs of parapodia are developed

Nectonema [něk′·tō·nēm″·ə] a genus of marine Nematomorpha. The male of *N. agile* [ăj″·əl·ē] may reach a length of 3 inches but the female is rarely more than an inch long. The larva is parasitic in shrimps

nectophore [něk′·tō·fôr″] the gelatinous swimming cells of a siphonophoran coelenterate

nectopod [něk″·tō·pŏd′, něk″·tə·pŏd] a limb adapted for swimming

nectosac [něk′·tō·săk″, něk′·tə·sak″] that portion of the nectophore of siphonophoran medusans which contains the radial canals

nectosome [něk′·tō·sōm″, něk·tə·sōm] that portion of the stem of a siphonophoran hydrozoan which bears swimming cells

Necturus [něk·tyōōr″·əs] a genus of completely neotenous urodele Amphibia, retaining their external gills throughout life. *N. maculosus* [măk′·yū·lōs″·əs] is the common mud puppy

negative cytotaxis the tendency of cells to separate (*see also* positive cytotaxis)

-nekt- *comb. form* meaning "to swim" often misspelled -nect- (q.v.)

nekton [něk″·tən, něk″·tŏn] swimming organisms sufficiently strong to be independent of the water currents among which they live (*see also* plankton, benthon, edaphonekton)

-nema- *comb. form* meaning "thread". In the form -neme- is frequently confused with -nema- (cf. -neme-). The following terms using this suffix are defined in alphabetic position:

CHROMONEMA   NUCLEOLONEMA
DIPLONEMA   PACHYNEMA
LEPTONEMA   ZYGONEMA

Nemalion [nə·māl″·ē·ən] a cosmopolitan genus of rhodophyte algae widely used in class teaching. They are thickly filamentous. One species (the Pacific *N. lubricum* [lyōō″·brĭk·əm]) reaches a length of about three feet

nemata plural of nema

Nemathelminthes an obsolete taxon of the animal kingdom once regarded as a phylum containing the Nematoda and the Acanthocephala and, from time to time, other thread-like worms

nematoblast = cnidoblast

nematocyst = cnidocyst

nematocyte = cnidocyst

Nematoda an enormous phylum of pseudocoelomatous animals called the threadworms. They are distinguished by the cylindrical body covered with a cuticle. There are paired nerve cords and paired excretory ducts. They occur as free living forms and as parasites of both plants and animals. In length they vary from a fraction of a millimeter to more than a foot and are nearly as ubiquitous and as numerous as the arthropods. The genera *Anguina, Ascaris, Dioctophyme, Dracunculus, Enterobius, Loa, Parascaris* and *Wuchereria* are the subjects of separate entries

Nematomorpha [nē′·măt·ō·môrf″·ə] a phylum of pseudocoelomatous bilateral animals commonly called the horsehair worms. This name is descriptive of the appearance. This group may also be regarded as a class of the phylum Aschelminthes. The genera *Gordius, Nectonema* and *Paragordius* are the subjects of separate entries

-neme- anglicized equivalent of -nema-. The following terms using this suffix are defined in alphabetic position:

DESMONEME   SPIRONEME

Nemertea [nēm·ûrt″·ē·ə] a phylum of acoelomate bilateral animals commonly called bootlace worms. They are distinguished by the possession of an eversible proboscis. The genus *Cerebratulus* is the subject of a separate entry

-nemic adjectival suffix from -nema- and thus literally "pertaining to a thread"

-nemo- *comb. form* meaning grassland, pasture, meadow (cf. -nema-)

-neo- *comb. form* meaning "recent"

neobiogenesis [nē′·ō·bī′·ō·jěn″·əs·ĭs] the theory that life may have been evolved several times and that there is a continuous recurring possibility of such synthesis taking place (*see also* biogenesis, abiogenesis)

neoblast [nē″·ō·blăst′] undifferentiated blastema cells which are used in the regeneration of invertebrate tissues

Neoceratodus [nē′·ō·sĕr·ăt″·ō·dəs] a monotypic genus of Dipneusti containing the Australian lungfish *N. forsteri* [fôrst″·ər·ī]. There is only one lung

neocerebellum [nē′·ō·sĕr·ə·bĕl″·əm] that part of the cerebellum which controls the integration of voluntary movements in limb-bearing vertebrates and which has its principal nervous connections with Variolus' bridge

neocortex [nē′·ō·kôr″·tĕks] the cerebral cortex of mammals

neodarwinism the view that explains the mechanics of evolution on the basis of mutation and selection. It differs from Darwinism in explaining the origin of species as a selective process from adaptive phenotypes produced by random mutations

neofemale [nē′·ō·fē″·māl] the female produced by sex reversal from the male

neogaea [nē′·ō·jē″·ə] the area comprising central and southern America, though sometimes used to include the whole "new world", particularly in regard to the distribution of plants

neomorph [nē″·ō·môrf′] a mutant allele influencing development in a manner markedly different from that of the ancestral allele

neomorphosis [*angl.* nē'·ō·môrf"·əs·ĭs, *orig.* nē'·ō·môr·fōs"·ĭs] the replacement, usually through regeneration, of one part by another different part

neopallium [nē'·ō·pal"·ē·əm] that portion of the cortex of the brain which covers the dorsal convexity of the cerebral hemispheres

Neopilina [nē'·ō·pĭl·ĭn"·ə] a genus of mollusks of doubtful affinities known as fossils from Cambrian times. A few living specimens were found in 1952 at very great depths off coast of Costa Rica and several more discoveries have since been made. The animals show a marked metameric segmentation and are thought by some to indicate vestiges of annelid ancestry

Neornithes [*angl.* nē·ôrn"·ĭth·ēz, *orig.* nē'·ôr·nith"·ēz] a doubtfully valid taxon erected to contain all birds except *Archaeopteryx* which was placed in the Archaeornithes

neoteny [nē·ŏt"·ən·ē] the condition of remaining indefinitely in an immature, or larval, state (*see* Ambystoma)

neotropical [nē'·ō·trŏp"·ĭk·əl] pertaining to that portion of the tropics which lies in the New World

neotype [nē'·ō·tĭp'] a type selected to represent the original type, usually collected from the original locality, when the original is no longer available

-nephr-, -nephro- *comb. form* meaning "kidney"

nephridiopore [nĕf·rĭd"·ē·ō·pôr'] the opening to the surface from either a protonephridium or a metanephridium

nephridium [nĕf·rĭd"·ē·əm] an individual excretory unit as contrasted with a nephron. Used without qualification the term is often synonymous with metanephridium. The following derivative terms are defined in alphabetic position:

HATSCHEK'S N. MICRONEPHRIDIUM
MEGANEPHRIDIUM PROTONEPHRIDIUM
METANEPHRIDIUM

nephrocoel [nĕf"·rō·sēl'] the cavity of a nephrotome

nephrocoel theory the ancestral coelom is the expanded inner end of a nephridium. Sometimes called Ziegler's theory

nephromixium [nĕf'·rō·mĭks"·ē·əm] a unit consisting of a coelomostome and a protonephridial tubule connected to the same duct

nephron [nĕf"·rŏn] a kidney unit. In mammals this consists of Bowman's capsule with its contained glomerulus, the proximal convoluted tubule, Henle's duct and the distal convoluted tubule

nephros a kidney. The following derivative terms are defined in alphabetic position:

ARCHINEPHROS METANEPHROS
HOLONEPHROS PRONEPHROS
MESONEPHROS

nephrostome [nĕf"·rō·stōm'] a peritoneal funnel forming part of, or directly derived from, a nephron

nephrotome [nĕf"·rō·tōm'] that portion of the embryonic mesoderm which lies between the myotome and the sclerotome and from which the nephric units are derived

nepionic [nĕp'·ē·ŏn"·ĭk] post-embryonic or larval

Nereis a genus of polychaete annelids commonly called clam worms. They have well developed tentacles and palps and protrusible jaws. Most live on or in sand but a few are free swimming. Frequently used in laboratory teaching (*see also* heteronereis)

Nereocystis a monotypic genus of laminariale Phaeophyta. *N. luetkeana* [lōōt'·kē·än·ə], the largest giant kelp, reaches a length of 120 feet

neritic [nə·rĭt"·ĭk] a term applied to those organisms that occur in more or less coastal waters in distinction from oceanic organisms

-nerv- *see* -neur-

nerve 1 (*see also* nerve 2, 3) a fiber, or more usually a bundle of fibers, which conducts impulses through organisms. The following derivative terms with this connotation are defined in alphabetic position:

ACCELERATOR NERVE EFFERENT NERVE
AFFERENT NERVE MOTOR NERVE

nerve 2 (*see also* nerve 1, 3) a specific group of fibers in a common sheath. The following derivative terms are defined in alphabetic position:

ABDUCENS NERVE OPTIC NERVE
ACCESSORY NERVE POSTTREMATIC NERVE
AUDITORY NERVE PRETREMATIC NERVE
COLLECTOR NERVE RECURRENT NERVE
CRANIAL NERVE SOMATIC NERVE
FACIAL NERVE SPINAL NERVE
GLOSSOPHARYNGEAL N. TRIGEMINAL NERVE
HYPOGLOSSAL NERVE TROCHLEAR NERVE
OCULOMOTOR NERVE VAGUS NERVE
OLFACTORY NERVE VISCERAL NERVE

nerve 3 (*see also* nerve 1, 2) used synonymously with vein in the description of insect wings and plant leaves

nerve cell any cell adapted to originate or transmit nerve impulses

nerve net a network of conducting fibrils running over the surface of many large invertebrates including the Protozoa; in the latter it consists of the kinetium

nervous system the sum total of all those structures in an organism which carry impulses. The following derivative terms are separately defined:

CENTRAL NERVOUS S. PARASYMPATHETIC N. S
CONCENTRATED N. S. PERIPHERAL NERVOUS S.
DIFFUSE NERVOUS S. SYMPATHETIC N. S.
ENTERIC NERVOUS S.

nervous tissue those elements of the body which are designed to perceive or originate stimuli or conduct impulses

net plankton those constituents of the plankton retained by a plankton net of any particular mesh

-neur- *comb. form* meaning "nerve". Properly, though very rarely -nevr-, frequently though quite improperly -nerv-

neural [nyūr"·əl] pertaining to nerves. The following derivative terms are defined in alphabetic position:

EPINEURAL MYONEURAL
HYPONEURAL

neural arch the arch that rises dorsally from the centrum of a vertebra and through which runs the spinal cord

neural canal 1 (*see also* neural canal 2) the cavity of the developing central nervous system

neural canal 2 (*see also* neural canal 1) the arch of ventral apodemes protecting the nerve cord in the thorax of arthropods

neural crest a band of cells just lateral to the line of closure of the neural folds in an embryo. It gives rise both to the spinal ganglia and the autonomic nervous system

neural plate the thickened cells on the mid-dorsal

surface of an early embryo that will later sink in to form the central nervous system

neural ridge the lateral margins of the neural plate as it begins to invaginate

neural tube a term applied to the vertebrate central nervous system at a stage when the neural groove is completely invaginated, but the neural folds have not yet fused

neurenteric canal [nyūr'·ĕn·tĕr"·ik] the connection, which is the remnants of the blastopore, between the archenteron and the neural canal

neurite [nyūr"·it] a general term for axons and dendrons, particularly in the cells of diffuse nervous systems where these are not clearly differentiated

neurobiotaxis [nyūr'·ō·bi'·ō·tăks"·ĭs] the growth of embryonic nerves in the direction of stimuli

neuroblast [nyūr'·ō·blăst"] those cells in the embryonic neural tube which will subsequently become neurons

neurochord cells [nyūr"·ō·kôrd'] unusually large unipolar ganglion cells lying between the ventral nerve chords of nemertines

neurocoel [nyūr'·ō·sēl"] the cavity of the central nervous system

neurocranium [nyūr'·ō·krān"·ē·əm] the part of the skull that encloses the brain in contrast to the splanchnocranium

neurofibrils [nyūr'·ō·fi·brĭlz] fine strands of protoplasm adapted to conduct impulses

neurogenic rhythm [nyūr·ō·jĕn"·ĭk] a rhythm of neural origin, particularly the rhythm of alary muscles of insects

neurohormone [nyūr'·o·hôr"·mōn] one produced in a nerve cell and which usually reaches its point of action by diffusion along a nerve

neurohypophysis [*angl.* nyūr'·ō·hi·pŏf"·əs·ĭs, *orig.* nyūr'·ō'·hi'·pō·fis"·ĭs] that portion of the hypophysis which is derived from the floor of the diencephalon as the infundidulum

neurolemma [nyūr'·ō·lĕm"·ə] the protoplasmic sheath of a nerve fiber

neuromast [nyūr"·ō·măst] a sensory cell; the term is applied both to the cells of invertebrate receptors and to those of the lateral line system of fish

neuron [nyūr"·ən] a unit of nervous structure consisting of the sum total of a nerve cell and all of its processes. The following derivative terms are defined in alphabetic position:

ADJUSTOR NEURON   SENSORY NEURON
INTERNUNCIAL N.
MOTOR NEURON

neuropodium [nyūr'·ō·pŏd"·ē·əm] the lower of the two lobes of a polychaete parapodium (*see also* parapodium, notopodium)

neuropile [nyūr'·ō·pil"] the central mass of medullary tissue of an invertebrate ganglion

neuroplankton [nyūr'·ō·plănk"·tən] that which is found only at certain seasons of the year

neuropore [nyūr"·ō·pôr'] the connection between the neural canal and the exterior in the neurula stage of a larva or embryo

Neuroptera [nyūr·ŏp"·tər·ə] an order of insects containing the ant lions, lace wings and dobson flies. All have complete metamorphosis, chewing mouth parts, four membranous wings and lack anal cerci (cf. Pseudoneuroptera)

neurosecretory cell [nyūr'·ō·sē·krē·tər·ē] a nerve cell that secretes a hormone

Neurospora [nyūr·ŏs"·pər·ə] a genus of saprophytic sphaeriale Ascomycetes. *N. crassa* [krăs"·ə] has been widely used for genetic experiments

neurula [nyūr"·yū·lə] that stage in the development of an embryo in which the whole of the neural plate is not yet invaginated

neuston [nyōō"·stŏn] organisms floating on the surface of water (*see also* pleuston, supraneuston)

neutrophil [nyōō"·trō·fil] a myelocyte with a trilobed nucleus the granules of which are neither eosinophilic nor basophilic

neutrophilic metamyelocyte [nyōō'·trō·fĭl"·ĭk] a descendant of a neutrophilic myelocyte in which the nucleus is "u" shaped

neutrophilic myelocyte a descendant of a promyelocyte in which the granules are apparent but the nucleus is not yet lobed

nexus [nĕks"·əs] the region of fusion of plasma membranes (cf. desmosome)

niacin [ni"·ə·sĭn] pyridine-3-carboxylic acid. A water soluble nutrient commonly regarded as a vitamin. Gross deficiency causes pelagra

niacinamide = nicotinamide

niche [nĭtch] variously used by ecologists to mean either the position occupied by an individual in an assemblage of individuals or to indicate a microhabitat or biotope (*see also* ecological niche)

Nicotiana [nĭk·ōsh'·ē·ăn"·ə] a genus of solanaceous flowering plants. *N. tobacum* [tō·băk"·əm] has a self explanatory name

nicotinamide [nĭk'·ō·tĭn"·əm·id'] niacinamide. Pyridine-3-carboxamide. Essentially similar to nicotinic acid but lacking some of the side effects when fed to humans

nicotinamide-adenine dinucleotide [nĭk'·ō·tĭn"·əm·id'-ăd'·ən·ēn" di'·nyōō·klē·ō·tĭd"] a frequent receptor in oxido-reductase systems. Usually abbreviated NAD

nicotinic acid = niacin

nictitating membrane [nĭk'·tĭt·āt"·ĭng] a transparent sheet which may be drawn horizontally across the front of the cornea from the inner angle of the eye, or beneath the lower lid

nidifugal [nid'·ē·fyōō"·gəl] nest-fleeing; commonly used as the antithesis of nidicolous

nipple any prominence on a compound gland (particularly the mammary gland) through which the product is excreted

Nissl granules granules demonstrable by optical microscopy in the cytoplasm of nerve cells by staining with an alkaline solution of methylene blue. Under the electron microscope they have a lamellar structure and are thought to contain a high concentration of transfer substance

Nissl body the aggregate of Nissl granules

Nitella [ni·tel"·ə] a genus of Charaphyta in which all the "branches" arising from the nodes remain free so that there is no cortex around the internodes

Nitrobacter [ni'·trō·bak"·tə] a genus of Pseudomonadales the members of which derive energy from the oxidation of nitrite to nitrate

nitrogen cycle a loose term covering the fixation of atmospheric nitrogen (either by biological or physical means), the synthesis of organic compounds from this nitrogen and the final return of the gas to the atmosphere as the end product of decomposition

Nitrosococcus = **Nitrosomonas**

Nitrosomonas [ni′·trō·sōm·ən·əs] a genus of pseudomonadale Schizomycetes that derives energy from the oxidation of ammonia to nitrite

-niva- *comb. form* meaning "snow" (cf. -niph-)

nival [ni′·vəl] pertaining to snow, particularly as a habitat for plants

NMN = nicotinamide mononucleotide

Nobelian bone [nō·bēl″·ē·ən] one of a pair of supporting rods in the intromittent organ of some anuran amphibia

nociceptor [nō″·sē·sĕp′·tər] one that is responsive to possibly injurious stimuli

Noctiluca [nŏk′·tĭl·yōō·kə] a monotypic genus of marine dinophycean algae. *N. scintillans* [sĭn·til″·ănz] is roughly spherical, about 1 mm in diameter and with a contractile tentacle replacing one of the flagella. A bloom of this species is responsible for the brilliant phosphorescence sometimes seen in oceans

nocturnal [nŏk·tûrn″·əl] pertaining to the hours of darkness either as a rhythmic activity or in opposition to diurnal and crepuscular

-nod- *comb. form* meaning "knot" or "knob"

node 1 [nōd] (*see also* node 2) a lump or knot of tissue in an organ, or strand, particularly the swollen portion of plant stems from which leaves arise. The following derivative terms are defined in alphabetic position:

HEART NODE	RANVIER'S NODE
HENSEN'S NODE	TAWARA'S NODE
PRIMITIVE NODE	

node 2 (*see also* node 1) in ants the nob between the thorax and posterior abdomen

nodose ganglion [nōd″·ōz] a ganglion on the intestinal branch of the vagus nerve

nodule [nŏd″·yōōl] diminutive of node. The following derivative terms are defined in alphabetic position:

LYMPHATIC NODULE   ROOT NODULE

-noma- *comb. form* meaning "wanderer", but hopelessly confused in biological literature with either or both meanings of -nomo-

nomen [nō″·mĕn] literally "name" (*see also* nym)

nomen conservandum [kŏn′·sûrv·ăn″·dəm] a name which is retained in virtue of its long established and wide usage

nomen dubium [dyōō″·bē·əm] a name which cannot be applied with certainty to a specific species

nomen inquirendum [ĭn′·kwir·ĕn″·dum] one under investigation by the International Commission on Nomenclature or other body

nomen novum [nō″·vəm] a new name applied to a form already having a name

nomen nudum [nyōō″·dəm] a name without status since its original publication was not accompanied by an adequate description or definition

nomen rejectum [rē·jĕk″·təm (but properly rē·yĕk″·təm)] an existing name which is no longer considered valid

-nomo- *comb. form* hopelessly confused from two homonymous roots, meaning respectively "custom" and "pasture"; ecologists have further confused -nomo- meaning "pasture" with -noma- meaning "wanderer". The following derivative terms are defined in alphabetic position:

AUTONOMOUS	HOMONOMOUS
HETERONOMOUS	

-nomy *comb. suffix* properly meaning "the application of laws" but in current usage "the science" or "study of"

non- a negative prefix. Such obvious compounds as "non-vascular" are not defined in this dictionary

-non-, -nono- *comb. form* meaning "nine"

non-homologous pairing the condition in which non-homologous chromosomes, or parts of chromosomes, pair in meiosis

noradrenaline = norepinephrine

norepinephrine [nôr′·ĕp′·ē·nĕf″·rin] a hormone similar in origin and function to epinephrine, but with a greater effect on arteriolar constriction

norleucine [nôr·lyōō″·sēn] α-amino-caproic acid. $CH_3(CH_2)_3CH(NH_2)CO_2H$. An amino acid not essential in rat nutrition. It is isomeric with leucine

normoblast [nôr″·mō·blăst′] the descendant of the polychromatophic erythroblast and the direct precursor of an erythrocyte. In mammals distinguished by a pyknotic nucleus which is subsequently extruded (*see also* pronormoblast)

norvaline [nôr·vāl″·ēn] α-aminovaleric acid. $(CH_2)_2CH(NH_2)COOH$. An amino acid, not essential to the nutrition of rats, isomeric with valine

nostril one of the paired apertures leading to the nasal cavity of mammals. In insects the term is applied to the rhinarium

Nostoc [nŏs″·tŏk″] a cyanophyte alga forming bead-like chains or filaments. Some species produce heterocysts contained in a gelatinous case that may reach an inch in diameter. The individual cells in the chains are about 5 µ in diameter

-noto- *comb. form* meaning "back", or "dorsal"

notocentrous vertebra [nō′·tō·sĕn″·trəs] one in which the centrum is derived entirely from the neural arches

notochord [nō′·tō·kôrd″] a turgid rod of cells lying immediately beneath, and parallel to, the nerve cord in Cephalochordata, and in vertebrate embryos. The centrum of the vertebra is derived from it

notochordal vertebra [nō′·tō·kôrd″·əl] one in which the centrum is hollow

notogaea [nō′·tō·jē″·ə] the combined neotropical and Australasian regions

Notonecta [*angl.* nŏt·ŏn″·ĕk·tə, *orig.* nō′·tō·nĕk″·tə] a genus of notonectid hemipteran insects that, unlike their relatives Corixa, swim upside down and are therefore called backswimmers

Notonectidae a family of aquatic hemipteran insects commonly called backswimmers. This name is adequately descriptive

Notophthalmus [nŏt′·ŏp·păl·ə·məs] the name now used for the common North American newts previously called *Triturus*. *N. viridescens* [vĭr′·ĭd·ĕs″·ĕnz], the red spotted newt, was the object of much research on regeneration

notopodium [nō′·tō·pōd″·ē·əm] the upper of the two lobes of polychaete parapodium (*see also* parapodium, neuropodium)

Notoryctes [nō′·tôr·ĭk″·tēz] a genus containing the "marsupial moles", the best known being *N. typhlops* [tif″·lŏps″]. The resemblance of this metatherian to eutherian moles is an astonishing example of convergent evolution

Notostraca [*angl.* nŏt·ŏst″·rə·kə, *orig.* nō′·tō·străk″ə] an order of crustacean arthropods commonly called tadpole shrimps. They are distinguished from other branchiopods by the fact that they have a broad shield-like carapace on the dorsal surface, from which the many segmented body projects. The genus *Apus* is the subject of a separate entry

notum [nō·təm]   the dorsal portion of an arthropod segment (= tergum)

-nov-   *comb. form* meaning "new"

nov-, -novem-   *comb. form* meaning "nine"

noviform [nō″·vē·fôrm′]   a cultigen of recent origin

nucellus [nyōō·sĕl″·əs]   the central portion of a plant ovule, and therefore, technically, the megasporangium

nuchal [nyōō″·kəl]   adjective from "nape" and therefore pertaining to the dorsal area, immediately behind the head or back of the neck of mammals, etc. also applied to a similar area of some insects

nuchal crest   the transverse, crescentic crest drawn out from the supraoccipital and exoccipital bones above the foramen magnum

nuchal flexure   the bend of the embryonic brain in the region of the medulla oblonga

nuci-   *comb. form* meaning "nut"

nuclear [nyōō″·klē·ər]   pertaining to the nucleus

nuclear membrane   the limiting membrane of the nucleus. It is thought by many to be continuous with the endoplasmic reticulum

nuclear net   the network of threads produced as an artifact by the acid fixation of nuclei

nuclear pore   a hole in the nuclear membrane

nuclear stains   those dyes that stain the nucleus differentially. The most commonly used is hematoxylin

nucleolus [nyōō·klĕ″·əl·əs, nyōō′·klē·ōl″·əs]   a discrete mass in the nucleus containing both fibrous and particulate components in an amorphous matrix. It is thought to be the site of RNA synthesis (*see also* lateral nucleolus)

nucleoprotein [nyōō′·klē·ō·prō″·tēn]   a compound protein, mostly found in the nucleus, combined with nucleic acids

nucleosidase [nyōō″·klē·ō·sid′·āz]   an enzyme catalyzing the hydrolysis of n-ribosyl-purines into purines and D-ribose

nucleoside diphosphatase [nyōō″·klē·ō·sĭd′ dĭ·fŏs″·fə·tāz]   an enzyme that catalyzes the hydrolysis of nucleoside diphosphate to the nucleotide and orthophosphate

nucleotidase [nyōō″·klē·ō′·tĭd·āz]   an enzyme catalyzing the hydrolysis of nucleotides to yield ribonucleosides and phosphoric acid

nucleotide pyrophosphatase [nyōō″·klē·ō′·tĭd]   an enzyme which catalyzes the hydrolysis of dinucleotides to two mononucleotides

nucleus 1 [nyōō″·klē·əs]   within non-dividing cells of animals and plants above the bacterial level, the membrane-bound vesicle containing, principally, the chromosomes and associated structures characteristic of the species. The following derivative terms are defined in alphabetic position:

DEFINITIVE N.	PARANUCLEUS
GENERATIVE N.	PRONUCLEUS
GERM NUCLEUS	RESTING NUCLEUS
INTERPHASE N.	RESTITUTION NUCLEUS
LATERAL NUCLEUS	SOMATIC NUCLEUS
MACRONUCLEUS	TUBE NUCLEUS
MICRONUCLEUS	VEGETATIVE NUCLEUS

nucleus 2 (*see also* nucleus 1) any more or less spherical mass embedded or contained in anything, including the central portion of a seed, but particularly aggregations of nerve cells in the brain. The following derivative terms are defined in alphabetic position:

HABENULAR NUCLEUS	PANDER'S NUCLEUS
MOTOR NUCLEUS	

Nudibranchiata   a suborder of gastropods commonly called sea slugs. The name derives from the dorsal respiratory extension of the mantle, often brilliantly colored. The genera *Dendronotus*, *Eolis* and *Hermaea* are the subjects of separate entries

numerical aperture (N.A.)   a controlling factor in the resolution of microscope objectives having the relationship

$$N.A. = \Delta \sin \theta$$

where Δ is the optical density of the environment and θ is the apical angle of the cone of light entering the objective. As $\Delta = 1$ for air it follows that the lens must be immersed in oil ($\Delta = 1.6$) if the N.A. is to be greater than 1

nurse cell   one cell which specifically provides nourishment to an adjacent cell as the follicle cells in an ovary

-nut-   *comb. form* meaning "nod"

-nychium   *comb. suffix* from onychium meaning a "claw" or "finger nail" (*see* hyponychium)

-nyct-   *comb. form* meaning "night" or "nocturnal"

nyctipelagic [nĭk·tē·pĕl·ăj″·ĭk]   those pelagic organisms which seek the surface at night

nyctipelagic plankton   that portion of a diurnal migrating plankton which rises at night

nyctitropism [nĭk′·tē·trōp″·ĭzm]   the condition of a leaf which changes its position at night

Nyctotherus [*angl.* nĭk·tōth″·ər·əs, *orig.* nĭk′·tō·pēr″·əs]   a genus of spirotrichan ciliates parasitic in the gut of many animals. *N. cordiformis* [kôr′·dē·fôrm″·ĭs] occurs in *Rana* where it is often confused with *Balantidium* from which it may be distinguished by the lateral "mouth"

-nym-   *comb. word* meaning "name". The following terms using this suffix are defined in alphabetic position:

BASONYM	METANYM
CACONYM	TAUTONYM
HOMONYM	TYPONYM
HYPONYM	

nymph [nĭmf]   in the United States this term is confined to those immature stages of arthropods which inhabit much the same environment as the parent which they more or less resemble in form. In Europe the word is used for the penultimate and sometimes the antipenultimate stage of any arthropod showing metamorphosis

nymphiparous   the bearing of young in the nymph stage

nymphophan [nĭm″·fō·făn]   an inactive stage, corresponding to an insect pupa, in the life history of some acarina

O

-ob- *comb. form* meaning "opposite" or "reverse"

obconic [ŏb·kŏn″·ĭk] in the shape of a cone standing on its thin end

Obelia [ō·bēl″·ē·ə] a genus of hydroid coelenterates. *O. campanulata* [kăm′·păn·nyōō·lā″·tə], a common cosmopolitan form, is widely used in class teaching

objective the front lens system of a microscope (*see also* resolution, numerical aperture, immersion objective)

obligate gamete [ŏb″·lĭg·āt] a gamete that cannot develop parthenogenetically

obligate parasite one which cannot exist in any other form

obverse [ŏb″·vûrs] one side of anything, the opposite side being the "reverse"; which is obverse and which is reverse is a matter of convention. In insects, obverse refers to the "head on" aspect

occipital [ŏk·sĭp″·ĭt·əl] pertaining to that region of the chordate head that lies in the dorsal third, or in the general vicinity of the occipital bone (*see also* otioccipital)

occipital arch a series of paired cartilages forming the side of a foramen magnum in chondrocranium

occipital bone any of several bones lying about the foramen magnum (*see* basioccipital bone, exoccipital bone, supraoccipital bone)

occipital condyle in insects, the lateral sclerite which articulates the head to the thorax

occipital lobe the most anterior lobe of the cerebral hemisphere

ocellus 1 (*see also* ocellus 2) one of the elements of a compound eye. The following derivative terms are defined in alphabetic position:
CIRCULAR OCELLUS   INVERSE OCELLUS
COMPOUND OCELLUS   PIGMENT-SPOT OCELLUS
CONVERSE OCELLUS

ocellus 2 (*see also* ocellus 1) a very small, simple, eye found in many invertebrates

-ochthon- *comb. term.* chopped off, without justification, from the Greek autochnon (native inhabitant) and reattached to various prefixes (*see* allochthonous, autochthonous)

-octa- *comb. form* meaning "eight"

oculomotor nerve [ŏk′·yū·lō·mō″·tôr] cranial III, arising from the ventral surface of the mesencephalon, and supplying the superior medial, inferior rectus, and inferior oblique muscles of the eye

Oddi's sphincter the sphincter at the aperture of the bile duct at its entrance to the duodenum

Odoceoleus [*angl.* ō·dōs″·əl·əs, *orig.* ō′·dō·sēl″·əs] the genus of artiodactyl mammals that contains the "white-tailed deer" and the "mule deer"

Odonata [ō′·dŏn·āt″·ə] the order of insects that contains the dragonflies and demoiselle flies. They are distinguished by the four large membranous wings and a freely moveable head bearing chewing mouth parts. The metamorphosis is incomplete

odontoblast [ō·dŏnt″·ō·blăst′] those mesodermal cells which secrete the dentin in the formation of a tooth

Odontoceti [ō·dŏnt′·ō·sēt″·ē] a suborder of Cetacea (whales) distinguished from the Mysticocetae by having teeth but no whale bone. The genera *Delphinus, Monodon, Orcinus, Phocaena* and *Physeter* are the subjects of separate entries

odontophore [ō·dŏnt′·ō·fôr′] the supporting structure of a molluscan radula

-oec- *comb. form* meaning "house" (*see* oecium)

-oecious adjectival form derived from oecium. The following terms using this suffix are defined in alphabetic position:
ANDROECIOUS   MONOECIOUS
DIOECIOUS   SYNOECIOUS
GYNOECIOUS

-oecism *comb. suffix* meaning the condition of having a special type of dwelling. The termination -oecy is sometimes used in the same sense. The following terms using this suffix are defined in alphabetic position:
ANDROECISM   TRIMONOECISM
GYNOECISM

-oecium- Latinized form, unknown to the Romans, of Greek *oikos*, a house. In biology usually, but not invariably, used for the housing of reproductive organs. The adjectival form -oecic, -oecious, and -oecial occur, the last two only very rarely. The substantive form oecism is used in the sense of "pertaining to a type of house" but the substantive oekete appears invariably with the k spelling. The general abbreviation -oeco- is nowadays invariably spelled -eco- (as in ecology, ecosystem). The following derivative terms are defined in alphabetic position:

ANDROECIUM   GYNOECIUM
GAMOECIUM   ZOOECIUM

Oedogonium [ē'·dō·gōn"·ē·əm] a genus of fresh-water filamentous chlorophyte algae with net-like chloroplasts. Asexual reproduction is by a single large zoospore. In sexual reproduction a large cell modified as an oogonium is fertilized by a very small zoospore produced in a number of small antheridia sometimes referred to as "dwarf males"

-oeko- *comb. form* meaning dwelling place, but almost universally replaced in biological literature by -eco-

oenocytoid [ē'·nō·sīt"·oid] an insect blood cell

oesophagus *see* esophagus

oestrone *see* estrone

oestrous *see* estrous

-ogen a termination once widely used for precourser compounds (e.g. pepsinogen, fibrinogen). The prefix pre- is currently preferred

-oidea suffix indicating superfamilial, or occasionally subordinal, rank in animal taxonomy

-oideae suffix indicating sub-familial rank in plant taxonomy. The zoological equivalent is -inae

Oikomonas [oi'·kō·mō"·nəs] a genus of zoomasti-gophoran protozoans with a plastic body but not forming pseudopodia. It may occur either free swimming or as part of a sessile colony

-oikos- *comb. form* meaning "house" usually replaced by oecium

Oleaceae [ōl'·ē·ās"·ē] the family of dicotyledons that contains, forsythia, the ash, the jasmines, the privits, the olives and the lilacs. The numerical plan of four with a superior ovary is distinctive of the family. The genus *Fraxinus* is the subject of a separate entry

olfactory [ōl·făk"·tər·ē] pertaining to the sense of smell

olfactory bulb one of a pair of evaginations from the antero-dorsal end of the telencephalon which contributes the sensory portion of the nasal organs

olfactory capsule = nasal capsule

olfactory cell a bipolar nerve cell in the nasal epithelium

olfactory cortex the combination of the hippo-campal cortex and the pyriform cortex

olfactory nerve cranial I, conducting impulses from the nerve organ

olfactory receptor a chemoreceptor activated by gasses

-oligo- *comb. form* meaning "few"

Oligocene epoch [ōl"·ig·ō·sēn'] one of the tertiary geologic epochs extending from about 35 million years ago to 25 million years ago. It was preceded by the Eocene and followed by the Miocene. Most modern mammals were in existence at the end of the epoch

Oligochaetae [ōl'·ĭg·o·kē"·tē] a class of terrestrial and freshwater annelid worms containing the earth-worms and their allies. They are distinguished from the Polychaetae by the absence of cephalic appendages and parapodia, and from the Hirudinia by the absence of suckers. The genera *Aelosoma, Chaetogaster, Diplocarida, Enchytraeus, Lumbricus, Nais* and *Stylaria* are the subjects of separate entries

oligodont [ōl"·ig·ō·dont'] having few and widely separated teeth

oligodynamic [ōl'·ĭg·ō·di·năm"·ĭk] said of waters containing sufficient impurities to kill delicate, but not sufficient to kill tough, organisms

oligogene [ōl"·ĭg·ō·jēn'] one having a major effect on qualitative phenotypic characters, as distinct from a polygene

oligoglia [ōl'·ig·ō·glē"·ə] macroglial cells having only a few fibers (= oligodendrocyte)

oligophyodont [ōl'·ĭg·ō·fī"·ō·dont'] having succes-sive series of teeth replacing each other

oligopod larva [ōl"·ĭg·ō·pŏd'] an insect instar with functional thoracic limbs

oligosaprobe [ōl'·ĭg·ō·săp"·rōb] an organism which flourishes in oligodynamic waters (*see also* polysaprobe)

oligotrophic 1 [ōl'·ĭg·ō·trōf"·ĭk] (*see also* oligo-trophic 2) said of plants that will grow on poor soil, of any organism that requires little food, and of those that are restricted to a narrow range of nutrients

oligotrophic 2 (*see also* oligotrophic 1) said of environments poor in nutrition, particularly waters and bogs

oligotrophic lake a weakly productive lake with little plankton

olynthus larva [ōl·ĭn"·pəs] that stage in the development of a sponge, in which the parenchymula becomes hollowed out to form a spongocoel

-omalo- *comb. form* meaning "equal". An initial "h" is frequently added

omasum [ō·mās'·əm] the division of the artio-dactyl stomach between the abomasum and the reticulum

-ombro- *comb. form* meaning "rain storm", but just as commonly used in the sense of "shade"

omentum [ō·měnt"·əm] that portion of the perito-neum which surrounds and supports the viscera (*see also* greater omentum)

-omma- *comb. form* meaning "eye", or "what the eye sees"

ommatidium [ŏm'·ə·tĭd"·ē·om] one of the units of a compound eye

ommochrome [ŏm'·ō·krōm"] a brown visual pig-ment found in insects

-omni- *comb. form* meaning "all"

omnivorous [ŏm·nĭv"·ər·əs] eating all kinds of thing

omphalo-mesenteric arteries [ŏm·fā'·lō·měs'·ĕn·tĕr"·ĭk] a major artery arising near the central region of the dorsal aorta and supplying blood to the principal viscera

omphalo-mesenteric vein a continuation of the subintestinal vein on both sides of the liver, draining directly into the sinus venosus

-on a terminal symbol, without meaning, used to indicate a group of organisms associated in an aquatic habitat (*see* benthon, herpon, neuston, plankton, etc.)

Onchorhynchus [ŏn'·kō·rĭn"·kəs] a genus of tele-ost fish containing the Pacific "salmon". *O. tshawytsha* [chä·wĭt'·shə] (the Chinook salmon), sometimes reaches a weight of 100 lbs

onchosphere larva [ŏn"·kō·sfēr'] the first, ciliated, spherical larva of a cestode platyhelminth

-onko- *comb. form* meaning "hook", frequently transliterated -onco- or -oncho-

ontogeny [*angl.* ŏn·tŏj"·ə·nē, *orig.* ŏn'·tō·jen"·ē] properly the "production of eggs" but used in the classic phrase "ontogeny repeats phylogy" in the sense of the successive developmental stages through which an egg develops into an adult

Onychophora [ŏn'·ē·kŏf"·ər·ə]   a group erected to contain the velvet worms as the members of the genus *Peripatus* are often called. They were at one time a separate phylum but are now usually regarded as a class of arthropoda. They are caterpillar-like and have from 17 to 40 pairs of ambulatory appendages, papilate skin, and antennae. The genera *Aysheaia* and *Peripatus* are the subjects of separate entries

-oo-   *comb. form* meaning "egg" Words with this prefix were at one time, and occasionally still are, written oö-

oocyst [ō'·ō·sĭst"]   an encysted zygote, particularly of a sporozoan protozoa

oocyte [ō'·ō·sit"]   a cell which is a precursor of an egg (*see also* secondary oocyte)

ookinete [ō'·ō·kin·ēt"]   a motile zygote, particularly of a sporozoan protozoa, in contrast to an oocyst

oolitic sand [ō'·ō·lĭt"·ik]   sand composed principally of granules of calcium cabonate

oogamy [*angl.* ō·ŏg"·ə·mē, *orig.* ō'·ō·găm"·ē]   a term, usually confined to the discussion of algae, which involves reproduction by anisogametes sufficiently different to be distinguished as sperm and eggs

ooplasm [ō'·ō·plăzm]   a yolk free area of cytoplasm immediately surrounding the nucleus in isolecithal eggs

oosphere [ō'·ō·sfēr"]   an unfertilized female gamete

oospore 1 [ō'·ō·spŏr"]   (*see also* oospore 2) a durable resting zygote resulting from the "fertilization" of an oogonium by a hyphal ingrowth

oospore 2 (*see also* oospore 1) the product of the fusion of male and female gametes

ootid [ō'·ō·tid"]   the cell which is produced at the end of the meiotic division of oocytes and which metamorphoses directly into an egg

ooze   that upper layer of mud which is sufficiently fluid to be subject to slow flow

Opalina [ōp·ə·lĭn"·ə]   a genus of multinucleate protociliate protozoans found in the colon of *Rana*. It lacks a mouth and is only doubtfully thought to be a degenerate mastigophoran

open association   one with ample room for further growth

open circulatory system   a circulatory system in which there is no closed connection between the end of the artery and the beginning of a vein or in which a hemocoel replaces a venous system (*see also* closed circulatory system)

open community   one in which the competition for space is not so severe as to render difficult the invasion by a new species

opercular [ō·pûr"·kyū·lər]   possessing a lid

opercular bone   one of the dorsal pair of the two large membrane bones in the operculum of fish (*see also* interopercular bone, preopercular bone, subopercular bone)

operculum [ō·pûr·kyū·ləm]   literally a lid and used in biology for almost any structure serving this function including the gill covers in teleost fish and amphibian larvas, the post-pedal plate in many gastropod mollusks, the shell plates of barnacles, the first pair of abdominal appendages in Xiphisura, the plate covering the book lungs in many Arachnida, the lid of a moss

capsule, the hinged chitinous lid of an Ectoproct zoecium, the protoplasmic covering over the open end of the tube of a nematocyst, and the cover of an ascus

operon [ŏp"·ər·ŏn]   a group of cistrons (q.v.) that act as a coordinated unit and have a common operator gene

Ophioderma [ŏf'·ē·ō·dûrm"·ə]   a genus of ophiuroid Echinodermata with oral papillae fringing the mouth. Most are West Indian but *O. brevispinosa* [brĕv'·ē·spin·ōs"·ə] occurs on the coast of Maine

Ophioglossales [ŏf'·ē·ō·glŏs·ăl"·ēz]   a small order of primitive pterophyte plants represented today by the forms known as serpents' tongues. They are distinguished from all other ferns by having the sporangia borne on a special fertile spike. The genus *Botrychum* is the subject of a separate entry

ophiopluteus larva [ŏf'·ē·ō·plŏō"·tē·əs]   a pluteus larva of ophiuroid echinoderms, distinguished by the fact that one pair of arms is much longer than the others

Ophiosaurus [ŏf'·ē·ō·sôr"·əs]   the genus of Squamata containing the American glass "snakes" which are in fact limbless lizards

Ophiuroidea [ŏf'·ē·yŏōr·oid"·ē·ə]   the class of Echinodermata that contains the brittle stars. They are distinguished by their long, flexible, sometimes branched arms. The genera *Gorgonocephalus* and *Ophiodesma* are the subjects of separate entries

Ophrys [ō"·frĭs, ŏf"·rĭs]   a genus of orchids the pollinium of which bears so striking a resemblance to the abdomen of some female insects that males of the species endeavour to copulate with it and thus transfer pollen

-ophthalm-   *comb. form* meaning "eye"

ophradium [ŏf·rād"·ē·əm]   a chemo-receptor gland in the incurrent siphon of many Mollusca

Opiliones = Phalangida

-opistho-   *comb. form* meaning "posterior"

opisthocoelous vertebra [ŏp'·ĭs·bō·sēl"·əs]   one in which the centra are convex in front and concave behind

opisthoglyph snake [ŏp'·ĭs·bō·glĭf"]   those snakes having enlarged grooved teeth at the rear of the maxillary series

Opisthorcis [ŏp·ĭs·bōr"·kĭs]   a genus of trematode platyhelminthes closely resembling *Clonorchis* in structure and life history. *O. viverrina* [viv'·ər·in"·ə] is a common human parasite in many parts of the USSR and in the Far East

opisthospondylous ring [ŏp'·ĭs·bō·spŏnd"·əl·əs]   a ring of cartilage around the centrum marking the posterior end of the vertebra (*see also* prospondylous ring)

opisthotic bone [ŏp'·ĭs·bot"·ik]   a chondral bone of the posterior end of the cranium of Reptilia and Amphibia (some), lying lateral to the basioccipital to which it is connected by a ventral process

opium   the dried, milky, juice extracted from the unripe seed pods of *Papaver sominiferum*. It is the source of morphine, heroin and other narcotic alkaloids

opportunism   in biology, the adaptation of an organism to the most pressing factors in its environment

-opse-, -opseo-   *comb. form* meaning "sight"

-opsin   a suffix denoting a retinal pigment. The following terms with this suffix are defined in alphabetic position:

CYANOPSIN
IODOPSIN
LUMIRHODOPSIN

METARHODOPSIN
RHODOPSIN
SCOTOPSIN

-opsis *comb. suffix* meaning "having the appearance of"

optic pertaining either to the eye or to sight (*see also* dermatoptic)

optic capsules the thin cartilaginous capsule which encircles, or lies above, the orbit in chondrocrania

optic chiasma the external cross-over of the optic nerves under the lower surface of the brain

optic lobe one of a pair of thickened outgrowths from the dorsal wall of the mesencephalon. In mammals, there are two pairs of such lobes, known as the corpora quadrigemina. In insects the term is applied to a lateral lobe of the protocerebrum

optic nerve cranial II, running from the retina to the floor of the diencephalon

optic pedicel a cartilaginous cup in the orbit of chondrichthian fish

optic stalk the connection in the embryo between the optic vesicle and the prosencephalon

optic tract that part of the optic nerve that lies between the optic chiasma and the geniculate body

optical microscope *see* microscope

Opuntia [ō·pŭnt"·ē·ə] a genus of cactus that lacks leaves and is therefore particularly adapted to dry environments. Members of the genus are commonly called prickly pears after the fruit. The genus is also a principal food plant for *Coccus cacti*

oral pertaining to the mouth (*see* aboral, preoral)

oral arm a lengthy perradial lobe descending from the corners of the manubrium in some Scyphozoa

oral disk the flattened upper surface of anthozoan coelenterates

oral funnel the cavity anterior to the velum in *Branchiostoma*

oral groove a depressed area leading to the cytopharynx in some protozoans

oral hood the bell-shaped opening at the anterior end of *Branchiostoma* (= oral funnel)

oral pharynx that portion of the amniote pharynx which lies below the plane of the soft palate

orangutan the pongine primate *Pongo satyrus*. It is superficially distinguished from the gorilla by its unpigmented skin and from the chimpanzee by its long hair

orbit a cavity housing the eye in the skull of chordates

orbital cartilage a cartilage which in the embryonic chondrocranium, lies just dorso-lateral to the optic nerve and which forms the dorsal region of the orbit in the cyclostome chondrocranium (*see also* interorbital cartilage)

orbital crest the horizontally flat ridge of cartilage projecting above the orbit

orbital fissure the gap between the alisphenoid and the orbitosphenoid bones

orbitosphenoid bone [ôr'·bĭt·ō·sfēn"·oid] one of a pair of chondral bones of the skull lying in the posterior region of the orbit, between the alisphenoid and the frontal (*see also* sphenoid bone, alisphenoid bone, basisphenoid bone, laterosphenoid bone, parasphenoid bone)

orcein [ôr'·sēn] a dye extracted from various lichens, particularly *Leucanora tinctoria* and *Rocella tinctoria*, used to demonstrate chromosomes in squashes

Orcinus [ôr·sin"·əs] the genus of odontocetid

cetaceans containing *O. orca* [ôr"·kə] (the "killer whale"). This is the largest extant carnivorous animal

-orcul- *comb. form* meaning "cask"

order a taxon ranking immediately below a class. For example, the Lepidoptera are an order of the class Insecta (*see also* suborder, superorder)

Ordovician period one of the Paleozoic periods extending from about 400 million years ago to about 350 million years ago. It was preceded by the Cambrian and followed by the Silurian. There were numerous invertebrates and many coral reefs, with pelecypod and cephalopod mollusks

organ a discrete mass of cells of specific function, or functions. The following derivative terms are defined in alphabetic position:

BIDDER'S ORGAN
BOJANUS' ORGAN
CORTI'S ORGAN
ELECTRIC ORGAN
EPIPHYSEAL ORGAN

HANSTRÖM'S ORGAN
INTROMITTENT ORGAN
JACOBSON'S ORGAN
STIRN ORGAN
WEBER'S ORGAN

organelle [ôr'·găn·ĕl"] a discrete portion, with a specific function, of a cell. Cilia and mitochondria are typical of what is termed an organelle

organicism [ôr·găn"·ĭs·ĭzm] the doctrine, closely allied to vitalism, that organisms can only be interpreted by a study of the whole, and that there may be undiscovered biological forces acting on the whole, as important in their way as are the physical and chemical forces which can be understood from examination of the separated parts

organiser an area of an embryo that influences the organization of other areas (*see* Spemann's organiser)

organogenesis [ôr'·găn·ō·jĕn"·əs·ĭs] the development of organs and organ systems

ornithine [ôrn'·ĭþ·ēn"] α,δ-*diaminovaleric acid*. $NH_2(CH_2)_3CH(NH_2)COOH$. An amino acid not essential to the nutrition of rats. Apparently derived from arginine and forming part of a cyclic reaction with arginine and citrulline

ornithine cycle *see* urea cycle

Ornithischia [*angl.* ôrn'·ē·þĭsh"·ē·ə, *orig.* ôrn'·ĭþ·ĭsh"·ē·ə] an order of extinct reptiles containing those dinosaurs with tetraradiate pelves

ornithogaea [ôrn'·ĭþ·ō·jē"·ə] the Polynesian region including New Zealand

Ornithorhynchus [ôrn'·ĭþ·ō·rĭn"·kəs] one of the two extant genera—the other is *Echinda*—of monotreme mammals. *O. anatinus* [ăn'·ə·tĭn"·əs] (the "duck-billed platypus"), the only species, is a semi-aquatic, flattened, elongate animal with a duck-like bill. It is confined to E. Australia

oronasal groove [ôr'·ō·nāz"·əl] a groove leading from the corners of the angle of the mouth to the internal nares in Elasmobranch fish, precursor to nasal passages of higher forms

ortet [ôr'·tĕt] the original ancestor of a clone (*see also* ramet)

-ortho- *comb. form* meaning "upright", though commonly used in the form of "straight"

orthograde [ôr'·þo·grād'] to walk with the body vertical

orthoploid [ôr'·þo·ploid'] a polyploid, the chromosomes in which are an exact multiple of the haploid number of chromosomes of the organism from which it is derived

Orthoptera [ôr·thŏp"·tə·rə] an order of insects

that today includes only the grasshoppers, locusts and crickets but at one time also contained those forms now more usually placed in the Dictyoptera (cockroaches and mantids), Grylloblattodea, and Phasmida (stick insects). The Orthoptera possess eleven apparent abdominal segments, two pairs of wings and have the third pair of legs adapted for jumping

Oryctolagus [angl. ŏr·ĭk·tŏl"·ə·gəs, orig. ŏr'·ĭk·tō·lā"·gəs] a genus of lagomorph rodents containing O. cuniculus [kyōō'·nĭk"·yōō·ləs], the Old World burrowing rabbit commonly miscalled "Lepus"

Oscillatoria [ŏs'·sĭl·ə·tôr"·ē·ə] a genus of cyanophyte Algae consisting of a filament of cells which slowly oscillates for reasons not yet explained

osculum [ŏs"·kyōō·ləm] literally a "little mouth" but most commonly used in biology for the main exhalent opening of a sponge

-osm- see -osme- and -osmo-

-osme- comb. form meaning "odor"

-osmo- comb. form meaning to "thrust" from the Greek osmos, but frequently confused with "odor" from Greek osme

osmosis [ŏz·mōs"·ĭs] a general term referring to the passage of a solvent through a selectively permeable membrane. The passage is more rapid from a region of low solute concentration to one of high solute concentration than in the reverse direction. Thus differential flow rate therefore produces "osmotic pressure" against the membrane (see also endosmosis, exosmosis)

osmotic equilibrium [ŏz·mŏt·ĭk] the condition of an aquatic organism that neither gains nor loses fluid

Ostracoda [angl. ŏs·trăk"·ə·də, orig. ŏs'·trä·kōd"·ə] a subclass of crustacean Arthropoda. They are distinguished by being totally enclosed in a sometimes calcareous bivalved shell

Ostracodermi [ŏs'·trăk·ō·dûrm·ĭ] a primitive group of very early (Ordovician to Silurian) agnath vetebrates lacking both paired appendages and girdles. The anterior end was covered with bony plates, thus differentiating them from the allied Marsippobranchia, which includes the extant hagfishes and lampreys

ostracum a molluscan shell. The following derivative terms are defined in alphabetical position:

EPIOSTRACUM   PERIOSTRACUM
HYPOSTRACUM   PROOSTRACUM

Ostrea [ŏs·trē"·ə, ŏs·trā"·ə] a genus of pelecypod mollusks usually called oysters. They attach to the substrate by the left valve of the shell which is irregular and variable. There is no foot. O. virginica [vûr·jĭn"·ĭk·ə] is the oyster of the Atlantic coasts and O. lurida [lyōōr'·ĭd·ə] that of the Pacific

otic [ōt"·ĭk] pertaining to or in the vicinity of the ear

otic capsule the postero-lateral capsule of a chondrocranium containing the semicircular canals

otic process the postorbital articular portion of the palatoquadrate in chondrichthian fishes

otioccipital [ōt'·ē·ŏk'·sĭp"·ĭt·əl] a general term for the posterior division of the endocranium in fish

otoconium [ōt'·ō·kōn"·ē·əm] a minute calcareous particle, analogous, or possibly homologous, to an otolith, found in the semicircular canals of higher vertebrates

otocyst [ōt'·ō·sĭst"] a sensory organ of doubtful function found in many invertebrates consisting of a fluid-filled cavity containing an otolith and sensory hairs

otolith [ōt"·ō·lĭþ'] a solid particle within a balancing organ

outer plexiform layer that layer of the retina in which the axons of the rods and the cones synapse with the dendrities of the bipolar cells (see also inner plexiform layer)

ovary 1 [ō"·və·ē] (see also ovary 2) the animal organ from which, or within which, eggs are developed

ovary 2 (see also ovary 1) the fertile base of a carpel within which the ovules are developed. After fertilization, the ovary usually gives rise to a fruit. The following derivative terms are defined in alphabetic position:

CHAMBERED OVARY   SUPERIOR OVARY
INFERIOR OVARY

overdominance a condition in which a heterozygote is dominant to the homozygous dominant

overlapping inversion one which is superimposed on a previous inversion

overturn (= fall overturn) the mixing of zones in a lake, commonly occurring in the Autumn, through the generation of rotary currents by wind action

-ovi- comb. form meaning "egg"

oviduct [ōv·ē·dŭkt'] the duct carrying eggs to the exterior. It is frequently divided into distinct portions (see also common oviduct, uterus, vagina)

oviductule [ōv'·ē·dŭk·tyōōl'] small collecting ducts, coming from a follicular (in the sense of scattered) ovary

oviparous [angl. ōv·ĭp"·ər·əs, orig. ōv'·ē·pâr"·əs] said of females which produce eggs

ovipore [ōv·ē·pôr'] the external opening of the female genital duct in many invertebrates

ovipositor [ōv·ē·poz"·ĭt·ər] a more or less elongate projection from the posterior end of many insects which is adapted to placing eggs at a distance from the body

ovisac [ōv·ē·săk'] either a membrane within which many eggs are extruded from the body, or that portion of a female reproductive system filled with ripe eggs (= egg sac)

ovocoel [ōv·ō·sēl"] the gonocoel of the ovary

ovogonium [ōv·ō·gōn"·ē·əm] the animal oocyte or, rarely, the female gametangium of plants (= carpogonium, oogonium)

ovotheca [ōv·ō·þē"·kə] a terminal expanded portion of an oviduct in which eggs are accumulated before laying as one mass (see also ootheca)

ovovitelline duct [ōv'·ō·vĭt·ĕl"·ĭn] a duct coming from a germovitellarium

ovoviviparous [ōv'·ō·vĭv·ĭp"·ər·əs] said of females which produce eggs that hatch inside the female immediately preceding birth

ovule [ōv·yūl] ovoid bodies developing from the placenta in the angiosperm ovary. Each contains a nucleus. After fertilization, the ovule gives rise to the seed

ovum [ō"·vŭm] the Latin for egg

oxbow an almost closed loop in a stream or inlet

oxbow lake a crescentic lake formed from an oxbow which has been cut off from its parent waters

oxidase [ŏks"·ĭd·āz'] the trivial name of a group of oxidoreductase enzymes in which oxygen is used as a hydrogen receptor and hydrogen peroxide is therefore a customary by-product of the reaction

oxidoreductase [oks'·ĭd·ō·rə·dukt"·āz] the preferred term for all enzymes entering into oxidation-reduction reactions and therefore including all those enzymes previously known as oxidases, reductases and dehydrogenases

-oxo-   *comb. form* meaning "sour", almost hopelessly confused with -oxy- meaning "sharp"

oxyginases   [ŏks'·ē·jĭn"·āz·əz]   a group of enzymes that catalyze the direct transfer of molecular oxygen to a molecule. They therefore differ from the oxidases in that neither water nor hydrogen peroxide are by-products of the reaction

oxyphile cell   [ŏks"·ē·fil']   technically, a cell that stains well in acid dyes, but specifically a heavily granulated cell in the parathyroid gland

oxytocin   [ŏks'·ē·tōs"·ĭn]   a smooth muscle contraction-stimulating hormone found in the neurohypophysis

-pachy- *comb. form* meaning "thick"
pachynema [păk'·ē·nēm"·ə] that stage in the process of meiosis immediately following the zygonema; synapsis is complete and chromosomes are contracted
Pacini's corpuscle= Vater's corpuscle
paddle a limb adapted to swimming (cf. nectopod)
-paed- *comb. form* meaning "child" often transliterated -ped- and thus confused with "foot"
-paedic *comb. suffix* meaning "pertaining to the young"
paedium as assemblage of young. The following derivative terms are defined in alphabetic position:
GYNOPAEDIUM   POLYGYNOPAEDIUM
PATROGYNOPAEDIUM   SYNCHOROPAEDIUM
PATROPAEDIUM
paedogenesis [pē'·dō·jen"·əs·ĭs] sexual reproduction by larval, or apparently immature, organisms. The term is used not only of animals but also of the flowering of plants, particularly trees, before they reach their full development
Pagurus [*angl.* păg"·ər·əs, *orig.* păg·yūr"·əs] a genus of decapod Crustacea containing most of the better known North American "hermit crabs", the soft abdomens of which are protected by being thrust into empty shells of gastropod mollusks. Many species show a remarkable commensalism with the hydroid *Hydractinia echinata* which grows on the edge of the shell and constantly increases its size, thus making it unnecessary for the crab to change houses after each molt
pairing-segment that portion of a sex chromosome which has an exact equivalent in the other sex chromosome
Palade's granule= ribosome
-palae- *comb. form* meaning "ancient"
palaearctic [păl'·ē·ärk"·tĭk] pertaining to those areas of the Old World which are not tropical, i.e. Europe, Africa N. of the Sahara, N. Arabia and Asia N. of the Himalayas
palaeogaea [păl'·ē·ō·jē"·ə] the region of relict faunas that includes the Arctic, Europe, parts of North Africa, India and the Australian region
palatal process those processes from the palatine bone which meet along the midline between the anterior palatine fenestrae

palate the sum total of those structures that separate the roof of the buccal, from the floor of the nasal, cavity of vertebrates
palatine bone [păl'·ə·tĭn"] one of a pair of dermal bones of the splanchnocranium, lying immediately behind the maxilla and forming the posterior part of the secondary palate of mammals
paleocerebellum [păl'·ē·ō·sĕr'·ə·bĕl"·əm] that part of the cerebellum which has its principal nervous connections with spinal fibers
paleostriatum [păl'·ē·ō·strī'·āt"·əm] that portion of the corpus striatum which is found in primitive forms and which is probably homologous to the globus pallidus of mammals
paleotropical [păl'·ē·ō·trŏp·ĭk·əl] pertaining to that portion of the tropics that lies in the Old World (*see* palaearctic)
Paleozoic era [păl'·ē·ō·zō"·ĭk] a geologic era extending from about 500 million years ago to 200 million years ago. It is divided into the Permian, Carboniferous, Devonian, Silurian, Ordovician and Cambrian periods. It was preceded by the Proterozoic and followed by the Mesozoic
-palin- *comb. form* confused from two roots and therefore meaning either "backwards" or "again"
palingenesis [păl'·ĭn·jĕn"·əs·ĭs] either the doctrine of simple descent or that part of the development of an individual which is supposed to repeat the phylogony
Palinurus [păl'·ĭn·yōōr"·əs] a cosmopolitan genus of marine decapod crustacea. *P. argus* [är"·gəs] is the Atlantic "sea crawfish" and *P. interruptus* [ĭn'·tûr·rŭp"·təs] the California "spiny lobster". The true lobster is *Homarus*
palisade cell an individual cell of palisade parenchyma
palisade layer one or more layers of palisade parenchyma forming the adaxial portion of the mesophyll of a leaf
palisade parenchyma parenchyma consisting of elongate, chloroplast containing, cells lying vertical to the surface of the blade of a leaf
pallial [păl'·ē·əl] pertaining to the mantle
pallial commissure the connection between the hippocampal areas on the two sides of the brain in lower

127

vertebrates, which in higher forms becomes differentiated into the corpus callosum, and the hippocampal commissure

pallium [păl'·ē·əm] literally a "mantle". The following derivative terms are defined in alphabetic position: ARCHIPALLIUM   NEOPALLIUM

palmar [păl'·mär, päm·ə] pertaining to the ventral surface of a hand or foot

palmiped [păl"·mē·ped', päm"·ē·ped'] said of birds having the three front toes fully webbed

palp [pălp] a sensory appendage (see also pedipalp)

palpebral bone one in the upper eyelid of some reptiles

paludal [pə·lōōd'·əl] pertaining to marshes

palustrine [pə·lŭs'·trĭn] pertaining to bogs

palynology [păl'·ē·nŏl"·ə·gē] the study of fossil pollens

pampa the South American equivalent of steppe, usually used in the plural

-pan- comb. form meaning all. -pan- is neuter, -pasis masculine and -pasa- feminine, a fact usually ignored by biological logotechnicians

Pan [păn] a genus of pongine primates containing P. troglodytes [trŏg'·lō·dĭt"·ēz], the chimpanzee (q.v.)

Panaeolus [păn'·ē·ōl"·əs] a common North American genus of agaricale fungi. Several species, particularly P. campanulatus [kăm'·păn·yōō·lāt"·əs] and P. papilionaceus [păp·ĭl·ē·ŏn·ās"·ē·əs] are alleged to be hallucinogenic

pancreas [păn'·krē·əs] a gland which discharges digestive enzymes into the intestine and also houses the insulin-secreting islets of Langerhans. The term is also applied to many invertebrate glands primarily concerned with the secretion of digestive enzymes (see also hepatopancreas)

pancreozymin [păn'·krē·ō·zĭm"·ĭn] a hormone secreted in the upper intestinal mucosa. It is active in controlling enzyme secretion by the pancreas

pandemic [păn·dĕm"·ĭk] epidemic over a wide area

Pander's nucleus the area of yolk free cytoplasm immediately under the blastodisc of telolecithal eggs

Pandorina [păn'·dôr·ĭn"·ə] a flagellate organism variously regarded as a flagellate protozoan (see Phytomastigophora), when it exists as single cells or as a green algae (of the order Volvocales) when it occurs as a multicellular flattened plate

Paneth cells large cells found typically in the bottom of the glands of Leiberkuhn in the small intestine

panformation a plant community of several combined formations, each, or at least one of which, is dominated by a genus or family

pangenesis [păn·jĕn"·əs·ĭs] the theory of evolution which held that heritable characters are transmitted through minute gemmules that develop in each cell, and lie dormant until the moment of reproduction. This was popular in Darwin's time

Panizza's foramen an opening between the two sides of the aortic trunk in Crocodilia

panmictic population [păn·mĭk"·tĭk] one which results from prolonged, random, cross-breeding

panmixis= random mating

Panthera [păn'·thə·rə] the genus of fissipede carnivores that contains the great cats, at one time placed in the genus Felis. P. leo [lē"·o] (the lion), P. tigris

[tĭ"·gris] (the tiger), P. pardus [pär'·dəs] (the leopard) and P. onca [ŏn'·kə] (the jaguar) are the best known species

-pantin- comb. form meaning "tendril"

pantothenic acid [păn'·tō·thĕn"·ĭk] D(d)-N-(α,γ-dihydroxy-β,β-dimethylbutyryl)-β'-alanine. A water soluble nutrient additive. Definitely required by birds as an antidermatitis agent. The requirements for other forms are not clear. Frequently regarded as a vitamin

papain [pə·pā"·ĭn] an enzyme found in the papaya which catalyzes the hydrolysis of numerous compounds at basic amino acid bonds

Papaver [păp'·ə·vûr] a genus of dicotyledons containing the poppies. It includes not only the well known ornamentals but also P. somniferum [sŏm·nĭf"·ĕr·əm] from which opium is derived

papilla [pə·pil"·ə] a nipple, or any small conical protuberance. The following derivative terms are defined in alphabetic position:

ADHESIVE PAPILLAE   RENAL PAPILLA
FILIFORM PAPILLA   URINARY PAPILLA
FUNGIFORM PAPILLA   VATER'S PAPILLA

papillary muscles muscular braces for the valves of the heart

-para- comb. form meaning "alongside"

parabiosis [pär'·ə·bi·ōs"·ĭs] either the condition of living together applied variously to mixtures of species of similar habit (as man and the rat) or the union of two individual animals (see next entry) or to the condition of symbiosis between two species of ant in which colonies of neighboring nests are contiguous but do not mingle

parabiotic twin [pär'·ə·bi·ŏt"·ĭk] a laboratory produced equivalent of a Siamese twin in a non-human animal

paracentric inversion [pär'·ə·sĕn"·trĭk] one that does not involve the centromere

parachordal cartilage [pär'·ə·kôrd"·əl] one of a pair of cartilages in an embryonic chondrocranium which lies immediately lateral to the anterior end of the notochord

paraglossa [pär'·ə·glŏs"·ə] one of a pair of structures on the insect labium corresponding to the galea of the insect maxilla

paraglossum [pär'·ə·glŏs"·əm] the anterior skeletal element of the tongue

paragnath [angl. pär·ăg"·năþ, orig. pär·ə·năþ] properly anything that lies alongside a jaw and applied to various palps and the like on the jaws of various arthropods and annelids

Paragonimus [pär'·e·gŏn"·ĭm·əs] a genus of trematodes the adults of which encyst in the lungs of mammals. The first intermediate host is a freshwater snail and many species are involved. The tiny leech-like cercariae liberated from the snail enter a decapod crustacean. In North America this is of necessity the crayfish but in the Orient many freshwater crabs are available to it. Animals eating raw freshwater decapods become parasitized by the adult. In North America the mink is the commonest host though occasional human infections have been recorded. In parts of Japan and Korea 40% to 50% of the population are parasitized.

Paragordius [pär·ə·gôrd"·ē·əs] a genus of Nematomorpha differing from Gordius in that the female lacks a trilobed tail. P. varius [vär"·ē·əs] is the commonest American nematomorph

parahormone [pär'·ə·hôr"·mōn] a substance functioning as a hormone but not produced by a gland. Carbon dioxide is a well known example.

Paramoecium [pär'·ə·mēs"·ē·əm] a genus of holotrichan ciliate protozoans widely used in class teaching. There are numerous species but almost everything shown to a student is called P. caudatum [kôr·dāt"·əm]

paranasal cartilage [pär'·ə·nāz"·əl] one of a pair of precursors of the nasal capsules in the embryonic chondrocranium

paranotum [pär'·ə·nōt"·əm] a lateral outgrowth, precursor of a wing, from the dorsal plates of the insect thorax

paranucleus [pär'·ə·nyōō"·klē·əs] the second, usually smaller, of two nuclei in a cell

parapancreas [pär'·ə·pǎn"·krē·əs] secretory cells of unknown function adnexed to the pancreas in many reptiles

paraphysis [angl. pə·rǎf"·əs·ĭs, orig. pär'·ə·fis"·ĭs] literally an "outgrowth alongside" but specifically used for the roof portion of the developing forebrain which lies between the cerebral hemispheres, and for a dorsal outgrowth of the brain at the junction of the telencephalon and diencephalon

parapineal [pär'·ə·pĭn"·ē·əl, pär·ə·pĭn"·ē·əl] a body similar to the pineal but having a nervous connection to the right posterior commissure. Some cyclostomes have both a pineal and a parapineal. The function is unknown

Parapithecus [pär'·ə·pĭþ"·ə·kəs] a genus of extinct primates found in the Egyptian early Oligocene. It was a very small monkey usually placed in a separate division (Parapithecoidea) of the Catarrhini

parapodium [pär'·ə·pōd"·ē·əm] a lateral extension from each segment of a polychaete annelid (see also notopodium, neuropodium). It may be an organ of locomotion or of respiration

parapophysis [pär·ə·pǒf"·əs·ĭs] the vertebral prominence with which the lower branch of a two headed rib articulates (see also apophysis, anterior apophysis, gonapophysis, zygapophysis)

parapsid [angl. pə·ǎp"·sĭd, orig. pär·ǎp"·sĭd] pertaining to a reptile skull in which there is only a single temporal opening

Parascaris [angl. pə·rǎs·kə·ĭs, orig. pär'·ǎs·kǎr"·ĭs] a genus of ascarid nematodes containing the horse worm P. equorum [ĕk·wôr"·əm] commonly miscalled "Ascaris megalocephala". A variety P. equorum univalens [yōō'·nē·vǎl"·ěnz] is distinguished by having only two chromosomes

paraseptal cartilage [pär'·ə·sĕpt"·əl] the precursor of Jacobson's organ in the embryonic chondrocranium

parasexual [pär·ə·sĕks"·yōōl] said of the activities of those organisms, such as the fungi imperfecti, in which protoplasmic fusion, nuclear fusion and chromosome reduction occur haphazardly and unpredictably rather than in an orderly sequence

parasite one organism which lives on another to the detriment of its host. The following derivative terms are defined in alphabetic position:

AUTOECIOUS P.   SOCIAL PARASITE
ECTOPARASITE   SUPERPARASITE
HETEROECIOUS P.   WATER PARASITE
METOECIOUS P.   XENOPARASITE
OBLIGATE PARASITE

parasite-saprophyte a parasitic plant that kills its hosts and then lives on the decomposing remains

parasphenoid bone [pǎ'·rĕ·sfēn"·oid, pär'·ə·sfēn"·oid] a bone forming the floor of the cranium of Amphibia and part of the floor in some other animals (see also alisphenoid bone, basisphenoid bone, orbitosphenoid bone, sphenoid bone)

parasternalia [angl. pə·ǎst·ûrn"·āl·ē·ə, orig. pär'·ə·stûrn·āl"·ē·ə] sternal rib-like bones found in the ventral abdominal wall between the last true rib of the pelvis in crocodilia, Sphenodon, and some fossil reptiles (= gastralia)

parasymbiosis [pär'·ə·sĭm·bī·ōs"·ĭs] the condition when one symbiont damages another. It is difficult to distinguish this from parasitism

parasympathetic ganglion [pär'·ə·sim·pǎþ·ět"·ĭk] a ganglion which receives its impulses from autonomic components of cranial nerves. They are usually embedded in the walls of organs

parasympathetic nervous system that portion of the autonomic nervous system consisting principally of cholinergic fibers which controls the involuntary responses of the head, thorax, and upper abdomen through its cranial parts and of the lower abdomen and reproductive system through its sacral portion

parathormone [pär'·ə·bôr"·mōn] = parathyroid hormone

parathyroid gland [pär·ə·þir"·oid] an endocrine gland arising from an anterior outpocketing from the pharyngeal pouches (see also thyroid gland)

parathyroid hormone one secreted by the parathyroid gland. It is active in the control of the calciumphosphorus ration in the blood

paratonic [pär'·ə·tōn"·ĭk] applied to movements in plants caused by external stimuli. These are also known as tropic responses or tropisms

paratype [pǎr'·ə·tīp"] a type collected at the same time as, thought by the collector of the holotype to be identical with it

paraxonic foot [pər·ǎks"·ōn·ĭk] one in which the axis of symmetry passes between the third and fourth digit. This situation pertains in the even-toed ungulates or Artiodactyla

parenchyma [pə·rĕn"·kə·mə] fundamental, or supporting, tissue of a plant. The following derivative terms are defined in alphabetic position:

DIFFUSE PARENCHYMA   SCLEROTIC PARENCHYMA
PALISADE P.   SPONGY PARENCHYMA
PHLOEM PARENCHYMA

parenchyma ray parenchyma cells in horizontal phloem

-parie- comb. form the "wall of a house"

parietal [pə·ri'·ə·təl] literally pertaining to the wall or periphery. Specifically said of an ovule attached to the wall, rather than to the axis, of a plant ovary, of certain bones forming the central portion of the roof of the skull of vertebrates and of structures, or organs, in the vicinity of these bones

parietal bone one of a pair of membrane bones, forming the posterior portion of the roof of the skull, lying immediately above the squamosal and posterior to the frontal (see also interparietal bone, postparietal bone)

parietal cells granula-free cells in stomach glands

parietal eye an eye developed from the anterior of the

two structures known as epiphysis and commonly miscalled pineal eye

parietal lobe   the lobe of the cerebral hemisphere that lies between the occipital and the frontal

parietal plate   a pair of cartilages in a chondrocranium immediately behind and above the otic capsules. In most embryonic forms, they are connected by the synotic tectum

-pario-   *comb. form* meaning "produce"

paroccipital process   [pär'·ŏk·sĭp"·ĭt·əl]   one of a pair of processes projecting around the dorsal and posterior margins of the otic capsule in chondrocrania

paroophoron   [păr'·ō·ə·fôr"·ŏn]   a remnant of the embryonic mesonephros remaining in the ovarian region of some mammals

-parous   *comb. suffix* meaning "to produce", usually in the sense of "giving birth to". The following derivative terms are defined in alphabetic position:

LARVIPAROUS	OVOVIVIPAROUS
MULTIPAROUS	PUPIPAROUS
NYMPHIPAROUS	VIVIPAROUS
OVIPAROUS	

-parthen-   *comb. form* meaning "virgin"

parthenita   [pär'·þěn·ē"·tə]   a stage in the development of a trematode platyhelminth (e.g. sporocyst, redia) which themselves reproduce parthenogenetically (cf. adolescaria and marita)

parthenogamete   [pär'·thěn·ō·găm"·ēt]   a gamete capable of parthenogenesis

parthenogamy   [*angl.* pär·þěn·ŏg"·əm·ē, *orig.* pär'·thěn·ō·gam"·ē]   parthenogenetic development of a diploid cell

parthenogenesis   [pär'·thěn·ō·jen"·əs·ĭs]   the reproduction of an organism from one gamete. The following derivative terms are defined in alphabetic position:

AMEIOTIC P.	HETEROPARTHENOGENESIS
ARTIFICIAL P.	MEIOTIC P.
AUTOMICTIC P.	ZYGOID PARTHENOGENESIS
FEMALE P.	
GENERATIVE P.	
HEMIZYGOID P.	

parthenogonidium   [pär'·thěn·ō'·gŏn·id"·ē·əm]   a protozoan cell that gives rise to a colony by continued fission

partial dominance   a condition in which the phenotype of the heterozygote is intermediate between the two parents

partial penetrance   a condition in which the phenotypic expression overlaps that of the wild type

partial polyploid   a polyploid in which segments of chromosomes but not entire genomes are replicated

-partur-   *comb. form* meaning to "bring forth young"

parturition   [pär'·tyŏōr·ĭsh"·ən]   the separation of an embryo mammal from its mother

-parv-   *comb. form* meaning "small"

passage association   an association in the course of changing from one type to another

passage cell   a thin-walled cell in a root through which water passes readily

passive immunity   that which results from the injection of serum or the like

Pasteurella   [păs'·tyŏōr·el"·ə]   a genus of eubacteriale Schizomycetes. All species are pathogenic but *P. pestis* [pěs'·tĭs] (causing plague) and *P. tularensis* [tyū·lär·ěn"·sis] (causing tularaemia) are the only

disease-causing species in man. *P. multocida* [mŭl'·tō·sē"·də] is, however, of great economic importance as the causative agent of fowl cholera and swine plague

patagium 1   [păt·āj"·ē·əm]   (*see also* patagium 2) a flap of skin protruding from a mammalian body and, supported by limbs, used for gliding or flying

patagium 2   (*see also* patagium 1) one of a pair of lateral expansions of the insect prothorax immediately in front of the base of the wing

-patell-   *comb. form* meaning a "small dish" or "saucer"

patella   [pə·tĕl"·ə]   a small dish and applied to almost any structure, plant or animal, of this shape including the joint between the femur and the tibia in the leg of chelicerate arthropods, the tibia of acarine arthropods, the adhesive disc on the foreleg of some male dytiscid beetles and a bone

patella bone   the cup shaped bone which protects the knee (= knee cap)

pathogen   [păth'·ō·jěn"]   parasitic organisms, particularly microorganisms, that cause damage to their host

-patr-   *comb. form* meaning "father"

patrogenesis   [păt'·rō·jen"·əs·ĭs]   the development of an enucleated egg induced by fusion with a normal sperm

patrogynopaedium   [păt'·rō·jin'·ō·pēd"·ē·əm]   a group in which both parents remain with their immediate offspring

patropaedium   [păt'·rō·pēd"·ē·əm]   a group consisting of a male and his immediate offspring (*see also* gynopaedium)

-pauc-   *comb. form* meaning "few"

paurometabolous   [pôr'·ō·mĕt·ăb"·əl·əs]   said of those insects in which the egg hatches into a nymph living in the same environment as the adult, but which usually lacks wings found in the adult animal

Pauropoda   [pôr'·ō·pŏd"·ə]   an order of minute arthropods usually with eleven legs of which the first pair are greatly reduced. The most typical character is the antenna consisting of four basal joints bearing at the end a pair of styli, the anterior of them with two flagella and the posterior with one flagellum

Pauropus   [pôr·ōp"·əs]   a typical genus of Pauropoda with the characteristics of the order. *P. huxleyi* [hŭks"·lē·ĭ], just over a millimeter long by a third of a millimeter wide, is of cosmopolitan distribution

pavement epithelium   squamous epithelium

peat   partially decomposed vegetable matter (*see also* mesotrophic peat)

peck order   the social stratum of animals so called from its expression in the behavior of the domestic hen but applied to other forms

-pect-   *comb. form* meaning "coagulation"

pecten   [pěk"·tən]   any comb-like structure particularly 1. a projection into the vitreous humor from the retina of the eye of birds. 2. A similar structure in teleost fish. 3. A comb-like structure of unknown function immediately behind the legs in scorpions

Pecten   [pěk"·těn]   the genus of pelecypod mollusks containing the scallops. Of the two valves the lower is usually strongly convex and the upper flat. The surface is ribbed and the margin scalloped. *P. irradians* [ĭr·rād"·ē·ănz] of the east coast and the almost identical *P. aequisulcatus* [ēk'·wē·sŭl·kät"·əs] of the Pacific coast are intensively fished, the large adductor muscles being removed and sold for food. In Europe *P. maximus* [măks"·ĭm·əs] is similarly collected but the whole animal, either raw or cooked, is eaten as are oysters in the United States. The edge of the mantle bears numerous eyes.

-pectin-   *comb. form* meaning "comb-like"
Pectinatella [pĕk'·tĭn·ə·tĕl"·ə] a relatively common genus of freshwater Ectoprocta forming rosette-shaped colonies on a gelatinous base. In *P. magnifica* [măg·nĭf"·ĭk·ə], a common United States species, this base may attain a thickness of several inches. The statoblasts are large and flat with a marginal ring of anchor-shaped hooks.

pectine [pĕk"·tĭn] used interchangeably with pecten, particularly as regards the appendages of scorpions

pectoral [pĕk"·tə·əl] pertaining to the chest

pectoral girdle the sum total of the skeletal elements to which the anterior limbs of vertebrates are articulated. It generally consists of a dorsal scapular bone and ventral coracoid and precoracoid bones. In man the precoracoid is replaced by the clavicle

-ped- *comb. form* meaning "foot". Also a very frequent and confusing mis-spelling of "-paed-"

pedal disc [pĕd'·əl] the adhesive base of a coelenterate polyp

-pede *comb. term.* synonymous with -ped, the form preferred in this work. -pede, however, is still used in such English words as centipede and millipede

pedicel [pĕd"·ĕ·sĕl] any slender stalk connecting two larger objects. Apart from this general use, which pertains equally to plants and animals, it is specifically applied to that section of a plant stem which is directly connected to a single flower and to the tube foot of an echinoderm (*see also* optic pedicel)

pedicellaria [pĕd'·ĕ·sĕl·âr"·ĭ·ə] one of numerous, minute, usually two or three "fingered" pincer-like organs found on the surface of many echinoderms

Pedicellina [pĕd'·ĕ·sĭl·ĭn"·ə] a genus of colonial marine Entoprocta. The zooids rise from a creeping stolon. The cup-shaped calyx with a score of tentacles is carried on a relatively long (several mms) stalk from which it is separated by a diaphragm

pedicular cartilage [pĕd·ĭk·yū"·lə] a cartilage rising from the side of the nasal septum, in some chondrichthian fish

pedipalp [pĕd'·ē·pălp"] the second appendage of the Arachnida

Pedipalpi [pĕd'·ē·pălp"·ĭ] at one time an order of arachnid Arthropoda. They are now regarded as two distinct orders the Uropygi and the Amblypygi

pedium *see* paedium

peduncle [pĕd"·ŭnk·əl] with great difficulty to be distinguished from pedicel, but usually used in the sense of a stalk bearing one part rather than a connection between two parts and thus applied to the stalk of a barnacle, or the main stem of an inflorescence (*see also* anterior peduncle, caudal peduncle)

peep order the equivalent, in some frogs, of the peck order of fowls

-pel- *comb. form* meaning "clay" or "mud" (*see* leptopel, sapropel)

-pelag- *comb. form* meaning "the surface of the sea"
pelagic [pə·lăj"·ĭk] properly, "inhabiting the open ocean" as distinct from shores and estuaries; however, frequently in biology assigned wider meanings and sometimes even used as synonymous with "free-swim-

ming". The following derivative terms are defined in alphabetical position:

Pelecypoda [pə·lĕs'·ē·pōd"·ə] a very large class of Mollusca containing the clams, oysters, scallops, mussels and their allies. The bilaterally symmetrical body is enveloped by the mantle which secretes the bivalved shell on the outside and protects the sheet-like gills on the inside. There is no distinct head and in most forms the alimentary canal is encased in the muscular, plow-shaped, foot. The genera *Anodonta, Mytilus, Ostraea, Pecten, Pholas,* and *Teredo* are the subject of separate entries

pellicle [pĕl·ĭk·əl] a thin skin, particularly an outer noncellular coat of an organism

Pelmatohydra [pĕl'·măt·ō·hĭd"·rə] a genus of freshwater hydroid coelenterates differing from *Hydra* in the variety of its nematocysts. *P. oligactis* [ŏl'·ĭg·ăk"·tĭs] is the common large brown hydra

pelogloea [*angl.* pə·lŏg"·lē·ə, *orig.* pĕl'·ō·glē"·ə] the inorganic material adhering to aquatic organisms and submerged objects

Pelomyxa [pĕl'·ō·mĭks"·ə] a genus of amebid protozoans. They are always multinucleate and it is difficult to know what is an individual since the giant (3 mms or more) *P. carolinensis* [kăr'·ō·lin·ĕns"·ĭs] may break up in many pieces that subsequently join together, or partially together. Some people consider *P. carolinensis* to be identical with *Chaos chaos*

-pelos- *comb. form* meaning "clay" or "mud"

-pelv- *comb. form* meaning "basin"

pelvic girdle [pĕl'·vĭk] the sum total of skeletal units to which the hind limbs of vertebrates are attached. It consists of the fused ilium, ischium and pubis

pelvis 1 [pel'·vĭs] (*see also* pelvis 2) the boney mass formed, in many vertebrates, by the fusion of the pelvic girdle with elements of the vertebral column

pelvis 2 (*see also* pelvis 1) the swollen end of the ureter within the kidney

-pemphi- *comb. form* meaning "bubble" or "blister"

-pencill- *comb. form* meaning a small, pointed "brush"

penetrance the extent or regularity with which a gene alteration appears phenotypically (cf. expressivity). The following derivative terms are defined in alphabetical position:

penetration path the route followed by the sperm immediately after penetrating the egg and, in some Amphibia, delineated by the movement of pigment from the exterior, carried by the male gamete through the egg

penicillinase [pen'·ē·sĭl"·ĭn·āz'] an enzyme catalyzing the hydrolysis of penicillin to penicilloic acid

Penicillium [pĕn'·ə·sĭl"·ē·əm] a genus of ascomycete fungi commonly called "green molds". *P. notatum* [nōt·ăt"·əm] was originally used in the production of penicillin but has now largely been replaced by highly selected mutants of *P. chrysogenum* [kri·sŏj"·ĕn·əm]. *P. roqueforti* [rōk·fôrt·ĭ] and *P. italicum* [ĭt·ăl"·ĭk·əm] are respectively the green veins in roquefort and gorgonzola cheese

penis the muscular male intromittent organ of most animals. Its chitinous equivalent is the aedeagus

penis stylet   a hard spine which in some invertebrates substitutes for a penis

-penn-   *comb. form* meaning "feather"

penniculus   [pə·nĭk"·yū·ləs]   the closely adpressed
basal granules of the cilia of the cytopharynx in some
ciliates or the rows of cilia themselves

-penta-   *comb. form* meaning "five"

Pentastomida = Linguatulida

pepsin   [pĕp"·sĭn]   an enzyme catalyzing the hydrolysis of peptides

peptidase   [pĕpt"·ĭd·āz']   a group of enzymes
catalyzing the hydrolysis of peptides by splitting off
residues

Perca   [pĕr"·kə]   a genus of freshwater teleost fish
containing the perches. *P. flavescens* [flăv·ĕs·ĕnz] is
the North American yellow perch. The genus is closely
allied to *Stizostedion*

-pereio-   *comb. form* meaning "to transport". Often
confused with -peri-

pereion   [pĕr"·ē·ŏn]   the thoracic region of isopod
crustacea, and sometimes used as synonymous with
prothorax

pereiopod   [pə·rē"·ō·pod']   in insects the second or
third pair of thoracic legs of the larva, or the second pair
of adults. In Crustacea, any thoracic limb

Perenema   [pĕr'·ə·nēm"·ə]   a genus of euglenid
protozoans with a very plastic elongate drop-shaped
body and a single flagellum of which only the tip
moves

perennial   that which occurs in more than two successive years, particularly a plant that blooms in this
manner (*see also* biperennial)

perfect flower   one which has both male and female
sex organs (*see also* imperfect flower)

-peri-   *comb. form* meaning "about" in the sense of
"surrounding"

perianth   [per"·ē·ănþ]   those parts (e.g. calyx,
corolla) which surround the petals of a flower. Also used
for the sheath which, in some lower plants, encase the
reproductive organs (*see also* pseudoperianth)

peribiliary body   [pĕr'·ē·bĭl"·yər·ē]   a lysosome of
the liver

periblast   [pĕr"·ē·blăst]   a mass of incompletely
separated yolk cells which unite the blastodisc with the
yolk mass in telolecithal eggs

periblastula   [pĕr'·ē·blăst'·yōō·lə]   a blastula consisting of a sheath one cell thick, enclosing the central
yolk

peribranchial chamber = atrial cavity

peribranchial groove   [pĕr'·ē·brănk"·ē·əl]   a
groove connecting the hypobranchial to the epibranchial grooves (*see also* epibranchial groove, hypobranchial groove)

pericardial cavity   [pĕr'·ē·kärd'·ē·əl]   that portion
of the coelom which immediately surrounds the heart

pericentric inversion   [pĕr'·ē·sĕn"·trĭk]   chromosome inversion around the centrosome

perichondral bone   [pĕr'·ē·kŏn"·drəl]   one in which
the first ossification takes place between the perichondrium and the cartilage

perichondrium   [pĕr'·ē·kŏn"·drē·əm]   the connec-
tive-tissue envelope of cartilage

perichordal tube   [pĕr'·ē·kôrd"·əl]   a thin layer of
fusiform cells derived from the sclerotome, lying
immediately around the notochord

periderm   [pĕr"·ē·dûrm']   the outermost layer of the
epidermis in the mammalian embryo, from which the
cornified layer derives, and also the chitinous tube
protecting the hydranths in some hydrozoan coelenterates

peridural space   [pĕr'·ē·dūr"·əl]   the perimeningeal
space in those forms having a dura spinalis (*see also*
subdural space)

perilymph   [pĕr'·ē·lĭmf"]   the fluid between the
semicircular canals of the inner ear and their skeletal
supports

perimysium   [pĕr'·ē·mĭs"·ē·əm]   white, fibrous connective tissue separating muscle bundles

perineural satellite   [pĕr'·ē·nyūr"·əl]   oligodendroglia closely pressed to the body of a nerve cell in the
brain (*see also* perivascular satellite)

perineurium   [pĕr'·ē·nyūr"·ē·əm]   the outer connective tissue coat of a nerve

period   a secondary division of geological time. The
primary divisions are eras. Thus the Mesozoic era is
divided into the Triassic, Jurassic and Cretaceous
periods. Periods are defined in alphabetic position and
listed under the eras of which they are divisions

periosteal bud   [pĕr'·ē·ŏst"·ē·əl]   a mass of invading
osteogenic cells, osteoblasts and capillaries appearing
in the course of endochondral formation of bone

periosteum   [pĕr'·ē·ŏst"·ē·əm]   the membrane
around the outside of bone

Peripatus   [pə·rĭp"·ə·təs]   a genus of Onycophora
(q.v.) with the characteristics of the phylum. Most are
neotropical

peripheral nervous system   the sum total of all
those parts of the nervous system which are not
included under the term central nervous system

peripheral stele   stele from which the vessels of
adventitious roots take their origin

periphyton   [pĕr'·ē·fĭt"·ən]   the surface population
of submerged objects or substrates

periportal connective tissue   [pĕr'·ē·pŏrt"·əl]   the
dense connective tissue of the lobules of the liver of
higher vertebrates

perisarc   [pĕr'·ē·särk"]   the sheath of hydroid
coelenterates. Sometimes used as synonymous with
periderm

perisperm   [pĕr'·ē·spûrm"]   storage tissue in the
seed derived from the remnants of the nucellus

-perisso-   *comb. form* meaning "uneven"

Perissodactyla   [pə·rĭs"·ō·dăk·tĭl'·ə]   an order of
placental mammals at one time fused with the Artiodactyla into the group Ungulata. It contains the horses,
rhinoceroses, and tapirs, in which there is an odd
number of digits on the foot with the axis of symmetry
passing through the centre digit. The family Equidae,
and the genera *Cerathotherium, Diceros, Rhinoceros* and
Tapiros are the subjects of separate entries

peristalsis   [pĕr'·ē·stăls"·ĭs]   wave-like contractions
passing along the length of a tube

peristome   [pĕr'·ē·stōm"]   literally "around the
mouth" and applied to the region surrounding entrance
to the cytopharynx in protozoa, the area surrounding the
aperture of a gastropod shell, that segment of an annelid
which immediately follows the prostomium and the
structures immediately surrounding the orifice of a
moss-capsule

perithecium   [pĕr"ē·thĕs"·ē·əm]   a hollow, usually
flask-like, ascocarp bearing asci over its inner surface

peritoneal funnel = coelomic funnel

peritoneum [pĕr'·ē·tŏn·ē·əm]   the lining of the abdominal cavity (cf. omentum) in vertebrates

Peritricha [pĕr'·ē·trĭk"·ə, pĕr'·ē·trĭk"·ə]   an order of ciliate Protozoa distinguished by the possession of an adoral row of cilia which passes from left to right round the peristome and by the absence of any other cilia from the body. The genus *Vorticella* is the subject of a separate entry

perivascular satellite   oligodendroglia surrounding capillaries in the brain (*see also* perineural satellite)

perivitelline membrane [pĕr'·ē·vĭt·ĕl"·ĭn]   a membrane formed beneath the vitelline membrane immediately after fertilization

perivitelline space   the space between the vitelline and perivitelline membrane

permeant [pûrm"·ē·ənt]   a roving member of a terrestrial community frequently passing from one community to another

permiable membrane   one which permits the passage of substances from one side to the other (*see also* differentially permiable membrane and semipermeable membrane)

Permian period [pûrm"·ē·ən]   the most recent of the Paleozoic periods extending from about 220 million years ago to 200 million years ago. It was preceded by the Carboniferous and followed by the Triassic. Much land was desert but a few coniferous trees were appearing. Reptiles and insects were the dominant land animals

-pero-   comb. form meaning "maimed"

Perodicticus [pĕr'·ō·dĭkt"·ĭk·əs]   a genus of West African lorisoid prosimians containing the potto (*P. potto* [pŏt"·ō]). They have soft fur, small ears and a vestigeal tail

Peromyscus [pĕr'·ō·mĭs"·kəs]   a very large genus of New World myomorph rodents. The common deer-mouse is *P. maniculatus* [măn·ĭk·yōō·lăt"·əs], of which more than 60 subspecies have been described

peroneal artery [pĕr'·ŏn·ē"·əl]   a branch of the popliteal artery supplying blood to the muscles of the calf

Peronosporales [pĕr'·ŏn·ō·spôr·āl"·ēz]   an order of biflagellate oomycete fungi distinguished by the production of sporangia. They differ from the Saprolegniales by producing two oospheres in each oogonium. The genera *Pythium* and *Phytophora* are the subjects of separate entries

peroxidase [pə·ŏks"·ĭd·āz]   a group name for enzymes catalyzing oxidations in which peroxide acts as the oxygen donor and water is therefore a product of the reaction. Peroxidase, without qualification, is used to describe a large group of haemoprotein enzymes oxidizing a great variety of materials by the indicated reaction

perpendicular plate   a projection of the ethmoid bone which separates the right and left nasal passages

perradius [pə·rād"·ē·əs]   that radius of a radially symmetrical organism on which organs are grouped

pessimum [pĕs"·ĭm·əm]   literally "the worst" and used in ecology to indicate the worst conditions under which a specific organism can just survive

petiolated abdomen [pĕt'·ē·ō·lāt"·ĕd]   an insect abdomen that, as in many Hymenoptera, is attached to the thorax by a stalk or petiole (*see also* abdomen, sessile abdomen)

petiole [pĕt"·ē·ōl]   any slender stalk, particularly that

which connects the leaf to a stem, or the swollen abdomen of a thin-waisted wasp to the thorax

-peto-   comb. form meaning "seek"

-petr-   comb. form meaning "rock" or "stone"

Petromyzon [pĕt'·rō·mī"·zŏn]   a genus of Marsipobranchii containing the Atlantic lamprey *P. marinus* [mə·rēn·əs]. This is a devastating parasite of fishes and a land-locked race of it has proved disastrous to the commercial fisheries of the Great Lakes. *Petromyzon* differs from *Myxine* in having a well developed branchial basket and a dorsal nasal aperture

petrosal bone [pə·trŏs"·əl]   one of a pair of chondral bones of the posterior end of the skull running from the supraoccipital to the tympanic bulla

petrosal ganglion   a ganglion on the ninth cranial nerve which in lower forms receives Jacobson's commissure

peyote [pā·ō'·tā]   a crude hallucinogenic extract from the dried tops of the cactus *Lophophorus williamsii*. The active principle is mescaline

peyotl [pā·ot'·əl]   the cactus *Lophophora williamsii* that yields a hallucinogenic mixture of alkaloids widely used by the Aztecs, a few of their descendants and many of their contemporary imitators

Peziza [pĕz·ēz"·ə]   a genus of pezizale fungi with large, cup-shaped, apothecia

Pezizales [pĕz'·ēz"·āl·ēz]   an order of ascomycete fungi distinguished by a cup-shaped or saucer-like ascocarp. They are often, from this fact, mistaken for Basidomycetes. The genera *Monilinia* and *Peziza* are the subjects of separate entries

pH   a measure of acidity or alkalinity based on the hydrogen ion concentration in terms of a reciprocal logarithmic scale. pH7 is neutral, pH's below 7 indicate increasing acidity and those above 7 indicate increasing alkalinity

-phaen-   comb. form meaning "to appear" or "appearance"; almost invariably transliterated -pheno-

-phaeo-   comb. form meaning "brown"

Phaeophyta [fē'·ō·fit"·ə]   a sub-phylum of Algae containing the brown algae. They are distinguished by the yellowish-brown chromatophores. The genera *Fucus, Laminaria* and *Sargossum* are the subjects of separate entries

-phaet-   comb. form meaning "main"

-phag-   comb. form meaning "a glutton", though more commonly used in the sense of "to eat"

-phage   comb. suffix indicating something that devours another. The following terms with this suffix are defined in alphabetic position:

BACTERIOPHAGE   MICROPHAGE
COLIPHAGE   PROPHAGE
MACROPHAGE   VITELLOPHAGE

phagocyte [făg"·ə·sit']   a cell that engulfs solid particles

phagocytosis [făg'·ō·sĭt·ōs"·ĭs]   the fact that, or method by which, solid particles may be invaginated by cell wall and carried into the interior of the cell

-phagous   comb. suffix used in the sense of "eater of". The suffix -vorous means the same thing. No compounds are listed in this dictionary since the meanings are obvious from the roots: for example, entomorphagous, insect-eating; myrmecophagous, ant-eating, etc.

phalanges [fā·lăn"·jēz]   plural of phalanx

Phalangida [fā'·lăn·jĭd"·ə]   an order of arachnid

arthropods containing those forms known in the United States as daddy-longlegs and in Britain as harvestmen. They are characterized by the possession of four pairs of walking legs and a segmented abdomen broadly joined to the cephalothorax

phalanx [fəl·ănks"] literally "an array of soldiers", but used specifically in botany for rays of stamens and in zoology for a bone in the vertebrate skeleton distal to the metacarpal or metatarsal. In man, a bone of the finger or toe

-phall- comb. form meaning "penis"

phaner- comb. form meaning "obvious" or "manifest"

phanerogam [făn"ər·ō·găm] a flowering plant that reproduces itself from seeds (cf. cryptogam)

phaoplankton [fā'·ō·plănk"tən] that which occurs only in depths to which enough light penetrates to permit photosynthesis

-pharo- comb. form confused between four Greek roots and thus variously meaning "a light, or lighthouse", "a garment", "a plough" or "to possess"

pharotaxis [făr'·ō·tăks"·is] the recognition and use of landmarks in bionavigation

pharyngeal [făr'·ĭn·jē"·ə] pertaining to the pharynx

pharyngeal basket a food gathering passage held permanently open by a girdle of trichites in some protozoa

pharyngesophagus [fə·rĭnj'·ē·sŏf"·ə·gəs] the dorsal of the two tubes into which the pharynx of the Marsippobranchia divides. The ventral is the respiratory passage

-pharyngo- comb. form from "pharynx"

pharynx [făr"·ĭngks] in vertebrates the part of the alimentary canal that lies between the oral cavity and the esophagus. In invertebrates a muscularized portion of the alimentary canal lying immediately behind the buccal tube or eosophagus. The following derivative terms are defined in alphabetic position:

CYTOPHARYNX	NASOPHARYNX
HYPOPHARYNX	ORAL PHARYNX

Phascolarctus [angl. făs·kōl"·ärk·təs, orig. făs'·kō·lärk"·təs] a monotypic genus of phalangid marsupia containing the "koala" or "kaola Bear" (*P. cinereus* [sĭn·ēr"·ē·əs]). These archityes of the teddy bear are arboreal, eating only the leaves of various species of *Eucalyptus*

-phase- comb. form meaning "appearance"

phase a transitory stage, particularly in cell division. The following derivative terms are defined in alphabetic position:

ANAPHASE	PROPHASE
INTERPHASE	TELOPHASE
METAPHASE	

phase contrast microscope a microscope in which light passing through the object is directed through a phase plate that causes it to become 1/4 wave out of phase with the background light which does not go through the object. Since colorless biological specimens in water usually shift the light passing through them by 1/4 wavelength, a total phase shift of 1/2 wavelength is produced in a phase contrast microscope. This results in maximum contrast between areas of varying optical density in the specimen

Phelebotamus [flə·bŏt'·ə·məs] a genus of blood sucking dipteran insects well known as vectors in the transmission of *Leishmania*

phellem [fĕl'·əm] cork tissue

-phello- comb. form confused between two Greek roots and therefore meaning either "cork" or "stony ground"

phellogen [fĕl"·ə·jĕn'] the type of cambium that produces cork

-pheno- comb. form derived from phaeno-, meaning "appearance". Sometimes used as though derived from -phaner- meaning "evident"

phenocopy [fēn"·ō·kŏp'·ē] a copy, resembling a gene mutant, but produced by the action of the environment

phenome [fē"·nōm] the sum total of all the phenotypic characteristics of an organism (cf. genome)

phenotypic [fēn'·ō·tĭp"·ĭk] used to designate the physical appearance of an organism as distinct from its genetic constitution (*see also* ecophenotype and genotype)

phenylalanine [fĕn'·əl·ăl"·ə·nēn] α-amino-β-phenylpropionic acid. $CH_6H_5CH_2CH(NH_2)COOH$. An amino acid essential to the nutrition of rats. It is apparently converted to tyrophenase in higher forms

-phero- comb. form meaning "to bear" in the sense of "carrying" (cf. -phore-)

pheromone [fēr"·ō·mōn'] a substance produced by one organism for purposes of chemocommunication with another (cf. hormone, from which this word is derived)

-phil-, -philo- comb. form meaning "loving" in the sense of being attracted to and hence used in botany in the sense of "fertilized through the agency of", and in ecology in the sense of "preferring to live on or among". The *subs. suffix* -phil is usually confined to leucocytes, -phile being used for most other forms, though -phila pertains to plant associations. The adjective suffix -philic is synonymous with, but less usual than -philous. The following terms with -phil as a suffix are defined in alphabetic position:

BASOPHIL	NEUTROPHIL
EOSINOPHIL	POLYCHROMATOPHIL
HETEROPHIL	

-phila see -phil- for meaning and geophila, rheophila for derivative terms

-phile see -phil- for meaning and basophile and symphile for derivative terms

-philic comb. term. synonymous with -philous, the form preferred in this work except for the terms below, defined in alphabetic position:

BASOPHILIC	CRYOPHILIC	THERMOPHILIC

Philodina [fil'·ō·dĭn"·ə] a very common genus of rotifers with the corona composed of two distinct wreaths of cilia

philopatry [fi'·lō·păt"·rē] the custom of returning from a distance to last year's breeding place

-philous termination identical in meaning to -philic but usually preferred for some terms, particularly in the sense of "dwelling in". The very numerous derivative terms are not separately defined since the meanings are obvious from the roots

-phloe- comb. form meaning "bark"

phloem [flō"·əm] the main food conducting tissue of vascular plants

phloem parenchyma parenchyma cells lying in logitudinal parenchyma

-pho- comb. form meaning "a light"

-phob- comb. form meaning "to fear" or "flee

from"; it is sometimes combined with the termination tropism to indicate negation of the operative word; thus, phobophototropism is negative phototropism

-phobic *adjective suffix* meaning "shunning" or "fleeing from". The meaning is opposite to -philous

Phocaena [fə·kēn"·ə] a genus of odontocetid cetaceans containing the porpoises. *P. phocaena* is the common "harbor porpoise" of both United States coasts. The porpoises differ from the dolphins in having compressed teeth with spade shaped crowns

-phol- *comb. form* meaning a "horny scale"

Pholas [fō"·ləs] a genus of rock-boring pelecypod mollusks. They are fusiform in shape with greatly reduced shells

Pholidota an order of eutherian mammals containing the pangolins and scaly anteaters. They are distinguished by the peculiar scales—actually plates of fused hair—which cover them. The genus *Manis* is the subject of a separate entry

-phon- *comb. form* confused between two Greek roots and therefore meaning either "sound" or "murder"

phonoreceptor [fōn'·ō·rē·sĕp"·tôr] a receptor distinguishing variable vibrations and therefore an organ of hearing (cf. scolopophore)

-phore- *comb. form* confused between four Greek roots and therefore meaning "bearer" (and, by extension, the vessel born), "movement", "thief" or "detector". The first of these is the most usual. The following derivative terms are defined in alphabetic position:

CHONDROPHORE	PHOTOPHORE
CONIDIOPHORE	PNEUMATOPHORE
GAMETOPHORE	SCOLOPOPHORE
GONOPHORE	SPERMATOPHORE
LOPHOPHORE	SPERMOPHORE
MELANOPHORE	SPORANGIOPHORE
NECTOPHORE	SPOROPHORE
ODONTOPHORE	TRICHOPHORE

phoresy [fŏr'·ə·sē] the condition of being carried about by another. Some pseudoscopions, for example, are phoretic, as distinct from parasitic, on houseflies

-phorous *adjective suffix* from -phore-

phosphoketolase [fŏs'·fō·kĕt"·ō·lāz"] an enzyme catalyzing the production of acetylphosphate and glyceraldehyde 3-phosphate from D-xylulose 3-phosphate and orthophosphate

phospholipase [fŏs'·fō·li'·pāz"] catalyzes the hydrolysis of lecithin into lysolecithin and an unsaturated fatty acid

phosphoprotein [fŏs'·fō·prō"·tēn] a protein which yields phosphorus on hydrolysis

phosphorylase [fŏs'·fō·rĭl·āz"] a group of transferase enzymes catalyzing reactions involving orthophosphate

phosphotase [fŏs'·fō·tāz"] group name for enzymes that hydrolyze phosphate ester linkages

photic pertaining to light. The following derivatives with this suffix are defined in alphabetic position:

APHOTIC	EUPHOTIC
DIPHOTIC	STENOPHOTIC

Photocorynus [*angl.* fō·tŏk"·ər·in'·əs, *orig.* fō'·tō·kôr·in"·əs] a deep-sea teleost fish in which the dwarf male is a permanently attached parasite of the female

photohorotaxis [fō'·tō·hōr'·ō·taks"·ĭs] response to a color or light pattern. The term should properly be photohoramotaxis. As here spelled it means "response to the time of day"

photolithotrophic [fō'·tō·liþ·ō·trōf"·ĭk] a type of microbial nutrition that requires inorganic hydrogen donors in the medium

photolysis [*angl.*fōt·ŏl"·əs·ĭs,*orig.*fō'·tō·lĭs"·ĭs] the arrangement of chloroplasts within the cell under the action of light

photometabolic [fō'·tō·mĕt'·ə·bŏl"·ĭk] pertaining to the utilization in biosynthesis of photon energy

photomicrograph [fō'·tō·mik"·rō·grăf] a photograph taken through a microscope (*see also* macrophotograph, microphotograph)

photonastic [fō'·tō·năst"·ĭk] said of the bending of a flattened plant structure, such as a leaf, caused by light-induced differential growth

photoperiodism [fō'·tō·pēr"·ē·ŏd'·ĭzm] the response to periodic, usually rhythmic variations in the light

photophore [fō"·tō·fôr'] a cell or organ used for the production of light by a living organism

photoreceptor [fō'·tō·rē·sĕpt·ər] a receptor capable of perceiving photon energy

photosynthesis [fō'·tō·sĭn"·thə·sĭs] the utilization of photon energy by chlorophylls in the synthesis of carbohydrates and the liberation of oxygen. In higher plants the chlorophyll is localized within chloroplast membranes (*see also* biosynthesis)

photosynthetic autotroph [fō'·tō·sĭn·thet"·ĭk] an organism capable of living in a simple inorganic medium obtaining energy from light (*see also* chemosynthetic autotroph)

phototaxis [fō'·tō·tăks'·ĭs] movement in response to light (*see also* aphototaxis, apophototaxis)

phototropism [fō'·tō·trōp"·ĭzm] movement stimulated by light. Without qualification usually means positive phototropism (*see also* diaphototropism, skototropism)

phragmoplast [frăg"·mō·plast'] a fibrous spindle, precursor to the cell plate and hence a spindle between two nuclei in the same cell, not in the process of division

-phret- *comb. form* meaning a "tank" or "cistern"

Phrynosoma [*angl.* frĭn·ôs"·əm·ə, *orig.* frĭn'·ō·sō"·mə] a genus containing the "horned toads" which are actually lizards (Squamata)

phthis- *comb. form* meaning "to waste away"

-phyco- *comb. form* confused between two Greek roots and therefore meaning either "alga" or "painted"

phycomycetes [fi'·kō·mi·sēt"·ēz] a no longer valid fungus taxon the members of which are now distributed between the Chytriomycetes, Oomycetes and Zygomycetes

phycobilosome [fi'·cō·bi'·lō·sōm"] granules, thought to contain supplementary photosynthetic pigments, on the chloroplast lamellae of Rhodophyta

-phyko- *comb. form* meaning "alga" or "seaweed", almost invariably transliterated -phyco-

-phyl- *comb. form* meaning "race" or "tribe"

phyletic [fil·ĕt"·ĭk] pertaining to race or tribe (*see* monophyletic, polyphyletic)

-phyll- *comb. form* meaning "leaf". The following derivative terms are defined in alphabetic position:

CATAPHYLL	MESOPHYLL
CHLOROPHYLL	MICROSPOROPHYLL
HYDROPHYLL	SPOROPHYLL
MEGASPOROPHYLL	

phyllode [fĭl″·ōd] in botany a flattened petiole having the appearance of a leaf blade (cf. cladode)

phyllorhiza [fĭl′·ō·rĭz″·ə] a leaf functioning as a root, as in some water plants

phyllospondylous vertebra [fĭl′·ō·spŏnd″·əl·əs] a shell-like vertebra not articulated with other vertebra

Phyllostomus [*angl.* fĭl·ŏst·əm·əs, *orig.* fĭl′·ō·stōm″·əs] a typical genus of leaf-nosed bats (Chiroptera). All have leaf-, or spear-shaped projections arising from the nose

phyllotactic fraction [fĭl′·ō·tăk″·tĭk] one in which the numerator represents the number of times the stem must be encircled before reaching a leaf vertically under the one from which the count starts and the denominator indicates the number of leaves which must be passed in so doing

-phyllous *adjectival suffix* from -phyll

phylogeny [*angl.* fĭl·ŏj″·ən·ē, *orig.* fĭl·ō·jĕn″·ē] properly the development of races but used in contrast to ontogeny in the sense of the successive evolutionary forms through which an organism has evolved from a remote ancestor

phylum a taxon representing a main division of the plant or animal kingdom. A phylum is divided into classes (*see also* sub-phylum)

Phymatotrichium [fĭ′·măt·ō·trĭk″·ē·əm] a genus of soil deuteromycete fungi, pathogenic to plants. *P. omnivorum* [ŏm′·nē·vôr″·əm] causes root rot of many commercial crops

-phys- *comb. form* meaning "a growth", or "outgrowth"

Physalia [fĭ·sāl″·ē·ə] a genus of planktonic siphonophoran coelenterates with a very large pneumatophore to the underside of which numerous specialized zooids are attached. *P. pelagica* [pə·lăj″·ĭk·ə], the "Portuguese man-of-war" has nematocyst-bearing tentaculozooids which may reach 50 feet in length

Physeter [fĭ·sēt′·ə] the genus of odontocentid whales containing the "sperm whale" or "cachalot" (*P. catodon* [kăt″·ō·dŏn′]) one of the two extant species of a once large group

-physis *substantive suffix* from -phys-. The following terms using this suffix are defined in alphabetic position:

ADENOHYPOPHYSIS	METAPHYSIS
ANTERIOR APOPHYSIS	NEUROHYPOPHYSIS
APOPHYSIS	PARAPHYSIS
EPIPHYSIS	PARAPOPHYSIS
GONAPOPHYSIS	SYMPHYSIS
HYPOPHYSIS	ZYGAPOPHYSIS

-physo- *comb. term* of confused origin properly meaning a "bubble" or "bladder" but often used in the sense of "swollen"

physoclystous [fĭs′·ō·klĭst″·əs] the condition of a fish in which the swim bladder is not connected to any portion of the alimentary canal

physostomous [*angl.* fĭs·ŏs″·tə·məs, *orig.* fĭs′·ō·stōm″·əs] the condition of a fish in which the swim bladder is connected to the esophagus

-phyt- *comb. form* meaning "plant"

phytase [fĭ′·tāz″] an enzyme catalyzing the hydrolysis of *myo*-inositol hexaphosphate to yield *myo*-inositol and orthophosphate

-phyte *substantive suffix* meaning "a plant". The meaning of most compounds are evident from their roots but the following terms are defined in alphabetic position:

AUTOPHYTE	SPERMATOPHYTE
CORMOPHYTE	SPOROPHYTE
EPIPHYTE	THALOLPHYTE
HOLOPHYTE	XEROPHYTE

phytochrome [fĭt″·ō·krōm′] a chromoprotein that regulates germination, growth and flowering in higher plants

Phytomastigophora [fī′·tō·măst′·ĭg·ŏf″·ər·ə] a subclass of mastigophoran Protozoa distinguished by their ability to synthesize chlorophyll and therefore considered by most to be algae. This zoological taxon does not, however, correspond to any acceptable botanical taxon. See, however, the orders Chrysomonadida, Cryptomonadida, Dinoflagellida, Euglenida and Phytomonadida

Phytomonadida [fī′·tō·mōn·ăd″·ĭd·ə] an order of phytomastigophorous protozoa distinguished by a large cup-shaped chloroplast and a cellulose wall. Most people consider them to be algae, in which case they are placed in the phylum Chlorophyta as the order Volvocales of the class Chlorophyceae

Phytophora [*angl.* fĭ·tŏf″·ər·ə, *orig.* fĭ′·tō·fôr″·ə] a genus of peronosporale Oomycetes, mostly parasitic. *P. infestans* [in·fĕst″·ănz] is the historically notorious "potato blight"

phytoplankton the plant constituents of the planktonic population (cf. phytopleuston)

pia mater [pē′·ə māt′·ə] the inner of the two meninges in higher forms

-pic- *comb. form* meaning "magpie" but now used almost entirely for "woodpecker"

Picea [pĭs′·ē·ə] the genus of Coniferales containing the spruces. They are evergreens with keeled aciculate leaves and reflexed cones

pigeon's milk a fluid produced in the crop of a pigeon and fed to its young

pigment a substance which produces the appearance of color, though often serving other functions (*see* respiratory pigment, visual pigment)

pigment cup the protuberant pigmented area surrounding the eye of many Mollusca

pigment-spot ocellus a non-protuberant ocellus, containing pigment and photoreceptors

-pil- *comb. form* meaning "hair"

-pile- *comb. form* meaning "cap"

pileus [pĭl″·ē·əs] the fleshy "cap" of agaricale fungi

pilidium larva [pĭl·ĭd″·ē·əm] a free-swimming stage in the development of many nemerteans. It is distinguished by its conical shape, large apical tuft of cilia and a pair of flap-like projections hanging down on each side

piliferous layer [pĭl·ĭf″·ĕr·əs] the layer from which root hairs are produced

pillar cartilage one of a pair of cartilages, running forward from and inside the base of the styloid cartilage in marsippobranchia

pillar cells supporting cells in the organ of Corti

pilomotor [pĭl″·ō·mōt″·ôr] pertaining to the movement of hair

pilus [pĭl″·əs] a fine, hair-like, protuberance from a bacteria

-pimel- *comb. form* meaning "fat"

pin feather a juvenile contour feather

-pinac- *comb. form* meaning "plank" and thus, by extension, "structural"

Pinaceae the pine family of Gymnospermae, containing the cone-bearing trees except the yews and Gnetales. The dry woody cones with winged seeds between the scales are distinctive. The genera *Abies* and *Pinus* are the subjects of a separate entry

pinacocyte [pĭn″·ə·kō·sĭt″]   the basic structural cell in the body wall of sponges

pineal [pĭn″·ē·əl, pĭ·nē″·əl] an organ of doubtful function arising from the epithalamus and retaining a nervous connection with the left habenular ganglion (*see also* parapineal)

pineal eye an eye-like structure, apparently functional in some cyclostomes and some lizards, derived from the pineal body

-pinn- *comb. form* meaning "wing" (cf. -penna-)

pinna [pĭn″·ə] literally "wing" but usually applied to the external lobe that surrounds the orifice of the ear in mammals, sometimes called the outer ear

pinnate leaf [pĭn″·āt] one with leaflets arranged on each side of a common petiole

Pinnipedia [pĭn″·ē·pēd′·ē·ə] the suborder of Carnivora which contains the marine forms in which flippers replace true feet. The genus *Phocaena* is the subject of a separate entry

-pino- *comb. form* confused between three Greek and one Latin root and therefore meaning "hunger", "dirt", "drink" or "pine tree"

pinocyte [pĭn′·ō·sĭt″, pĭn′·ō·sĭt″] a cell which engulfs fluids as a phagocyte engulfs solids

pinocytosis [pĭn′·ō·sĭt·ōs″·ĭs, pĭn′·ō·sĭt·ōs″·ĭs] the engulfing of fluids by a cell (*see also* emeiocytosis)

Pinus [pĭn″·əs] the genus of coniferales distinguished by the deciduous primary leaves and the large woody cones with winged leaves

Pipa [pēp″·ə] a genus of tongueless frogs containing the famous surinam toad (*P. pipa*). The male deposits the female's eggs in pouches on the female's back which swell and totally enclose them until the entire larval development is completed

Piroplasma *see Babesia*

-pis- *comb. form* meaning either "pea" or "meadow"

Pisces [pĭs″·ēz] a superclass of craniate chordates containing all those forms which are generally called fish including the lampreys and sharks. The classes Agnatha, Chondrichthyes, and Osteichthyes are the subject of separate entries

pisiform bone [pĭs″·ē·fôrm′] that bone which lies beneath, and usually laterally to, the triangular bone, between the fifth metacarpal and the ulna

pisiform process a bony bump on the under surface of the carpometacarpus of birds

pistil [pĭs″·təl] the ovary, style and stigma of a mature flower. In flowers having a syncarpous gynoecium, the whole structure is popularly referred to as a pistil

piston cartilage a long, unpaired cartilage running backward from the apical cartilage in the skeleton of the rasping organ in Marsippobranchia

Pisum [pĭs″·əm] a genus of leguminous plants containing the edible pea (*P. sativum*) [săt·ēv″·əm]

pit any open top cavity. Without qualification, usually a cavity in a plant secondary cell wall. The cavities in the primary cell wall are called primary pit fields. The following derivative terms are defined in alphabetical position:
ANAL PIT   NASAL PIT
BORDERED PIT   RAMIFORM PIT

GASTRIC PIT   SIMPLE PIT
KOLLICKER'S PIT   VENTILATING PITS

pith 1 (*see also* pith 2) that part of the fundamental supporting system of a plant which lies in the center of the stem or root

pith 2 (*see also* pith 1) a verb used to describe the method of killing an animal by the destruction of the spinal cord

Pithecanthropus [pĭþ′·ə·kăn″·þrō·pəs] the fossil "Java man" now thought to be a race of *Homo erectus* (see Homo)

pitocin = oxytocin

pitressin [pĭt·rĕs″·ĭn] a hormone found in the neurohypophysis, acting on the contraction rate of smooth muscle in the walls of the blood vessels

pituitary gland a compound endocrine gland, derived in part from the stomodeum (adenohypophysis) and in part from the diencephalon (neurohypophysis)

Pituophis [*angl.* pĭt·yōō″·əf·ĭs, *orig.* pĭt′·yōō·ō″·fĭs] a genus of Squamata containing the North American bullsnakes

-placent- *comb. form* meaning a round, flat or slightly domed, "cake"

placenta 1 (*see also* placenta 2) a compound organ, derived in part from the wall of a mammalian uterus and in part from the embryonic chorion and amnion, which provides a mechanism for osmotic exchange between maternal and foetal blood. The following derivative terms are defined in alphabetic position:
DIFFUSE PLACENTA
DISCOIDAL PLACENTA
ENDOTHELIOCHORIAL PLACENTA
EPITHELIOCHORIAL PLACENTA
HEMOCHORIAL PLACENTA
SEMIPLACENTA
SYNDESMORCHORIAL PLACENTA
ZONORY PLACENTA

placenta 2 (*see also* placenta 1) the tissue which carries the ovules in a plant ovary, or which, in cryptogams, bears sporangia

Placobdella [plăk′·ăb·dĕl″·ə] a genus of leeches with a large body bearing a row of dorsal tubercles. *P. parasitica* [păr·ə·sĭt″·ĭk·ə] is a common United States species, 3 to 4 inches long, parasitic on turtles, particularly the snapping turtle (*Chelydra serpentina*)

placoid scale a scale found in elasmobranch fish having a plate-like bony foot from which rises, above the surface of the skin, a tooth-like surface plate

-plagio- *comb. form* meaning "oblique" , and by extension "side" or "flank"

plague a frequently fatal disease of man caused by *Pasteurella pestis* transmitted by the flea *Xenopsylla cheopsis*. The disease is also fatal to many rodents including rats, squirrels, prairie dogs and marmots; others, particularly voles and gerbils, harbor and transmit the disease without themselves showing symptoms. Plague is therefore endemic in many areas but serious epidemics are usually started by population explosions among rats

-plan- *comb. form* meaning "motile" or "flat"

Planctosphaeroidea [plănk′·tō·sfēr·oid″·ē·ə] a class of the phylum Hemichordata known only from their peculiar transparent pelagic larvas

plane a flat surface, in biology usually a section cut across a solid body. The following derivative terms are defined in alphabetic position:

HORIZONTAL PLANE   SAGITTAL PLANE
LONGITUDINAL PLANE   TRANSVERSE PLANE

plankt- *comb. form* meaning "wandering"
plankton [plănk″·tŏn, plănk″·tən] organisms living in the upper part of any body of water and which drift with the current (*see also* neuston). The following derivative terms are defined in alphabetical position:

ADVENTITIOUS P.	MONACMIC PLANKTON
AEROPLANKTON	NANNOPLANKTON
AUTOPELAGIC P.	NET PLANKTON
BATHYPELAGIC P.	NEUROPLANKTON
CRYOPLANKTON	NYCTIPELAGIC PLANKTON
DIACMIC PLANKTON	PHYTOPLANKTON
EPIPLANKTON	POTAMOPLANKTON
EUPELAGIC P.	RHEOPLANKTON
KNEPHOPLANKTON	SKOTOPLANKTON
MACROPLANKTON	THALASSOPLANKTON
MEGAPLANKTON	TYCHOPLANKTON
MEROPLANKTON	ZOOPLANKTON

Plannipennia [plăn′·ē·pĕd″·ē·ə] the order of insects, once included in the Neuroptera, which contains the ant lions. The larvas dig pits in loose soil to entrap small insects
-plano- *comb. form* meaning "roaming"
planogamete [*angl.* plăn·ŏg″·ə·mēt, *orig.* plăn′·ō·găm″·ēt] a mobile gamete
Planorbis [plăn·ôrb″·ĭs] a genus of freshwater pulmonate gastropod mollusks with the shell wound in a flat, sinistral, spiral. The genus is so large that it has been divided into a score of sub-genera. One of these, *Biomphalaria*, is a vector of *Schistosoma* in Africa
plant hair a hair-like projection, which may be either unicellular or multicellular, from the surface of a plant. Better called a trichome
plant hormone a group of plant compounds analogous to animal hormones (*see* hormone) in that they are synthesized in one part of the plant and transported to other parts where they exercise their effect. The three principal groups are auxins, gibberellins and cytokinins
Planta [plănt″·ə, plănt″·ə] the kingdom of organsms that contains the plants. No single characteristic distinguishes plants from animals though, among plants, only the fungi and a few parasitic higher forms lack chlorophyll. In general, the presence of chlorophyll, the presence of cellulose cell walls and a relatively slow response to external stimuli distinguish plants from animals
plantar [plăn′·tər] pertaining to the sole of the foot or the palm of the hand
plantigrade [plănt″·ē·grād′] pertaining to those animals that walk on the flat of the foot with the heel touching the ground
planula larva [plăn″·yōō·lə] a free-swimming ovoid or pear-shaped post-blastula stage in the development of many coelenterates
-plasm- *comb. form* properly meaning "molded" but by extension "layer", "substance" or even "organelle". These meanings shade into each other and are not separated in the following list of derivative terms that are defined in alphabetic position:

CYTOPLASM	ENDOPLASM	PROTOPLASM
ECTOPLASM	OOPLASM	SARCOPLASM

plasma [plăz″·mə] blood from which the corpuscles have been removed (cf. serum)
plasma membrane the surface boundary of a cell
plasma stains dyes that are used to stain the cytoplasmic constituents of cells. The most commonly used

in histology are various fluorescein halides known as "eosins"
plasmablast [plăz″·mə·blăst′] a large leucocyte thought by some to be the site of antibody production
plasmagene [plăz′·mə·jēn″] a genetic unit found in the cytoplasm (*see also* episome)
plasmid [plăz″·mĭd] a term coined to include both intrinsic plasmogenes and extrinsic factors such as viruses the effect of which may be mistaken for that of a true plasmogene
plasmin [plăz″·mĭn] an enzyme hydrolyzing peptides and esters of arginine and lysine and thus converting fibrin into soluble products
plasmodesma [plăz″·mō·dĕz′·mə] a strand of protoplasm running into, and possibly through, a plant cell wall
Plasmodiophora [*angl.* plăz′·mōd·ē·ŏf″·ər·ə, *orig.* plăz′·mōd·ē·ō·fôr″·ə] a genus of Myxomycophyta, many of which are plant pathogens. *P. brassicae* [brăs″·ĭk·ē] is responsible for the disease "club root" in cole crops
Plasmodium [plăz·mōd″·ē·əm] a genus of sporozoan protozoans parasitic in the blood cells of many vertebrates. Mosquitoes of the genus *Anopheles* are the vector in most mammals and the genus *Culex* in most other forms. In man *P. vivax* [vī′·văks] causes "tertian malaria", *P. malariae* [măl·är″·ĭ·ē] "quartan malaria" and *P. falciparum* [făl″·sē·pâr″·əm] "malignant tertian malaria"
plasmodium a multinucleate mass of protoplasm, particularly that formed from the fusion of zoospores in the life history of myxomycophytes and sporozoans. The following derivative terms are defined in alphabetic position:

AGGREGATE P.	FUSED P.	PSEUDOPLASMODIUM

plasmogamy [*angl.* plăz·mŏg″·əm·ē, *orig.* plăz′·mō·găm″·ē] the condition of two cells of which the cytoplasm but not the nuclei have fused, particularly, in rhizopod Protozoa
plasmolysis [*angl.* plăz·mŏl″·ə·sĭs, *orig.* plăz′·mō·lĭs″·ĭs] the shrinking of plant protoplasm from the cell wall
plasmotomy [*angl.* plăz·mŏt″·əm·ē, *orig.* plăz′·mō·tōm″·ē] the division of a multinucleate protozoan into many parts, without mitosis of any of the nuclei
plasmotrophoderm [plăz′·mō·trōf″·ō·dûrm] the outer layer of a mammalian embryo before the formation of the chorion (*see also* trophoderm, cytotrophoderm)
-plast- *comb. form* indicative of "formation", of "shaping", in the sense, which is the original meaning, of "molding". See -blast- for remarks on further confusion. The following terms with this suffix are defined in alphabetic position:

AMYLOPLAST	HEMOCYTOPLAST
BLEPHAROPLAST	KINETOPLAST
CHLOROPLAST	PHRAGMOPLAST
ELAIOPLAST	POLYPLAST

-plastic- *comb. form* from Greek having precisely the same meaning as the English word plastic, but used in biology as an adjectival form from both -plast- and plastid
plastid [plăs′·tĭd] any discreet organelle particularly those in plant cells
plastogene [plăst′·ō·jēn″] a non-nuclear genetic locus associated with the plastids of plants and in part responsible for the traits of these plastids
plastospecies [plăst″·ō·spēsh′·ēz] a population of

plastospecies can, but usually does not, interbreed with another population of plastospecies

plastron [plăst′·rən]   the central plate of a tortoise's shell

-plasy *comb. suffix* used in suffix to denote alteration in structures

plate any flat structure of limited area. The following derivative terms are defined in alphabetic position :

ALAR PLATE
ANAL PLATE
BASAL PLATE
BASITRABECULAR P.
CARDIOGENIC PLATE
CELL PLATE
CLOSING PLATE
CRIBIFORM PLATE
DERMAL PLATE
FLOOR PLATE
GENITAL PLATE
HYPOPHYSEAL PLATE
LATERAL PLATE
MADREPORIC PLATE
NEURAL PLATE
PARIETAL PLATE
PERPENDICULAR PLATE
PRECHORDAL PLATE
ROOF PLATE
SCLEROTIC PLATE
SIEVE PLATE
SKELETOGENOUS PLATE
TARSAL PLATE

platelet detached portions of megakaryocyte cytoplasm found in blood

-platy- *comb. form* meaning ''broad'', but widely used in biology as though it meant ''flat''

Platydorina [*angl.* plə·tïd″·ər·ïn·ə, *orig.* plăt′·ē·dôr·ïn″·ə]   a genus of colonial volvocale algae. The colony of 16 to 32 cells is housed in a flattened, horseshoe-shaped, test

Platyhelminthes [plăt′·ē·hĕl·mïn″·thēz]   the phylum of the animal kingdom that contains the flatworms. These are usually regarded as the most primitive phylum in the group Bilateralia and are distinguished by the presence of a mouth but no anus and by the fact that the space between the endoderm and ectoderm is filled with parenchymatous tissue. The classes Turbellaria, Trematoda (flukes) and Cestoda (tapeworms) are the subjects of separate entries

Platyrrhina [plăt′·ē·rïn″·ə]   a division of the anthropoidea containing the New World as distinct from the Old World forms ; they derive their name from the fact that the nostrils point either upwards or forwards; this division therefore comprises the sub-orders Hapaloidia and Ceboidia. The genera *Alouetta* and *Ateles* are subjects of separate entries

Plecoptera [plĕk·ŏpt″·ər·ə]   the order of insects that contains the stoneflies. They are distinguished by four membranous wings lying flat on the body and by the presence of segmented cerci on the last abdominal segment

-plect- *comb. form* meaning ''woven'', ''twisted'' or, by extension, ''folded'' (cf. -ploc-)

Plectascales [*angl.* plĕkt·ăs″·kə·lēz, *orig.* plĕk′·tă·skăl″·ēz]   an order of ascomycete fungi distinguished by possessing ascogenous hyphae. *Penicillium* and *Aspergillus* are the best known genera and the subject of separate entries

plectenchyma [*angl.* plĕkt·ĕn″·kə·mə, *orig.* plĕkt′·ĕn·kïm″·ə]   a tissue composed of hyphae

-pleio- *comb. form* meaning ''more'' and actually the comparative form of -poly-. The spelling -pleo- leads to unnecessary confusion

Pleistocene epoch a geologic epoch covering roughly the last million years. There are varying views as to whether there is any difference between Pleistocene and Recent

-pleo- *comb. form* meaning ''swim''. It is frequently confused with -pleio-

Pleodorina [*angl.* plē′·ə·dŏr″·ən·ə, *orig.* plē′·ō·dŏr·ïn″·ə]   a volvocale Chlorophyceae. The more or less spherical colony contains both reproductive and vegetative cells

pleomorphism [plē′·ō·môrf″·ïzm]   polymorphism, but by some restricted to polymorphism exhibited as successive stages in a life cycle

pleon [plē′·ən]   the abdominal region of those crustacea (e.g. crayfish, lobster) in which the whole abdomen is used for swimming. The term is also used as synonymous with abdomen in some other crustacea

pleopod [plē′·ō·pŏd′]   a crustacean abdominal leg, usually one adapted for swimming

-pler- *comb. form* meaning ''full''

plerocercoid larva [plûr′·ō·sĕr″·koid]   a larval stage in the development of a tapeworm consisting of a scolex, some strobila, and sometimes even genital organs (cf. cysticercus)

-plesio- *comb. form* meaning '' near'', ''approximate'' or ''recent''

plesiotype [plēs″·ē·ō·tïp′]   any specimen which is thought to be identical with a holotype or paratype by an individual other than the original describer

-pleth- *comb. form* meaning to ''swell'' or ''fill up''

Plethodon [plĕþ′·ə·dŏn]   a very large genus of lungless terrestrial salamanders (Urodela) widely distributed in moist habitats in the United States. *P. glutinosus* [glōō′·tïn·ōs″·əs] (the slimy salamander) is found all over the east, south and middle west

-plethysmo- *comb. form* meaning a ''swelling'' or ''enlargement''

-pleur- *comb. form* meaning ''side'' or ''rib'' (*see also* -pleura-)

-pleura- *comb. form* meaning ''rib cage'' or ''side'' in the sense of a ''side of beef''

pleura [plŏŏr′·ə]   the lining of the thoracic cavity of mammals

pleural rib [plŏŏr′·əl]   a rib lying in the horizontal septum which separates the body musculature into dorsal and ventral divisions

pleural sac the fold of peritoneum containing the lung of a mammal

pleurite [plŏŏr′·ït]   used both in the sense of sclerite and for flexible, intersegmental, membranes of arthropods

Pleurobrachia [plŏŏr′·ō·brăk″·ē·ə]   a cosmopolitan genus of ctenophorans admirably described by their common name of ''sea gooseberries''

pleurocentrum [plŏŏr′·ō·sĕn″·trəm]   that part of the centrum which is derived from the inner portions of the dorsals and ventrals; in amniotes, the whole centrum is derived from the pleurocentrum

pleurodont [plŏŏr″·ō·dont′]   said of a dentition in which the teeth rise from the top of a bony ridge in the jaw or are attached to the outer wall of the alveolar groove

pleuron [plŏŏr·ən]   the lateral region of the segment of an arthropod body (*see also* epipleuron)

pleuston [plyū′·stən]   organisms that float in virtue of their low specific gravity (cf. neuston)

plexiform layer [plĕks″·ē·fôrm]   those layers of the retina containing synapses (*see* inner plexiform layer, outer plexiform layer). Also the peripheral layer of the cerebral cortex

plexus [plĕks″·əs]   a network of nerves or blood

vessels. The following derivative terms are defined in alphabetic position:

AUERBACH'S PLEXUS	MEISSNER'S PLEXUS
CHORIOID PLEXUS	SOLAR PLEXUS

-plic- *comb. form* meaning "fold" or to "fold", or occasionally to "plait"

Pliocene epoch [plīʺ·ō·sēn´] the most recent of the Tertiary geologic epochs extending from about 10 million years ago to 1 million years ago. It is preceded by the Miocene epoch and followed by the Pleistocene epoch of the Quaternary period

Pliohippus [plīʹ·ō·hĭpʺ·əs] the first one-toed horse and probably the ancestor of the modern *Equus* in the Pliocene epoch

-ploc- *comb. form* meaning to "weave", or "twist together" (cf. -plect-)

-ploid *comb. suffix* which has come to mean "replicate" by extension from the Greek root meaning "doubled". The following terms using this suffix are defined in alphabetic position:

ALLOPOLYPLOID	DYSPLOID
ALLOTETRAPLOID	EUPLOID
ALLOTRIPLOID	HAPLOID
AMPHIPLOID	HETEROPLOID
ANEUPLOID	HYPERDIPLOID
ARTIOPLOID	HYPOPLOID
AUTOPLOID	ORTHOPLOID
AUTOPOLYPLOID	PARTIAL POLYPLOID
AUTOTETRAPLOID	POLYPLOID
AUTOTRIPLOID	SECONDARY POLYPLOID
DIDIPLOID	SYNDIPLOID
DIPLOID	TETRAPLOID
DOUBLE HAPLOID	TRIPLOID
DOUBLE TETRAPLOID	

ploidy [ploid´·ē] the condition of having the chromosome sets replicated

plumule [plŏŏmʺ·yōōl] the rudimentary shoot containing a leaf primordium lying immediately above the cotyledons (cf. epicotyl, hypocotyl)

plumule bulb a bulb arising directly from a seed

-plur- *comb. form* meaning "many" or "several"

pluteus larva [plŏŏʺ·tē·əs] a general term for the bilaterally symmetrical free-swimming echinoderm larva in which the ciliated bands are prolonged over anteriorly directed arms

-pneum- *comb. form* meaning ' wind" and, by extension, "breathing"

pneumatic duct the connection of the swim bladder of a fish to the alimentary canal

Pneumococcus not a recognized genus but a term at one time applied to any organism involved in pulmonary pneumonia (*see Diplococcus*)

pneustic [nyūsʺ·tĭk] pertaining to breathing (*see* amphipneustic)

-pocul- *comb. form* meaning a "cup", deeper than is indicated by -calyc- but shallower than is indicated by -cyath-

-pod- *comb. form* meaning "foot"

pod 1 (*see also* pod 2) *comb. form* meaning "foot" or "limb" (*see also* podite). The following derivative terms are defined in alphabetic position:

CERCOPOD	PLEOPOD
GNATHOPOD	PYGOPOD
GONOPOD	TELEPOD
NECTOPOD	TETRAPOD
PEREIOPOD	

pod 2 (*see also* pod 1) as a synonym of pseudopodium. The following derivative terms are defined in alphabetic position:

AXOPOD	RETICULOPOD
FILOPOD	RHIZOPOD
LOBOPOD	

-podial adjectival suffix meaning "footed" and therefore synonymous with -podous. The alternate form -podal is rarely used. The following terms using this suffix are defined in alphabetic position:

MONOPODIAL	SYMPODIAL

podial canal [pō´·dē·əl] a branch of the watervascular system of holothurian echinoderms, which supplies the podia

-podite an arthropod limb or any part of it. The following derivative terms are defined in alphabetic position:

BASIPODITE	ISCHIOPODITE
CARPOPODITE	MERIOPODITE
COXOPODITE	MEROPODITE
DACTYLOPODITE	PROPODITE
ENDOPODITE	PROTOPODITE
EPIPODITE	TELEOPODITE
EXOPODITE	

podium literally, foot. Specifically applied to the tube feet of echinoderms. The following derivative terms are defined in alphabetical position:

METAPODIUM	PARAPODIUM
NEUROPODIUM	PSEUDOPODIUM
NOTOPODIUM	

podocyte [pō´·dō·sĭt´] a cell lying on the outer surface of the glomerulus of vertebrate kidneys

podosoma [pō´·dō·sōmʺ·ə] that region of the acarine body which carries legs

-podous pertaining to -pod, or -podium. The form -podial is synonymous

-poecil- *see* -poikel-

-poes- *see* -poies-

-poies- *comb. form* meaning "to produce". The following terms using this suffix are defined in alphabetic position:

ERYTHROPOIESIS	HEMOPOIESIS
GALACTOPOIESIS	

-poikel- *comb. form* meaning "variegated" or "various", frequently transliterated -poicel-

poikelotherm [poi·kēlʺ·ō·thûrm] an animal whose body temperature varies according to the temperature of the environment

poikelosmotic [poi·kēl´·ŏz·mŏtʺ·ĭk] the condition of an organism the body fluids of which are continuously in osmotic equilibrium with its environment

point mutation one which appears to involve only a single locus

pointer cell = deuter cell

polar pertaining to the poles either of the planet earth or of a spherical egg (*see also* heteropolar)

polar body the nuclear material, with a minute amount of accompanying cytoplasm, extruded from the egg in the course of meiosis

polar cap a solid mass formed at the poles of the nuclei in some Protozoa undergoing mitosis

polar cartilage one of a pair of cartilages in the embryonic chondrocranium lying immediately external to, and on the other side of the carotid artery from, the hypophyseal cartilages

polar cell = polar body

polar field the sensory floor of a statocyst

polarity in biology, the existence of a difference between the two ends of an organism. In developing eggs the more active is known as the animal and the less active as the vegetal pole

Polian vesicle [pŏl′·ē·ən] an elongated sac hanging from the ring canal of the water-vascular system of holothurians and projecting into the coelom. In other echinoderms, they are smaller, blindly ending sacs, apparently serving as additional water reservoirs

-pollac- *comb. form* meaning "frequent" or "numerous"

pollen the male gametophyte of flowering plants that produces two spermid cells

pollen basket = corbiculum

pollen chamber a cavity in the apex of a style in which pollen accumulates

pollen comb = scopa

pollen mass a coherent mass of pollen

pollen sac that cavity in the stamen in which the microsporangia are formed

pollen tube the tube that develops from the pollen in the course of fertilization

pollex [pŏl″·ĕks] the first digit of the fore limb (= thumb in primates). Also any finger-like spine on an arthropod limb. The term was once synonymous with inch

pollination the act, or method, of transferring pollen to the female portion of the flower. The names for many methods terminate in -philous

-poly- *comb. form* meaning "many"

Polychaetae the class of annelid worms commonly called bristle worms. They are distinguished by the presence of parapodia, bearing numerous chaetae. All but a very few are marine. The free living forms (polychaetae errantia) usually have chitinous jaws and the sedentary forms (polychaetae sedentaria) frequently have prostomial and peristomial gills. The genera *Amphitrite, Aphrodite, Arenicola, Chaetopterus, Eunice, Neanthus, Nereis* and *Sabella* are the subjects of separate entries

polychromatophilic erythroblast [pŏl′·ē·krōm′·ăt·ō·fĭl″·ĭk] a descendant of a basophilic erythroblast and precursor of a normoblast. It is distinguished by the fact that it takes up both the acidic and basic components of blood stains

polyclimax theory [pŏl′·ē·klī″·măks] there may be several climax communities, each influenced by some feature of the climate

polydemic [pŏl′·ē·dĕm″·ĭk] occurring in several, separated areas (*see also* endemic, epidemic, pendemic)

Polyergus [pŏl′·ē·ĕrg″·əs] a genus of ants known as "amazons" that are obligatory slave makers. The colonies conduct organized raids on nests of smaller ants, usually species of *Formica*, from which they take the pupae. These they hatch in their own nests and train the ants as slaves

polyestrous [pŏl′·ē·ĕst″·rəs, pŏl·ē·ĕs″·trəs] said of mammals which have numerous oestrous cycles in the course of one breeding season

polygalacturonase [pŏl′·ē·găl·ăkt·yōōr″·ən·āz″] an enzyme catalyzing the break down of pectate through the hydrolysis of α-1,4-d-galacturonide links

polygamy [*angl.* pŏl·ig″·əm·ē, *orig.* pŏl′·ē·găm″·ē] the condition of having many sexual partners

polygene [pŏl′·ē·jēn″] one of several genes which in combination exercise a relatively minor effect

polygenesis [pŏl′·ē·jĕn″·əs·ĭs] the origin of something at several places either in time or space

Polygordius [pŏl′·ē·gôrd″·ē·əs] a genus of archannelids with a filiform body indistinctly segmented at the anterior end. There are neither parapodia nor setae

polygynopaedium [pŏl′·ē·jĭn′·ō·pēd″·ē·əm] an assemblage consisting of one female together with her offspring and the parthenogenetic descendants of her offspring

polykaryotic [pŏl′·ē·kâr′·ē·ŏt″·ĭk] said of an organism that multiplies only by binary fission

polykont [pŏl′·ē·kŏnt″] having many flagella

polylobular kidney [pŏl′·ē·lŏb″·yū·lə] a metanephros which is divided into lobes, each entering the pelvis through a separate branch

polymely [*angl.* pŏl·ĭm″·əl·ē, *orig.* pŏl′·ē·mēl″·ē] the abnormal condition of a tetrapod organism having more than four limbs

polymeric [*angl.* pŏl·ĭm″·ər·ĭk, *orig.* pŏl′·ē·mĕr″·ĭk] having many parts and particularly having many members of each series of a flower

polymetaphosphatase [pŏl′·ē·mĕt′·ə·fŏs″·fə·tāz′] general term for enzymes that catalyzes the hydrolysis of polyphosphates into pentaphosphates

polymorph [pŏl′·ē·môrf″] that which exhibits several forms (also a common abbreviation for polymorphocyte)

polymorphism [pŏl′·ē·môrf″·ĭzm] the condition of having one organism occur in several forms—as the polyp and medusoid stage of coelenterates

polymorphocyte = neutrophil

Polyodon [pŏl′·ē·ō″·dŏn] a genus of chondrostean Osteichthyes distinguished by a paddle-like extension of the upper jaw. *P. spathula* [spăþ·yōō·lə] is the paddlefish of the Mississippi River

polyp literally an aquatic animal but specifically used to distinguish the sessile form of a polymorphic coelenterate in contrast to the medusa

polyphenism [pŏl′·ē·fēn″·ĭzm] the exhibition of several forms by one individual. It applies equally to larval changes, social castes in insects, or polymorphism and cyclomorphosis

polyphyletic [pŏl′·ē·fĭl″·ĕt·ĭk] descending from many stocks or individuals

polyphyodont [pŏl′·ē·fī″·ō·dŏnt] having successive generations of teeth replacing each other

Polyplacophora [pŏl′·ē·plăk·ŏf″·ər·ə] an order of amphineuran mollusca commonly referred to as chitons. They are distinguished by having a shell of transverse separated plates and in the fact that the mouth and anus are at opposite ends of the elongate body. The genera *Chiton* and *Ischnochiton* are the subjects of separate entries

polyplast [pŏl′·ē·plăst″] in botany a multicellular, undifferentiated embryonic stage, roughly corresponding to the morula of the zoologist

polyploid [pŏl′·ē·ploid″] a cell, or organism, having more than the usual number of chromosomes (*see also* haploid, diploid, triploid). The following derivative terms are defined in alphabetic position:

ALLOPOLYPLOID   PARTIAL POLYPLOID
AUTOPOLYPLOID   SECONDARY POLYPLOID
ENDOPOLYPLOID

polypod larva an insect instar with a completely segmented abdomen, each segment bearing a set of functional legs

Polypodiaceae [pŏl'·ē·pŏd·ē·ās"·ē] the family of pteridophytes containing the great majority of ferns. They are distinguished by the possession of a vertical incomplete annulus associated with a lenticular long-stalked sporangium. The type genus *Polypodium* has numerous species among which *P. vulgare* [vŏŏl·gär"·ē], the adder fern, is probably the best known

Polyporus [pŏl'·ē·pōr"·əs] a genus of aphyllophorale Basidiomycetes forming an annual fleshy bracket on wood (*see also* Fomes)

Polypterus [pŏl·ĭp"·tĕr·əs] a genus of chondrostean fish with lobed fins, ganoid scales, spiracles and non-alveolated lungs. There are about a dozen African species called the "African lung fishes"

polysaprobe [pō'·li·săp"·rōb] an organism that is adapted to live in heavily contaminated water

Polysiphonia [pŏl'·ē·si·fōn"·ē·ə] a very large genus of rhodophyte algae with a much branched, almost feathery, thallus. They are heterothallic, the spermatangia forming an ovoid mass on only a single branch. The carpogonia are four-celled

polytene chromosome [pŏl"·ē·tēn'] a giant chromosome resulting from the endoreduplication of numerous parallel chromatids

polytrophic [pŏl'·ē·trōf"·ĭk] said of organism deriving food from a wide area or of bees that range over a large number of flowers or of organisms having an unusually varied diet

polyzygosis [pŏl'·ē·zïg·ōs"·ĭs] the fusion of more than two gametes

pome [pōm] a fleshy fruit consisting of a thin skin and an outer zone of edible flesh. The apple and pear are typical examples

pond a standing body of fresh, brackish or, rarely, sea water too small to be called a lake

Pongidae [pŏn"·jĭd·ē] a family of anthropoids containing the subfamilies Ponginae and Hylobatinae. Some workers add the subfamily Homininae. The subfamilies Ponginae and Hylobatinae, and the genera *Dryopithecus* and *Pliopithecas* are the subjects of separate entries

Ponginae [pŏn·jïn"·ē] the subfamily of apes that contains the chimpanzee, gorilla, orangutan, and, the opinion of many, man. These are often called the great apes leaving the term lesser apes to apply to the Hylobatinae. The genera *Gorilla, Pan,* and *Pongo* are the subjects of separate entries (*see also* Homo)

Pongo [pŏng"·gō] the generic name now customarily used for the orangutan (*P. satyrus*) previously called Simia. It is distinguished from the gorilla by its light brown skin and from the chimpanzee (Pan) by its long hair

-pont- *comb. form* meaning "sea", and more particularly the Black Sea, but used by some ecologists in the sense of "deep sea"

pontal flexure [pŏnt"·əl] the flexure of the developing brain in the region of the cerebellum which is in the reverse direction to the primary and nuchal flexures

popliteal artery [pŏp'·le·tē"·əl] that portion of the sciatic artery which lies in the lower leg

-por- *comb. form* confused between four Greek roots and therefore meaning "blind", "soft (or by extension "porous") stone", "a hole" or "blind"

pore a small hole. The following derivative terms are defined in alphabetic position:

ASCOPORE   NEUROPORE
ATRIOPORE   OVIPORE
BLASTOPORE   TASTE PORE
GONOPORE   UROPORE
NEPHRIDIOPORE

pore cork cork cells in lenticels

pore passage the connection between the inner and outer surfaces of a stoma

porenchyma [pôr·ĕn"·kə·mə] a tissue of pitted cells

Porifera [pôr·ïf"·ər·ə, pər·ïf"·ər·ə] the phylum of the animal kingdom that contains the sponges, distinguished by the presence of choanocytes and, almost invariably, a skeleton of fibers or spicules. The classes Calcarea and Hexactinellida, and the genera *Euspongia, Spongilla* and *Stylotella* are the subjects of separate entries

porocyte [pŏr'·ə·sit'] the tubular cell of a sponge, the outer end of which terminates in an ostium

Porphyra [angl. pôr'·fər·ə, orig. pôr'·fēr'·ə] a genus of rhodophyte algae several species of which are cultivated for food in Japan

porphyropsin [pôr'·fə·ŏps"·ïn] a visual pigment of the rod cells of the retina closely allied to rhodopsin

portal vein [pôr·təl] a vein which carries blood to an organ other than the heart

position effect the circumstance that a gene may exercise a different phenotypic effect in relation to its position in the chromosome

positive cytotaxis the tendency of cells to aggregate (*see also* negative cytotaxis)

-positor- *comb. form* meaning "to place" (*see* larvipositor, ovipositor)

-post- *comb. form* meaning "behind" either in time or place

post spiracular bone one of a pair of membrane bones in the dermocranium of Crossopterygian fish, lying immediately dorsal to the junction of the opercle and squamosal, and immediately below the junction of the tabular and extrascapular

postcleithrum bone one of a pair of bones in the pectoral girdle of Crossopterygian fish, lying between the supracleithrum and the cleithrum (*see also* cleithrum bone, supracleithrum)

postclimax *either* the passing, or alternation. of an existing climax owing to a climatic change *or* a situation which exists in a region contiguous to a true climax but in which the climate is more favorable and the situation therefore more advanced (*see* preclimax)

postclisere proceeding from a lower to a higher climax

postemporal bone one of a pair of bones articulating anteriorly with the extrascapular and fused posteriorly with the supracleithrum in the pectoral girdle of Crossopterygian fish (*see also* temporal bone, intertemporal bone, supratemporal bone)

posterior cardinal vein the principal vein draining the posterior region of the body of a vertebrate embryo; it unites with the jugular at the Cuvierian duct or sinus

posterior chamber that portion of the cavity between cornea and lens which is internal to the iris of the eye

posterior horn one of the ventral (posterior) branches of the H-shaped mass of gray matter seen in a transverse section of the spinal cord

posterior lateral cartilage one of a pair of cartilaginous plates forming the side of the chondrocranium of Cyclostomes

posterior pleopod   the anal claspers, or clasping legs, of lepidopteran larvas

posterior tectal cartilage   one of a pair of cartilages in the roof of the chondrocranium of Cyclostomes, occupying much the same position as the parietal bone of higher forms

postfrontal bone   a small bone at the posterior dorsal edge of the orbit, lying across the anterior end of the postorbital, and usually running a short distance parallel to the frontal, in many vertebrates other than mammals (*see also* frontal bone, prefrontal bone)

postglenoid process   a downwardly directed process from the base of the zygomatic arch

postorbital bone   one of a pair of membrane bones lying immediately posterior to the postfrontal and connected with this bone in front, the jugal below and the squamosal behind, in many vertebrates other than mammals. In some fish there are several pairs of postorbitals

postparietal bone   a membrane bone of the rear end of the skull lying immediately above the supraoccipital and between the two parietals (*see also* parietal bone)

posttrematic nerve   [pŏst'·trə·măt"·ĭk]   those branches of Cranial IX and X which pass behind the gill crest

-potam-   *comb. form* meaning a "river"

potamoplankton   [pŏt'·ə·mō·plănk"·tən]   that which occurs in rivers

potency   in embryology the potential of a part if its fate is overridden (e.g. the production of a whole organism from either of two separated blastomeres). Also applied to the genetically controlled ability of an egg to develop (*see also* totipotency)

pouch   a hollow protuberance, too small to be called a lobe, or an inwardly directed sac (*see also* pocket). The following derivative terms are defined in alphabetic position:

AMNIOTIC POUCH   RATHKE'S POUCH
BRANCHIAL POUCH

prae-   adjectival prefix meaning "before", either in time or place

prairie   loosely applied to almost any type of grassland habitat, usually that in which tall grasses predominate. The term has also been applied to open marshland in swamps

pre-   *comb. prefix* meaning "before", either in time or place

pre-adaptation   the possession of a mutation not advantageous in the organism's present environment but which may adapt it to a new environment

prearticular bone   one of a pair of dermal bones in the posterior region of the lower jaw, immediately ventral to the articular and posterior to the angular, found in many vertebrates other than mammals (*see also* articular bone, retroarticular bone)

precartilage   an aggregate of mesenchyme cells with indeterminate cell boundaries and lamellar ground substance

prechordal plate   [prē·kôrd"·əl]   the anterior, undifferentiated mesoderm in the early embryo of elasmobranch fish

preclimax   *either* the vegetation that immediately precedes a climax *or* a situation which exists in an area contiguous to a true climax but in which the climate is less favorable, and the situation is therefore less advanced (*see* postclimax)

preclisere   one which proceeds from a higher to a lower climax

precoxa   [prē·kŏks"·ə]   an additional segment, lying between the coxopodite and the body, in some crustacean appendages

preen   the actions of a bird in arranging its feathers (*see also* allopreening)

preen gland = uropygial gland

preethmoid bone   [prē·ĕth"·moid]   one of a pair of chondral bones in the anterior wall of the orbit of fish (*see also* ethmoid bone)

preformation theory   the medieval idea that the egg or sperm contains an adult in miniature, and that this adult develops purely by a process of "unfolding"

prefrontal bone   one of a pair of membrane bones lying in the antero-dorsal of the orbit between the frontal nasal and maxilliary bones, in many vertebrates other than mammals (*see also* frontal bone, postfrontal bone)

prehepatic hemopoiesis   the formation of blood cells from embryonic blood islands (*see also* hemopoiesis)

premaxilla bone   one of a pair of dermal bones of the splanchnocranium lying at the anterior end of the maxilla and bearing, in mammals, the incisor teeth (*see also* maxilla bone, septomaxilla bone)

premolar tooth   one which lies immediately anterior to the molar teeth in the jaw of a mammal

premutation   a term of early genetics used to account for the unexpected appearance of a phenotypic character which was held to have been "premutated" within the organism

prenasal process   one of a pair of prolongations of the premaxillae

preopercular bone   one of a pair of bones in the cheek skeleton of fish, lying immediately anterior to the junction of the opercular and subopercular (*see also* opercular bone, interopercular bone, subopercular bone)

preoptic recess   [prē·opt"·ĭk]   that slightly swollen portion of the embryonic forebrain from which the eye will subsequently develop

preoral   anterior to the mouth

prepattern   a term used to describe those chemical and physical factors that, by their non-random distribution in an undifferentiated tissue, account for the local initiation of differentiation

prepollex bone   [prē·pŏl"·ĕks]   the small bone lying between the outer margin of the radius and the outer margin of the navicular bone in some vertebrates and a few mammals

prepotence   [prē·pōt"·ənz]   a term of animal breeding reflecting genetic dominance of a male or, less usually, female parent

prepupa   [prē·pyōō"·pə]   a quiescent larval instar that has not yet pupated

presoma   the short, anterior division of the body of priapulids and acanthocephalans

pressure receptor   a mechanoreceptor stimulated by compression forces and thus one of the organs of touch

pretarsus   a joint distal to the tarsus on the appendages of some arachnids

pretrematic nerve   [prē'·trə·măt"·ĭk]   those branches of Cranial IX and X which pass in front of the gill crest

prevailing climax   a term used to describe a growth

form occupying the majority of sites in a given area even though no true climax has been formed

prevertebral ganglion   one of a series of ganglia, some single and some paired, lying ventral to the aorta and connected to the chain ganglia

Priapulida   [pri'·ə·pyū"·lid·ə]   a small group of marine worms variously regarded as a separate phylum, an order of the phylum Gephyrea, or a class of the Annelida. They are now usually regarded either as a separate phylum, or as a class of the pseudocoelomate phylum Aschelminthes. They are distinguished by having the appearance of a warty cylinder with an introversible presoma

Priapulus   [pri·ăp"·yū·ləs]   the type genus of Priapulida. The characteristics of the genus are those of the group

prickle cell layer   a layer of cells immediately above the Malphighian layer in the skin

primary cell wall   the original cellulose wall formed round a plant cell (*see also* cell wall, secondary cell wall)

primary flexure   the initial bend in the embryonic brain by which the procencephalon and its derivatives are bent at right angles to the remainder

primary growth   that growth of a plant which derives directly from apical meristem (*see also* secondary growth)

primary root   that which develops directly from the radicle

primary sensory cell   a cell an extension of which acts as a direct receptor of a stimulus rather than receiving the stimulus from another cell (*see also* sensory cell)

primary sexual character   those organs which are directly concerned with the establishment of sex i.e. the gonads (*see also* secondary sexual character, tertiary sexual character)

primary somatic hermaphrodite   a somatic hermaphrodite having the gonads of one sex while exhibiting either all or some part of the secondary sexual characters of both sexes (*see also* secondary somatic hermaphrodite)

primary xylem   xylem developed in the primary part of the plant body, directly from procambium

Primates   [pri·māt"·ēz, prim'·āts]   the order of placental mammals that contains the monkeys, baboons, and apes. They are distinguished by a pentadactyl pattern in the limb and an orbit completely encircled by bone. Most have pectoral mammary glands. The suborders Lemuroidea, Lorisoidea and Anthropoidea are the subjects of separate entries

primitive   properly, the first of its kind, and thus the ancestral form; it is also used to mean un-, or under-, developed, and also occasionally for "wild type" in genetics

primitive node = Hensen's Node

primitive streak   an area of concreted cells corresponding to the blastopore, at the posterior end of the blastodisc of a developing telolecithal egg

primordial follicle   a group of cells cut off from and sinking below the surface of a mammalian ovary

primordium   [prim·ôrd"·ē·əm]   the earliest stage at which the differentiation of an organ can be perceived; probably the best translation of the widely used German word "anlage"

-prion-   *comb. form* meaning "saw"

priority law   the proper name of an organism is,

subject to certain modifications of the international rules, the first under which it was described

Pristis   [pris"·tis]   a genus of batoidean elasmobranchs containing the sawfishes. The flat calcified elongated snout is furnished with teeth on either edge. This is one of the few genera of batoids that has considerable freshwater populations, that in Lake Nicaragua apparently being permanent

-pro-   *comb. form* meaning "before" and "early"

proamnion   [prō·ăm"·nē·ən]   the beginnings of an amniotic fold consisting entirely of ectoderm

proatlas   [prō·at"·ləs]   a neural arch, of unknown homologies, found anterior to the atlas in some reptiles

Proboscidea   [prō·bŏs·id"·ē·ə]   the order of placental mammals that contains the elephants. The elongation of the nose into a trunk or proboscis is typical of the order. The genera *Elaphas* and *Loxodonta* are the subjects of separate entries

procambium   [prō·kăm"·bē·əm]   partially differentiated apical meristem which will later give rise to the vascular system of the plant

Procavia   [prō·cāv"·ē·ə]   the genus of hyracoid mammals containing *P. capensis* [kăp·ĕn"·sis], the common "dassy"

Procellariiformes   [prō·sĕl·är"·ē·ə·fôrm·ēz]   the order of birds that contains the shearwaters and petrels. They are distinguished from other gull-like birds by the fact that the nostrils are in raised tubes

procentric synapsis   [prō·sen"·trik]   that which first occurs at the centromere and continues outwards

procercoid larva   [prō·sĕrk"·oid]   an elongate development of a coracidium or onchosphere in which the hooks pass through the posterior and form an organ of attachment

process   anything that projects from something. The following derivative terms are defined in alphabetic position:

ANGULAR PROCESS	OTIC PROCESS
ARTICULAR PROCESS	PALATAL PROCESS
CILIARY PROCESS	PAROCCIPITAL PROCESS
CLINOID PROCESS	PISIFORM PROCESS
CORONOID PROCESS	POSTGLENOID PROCESS
ILIAC PROCESS	PRENASAL PROCESS
ISCHIAL PROCESS	PUBIC PROCESS
MASTOID PROCESS	ZYGOMATIC PROCESS

processus alaris   [prō·sĕs"·əs ăl·âr"·is]   an anterolateral process of the polar cartilage

procoelous vertebra   [prō·sēl"·əs]   one in which the centra are concave in front and convex behind

Proconsul   [prō·kŏn"·səl]   an extinct genus of Miocene catarrhine primates thought by many to be ancestral to the Pongidae

procoracoid bone   [prō·kŏr"·ə·koid']   a bone lying ventro-medial to the coracoid in the skeleton of some reptiles and birds (*see also* coracoid bone)

Procotyla   [*angl.* prō·kŏt"·əl·ə, *orig.* prō'·kŏt·il"·ə]   a genus of freshwater turbellarians. *P. fluviatalis* [flōō·vē·ăt"·əl·is] is the "*Dendrocoelum lacteum*" of many elementary texts

procryptic color   [prō·kript"·ĭk]   a color designed to assist in concealing an organism (*see also* cryptic color, antecryptic color)

-proct-   *comb. form* meaning "hind gut" or "anus"

proctodeum   [prŏk'·tō·dē"·əm]   a posterior invagination, which subsequently becomes either the cloacal or anal aperture

Procyon [prō·sī″·ən] the genus containing the American racoon (*Procyon lotor*) [lō″·tôr]. It forms, together with the coatimundes and kinkajous, a family distinguished from other fissipede carnivores by having two molars in each half of both jaws

proerythroblast [prō′·ĕr·ĭth″·rō·blăst′] a descendant of a hemocytoblast giving rise to a basophilic erythroblast and distinguished in general by the large size of the nucleus in which two nucleoli are clearly visible

Profelis [prō·fēl″·ĭs] the genus of felid fissipede carnivores that contains the cougar, or mountain lion (*P. concolor*) [kŏn·kŭl′·ôr] and the clouded leopard (*P. nebulosa*) [nĕb′·yūl·ōs″·ə]

progamete [prō·găm′·ēt] that cell which gives rise directly to gametes either by a single division or by metamorphosis of itself

progamic cell [prō·găm″·ĭk] that cell in the pollen grain that has the sperm nucleus

progamous fission [prō·găm′·əs] the type of fission that, in some ciliate protozoans, divides the organism unequally into macro- and micro-conjugants

progesterone [prō′·jĕst″·ər·ōn′] a hormone secreted by the ovary when stimulated by luteinizing hormones. Active in the preparation of the uterine wall for the implantation of the embryo

proglottid [prō·glŏt″·ĭd] one of the divisions, improperly called segments, into which the body of a tapeworm is divided. They are budded off ("strobilated") from the scolex and become successively sexually mature and then distended with fertile eggs

proglottis [prō·glŏt″·ĭs] the tip of the tongue. This word is *not* the singular of proglottid

prognathous [prŏg·nāþ″·əs] with jaws directed forward, used particularly of insects with a horizontal head and projecting jaws

progressive association any association which is not stable

progressive metamorphosis the change of vegetative into sexual organs in plants

progynous [prō·jĭn″·əs] the condition of a hermaphrodite in which the female portions mature first (cf. protogynous)

prokaryotic [prō′·kâr·ē·ŏt″·ĭk] the condition, such as that in bacteria and Cyanophyceae, of having a nucleus lacking a nuclear membrane (cf. eucaryotic)

-prol- *comb. form* meaning "offspring"

prole [prōl, prō′·lē] offspring; progeny; the result of breeding. Used frequently in German and sometimes in English for "subspecies"

proleg [prō″·lĕg] an unjointed arthropod appendage used in locomotion as the abdominal appendages of many caterpillars

l-proline [el prō″·lēn] 2-pyrrolidinecarboxylic acid. An amino acid not essential to the nutrition of rats. It is a constituent of most plant proteins

promeristem [prō·mĕr′·ĭs·tĕm″] initiating cells and undifferentiated derivative cells

promitosis [prō′·mī·tōs″·ĭs] that form of division in which both centrioles and chromosomes come from the endosome

promycelium [prō′·mī·sēl″·ē·əm] the tube which projects from a germinating fungus spore

promyelocyte [prō·mī″·ĕl·ō·sīt′] a direct descent of a myeloblast, containing a few, not readily differentiated, granules. Directly ancestral to myelocytes

-pron-, -prono- *comb. form* confused from two Greek roots and therefore meaning either "to bend down" or "a prominence" or "headland"

prone lying flat on the face (*see also* supine)

pronephros [prō·nĕf″·rŏs] a kidney composed of units in which a coelomic funnel leads to a tubule but in which the glomerulus is free in the coelomic cavity and does not form part of the nephron

pronograde [prōn″·ə·grād′] to walk with the body horizontal

pronormoblast [prō·nôrm″·ō·blăst′] the second stage in the development of an erythrocyte

Pronuba a name often incorrectly used for *Tegeticula*

pronucleus [prō·nyōō″·klē·əs] the haploid nucleus of a gamete

prootic bone [prō·ŏt″·ĭk] one of a pair of chondral bones running from the basisphenoid to the latero-posterior angle of the skull in vertebrates other than mammals

propagatory cell in the early cleavage division of the trematode egg, one blastomere remains as a propagatory cell which contributes to the miracidium larvae. The other is a somatic cell which forms a vitelline membrane

prophage [prō″·fāj′] a factor in a bacterial genome that can, in combination with another factor, cause the production of a bacteriophage

prophase [prō″·fāz′] the first stage in mitotic division in which the chromosomes are visible within the nucleus

Propliopithecus [*angl.* prŏp′·li·ō·pĭth″·ə·kəs, *orig.* prō′·pli·ō·pĭth·ēk″·əs] an extinct, very primitive, pongid primate from the Egyptian Oligocene

propodite [prō″·pŏd·ĭt] the joint of a crustacean appendage which lies between the dactylopodite and the carpopodite

-proprio- *comb. form* meaning "proper" in the sense of "one's own"

proprioceptive system [prō′·prē·ō·sĕp″·tĭv] the sum total of those receptors which receive stimuli from muscles, tendons and joints (*see also* interoceptive system, exteroceptive system)

proprioceptor [prō′·prē·ō·sĕp″·tôr] a structure which enables a complex organism, such as man, to locate one part of his body in relation to another

-pros- *comb. form* meaning "towards"

prosencephalon [prŏs′·ĕn·sĕf″·əl·ŏn] the anterior region of the developing brain which subsequently divides into the telencephalon and diencephalon

prosere [prō″·sēr] a migratory community playing a temporary role in seral development

prosopyle [prŏs′·ō·pīl″] the pore between the incurrent and the radial canals in a syconoid type sponge

prospondylous ring [prō·spŏnd″·əl·əs] a ring of cartilage around the notochord marking the anterior end of a vertebra (*see also* opisthospondylous ring)

prostate gland [prŏs·tāt″] a male accessory sex gland of mammals, the secretion of which dilutes, and possibly activates, spermatic fluid

-prostho- *comb. form* meaning "before" in the sense of position

prosthomere [prŏs′·þō·mēr″] a preoral segment in an arthropod embryo

prostomium [prō·stōm″·ē·əm] that "which lies in front", usually immediately in front, "of the mouth".

Specifically that segment of an annelid which immediately precedes the mouth

-prot-, -proto- *comb. form* meaning "first" in time and by extension "primitive"

protandrous [prŏt·ăn"·drəs] the condition of a hermaphrodite in which the male portion develops first or an organism that is first male, and later sex reverses to a female. Also said of a flower in which the pollen matures before the stamen is receptive

protaspis larva [prŏt·ăs"·pĭs] the minute, megacelaphic, almost unsegmented, larva of trilobites

-prote- *comb. form* meaning "polymorph"

protective layer the layer immediately on each side of the separation layer in an abscission zone

protein any of vast number of organic compounds basically consisting of peptide-linked amino-acids. The following derivative terms are defined in alphabetic position :

CHROMOPROTEIN | LIPOPROTEIN
COMPOUND PROTEIN | NUCLEOPROTEIN
CONJUGATED PROTEIN | PHOSPHOPROTEIN
FIBROUS PROTEIN | SIMPLE PROTEIN
GLYCOPROTEIN | SOLUBLE PROTEIN
LECITHOPROTEIN

proterminal synapsis that which begins at the tips of the chromosomes and continues toward the center

-protero- *comb. form* meaning "before", either in time or space

proterogynous [prŏ·tə·ŏ·jĭn"·əs] having the female parts mature before the male parts

Proterozoic era [prŏ'·tər·ō·zō"·ĭk] a geologic era extending from about a billion years ago to half a billion years ago. It was preceded by the Archeozoic and followed by the Paleozoic. It showed the development of organized life represented by a few protozoans and sponges

Proteus [prŏ"·tē·əs] a genus of eubacteriale schizomycetes found in the alimentary tract of most vertebrates. A few occasionally cause pathogenic reactions

prothallus [prŏ·păl"·əs] the gametophyte of a fern or the initial stages of any thallophyte

prothorax [prŏ·þor"·ăks] the first of the three thoracic segments in insects

Protista [prŏ·tĭst"·ə] a taxon erected to contain those acellular (or "unicellular") organisms that in contrast to the Monera have a diploid nucleus bounded by a nuclear membrane

-proto- *comb. form* meaning "first"

protocercal [prŏ'·tō·sĕrk"·əl] said of a caudal fin in which the ray supported part of the fin is limited to the margin and in which there are both simple epichordal and hypochordal folds

Protociliata [prŏ'·tō·sĭl"·ē·ăt·ə] a group of protozoa, of which *Opalina* is typical, of doubtful affinities. They are sometimes regarded as a subclass of Ciliata but have by some been removed from the Protozoa into a phylum of their own

Protococcus [prŏ'·tō·kŏk"·əs] a genus of unicellular chlorophyta. *P. viridis* [vĭr"·ĭd"·ĭs] is often responsible for the green incrustation found on the leeward side of trees, walls and fences

protoconch [pro'·tō·kŏngk"] the shell of a larval gastropod

protocorm [prŏ'·tō·kôrm"] a plant embryo which is not yet morphologically differentiated

protoderm [prŏ'·tō·dûrm"] partially differentiated

meristem which will later give rise to the epidermal system of a plant

protogonia [pro'·tō·gŏn"·ē·ə] the terminal angle of the insect forewing

protogynous [prŏ'·tō·jĭn"·əs] frequently used as synonymous with progynous but also is used to describe the condition of an organism which is first female and later sex-reverses to male

protonephridium [prŏ'·tō·nĕf·rĭd"·ē·əm] a nephridium consisting of a blind tubule terminating in a flame cell or in a solenocyte

protonymphon larva [prŏ'·tō·nĭm"·fŏn] the larva which hatches from the egg of Pycnogonida. It has three pairs of appendages, of which the anterior are chelate, and a well-developed proboscis

protophloem [prŏ'·tō·flŏ"·əm] phloem elements differentiated before a plant organ completes elongation

protoplasm [prŏ'·tō·plazm"] the sum total of the contents of a living cell. The term includes all organelles together with the matrix of water-dispersed proteins in which they lie

protopod larva [prŏ'·tō·pod"] an arthropod instar in which appendages are not yet differentiated

protopodite [*angl.* prŏt'·ŏp·ə·dĭt", *orig.* prŏ'·tō·pōd"·it] in insects, the basal portion of the maxilla

Protopterus [prŏt·ŏp"·tə·əs] a genus of Dipnesti containing four species of African lung fishes. The best known is the W. African *P. annectens* [ăn·ĕk"·tĕnz]

protosclerenchyma [prŏ'·tō·sklĕr·ĕn"·kə·mə] collenchyma resembling sclerenchyma

Protospondyli [prŏ'·tō·spŏnd"·əl·ĭ] an order of holostean Actinopterygii distinguished from other Osteichthyes by the amphicoelous vertebrae. The only extant genus is *Amia*, the subject of a separate entry

protospore [prŏ'·tō·spŏr"] a spore which produces a promycelium

protostele [prŏ'·tō·stēl"] a most primitive type of stele consisting of a solid column of vascular tissue with xylem occupying the center and phloem forming the surrounding tissue

Protostomia [prŏ'·tō·stŏm"·ē·ə] a term coined to define those phyla of bilaterally symmetrical animals in which the embryonic blastopore becomes the mouth. This division therefore includes all phyla of the Lateralia except Chaetognatha, Echinodermata, Hemichordata, and Chordata (cf. Deuterostomia)

Prototheria [prŏ'·tō·ther"·ē·ə] a subclass of the phylum Mammalia containing the single order Monotremata. The characters of the order are typical of the subclass

prototroch larva a pre-trochophore larva possessing one to three equatorial bands of preoral cilia

prototroph [prŏ'·tō·trof"] a mutant microorganism which can grow in a medium lacking a factor necessary for the growth of the wild type

prototrophic [prŏ'·tō·trōf"·ĭk] said of microorganisms capable of growth on simple sugars and salts without specific nutrient requirements (cf. auxotrophic)

protoxylem [prŏ'·tō·zi"·ləm] xylem maturing before elongation commences or during elongation

Protozoa [prŏ'·tō·zō"·ə] a phylum of animals which are not obviously divided into cells. Many are solitary uninucleate organisms but they may also exist as multinucleate individuals or as large colonies. Some biologists prefer the term "acellular" for "unicellular" in describing this phylum. The following classes are the subject of separate entries: Ciliata, Mastigophora, Sarcodina, Sporozoa, Suctoria

protrichocysts [prō'·tō·trĭk'·ō·sĭst", prō'·tō·trĭk'·ō·sĭst] rod-like structures, which are never discharged like trichocysts, in the pellicle of certain ciliate protozoans (see also cnidotrichocyst)

protrochula larva [prō·trŏk"·yū·lə] a free-swimming larva of some Turbellaria, and which is supposed by some to be a precursor of the trochophore larva of other types

Protura [prō·tyūr"·ə] a small, anomalous, group of hexapod arthropoda excluded by many from the Insecta on the ground that they have fifteen postcephalic segments, some being added during postembryonic development

proventriculus 1 [prō'·vĕn·trĭk"·yū·ləs] (see also proventriculus 2, 3) any expanded area of the alimentary canal lying immediately in front of the stomach

proventriculus 2 (see also proventriculus 1, 3) in birds, a glandular sac between the crop and gizzard

proventriculus 3 (see also proventriculus 1, 2) a pouch, frequently with internal teeth, at the junction of the foregut and midgut in insects. Analogous to the gizzard of birds

proximal [prŏks"·əm·əl] "that which lies nearer to". In general biological usage proximity starts at the anterior end, or at the main axis, thus the esophagus is proximal to the stomach and the shoulder is proximal to the arm (cf. distal)

proximal chiasma one between an inversion loop and the centromere

psalterium [sŏl·tĕr"·ē·əm] the third of the four divisions of the stomach of artiodactyls (cf. abomasum, reticulum, rumen)

-psamm- comb. form meaning "sand" or, by extension, "beach"

-pseud- comb. form meaning "false"

pseudaposematic [sūd'·ăp·ō·səm·ăt"·ĭk] a protective or warning coloration borne by a defenseless species in imitation of one well able to defend itself (see also aposematic)

Pseudemys [sūd"·ēm·ĭs] a common genus of freshwater turtles (Chelonia) usually, with several other related genera, called terrapins. *P. scripta* [skrĭp"·tə] is the Middle Western red-eared terrapin

pseudepisematic color [sūd'·ĕp·ē·sēm"·ət·ĭk] colors which attract (see also sematic color, episematic color, aposematic color, antipseudepisematic color)

pseudoallele [sūd"·ō·əl·ēl'] a group of closely linked loci once thought to be a single locus. Pseudoalleles do not complement and recombine very rarely

pseudobranch [sūd'·ō·brănk"] the external gill of a mollusk as contrasted with the true gill or ctenidium. Also a gill-like structure thought to lack respiratory function

pseudobulb [sūd"·ō·bŭlb'] the thickened basal internode of an orchid

pseudocambium [sūd'·ō·kăm"·bē·əm] meristem resembling cambium

pseudochitin = tectin

pseudocoel [sūd'·ō·sēl"] a body cavity, such as that of a nematode, which is not a true coelom but is derived from the blastocoel

Pseudocoelomata [sūd'·ō·sēl'·ō·māt"·ə] a term coined to describe those phyla of the animal kingdom which are neither acoelomate nor coelomate. This

taxon therefore embraces Acanthocephala, Rotifera, Gastrotricha, Kinorhyncha, Priapulida, Nematoda, Nematomorpha, and Entoprocta. All these phyla except the first and the last are sometimes regarded as classes of the phylum Aschelminthes

pseudodominance the appearance of a recessive character in a heterozygous individual owing to the fortuitous absence of the dominant allele from that individual

pseudogamy [angl. sū·dŏg"·əm·ē, orig. sūd'·ō·găm"·ē] the situation which results when a spermatozoa enters and activates an egg but degenerates without its nucleus fusing with that of the egg

pseudoglobulin [sūd'·ō·glŏb"·yōō·lĭn] a simple protein soluble in ammonium sulfate solution but not in water (see also globulin)

pseudogyne [sū'·dō·jĭn"] a parthenogenetic female, particularly in insects

pseudohermaphrodite [sūd'·ō·hûrm·ăf"·rō·dĭt] an animal which produces functional gametes of one sex but exhibits some of the secondary sex characters of the other sex

pseudometameric theory [sūd'·ō·mĕt'·ə·mĕr"·ĭk] metameric segmentation arose in consequence of the impedance to free movement occasioned by periodic swelling of lobed gonads

pseudomitosis [sūd'·ō·mĭt·ōs"·ĭs] an early term for a type of nuclear division in which discrete chromosomes were not visible

Pseudomonadales [sūd'·ō·mŏn·əd·āl"·ēz] an order of schizomycetes which may be coccoid, straight, curved, or spiral, sometimes occurring in chains, but never in trichomes, and usually motile by means of polar flagella. The genera *Nitrobacter, Nitrosomonas, Pseudomonas, Thiobacillus* and *Vibrio* are the subjects of separate entries

Pseudomonas [sūd'·ō·mŏn"·əs] a genus of pseudomonale schizomycetes present in almost all meat. "Slimy" meat usually has a count of about 1×10^7 *Pseudomonas* per cm². Many members of the genus are very sensitive to oxygen and were used as oxidation-reduction indicators in early biological studies

pseudomycelium [sūd'·ō·mī·sēl"·ē·əm] short chains of fungal cells such as those produced by some yeasts and bacteria

pseudoplasmodium [sūd'·ō·plăz·mōd"·ē·əm] the result of the aggregation, but not fusion, of myxoamoebae

pseudopodium [sūd'·ō·pōd"·ē·əm] a protuberance, either temporary or permanent, from the body of a sarcodine protozoan. The various types of pseudopodia will be found under -pod

pseudopupa [sū'·ō·pyōō"·pə] a resting stage between insect instars which is not covered with a pupal case and in which histolysis and histogenesis do not take place

pseudosacral vertebra [sūd'·ō·sāk"·rəl] one which is attached to the pelvis by its transverse processes rather than by its ribs (see also sacral vertebra and urosacral vertebra)

Pseudoscorpionida [sūd'·ō·skôrp'·ē·ŏn"·id·ə] an order of arachnid arthropods containing the animals commonly called pseudoscorpions. Save for the very much smaller size they resemble scorpions in external appearance though they lack a sting on the last abdominal segment

Pseudosporochus [sūd'·ō·spôr·ōk"·əs] an extinct Devonian-Carboniferous plant, usually considered to be in the evolutionary series of pre-ferns

pseudostratified columnar epithelium columnar epithelium in which not all the columns reach from the basement membrane to the surface

-psil- *comb. form* meaning "nude", but commonly used in biology for "slender"; it is used by ecologists in the sense of "prairie"

Psilophyta [sil'·ō·fit"·ə] a no longer acceptable taxon of plants. The fossil forms have been distributed between the Zosterophyllophytina, the Trimerophytina and the Rhyniophytina. The extant genera Psilotum and Tmesipteris are now regarded as primitive ferns

Psilophyton [sil'·ō·fit"·ən] an extinct genus of plants of the group Trimerophytina. They appeared in the lower Devonian when the Rhyniophytina were well established. The lateral branches were terminated by clusters of dehiscent sporangia

Psocoidea [sō·koid"·ē·ə] an order of small insects commonly called "book lice" for the reason that several wingless members of the group are common in old books. The majority of psocids possess four functional wings. The order is distinguished by the possession of chewing mouth parts which include an unusually elongate and chisel-like lacinia of the maxilla

Psocoptera = Psocoidea

-psychro- *comb. form* meaning "cold"

-pteno- *comb. form* meaning "winged"

pterin [tə·rín"] a general term for a group of white, yellow and red biochromes closely allied to the purines. They are responsible for the body and wing colors of many insects

-pteris- literally "fern" but frequently abbreviated -pter- and thus confused with wing

Pteris [tĕr'·is] a genus of Pterophyta (ferns). The best known species is the bracken (*P. aquilina*) [ăk'·wil·in"·ə] which in many parts of the world is a menace to agriculture through its encroachment on cultivated land

-ptero- *comb. form* meaning a "wing"; frequently confused with pteris-

Pterobranchia a class of the phylum Hemichordata distinguished by the presence of a U-shaped digestive tract and the possession of secreted encasements resembling those of the Ectoprocta. The genus *Cephalodiscus* is the subject of a separate entry

Pterophyta [tĕr·ō·fit"·ə] a phylum of plants containing the ferns. They are distinguished from all other cryptogamic plants by the fact that the sporophyte is differentiated into stems, leaves, and roots. The genera *Azolla*, *Pteris* and *Salvinia* are the subjects of separate entries

Pterydophyta [tĕr'·id·ō·fit"·ə] a subkingdom of plants containing the clubmosses, horsetails, and ferns. They differ from the bryophytes and spermatophytes in the fact that both the gametophytes and the sporophytes are independent plants at maturity. Many taxa of pterydophytes are known only as fossils

-pterygium literally a wing but used in biology for fin (*see* mesopterygium, metapterygium)

pterygoid [tĕr'·ē·goid"] wing-shaped

pterygoid bone a pair of dermal bones of the splanchnocranium. It lies behind the palatine and forms the lateral walls of the nasal passage (*see also* ectopterygoid bone, epipterygoid bone, metapterygoid bone)

pteryla [tĕr·əl·lə] (*see also* tract) an area of bird skin from which contour feathers spring (*see also* apterium)

Pterypoda [*angl.* tə·rip"·əd·ə, *orig.* tĕr'·ē·pōd"·ə] a group of pelagic Mollusca once regarded as a separate order but now usually accepted as a sub-order of Gastropoda. The distinctive character is the foot, the sides of which are drawn out into two flap-shaped paddles

-ptil- *comb. form* meaning "wing" but frequently used in biology in the sense of "sheath" (*see* coleoptile)

pubic [pyōō'·bik] pertaining to the area of the genitalia (cf. pudendic)

pubic process an anterior process of the cartilaginous pectoral girdle of primitive fish

Puccinia [pŭk·sin"·ē·ə] a genus of uredinale fungi known as rusts. Many, such as *P. graminis* [grăm"·in·is] (wheat rust) and *P. punctiformes* [pungkt'·ē·fôrm"·is] exude a scented nectar which attracts insects

pudendic [pyōō·dĕn"·dĭk] pertaining to the external genitalia (cf. pubic)

pull root one which by its contraction draws a plant deeper into the ground

-pulmo- *comb. form* meaning "lung"

pulmonary circulation [pŭl·mŏn"·ər·ē, pŭl"·mŏn·ər·ē] the circulation of blood from the heart through the lung

Pulmonata [pŭl'·mŏn·āt"·ə] an order of gastropod Mollusca containing the land and freshwater snails and slugs. The respiration is by "lungs", highly vascularized sacs or folds of the mantle. The genera *Ampullaria*, *Helix*, *Lymnaea* and *Planorbis* are the subjects of separate entries

pulp cord the red pulp situated between adjacent sinusoids in the spleen. The term is also applied to strings of sex cells in the developing ovary

-pulvin- *comb. form* meaning "cushion"

-punct- *comb. form* meaning "puncture" or "sting"

pupa [pyōō·pə] The dormant, pre-imagal, stage of holometabolous insects; the term was at one time applied to the doliolaria larva of holothuria. The following derivative terms are defined in alphabetic position:

COARCTATE PUPA	PSEUDOPUPA
MASKED PUPA	SEMI-PUPA
PREPUPA	

puparium [pyōō·pâr"·ē·əm] the thickened larval case within which the pupa of many Diptera is formed

pupation hormone an insect neurohormone inducing pupation

pupil the contractile aperture in the iris of an eye

Pupipara [*angl.* pyūp'·əp·ər·ə, *orig.* pyū'·pē·pär"·ə] an assemblage of dipteran insects that are permanent ectoparasites on other animals and which either lack wings or have wings so poorly developed that they cannot pass from one host to another. The name derives from the fact that the larvas are developed within the mother and pupate immediately after being shed. Bat flies are a typical example

pure association an association completely dominated by a single species

pure forest one which consists of only a single species of tree

pure line = clone

Purkinje cell [pûr·kĭnj"·ē] a large glass-shaped nerve cell in the cerebellum, with much branched dendrites

Purkinje fiber an atypical, impulse-conducting, cardiac muscle fiber

-pycn- *comb. form* meaning "dense"

pycnium [pĭk″·nē·əm] the spermatogonium of *Puccinia*

Pycnogonida [pĭk′·nō·gŏn″·ĭd·ə] an aberrant class of marine arthropods called sea spiders. They are distinguished by a greatly reduced slender body to which are attached four, five, six, or twelve pairs of very long legs

pycnosis [pĭk·nŏs″·ĭs] the condition of having a contracted nucleus, usually in moribund cells

-pyg- *comb. form* indicating "rump" or "buttocks"

pygidium [pĭ·jĭd″·ē·əm] any terminal region of the body which cannot properly be called a tail for example, the terminal tergite of the insect's abdomen, the tail shield of the trilobite, and the terminal segments of an adult polychaete worm

pygopod [pĭ′·gō·pŏd″] the appendages on the tenth abdominal segment of insects

pygostyle [pĭg″·ō·stĭl′] a short squat "tail" of fused vertebrae found in birds

-pyle- *comb. form* properly meaning "entrance" but also misused in the sense of "hole" or "pore". The following terms using this suffix are defined in alphabetic position:

APOPYLE   MICROPYLE   PROSOPYLE

pyloric caecum [pī·lŏr″·ĭk] a caecum arising from the junction of the pyloric stomach and the intestine in some fish

pyloric stomach the region of the stomach adjacent to the intestine (cf. cardiac stomach)

pyloric valve that which closes the pyloric end of the stomach

pylorus [pī·lŏr″·əs] the opening between the stomach and the intestine

-pyr- *comb. form* confused from three Greek roots and therefore meaning either "pear", "fire" or "wheat"

pyramid a solid body the sides of which are triangles coming to a common point. Specifically applied, without qualification to the ventro-lateral portion of the cerebellum; and the expanded floor and walls of the medulla oblonga. The following derivative terms are defined in alphabetic position:

ANAL PYRAMID   RENAL PYRAMID
ELTONIAN PYRAMID

pyrenoid [pĭr″·ĕn·oid′] a dense proteinaceous mass in the center of a chloroplast

pyriform cortex [pĭr″·ē·fôrm′] that portion of the cortex of the brain which lies superficial and dorsal to the corpus striatum

pyriform lobe the backwardly directed lobe arising from the anterior region of the cerebral cortex

Pyrrophyta [pĭr′·ō·fit″·ə] a phylum of algae distinguished by the fact that the pigments in the chromatophores are brownish-green through the presence, in addition to chlorophyll, of carotenes and xanthophylls

Pyrus [pĭr·əs] a large genus of Eurasian trees and shrubs of the Rosaceae. *P. communis* [kom″·yū·nĭs] is the pear

Pythium [pĭth″·ē·əm] a genus of terrestrial peronosporale Oomycetes. Several species are responsible for the "damping off" of seedlings in seed pans

pythmic [pĭth″·mĭk] pertaining to lake bottoms (cf. bathile, chilile)

-pyxid- *comb. form* meaning "chest", in the sense of a box with a hinged lid

-quadr- *comb. form* meaning "four"

quadrat [kwŏd'·rət] a square area in which the organisms are intensively examined and which forms the basis for assessing the entire population of the area

quadrate [kwŏd"·rāt] more or less square

quadrate bone one of a pair of membrane bones found at the posterior outer angle of the cranium, in vertebrates other than mammals

quadratojugal bone [kwŏd'·rāt·ō·jōō"·gəl] one of a pair of chondral bones lying in the angle between the quadrate and the jugal where these exist, or fused with the latter in the crania of many vertebrates other than mammals (*see also* jugal bone)

quasiclimax [kwās"·ē·klī'·măks] a community, such as that of freshwater plants, which is a climax in the sense that it is relatively stable but not a climax in that it is more or less independent of climate

quaternary [kwăt·ûrn"·ĕr·ē] arranged in fours or of the fourth order

Quaternary period the most recent period of the Cenozoic era extending over about the last million years. It was preceded by the Tertiary. It is the first age of organized man

queen a regnant female, particularly the fertilized, egg-laying female of an insect colony

Quercus [kwûr·kəs] a genus of about 500 species in the family Fagaceae. All are called oaks and many are of economic importance. Many yield timber, particularly in the United States. *Q. alba* [ăl'·bə] (white oak), *Q. ilex* [ī'·lĕks] (holm oak) and *Q. petraea* [pĕt·rē"·ə] are a source of tan bark while *Q. suber* [syū'·bə] is the cork oak

quinary [kwĭ·nər·ē] arranged in fives or the fifth order

quincunx [kwĭn"·kŭngks"] a pattern of five objects, four placed at the corners of a square, and the fifth at its center. Serial repetition of quincunces produces a pattern of alternating rows to which the term quincunical is commonly applied

quinine the principal alkaloid derived from the bark of several species of trees of the genus Cinchona. Until quite recently it was the only specific for malaria, a disease that ran unchecked in the Old World until quinine became available through the discovery of the New World

quinone biochrome [kwĭn·ōn"] a common group of reds and yellows best known as the pigment of the cochineal insect

R

race   a geographic enclave of a species the gene pool of which differs from that of other similar enclaves of the same species. The term is also loosely used in the sense of a variety that breeds true. The following derivative terms are defined in alphabetic position:
ADAPTIVE RACE   ECOLOGICAL RACE
CLIMATIC RACE
-racem-   *comb. form* meaning a "bunch of berries"
racemase   [răs″·əm·āz′]   a group name for those enzymes which change L to D forms and vice versa. The names are all completely descriptive (e.g. glutamate racemase catalyzes the production of D-glutamate from L-glutamate) and are not listed separately
raceme   [răs″·ēm′]   an indeterminate centripetal inflorescence with a long axis; the term derives from the appearance of a bunch of grapes
rach-   *comb. form* meaning "cliff" or "crag" but almost universally used as a misspelling of -rhach-
rachion   [răk″·ē·ən]   the area of maximum wave turbulence on a shore
rachis   [răk″·ĭs]   a stem or shaft particularly of a feather
racial immunity   that which appears to be associated with a geographically limited community
radial canal   gastrodermal canals radiating from the coelenteron of medusoid coelenterates
radial cleavage   holoblastic cleavage in which the tiers of cells lie on top of each other parallel to the polar axis
radial symmetry   a pattern in which the parts are so arranged around a central point or shaft, that any vertical cut through the center would divide the whole into two identical halves. A radially symmetrical organism may be heteropolar
radiale bone   [rā′·dē·äl″·ē]   that carpal which lies next to the radius (= navicular)
-radic-   *comb. form* meaning "root"
radiobilateral symmetry   a variety of radial symmetry which is strongly modified, as in the Anthozoa, in the direction of bilateral symmetry
Radiolaria   [rā′·dē·ō·lâr″·ē·ə]   a large order of actinopodous protozoa usually producing a silicious skeleton of spicules fused into a radially symmetrical basket. These "shells", from 0.01 mm to 0.1 mm in diameter form immense fossil beds in several parts of

the world. The genus *Thalassicola* is the subject of a separate entry
radius   a straight line running from the center of a circle to the circumference, or a structure lying in the position of such a line. The following derivative terms are defined in alphabetic position:
ADRADIUS   PERRADIUS
INTERRADIUS
radula   [răd″·yōō·lə]   a ribbon-like band bearing transverse rows of teeth found about the mouth in many mollusks
rain forest   one composed of plants which require an unusually high level of precipitation (*see also* equitorial rain forest, tropical rain forest)
Raja   [rā′·yə]   a widely distributed genus of batoid elasmobranchs. *R. clavata* [klăv·āt″·ə] is the familiar thornback ray
-ram-   *comb. form* meaning "branch"
Ramapithecus   [rä′·mə·pĭþ″·ə·kəs]   an extinct anthropoid ape known only from a jaw fragment about fourteen million years old. It is thought by many people to have manlike characteristics and therefore to be the earliest known possible ancestor of man
ramet   [răm″·ət]   one individual of a clone (cf. ortet)
-rami-   *comb. form* meaning a branch
Rana   [rä′·nə]   a genus Salientia containing forms common in both the Old and New Worlds. The best known of the dozen American species are *R. catesbiana* [kătz′·bē·ān″·ə] (bullfrog), *R. clamitans* [klăm″·ĭt·ănz] (green frog) and *R. pipiens* [pĭp″·ē·ĕnz′] (leopard frog). There are more than 250 other species of which the best known are *R. temporaria* [tĕm′·pôr·âr″·ē·ə] (European common frog) and *R. esculenta* [ĕs′·kyū·lĕnt″·ə] (the water frog)
Ranunculaceae   [răn′·ŭn·kyūl·ās″·ē]   a very large family of dicotyledons that contains the aconites, the anemones, the columbines, the peonies, and the buttercups. Distinctive characters are the numerous hypogynous stamens and the spiral floral structure. The genus *Ranunculus* is the subject of a separate entry
Ranunculus   [rə·nŭn″·kyūl·əs]   the type genus of the Ranunculaceae, with the characteristics of the family. Most are called buttercups, the yellow flowers of which are familiar the world over

Ranvier's node an interruption in the myelin sheath of axons and neurons

-raph- *comb. form* meaning "needle" or "seam"

raphid [rā″·fĭd] literally a "little rod" but applied specifically to needle-like crystals of calcium oxalate found in some plant tissues

Rathke's pouch a dorsal evagination from the roof of the embryonic buccal cavity which develops into the adenohypophysis

Rattus [răt″·əs] a genus of murid rodents containing more than a hundred species of which two, *R. rattus* (the black rat) and *R. norvegicus* [nôr·vēg″·ĭk·əs] (the brown, or "Norwegian") rat have become commensals of man

Rauvolfia [rou·vŏlf″·ē·ə] a genus of woody plants of the family Apocynaceae. The genus yields large numbers of alkaloids of which reserpine, from *R. serpentina* [sûr′·pĕn·tēn″·ə], is the best known

ray 1 (*see also* ray 2, 3) woody tissue growing at right angles to the axis of the growth and therefore appearing as "rays" in transverse section. In this connotation the following are defined in alphabetic position:
MEDULLA RAY   VASCULAR RAY
PARENCHYMA RAY

ray 2 (*see also* ray 1, 3) in the sense of radius. In this connotation the following derivative terms are defined in alphabetic position:
ASTRAL RAY   MEDULLARY RAY

ray 3 (*see also* ray 1, 2) thin boney structures stiffening the fins, tail and gill chambers of fishes. In this connotation the following derivative terms are defined in alphabetic position:
BRANCHIOSTEGAL RAY   FIN RAY   HYOID RAY

ray initial a cell of the vascular cambium which gives rise to xylem and phloem lying at right angles to the direction of growth

recapitulation law = Baer's law

receptaculum seminalis [rē′·sĕp·tăk″·yū·ləm sĕm′· ĭn·ăl″·ĭs] a vessel used in hermaphrodite animals for the storage of sperm derived from another animal

receptor 1 (*see also* receptor 2, 3) an organ or cell designed to perceive external stimuli, and transmit them through nerves. The following derivative terms are defined in alphabetic position:
CHEMORECEPTOR   PRESSURE RECEPTOR
EQUILIBRIUM R.   RHEORECEPTOR
GRAVITY RECEPTOR   STATORECEPTOR
INTERORECEPTOR   STRETCH RECEPTOR
MECHANORECEPTORS   TANGORECEPTOR
OLFACTORY RECEPTOR   TASTE RECEPTOR
PHONORECEPTOR   THERMORECEPTORS
PHOTORECEPTOR   TOUCH RECEPTORS

receptor 2 (*see also* receptor 1, 3) the individual to which a tissue transplant is transferred from a donor

receptor 3 (*see also* receptor 1, 2) a molecule to which atoms are transferred in the course of an enzyme-catalyzed reaction

recessive a term applied to organisms having recessive characters (*see* bottom recessive, double recessive)

recessive allele one which produces the phenotypic expression only when in the homozygous state

recessive character the opposite of dominant character; therefore it appears only in individuals homozygous for this character

reciprocal hybrid one obtained from the same parents, but with the sexes transposed (i.e. male parent A times female parent B as against male parent B times female parent A) (*see also* sesquireciprocal hybrid)

reciprocal translocation one in which each chromosome receives a portion of the other

recombination 1 (*see also* recombination 2) the rejoining of a broken portion of a chromosome in such a manner as causes a genetic aberration (cf. restitution)

recombination 2 (*see also* recombination 1) the result of the exchange of linked genes through crossing over. In bacterial genetics the result of pairing opposite mating types but with a unidirectional transmission from donor to recipient with the recombinant then developing within the recipient cell

recon the smallest unit of recombination within a cistron

-rect- *comb. form* meaning "straight"

rectal gland a digitiform gland arising from the rectum of elasmobranch fish. It functions as a salt gland

rectum that portion of the large intestine immediately adjacent to the anus

red tide discolored water, sometimes red but more often brown, occasioned by a massive bloom of dinophycean algae. This may cause the death of many other forms either by deprivation of oxygen or the production of toxins (*see Goniaulax*)

redia larva [rē″·dē·ə] a larval stage of a trematode developed within a sporocyst. The redia possesses a distinct mouth and pharynx and produces cercaria by internal budding

redifferentiation the process by which dedifferentiated cells return to specific forms or patterns (*see also* differentiation, dedifferentiation)

reduced expressivity the condition of an individual showing a lesser phenotypic expression of the character than is customary in the species

reductase [rə·dŭkt″·āz] a group name for enzymes that catalyze a reduction reaction. In the great majority of cases NADP is the hydrogen receptor becoming NADPH₂ (*see also* oxidoreductase)

reduction body a degenerate, or dedifferentiated, mass of tissue formed in some invertebrates and urochordates from which a new individual can regenerate. In ectoprocts it is also called brown body

reduction division = meiosis

reef a rocky ridge, frequently of coral origin, projecting near or slightly above the surface of oceans adjacent to the shore (*see* barrier reef, fringing reef)

refugium [rə·fyū″·jē·əm] an area that remains unaffected by a general climatic change and which therefore contains a population of organisms typical of those once spread over the whole region

regeneration the ability of an organism to reproduce a part that has been lost or to reorganize dedifferentiated material (*see also* epimorphic regeneration, homoetic regeneration)

regressive evolution the appearance in a taxon of organisms having characters usually considered to be associated with more primitive forms

regressive metamorphosis in botany, when a female structure is metamorphosed into a male or when a reproductive structure is metamorphosed into a vegetative one

regulative egg one in which the developmental pattern is not established before cleavage so that induced modifications of cleavage do not modify subsequent development (cf. mosaic egg)

relaxin [rə·lăks"·ĭn] a hormone secreted by the ovary and producing relaxation of the pubic ligament and softening of the cervix in pregnancy

relict an organism or group of organisms belonging to an earlier time than the other members of the biopopulation in the area in which the relict occurs

remex [rĕm"·ĕks] a primary flight feather attached to the ulna of birds

remiges [rĕm"·ĭj·ēz] plural of remex

-ren- comb. form meaning "kidney"

renal [rē"·nəl] pertaining to the kidney

renal papilla the projecting fused end of numerous collecting tubules projecting into the pelvis of a mammalian kidney

renal pyramid a compact pyramid-shaped portion of the medulla in the kidneys of birds and mammals

renin [rĕn"·ĭn] an enzyme catalyzing the conversion of hypertensinogen to hypertensin

rennin [rĕn"·ĭn] an enzyme catalyzing the hydrolysis of peptides

Rensch's Laws in cold climates races of birds have larger clutches of eggs and mammals larger litters than races of the corresponding species in warmer climates; in warm climates, birds have shorter wings and mammals have shorter fur; in snails, races of land snails in cold climates have brown shells and those in hot have white shells; the thickness of the shell is positively associated with strong sunlight and arid conditions

repand [rĕp"·ənd, rē"·pănd] said of a bacterial colony on agar which has a wavy surface

reproduction the replication of an organism. The following derivative terms are defined in alphabetic position:

ASEXUAL REPRODUCTION   SEXUAL REPRODUCTION

reproductive isolation the isolation of species inhabiting the same environment but which are unable to interbreed

Reptilia a class of chordates containing the reptiles. They are distinguished by the presence of epidermal scales on the skin. The six sub-classes mentioned in this dictionary are the Anapsida, Archosauria, Euryapsida, Ichthyopterygia, Lepidosauria and Synapsida. Of these all are extinct except one order (Chelonia or turtles) of the Anapsida, one order (Crocodilia) of the Archosauria and two orders (Rhychocephalia, with a single extant species and Squamata containing the snakes and lizards) of the Lepidosauria

repulsion the condition of two loci on the same chromosome one being dominant and the other recessive

reserpine [rĕs·ə·pēn, rĕz·ə·pēn] the principal alkaloid extracted from *Rauvolfia serpentina*. Crude extracts of this plant were used as a tranquilizer by witch doctors for many centuries before it was introduced to Western medicine in the 1950's

reservoir lake a lake resulting from the artificial impoundment of water

residual meristem meristem from which the vascular system of a shoot develops

resin any water insoluble exudate of a tree. Water soluble exudates are properly gums

resolution the limiting factor in microscopy, in that it defines the smallest object that can be clearly seen. Resolution is dependent on the numerical aperture

N.A. (q.v.), the wavelength of the illuminating beam (λ) and the contrast (k) according to the relationship

$$R = \frac{kNA}{\lambda}$$

As the wavelength of a beam of electrons is one hundred thousand times smaller than that of a beam of photons it follows that the electron microscope can image objects one hundred thousand times smaller than can the optical microscope

respiration the sum total of the method or methods by which an organism secures oxygen (*see also* aerobic respiration, anaerobic respiration)

respiratory circulation the circulation of the blood from the heart through a respiratory organ either lung or gill

respiratory duct the passage through which air for respiration is drawn through the nose. In some urodele amphibia, it is separated from the olfactory duct, which is lined with sensory epithelium

respiratory passage the ventral of the two tubes into which the pharynx of marsippobranchs divides. The dorsal is the pharyngoesophagus

respiratory pigment one that carries oxygen. See heme-, hemo- and cruorin

respiratory tree large, dendritic extensions that run from the anterior part of the cloaca throughout most of the length of Holothuria. They are the main respiratory system of these forms

resting nucleus one which is not dividing

resting spore one with an unusually thick wall designed to survive adverse conditions.

restitution the rejoining of a severed part of a chromosome in such a manner as to cause no genetic effect (cf. recombination)

restitution nucleus one which is restored through the refusion of replicated chromosomes which have failed to separate at mitosis

-ret-, -rete- comb. form meaning "net"

rete mirabile [rā"·tā mēr·ä"·bĭl'·ā] the internal network of collecting ducts within the testes

reticular [rĕt·ĭk"·yū·lər] net-like

reticular cell a cell with numerous processes so that reticular cells with their processes joined make a sponge-like framework or stroma such as that found in lymphatic tissue

reticular fibers small branched fibers closely allied to collagen fibers but smaller in diameter and differing also from collagen in that reticular fibers can be demonstrated by silver staining techniques

reticular layer the layer of the dermis immediately underneath the papillary layer

reticulocyte [rĕt·ĭk"·yū·lō·sit'] the penultimate stage in the development of an erythrocyte

reticuloendothelial cell [rĕt·ĭk"·yū·lō·ĕnd'·ō·þĕl"·ē·əl] a descendant of a primitive reticular cell specialized for phagocytosis and lining blood passageways. The name derives from their ability to make networks of reticular fibers

reticulopod [rĕt·ĭk"·yū·lō·pŏd'] a thread-like pseudopodium which branches and anastomoses into networks

reticulum 1 [rĕt·ĭk"·yū·ləm] (*see also* reticulum 2) any net-like structure (*see* endoplasmic reticulum)

reticulum 2 (*see also* reticulum 1) the second of the

four divisions of the stomach of artiodactyls (see abomasum, psalterium, rumen)

retina　the photosensitive portion of an eye. It receives the image formed by the lens and transforms the components of this image into nervous impulses that are interpreted by the brain

retinacula　a rope or chain particularly one which restrains or holds something. This word is not the plural of retinaculum (a "little net"). The word has been given so many meanings in biology as to be almost worthless particularly as it is frequently confused with retinaculum. It is also misused, apart from its proper usage as a holdfast, to mean a retractile muscle in many invertebrates

retinaculum　a little net (plural retinaculi). Retinacula (plural retinaculae) derives from a different root. There are many meanings in biology but this term usually refers to one body which retains or tethers another; best known as applying to the gland by which pollen-masses are retained in orchids. The term is, in insects, used mostly for restraining parts, such as the ring which prevents the hymenopteran sting being pushed out too far. There seems little justification for its application to the muscular sheath of a nerve in Acanthocephalans

retinal rod = rhabdomere

retinene　[rĕt'·ĭn·ēn"]　one of the two breakdown products of metarhodopsin (the other is scotopsin) in the presence of ample supplies of vitamin A; retinene and scotopsin recombine into rhodopsin

retroarticular bone　[rĕt'·rō·är·tĭk"·yū·lər]　one of a small pair of membrane bones at the posterior medial corner of the lower jaw of actinopterygian fish (see also articular bone, prearticular bone)

retrogressive association　an association which has passed the stable phase

retroperitoneal　[rĕt'·rō·pĕr'·ĭt·ŏn·ē"·əl]　said of an organ covered by peritoneum, but not projecting into the coelom. The kidney of fish is an example

revehent vein　[rə·vē'·ənt, rĕv'·ē·ənt]　a vein carrying blood from the mesonephros to the posterior vena cava

reverse mutation　one which transforms a mutant back to its original state

reversion　a term for plant and animal breeding used to explain the appearance of a phenotype which has not appeared for several generations

-rhabd-　comb. form meaning "rod"

rhabdoid　[răb"·doid]　a solid rod or spicule in the epidermis of Turbellaria

rhabdomere　[răb"·dō·mēr']　that portion of a retinal cell that contains the photosensitive pigment

-rhach-　comb. form literally meaning "spine" or "backbone" but used in biology for any axial support. The almost universal misspelling -rach- causes much confusion

-rhag-　comb. form confused from five Greek roots and therefore variously meaning a "break", "berry" "violence" and "poisonous spider". An extension of this list provides the justification for the next entry

rhagon larva　[rā"·gən]　a larval stage of many sponges having the form of a cone with a single exhalant aperture at the tip

Rhea　[rē"·ə]　a genus of rheiform birds with the characteristics of the order. The "common rhea", often miscalled the American ostrich, is R. americana [ăm'·ĕr·ĭk·än"·ə]

Rheiformes　[rā'·ē·fôrm"·ēz]　an order of birds containing the South American rheas. They are distinguished from the Struthionideae (ostriches) by the three-toed feet and the fact that the head and neck are partially feathered

-rheo-　comb. form meaning "to flow" or pertaining to that which flows (e.g. a "stream") (cf. -rhya-)

rheoplankton　[rē'·ō·plănk"·tən]　that which lives in running water

rheoreceptor　[rē'·ō·rē·sĕp"·tər]　a receptor perceiving the direction of current flow in water

rheotaxis　[rē'·ō·tăks"·ĭs]　a movement in response to flow, either of air or water

Rhesus　see Macaca

-rhin-　(see also -rhynch- and -rostr-) comb. form confused between two Greek roots and therefore meaning either "nose" or "file"

rhinal fissure　[rī"·nəl]　the fissure which separates the olfactory lobes from the rest of the cerebrum in higher forms

rhinarium　[rin·är"·ē·əm]　a naked and moist glandular skin surrounding the nostrils in many mammals

Rhincodon　[rĭn"·kō·dŏn']　a monotypic genus of elasmobranchs containing the whale shark (R. typus) [tī·pəs]. This shark, spotted like a leopard and up to 45 feet long, lives on the surface and feeds on sardine-sized fish

Rhineura　[rī·nyōōr"·ə]　a genus of Squamata containing the Florida worm lizard R. floridana [flor'·id·än"·ə] (see Amphisbaena)

Rhinoceros　[ri'·nŏs"·ə·rəs]　there are only two extant representatives of this one horned genus. R. indicus [ĭn·dĭk"·əs] (the Indian rhinoceros) and R. sondanicus [sŏn·dăn"·ĭk·əs] (the Java rhinoceros). The other rhinocerosses are divided among several genera of which only Diceros and Ceratotherium are mentioned in this book

-rhinous　comb. term meaning pertaining to the nose

-rhiz-　comb. form meaning "root". The following derivative terms are defined in alphabetic position:
MYCORRHIZA　　　PHYLLORHIZA

Rhizobium　[riz·ō"·bē·əm]　a genus of nitrogen-fixing enbacteriale Schizomycetes occurring as symbiotes in root modules of leguminous plants

Rhizocephala　[rĭz'·ō·sĕf"·əl·ə]　an order of cirripede crustacea thought by some to be a separate subclass. All are parasitic on crabs in which the female degenerates to a tumor-like sac protruding from the abdomen of the host. The genus Sacculina is the subject of a separate entry

rhizoid　[rĭz"·oid]　a root-like filament of a non-vascular plant. The term is also applied to anchoring extensions of the stolon in Ectoprocta

rhizome　[rīz"·ōm]　a swollen underground root commonly used for reproduction and food storage. The term is also applied to the stolon of hydroid coelenterates

rhizopod　[rĭz"·ō·pŏd]　a narrow lobopod

Rhizopoda　[angl. rĭz·ŏp"·əd·ə, orig. rĭz'·ō·pōd"·ə]　the term is used as synonymous with Sarcodina or as a subclass of Sarcodina. In the latter case it comprises all of the order of Sarcodina except the Heliozoa and Radiolaria which are placed in the subclass Actinopoda. The Amebida are the subject of a separate entry

Rhizopus　[rĭz·ōp"·əs]　a genus of zygomycete fungi. R. stolonifera [stōl'·ŏn·ĭf"·ĕr·ə] and R. nigricans [nig·rĭk″·ănz] are the well-known "black bread-molds"

Rhizostoma [*angl.* rĭz·ŏst″·əm·ə, *orig.* rĭz′·ō·stōm″·ə] a genus of mostly large Scyphozoan coelenterates. There are no marginal tentacles and eight very large oral lobes, the sucking pores along the edges of which replace the mouth

-rhod- *comb. form* meaning "red"

Rhodophyta [rō′·dō·fī″·tə] a phylum of algae commonly called the red algae. They are the only algae to show a form of sexual reproduction in which non-flagellated male gametes are transported to the female sex organ. The genera *Chondrus, Gelidium, Polysiphonia* and *Porphyra* are the subject of separate entries

rhodopsin [rō·dŏp″·sĭn] the primary visual pigment that converts photon energy into nerve impulses in the rods of the retina (*see also* porphyropsin)

-rhomb- *comb. form* meaning either "lozenge-shaped" or "whirling"

rhombencephalon [rŏmb′·ĕn·sĕf″·əl·ən] the portion of the embryonic brain that subsequently divides into the metancephalon and myelencephalon

Rhus [rŭs] a genus of anacardiaceous shrubs many of which produce irritating non-volatile oils. *R. radicans* [răd′·ĭk·anz″] is the poison ivy, *R. toxicodendron* [tŏks′·ĭk·ō·dĕn″·dran] is poison oak and *R. vernix* [vûr″·nĭks] the poison sumac

-rhya- *comb. form* meaning "torrent" (cf. -rheo-)

-rhynch- *comb. form* meaning "beak"

Rhynchobdellida [rĭn′·kŏb·dĕl″·ĭd·ə] an order of hirudinian annelids with a protrusible proboscis and no jaws. The genera *Glossiphonia* and *Placobdella* are the subjects of separate entries

Rhynchocephalia [rĭn′·kō·sĕf·āl″·ē·ə] an order of lepidosaurian reptiles containing the single living representative *Sphenodon,* the tuatara of New Zealand. The order differs from other diapsids in that the quadrate bone is immovable and there are upper and lower temporal arcades. They occur from the Permian on. The genus *Sphenodon* is the subject of a separate entry

Rhynia [rĭ″·nē·ə] a genus of fossil plants of the group Rhyniophytina. The leafless stems and terminal sporangia are typical of this Devonian form

Rhyniophytina [rĭn′·ē·ō·fĭt″·ĭ·nə] a group of vascular plants known only from Silurian and Devonian fossils. This group was once included with the Psilophyta. The genera *Horneophyton* and *Rhynia* are the subjects of separate entries

-rhyss- *comb. form* meaning "wrinkle" but sometimes used in the sense of canal

rhythm a repeated cyclic change. The following derivative terms are defined in alphabetic position:

ENDOGENOUS RHYTHM   NEUROGENIC RHYTHM
EXOGENOUS RHYTHM   SEASONAL RHYTHM

rib 1 (*see also* rib 2) a cartilaginous or bony rod articulating with or dependent from the vertebrae. The following derivative terms are defined in alphabetic position:

ABDOMINAL RIB   PLEURAL RIB
FALSE RIB   SACRAL RIB
FLOATING RIB   TRUE RIB
HEMAL RIB

rib 2 (*see also* rib 1) a leaf vein that protrudes above the surface

riboflavin = Vitamin B$_2$

riboflavinase [rī′·bō·flav″·ĭn·āz′] an enzyme cata-lyzing the hydrolysis of riboflavin to ribitol and lumi-chrome

ribonuclease [rī′·bō·nyōō″·klē·āz′] an enzyme that catalyses the formation of cyclic nucleotide by the transfer of a pyrimidine nucleotide residue

ribonucleic acid (rī′·bō·nyōō·klā″·ĭk] this close analogue of deoxyribonucleic acid is replicated from the latter in the nucleus and transfers the genetic information to the sites of protein synthesis in the ribosomes (*see also* deoxyribonucleic acid, ribosome)

ribosome [rī″·bō·sōm′] a discrete particle (*circ.* 200 Å in diameter), not membrane bounded, on the walls of the endoplasmic reticulum (cf. lysosome). It is a primary site of protein synthesis (*see* ribonucleic acid)

Rickettsiales [rĭk′·ĕts·ē·āl″·ēz] an order of schizo-mycetes containing the forms commonly called rickett-sias. Structurally they resemble minute bacteria and are mostly below 0.1 microns in size. They are distinguished as vertebrate parasites transmitted by arthropod vectors and are the cause of, inter alia, epidemic typhus, Rocky Mountain spotted fever, and "Q" fever

ridge an elongate prominence rising from a flat surface. The following derivative terms are defined in alphabetic position:

GENITAL RIDGE   NEURAL RIDGE
HYPOBRANCHIAL RIDGE

-rim- *comb. form* meaning "fissure"

ring an annular structure. The following derivative terms are defined in alphabetic position:

ANNUAL RING   PROSPONDYLOUS RING
GROWTH RING   VERTEBRAL RING
OPISTHOSPONDYLOUS R.

ring canal the circumoral portion of the water-vascular system of echinoderms

ring-porous wood woods with the distinct rings of large and small vessle (*see also* diffuse porous wood)

ring speciation the condition that occurs when an extensive cline curves round on itself so that the extremes overlap but interbreeding does not occur

Rio-Hortega's glia = microglia

riparian [rī·pâr″·ē·ən] pertaining to the shores, or banks, of rivers

Rocella [rō·chĕl″·ə] a genus of lichens. *R. tinctoria* [tĭng·tŏr″·ē·ə] yields the dye orcein

rod cell a rod-shaped cell in the retina which is sensitive to light but which cannot perceive color

Rodentia [rō·dĕnt″·ē·ə] a large order of placental mammals containing the rats, mice, squirrels, porcupines, etc. They are distinguished by the large single pair of upper and lower incisors adapted to gnawing from which the name is derived. The genera *Castor, Cynomys, Microtus, Mus, Peromyscus, Rattus* and *Sciurus* are the subjects of separate entries

Romalea see Brachystola

roof plate a narrow area along the dorsal side of the brain stem separating the alar plates

root 1 (*see also* root 2) the portion, usually under-ground, of a vascular plant that is adapted both to absorb water and minerals and to anchor the aereal portions (cf. rhizoid, rhizome). The following derivative terms are defined in alphabetic position:

PRIMARY ROOT   STILT ROOT
PULL ROOT   TAP-ROOT

root 2 (*see also* root 1) the basal portion of an organ, or part of an organ. The following derivative terms are defined in alphabetic position:

AORTIC ROOT   DORSAL ROOT   VENTRAL ROOT

root hair   a hair-like extension of a root cell through the membrane of which both water and inorganic nutrients are absorbed

root nodule   without qualification usually refers to the bacteria containing nodules on the roots of leguminous plants (*see Rhizobium*)

rootcap   a cap of parenchyma cells lying immediately over the root peristem

Rosaceae   [rōz·ās"·ē]   one of the largest families in the plant kingdom. Apart from the roses and their flowering allies most temperate zone edible fruits and berries belong in this family. The family is clearly distinguished by the perigynous flower, the numerous cyclic stamens and the numerous cyclic carpels. Many of the flowers superficially resemble those of the Ranunculaceae but these have hypogynous flowers. The genus *Pyrus* is the subject of a separate entry

rostellum   [rŏs·tĕl"·əm]   diminutive of rostrum and applied to numerous plant and animal structures in the form of a short or small beak, such as the mobile, cone-shaped tip of the scolex of some tapeworms

rostral   [rŏs"·trəl]   pertaining to the rostrum but commonly used as pertaining to the beak

rostral bone   one of a pair of dermal bones in the orbital skeleton of crossopterygian fish, lying at the anterior end, immediately below the septomaxilla

rostral cartilage   one of a number of cartilages supporting the rostrum of chondrichthian fish

rostrum   [rŏs"·trəm]   literally a "beak" but applied to almost any pointed non-retractile prominence in organisms

-rota-   *comb. form* meaning "wheel"

Rotatoria   [rōt'·ə·tôr"·ē·ə]   a taxon of the animal kingdom which may either be regarded as a phylum of pseudocoelomate Bilateria or as a class of the phylum Aschelminthes. They are distinguished by the presence of an anterior ciliated corona from which the popular name "wheel animalcule" is derived. The term "rotifer" is also common. The genus *Philodina* is the subject of a separate entry

row   a number of objects placed side by side in a horizontal row, particularly in contrast to a vertical "rank" though this distinction is not universal (*see* axial row, comb row)

Rubiaceae   [rōōb'·ē·ās"·ē]   the family of shrubs and trees which contains the coffee shrub, the cinchona, the cape jasmine and the partridge berry. The family is closely related to the Caprifoliaceae but is distinguished from them by having stipules or whorled leaves. The genus *Cincona* is the subject of a separate entry

-rubu-   *comb. form* meaning "bramble"

rudiment   the first vestige of a developing organ, or a functionally useless organ

Ruffini's corpuscle   a heat-sensing end organ, containing a loose arborization of nerve fibers, ending in flattened expansions in a mass of granular tissue

rule   a term preferred to "law" by many contemporary scientists. "So-and-so's rule" is, to avoid repetition, given in this work as "So-and-so's Law". The various international bodies controlling nomenclature, and the like, promulgate rules having the force of laws

rumen   [rōō"·mĕn]   the first of the four divisions of the stomach of artiodactyls (cf. abomasum, psalterium, reticulum)

-rupes-   *comb. form* meaning "rock"

-rupt-   *comb. form* meaning "broken"

S

Sabella [sə·běl″·ə] a genus of tube-dwelling Polychaetae Sedentaria with parchment-like tubes and a large circular plume of prostomial palps modified as gills
-sabul- *comb. term* meaning "sand" (cf. -psamm-)
saccate [săk″·ăt] in the form of or possessing a pouch or sack; used specifically by microbiologists to describe the appearance of the liquified area in a stab culture when it is a blunt-nosed cone not reaching to the bottom of the tube
-sacchar- *comb. form* meaning "sugar"
Saccharomyces [săk′·ə·rō·mi″·sēz] a large genus of Endomycetales used by man since prehistoric times in the production of bread and alcoholic beverages. *S. cerevisiae* [sěr′·ə·vis″·i·ē] is the "top yeast" used for English type beers and in baking bread and *S. carlsbergensis* [kârlz′·bə·gěn″·sĭs] the "bottom yeast" used for lager type beers. More than a hundred species of "wine yeasts" have been identified
Saccoglossus [săk′·ō·glŏs″·əs] a genus of enteropneustan hemichordata. They are mostly burrowing forms making U-shaped tubes in sand or mud
Sacculina [săk′·yōō·lin″·ə] a typical rhizocephalan with the characteristics of the group
sacculus [săk″·yōō·ləs] the ventral of the two divisions into which the auditory vesicle becomes first divided (cf. utriculus)
sacral rib [sāk·rəl] a rib articulating with the pelvis
sacral vertebra one which is directly connected with or forms part of the sacrum (*see also* pseudosacral vetrebra and urosacral vertebra)
sacrum that part of the vertebral column which articulates with the pelvis (*see also* synsacrum)
-sagitta- *comb. form* meaning "arrow head"
Sagitta [săj·ĭt″·ə] the best known genus of Chaetognatha, with the characteristics of the phylum. The various species of *Sagitta* are not only a major component of the plankton in all the oceans of the world but are also found at great depths
sagittal axis [săj·ĭt″·əl] the axis of a biradially or a bilaterally symmetrical object which divides the object into equal halves
sagittal crest the median dorsal crest of the skull
sagittal plane a plane parallel to the sagittal axis. A sagittal section is therefore a vertical longitudinal section

Salientia [săl·ē·ěnt″·ē·ə] a large order, or subclass, of Amphibia distinguished by the absence of a tail in the adult, development of the hind legs for jumping, and the reduction or absence of ribs. Commonly called frogs and toads (= Anura). The genera *Ascaphus, Bufo, Hyla, Pipa, Rana* and *Xenopus* are the subjects of separate entries
saliva a secretion, accumulated in the buccal cavities of many mammals, from the salivary glands
salivary gland any of the several glands adjacent to the buccal cavity of mammals which produce saliva. The term is extended to glands of similar location and function in other organisms
salivary pump the organ at the base of the piercing stylet in biting insects through which the products of the salivary gland are pushed into the victim
Salmo [săl″·mō] a genus of teleost fish of great economic and recreational importance. *S. salar* [sā″·lär] is the Atlantic Salmon and *S. trutta* [trŭt″·ə] is the brown trout
Salmonella [săl′·mŏn·ěl″·ə] a genus of eubacteriale Schizomycetes of which more than 400 types have been identified. *S. typhi* [ti″·fi″] and *S. paratyphi* [pär′·ə·ti″·fi′] are groups, rather than species, producing typhus, paratyphus and other enteric fevers
-salp- *comb. form* meaning "trumpet"
Salpa a genus of thaliacean urochordates so common that the whole class are sometimes called salps. The test of the solitary forms is usually prolonged into two posterior spines. There are many species among which *S. democratica* [děm″·ō·krăt″·ĭk·ə] often occurs in large swarms on both sides of the Atlantic
salpinges [săl·pĭn″·jēz] plural of salpinx
salpinx [săl″·pĭngks] literally a trumpet, but used for any tube, such as the oviduct or Eustachian tube, which terminate in a trumpet-shaped orifice
salt gland a compound tubular gland found about the eyes in marine reptiles and birds and as an adanal structure in some fish. The gland excretes large quantities of sodium chloride thus permitting the animal to drink salt water or rid itself of adsorbed salt
salt marsh a marsh the waters of which are brackish
saltation [săl·tā″·shŭn] a type of evolution (= saltatory evolution) which proceeds by leaps and bounds through the production of mutants which differ grossly from the parent

157

saltatory evolution = saltation

Salvinia [săl·vĭn"·ē·ə] a genus of aquatic Pterophyta with the leaves a set in whorls of three, two floating and one submerged. The group is further distinguished by being heterosporous, the sporocaps containing either macro- or microsporangia

samara [sə·mâr"·ə] a single-seeded winged fruit like that of the maple or ash

Samia [săm"·ə·a] a genus of moths containing *S. cercropia* which shares with *Citheronia regalis* the distinction of being the largest United States lepidopteran

sand rock particles, frequently silicious, from .05 millimeters to 1 millimeter in size. However, particles of oolytic limestone form many of the "sandy" beaches in the Caribbean (cf. gravel, silt and also hydatid "sand")

"sand" granule a phosphate granule appearing in the pineal gland

Santorini's duct an accessory pancreatic duct opening posterior to the main duct

sap the fluid that circulates in plants. It carries dissolved gases and inorganic materials to the sites of photosynthesis and sugars away from them

-sapro- *comb. form* meaning "decomposed" in the sense of rotten or "putrid" (*see also* polysaprobe)

Saprolegnia [săp'·rō·lĕg"·nē·ə] a genus of saprolegniale fungi commonly called water molds. They are unusual in that some species have two strains, one saprophytic and the other parasitic. *S. parasitica* [păr'·ăs·ĭt"·ĭk·ə] is a major problem in aquaria and fish hatcheries

Saprolegniales [săp'·rō·lĕg'·nē·ăl"·ēz] an order of oomycete fungi producing two oospores. The other order is the Peronosporales members of which produce only one oosphere in each oogonium

sapropel [săp·rō"·pĕl, săp"·rə·pĕl] leptopel (q.v.) which has accumulated on the bottom under anaerobic conditions and is gas-producing

saprophyte a plant that cannot synthesise all its food from inorganic salts and therefore requires organic matter as a nutrient. The term saprobe is not infrequently used as a synonym. The following derivative terms are defined in alphabetic position:
HOLOSAPROPHYTE   SYMBIOTIC SAPROPHYTE
PARASITE-SAPROPHYTE

saprozoic nutrition [săp'·rō·zō"·ĭk] that type of nutrition that does not involve the ingestion of solids by a protozoan, but only the absorption of organic solutes

-sarc- *comb. form* meaning "flesh" (*see* coenosarc, perisarc)

sarcocyst [sär'·kō·sĭst'] a many-chambered cyst of a sporozoan parasite

Sarcodina [sär'·kō·dĭn"·ə] a class of Protozoa containing the amebas, Foraminifera and the Radiolaria. They are distinguished from the Mastigophora and Ciliata by lacking locomotor organelles and from the Sporozoa by their methods of reproduction. The orders Foraminifera, Heliozoa, Radiolaria, Rhizopoda and Testacea are the subjects of separate entries

sarcolemma [sär'·kō·lĕm"·ə] the sheath of a striated muscle fiber

sarcomere [sär"·kō·mēr'] the portion of a myofibril that lies between two Z-lines

Sarcophilus [*angl.* sär·kŏf"·əl·əs, *orig.* sär'·kō·fil"·əs] a marsupial (metatherian mammal) once called the

"Tasmanian devil" but now more usually the "Tasmanian badger". *S. harrisii* [hăr·rĭs"·ē·ī], the only extant species, is not unlike a badger with a short, partially hairless, tail

sarcoplasm [sär"·kō·plăsm'] the amorphous material in which myofibrils are embedded

Sarcoptes [sär·kŏp"·tēz] a genus of acarine arthropods distinguished by the minute globular body with short legs ending in two claws. They burrow in the skin of mammals which gives rise to their popular name of itch mites. *S. scabei* [skăb"·ē·ī] is native to the hog but attacks man as does also *S. canis* [kăn"·ĭs], the cause of mange in dogs. *S. cati* [căt·ī], the cause of cat mange, is not apparently transferable

sarcosine [särk"·ə·sēn] methyl-glycine. A decomposition product of the alkaline hydrolysis of creatine and sometimes, for this reason, classified among the amino acids

Sargassum [sär·găs"·əm] a genus of phaeophyte algae with a long branching thallus with both leaflike branches and pneumatocysts. Like most brown algae it normally grows in the intertidal zone but in many areas storm detached pieces have reproduced into enormous floating masses. The huge mass between the West Indies and the Azores, known as the "Sargasso Sea" has existed for so long that it has developed a unique fauna

satellite cell a cell in the capsule of a peripheral nerve ganglion

-sathr- an occasional variant of -saphr- (q.v.) and sometimes applied to humus in distinction from other forms of decaying matter

-saur- *comb. form* meaning "lizard", but widely used in the sense of "reptile"

Sauria [sôr"·ē·ə] a sub-order of Squamata containing the lizards as distinct from the snakes (Serpentes). It is a doubtfully valid taxon though most lizards can be distinguished by having legs (though many do not) and a moveable eye-lid (though some do not). The genera *Anolis, Chamaeleo, Draco, Eumeces, Heloderma, Phrynosoma, Scleroporus* and *Varanus* are the subjects of separate entries

Saurischia [sôr·ĭsk"·ē·ə] an order of extinct reptiles, in which the pelvis is triradiate when seen from the side since the pubic bone is directed forward at an angle to the ischium. They were mostly carnivorous and bipedal. The genus *Branchiosaurus* is the subject of a separate entry

Sauropsidia [sôr'·ŏp·sĭd"·ē·ə] a general term, not usually accepted as a taxon, for birds and reptiles taken as a whole

savannah a type of grassland habitat found in the tropics in which forb usually predominates over grass and on which there are scattered clumps of shrubbery and tall trees

-scalar- *comb. form* meaning "ladder"

scalariform conjugation [skăl·är'·ē·fôrm"] when two alga filaments conjugate as they lie side by side

scale 1 (*see also* scale 2, 3) a dermal (fish) or epidermal (Sauropsidia) protective platelet. The following derivative terms are defined in alphabetic position:
CTENOID SCALE   GANOID SCALE
CYCLOID SCALE   LOREAL SCALE
DERMAL SCALE   PLACOID SCALE

scale 2 (*see also* scale 1, 3) a modified, flattened, seta found on many arthropods, particularly insects

scale 3   (*see also* scale 1, 2) a modified, flattened, trichome (= peltate plant hair)

Scalopus [skăl″·ō·pəs]   the genus of insectivores containing the common eastern mole *S. aquaticus* [ə·kwăt″·ĭk·əs]

-scand-   *comb. form* meaning "climb"

-scaph-   *comb. form* meaning a "boat" ("skiff") or boat-shaped container

scaphognathite [*angl.* skăf·ŏg″·nə·thit, *orig.* skăf′·ŏg·năth″·ĭt]   the second maxilla of decapod crustacea which fans water over the gills

Scaphopoda [*angl.* skăf·ŏp″·əd·ə, *orig.* skăf′·ō·pŏd″·ə]   a class of small Mollusca with tusk-shaped shells from the mouth of which a lobed foot and cephalic tentacles emerge. Commonly called "elephant tooth shells". The genera *Clione* and *Dentalium* are the subjects of separate entries

scapula bone [skăp″·yōō·lə]   the shoulder blade. The bone in the pectoral girdle of vertebrates that lies dorsal to the vertebral column (*see also* extrascapular bone)

scapulocoracoid cartilage [skăp′·yōō·lō·kŏr″·ə·koid′]   a cartilage lying between the cleithrum and the clavicle in the pectoral girdle of some fish (*see also* supracoracoid cartilage)

Sceloporus [*angl.* sĕl·ŏp″·ər·əs, *orig.* sĕl′·ō·pôr″·əs]   a genus of Squamata containing the spiny lizards in which the scales are drawn out in spine-like points

schemochrome [skē″·mə·krōm′]   a color produced by structure, as the iridescent colors of butterfly wings

schindylesis [skĭn′·də·lē″·sĭs]   the type of joining in which one plate of bone lies in a groove between two plates of another bone or plates of two other bones

-schist-   *comb. form* meaning "split" (*see* isoschist)

Schistosoma [shĭst′·ə·sōm″·ə]   a genus of trematodes that cause schistosomiasis, a disease resulting from the presence of the flukes in the mesenteric and pelvic veins. The eggs pass out through the urine or feces, occasioning much damage while passing through the tissues. The miracidia that hatch from the eggs enter a snail, specific for each species, and produce sporocysts from which cercaria escape into water. They enter the adult host—man in the case of *S. hematobium* [hē′·mə·tōb″·ē·əm], *S. mansoni* [măn·sŏn″·ĭ] and *S. japonicum* [jăp·ŏn″·ĭk·əm]—by burrowing through the skin of bathers or waders. This sometimes produces a temporary irritation known as bather's itch

-schizo-   *comb. form* meaning "split" (cf. -schist-)

schizocoel [skĭz″·ō·sēl, skē″·zō·sēl′]   a coelom derived from the splitting of a mesodermal band or plate

schizogenic development [skĭz′·ō·jĕn″·ĭk, skĭz′·ō·jĕn″·ĭk]   development resulting from cell division as distinct from increase in cell size

schizogenous [skĭz·ŏj″·ĕn·əs, skĭz·ŏj″·ĕn·əs]   derived through splitting. The term is commonly applied to intercellular spaces in plants

schizogony [*angl.* skĭz(skĭz)·ŏg″·ən·ē, *orig.* skĭz′·(skĭz′)·ō·gŏn″·ē]   the process of multiple fission when the end products develop directly into adults, not into spores or gametes (cf. gamogony, sporogony)

Schizomycetes [skĭz′·ō·mi·sēt″·ēz, shĭz′·ō·mi·sēt″·ēz]   the "fungi that divide" is a class of akaryote organisms containing many varied forms such as the bacteria, spirochaetes, slime bacteria and Rickettsias. The following orders are the subject of separate entries: Pseudomonadales, Eubacteriales (bacteria proper),

Chalmydobacteriales (filamentous iron bacteria), Beggiatoales (filamentous sulfur bacteria), Myxobacteriales (slime bacteria), Spirochaetales, Mycoplasmatales, Rickettsiales. There is much divergence of opinion as to the relation of these orders and as to the proper inclusion of some in the Schizomycetes

schizont [skĭz″·ŏnt, skĭz″·ŏnt]   an organism derived by splitting or which will give rise to other organisms by splitting

Schwann's cell   a cell which produces the sheath round a myelinated nerve fiber. It becomes wrapped round the fiber in several layers each of which then becomes swollen with lipids

Schweiger-Seidel sheath   a sheath consisting of densely packed reticular cells round the arterioles in the red pulp of the spleen

sciatic artery [si·ăt″·ĭk]   an artery running down the posterior side of the leg. It derives from the internal iliac artery and becomes the popliteal artery

scion [si″·ən]   a "twig", particularly one that is used as a graft

-sciur-   *comb. form* meaning "squirrel"

Sciurus [si·yūr″·əs]   the genus to which most North American and all European squirrels belong. They are all bushy-tailed, moderate sized, arboreal forms

-scler-   *comb. form* meaning "hard"

sclera [sklir·ə]   the outer coat of the back of the eye

sclereid [sklir·ē·ĭd]   a short sclerenchyma cell (cf. fiber)

sclerenchyma [*angl.* sklir·ĕn″·kəm·ə, *orig.* sklĕr″·ĕn·kĭm″·ə]   coherent masses of skeletal, as distinct from conducting, tissue of a plant (*see also* girder sclerenchyma, protosclerenchyma)

sclerite 1 [sklĕr″·ĭt]   (*see also* sclerite 2) a general term for sponge spicules (*see also* megasclerite, microsclerite)

sclerite 2   (*see also* sclerite 1) rigid portions of the arthropod exoskeleton, usually bounded by sutures

scleroblast [sklĕr″·ō′blăst′]   a sponge mesenchyme cell which produces spicules, or other skeletal elements

sclerocoel [sklĕr″·ō·sēl′]   that portion of the coelom that forms the cavity in a sclerotome

scleroprotein [sklĕr″·ō·prō″·tēn]   any protein which forms a hard skeletal, or protective structure

sclerotic parenchyma [sklĕr·ŏt″·ĭk]   parenchyma cells with lignified secondary walls

sclerotic plate   one of a ring of bones surrounding the orbit

sclerotome [sklĕr″·ō·tōm′]   a mass of cells proliferated from the ventral inner corner of the myotome and from which the ribs and vertebrae are derived. The term is sometimes confined to the inner wall of the myotome as defined in above, in which case the outer wall becomes the dermatome (*see also* myosclerotome)

sclerotomic fissure [*angl.* sklĕr·ŏt″·əm·ĭk, *orig.* sklĕr′·ō·tōm″·ĭk]   a space free of cells that lies between successive sclerotomes

sclerotomite [*angl.* sklĕr·ŏt″·əm·ĭk, *orig.* sklĕr′·ō·tōm″·ĭk]   a division of the vertebral column, running from one intervetebral space to the next. It does not correspond to a somite which runs from the center of one vertebra to the next (*see also* cranial sclerotomite, caudal sclerotomite)

sclerotize [sklĕr″·ō·tiz′]   used specifically to describe the hardening of arthropod appendages by materials other than chitin, as in the calcareous lobster claw

-scol- *comb. form* confused from three Greek roots and therefore meaning a "thorn", a "worm" or "crooked"

-scolec- *see* "-skolex-"

scolex [skō″·lĕks] the anterior, strobilating proglottid of a cestode platyhelminth worm (tapeworm), often furnished with hooks, suckers or both; or the same structure, in a prestrobilating condition, in the cysticercus larva

-scolio- *see* -skolio-

Scolopendra [skŏl′·ō·pĕn″·drə] a cosmopolitan genus of Chilopoda. The members are generally broad and short (in comparison with *Geophilus*). They occur mostly in warm climates, the largest being *S. gigantea* [ji·gănt″·ē·ə] (up to a foot) in the East Indies. *S. heros* [hĕr″·ŏs] up to about 3 inches, occurs in most Southern states of the United States. The bite of all species is extremely painful but it is doubtful if even *S. gigantea* is fatal to man

scolopophore [*angl.* skəl·ŏp″·ə·fôr, *orig.* skŏl″· ŏp·ō·fôr] a sense organ perceiving continuous vibration as distinct from a tangoreceptor which perceives individual touches. A tuned scolopophore is an organ of hearing (cf. phonoreceptor)

-scop- *comb. form* meaning "watchman" and thus, by extension, "one who examines things" or the actual "process of examination"

scopa [skōp″·ə] a comb-like structure commonly called a pollen comb on the tibia of the hind leg of a bee; the scopa of one leg selects pollen which is transferred to the corbiculum of the other leg

-scopic pertaining to the appearance. The following derivative terms are defined in alphabetic position:
MACROSCOPIC　　ULTRA MICROSCOPIC
MICROSCOPIC

scopula [skŏp′·yōō·lə] literally, a small broom and used for any small brush-like structure particularly on the legs of arthropods, for compound cilia in some protozoa, and particularly for the adhesive disc on the posterior end of some peritrichous ciliates

Scorpionida [skŏrp′·ē·ŏn″·id·ə] an order of arachnid arthropoda containing the scorpions. They are clearly distinguished by the modification of the pedipalps into chelate appendages, the presence of a comb-like appendage on the third abdominal segment, and the segmented tail usually terminating in a sting

-scoto- *see* -skoto-

scotopsin [skŏt·ŏp·sĭn] one of the two breakdown products (cf. retinene) of metarhodopsin (q.v.)

Schrophulariaceae [skrŏf′·yōō·lär′·ē·ās″·ē] an enormous family of tubuliflorous dicotyledons containing, *inter alia*, the snapdragons, toadflaxes, and foxgloves. The peculiar irregular corolla and reduced number of stamens are distinctive of this easily recognizable family. The genus *Digitalis* is the subject of a separate entry

-scrot- *comb. form* meaning "pouch"

scrotum [skrō·təm] the sac in which the descended testes lie in those mammals in which they leave the abdominal cavity

scrub [skrŭb] a habitat consisting principally of low growing bushes and stunted trees

-scut- *comb. form* meaning an elongate quadrangular shield or buckler (cf. -clyp- and -pelt-)

scute [skyūt] a scale or shield. Specifically the large ventral scales on the lower surface of a snake, a

tergal plate of a chelopod arthropod, the chitinous plates on many larval insects and the scale-like notopodial cirri on some polychaete annelids (*see also* corneoscute)

scutellum [skyūt″·ĕl·əm] the upper portion of the thoracic skeleton of an insect

Scutigera [skyū·tĭj″·ər·ə] an unusual genus of Chilopoda easily recognized by the sixteen pairs of very long legs, the centers of 15 of which are crooked upwards as the creature scuttles. The terminal pair are longer than the body and trail behind. *S. forceps* [fôr″·sĕps], with a one inch body and a two inch pair of hind legs is sometimes called the "house centipede" since it is often found in damp cellars searching for cockroaches, its preferred food

Scutigerella [skyū′·tĭj·ər·ĕl″·ə] a typical genus of Symphyla with the characteristics of the order. *S. gratiae* [grăsh·ē], about a tenth of an inch long, is of world wide distribution

scutum (*see also* scutum 1, 2) literally, an oblong-shaped shield; the plural is scuti, *not* "scuta" as usually given

scutum the dorso-central sclerite of an insect notum

-scyph- *comb. form* meaning "cup" (cf. -calyc-, -cyath-, -pocul-)

Scypha = *Sycon*

scyphistoma larva [skĭf′·ĭst·ŏm″·ə, ski·fist″·am·ə] the sessile hydroid stage of scyphozoans. It strobilates into ephyra larvas

Scyphozoa [sif′·ō·zō″·ə] the class of Coelenterata that contains the jellyfish. The hydroid generation is greatly reduced or even entirely wanting. When it occurs it is known as a scyphistoma larva. The genera *Aurelia* and *Rhizostoma* are the subjects of separate entries

sea water the water of seas and oceans usually containing between 35 and 37 parts per thousand of dissolved salts. Parts of the Baltic Sea are only brackish while the salinity of the Red Sea may reach 41% (*see* haline, freshwater, brackish water)

seasonal cycle rhythmic cycles caused by, or associated with, the procession of the equinoxes

seasonal rhythm an endogenous rhythm of hibernators which makes it difficult to induce artificial hibernation by lowering the temperature out of season

sebaceous gland [səb·ās″·ē·əs] a small oil gland opening into the hair follicle

sebum [sē″·bəm] the oily material secreted by sebaceous glands

second maxilla in Crustacea the second of the two pairs of appendages so named but in insects the labium

second order bronchus a tube dividing off from the bronchus

secondary cell wall a wall of one or more layers, sometimes of varied composition, formed within the primary cell wall of a plant (*see also* cell wall)

secondary growth that plant growth which results in thickening as distinct from elongation (*see also* primary growth)

secondary metamorphosis an alteration in the life pattern and form of an adult the first metamorphosis of which was the change from the larval to the adult condition

secondary polyploid an allopolyploid with a predominance of genomes from one ancestor

secondary sexual characters those characters

which are directly concerned with the function of sex, i.e. intromittent organ, mammary gland, etc. (*see also* primary sexual character, tertiary sexual character)

secondary somatic hermaphrodite a somatic hermaphrodite having the gonads and secondary sexual characters of one sex together with parts of the secondary sexual characters of the other sex (*see also* somatic hermaphrodite)

secondary xylem xylem produced from vascular cambium

-secret- *comb. form* meaning "to part from" and, by biological extension "to secrete"

secretagogue [sə·krĕt″·ə·gōg] a material, not in itself endocrine, which initiates endocrine secretions

secretin [sə·krĕt″·ĭn] a hormone secreted in the upper intestinal mucosa. It acts in controlling the volume rate of the secretion of pancreatic enzymes

secretion the production by an organism of a desirable product as distinct from the excretion of an unwanted one

secretory granule one which appears in cells of exocrine glands that are actively secreting

section a thin slice, usually intended for microscopic study. Most sections are cut in the transverse, sagittal or frontal planes. Sections for examination under an optical microscope are usually cut about 10 μ thick, from material embedded in wax; those for the electron miscroscope are cut about 200 Å thick from material embedded in plastic (*see also* plane, serial section)

-secund- *comb. form* properly meaning "second" or "following" but widely used in biology to mean "directed in one sense"

Sedentaria [sĕd′·ĕn·târ″·ē·ə] an assemblage, sometimes regarded as an order, of those polychaete annelids which live in tubes or burrows (cf. Errantia)

seed the resting stage of a higher plant, containing a partially developed sporophyte and a food reserve called endosperm both enclosed in a seed coat or testa (*see also* albuminous seed, exalbuminous seed)

segment 1 (*see also* segment 2) in the sense of a discrete portion of an organism or appendage; in arthropods the term has become hopelessly confused with "joint" which is properly the connection between segments, or segments of limbs and appendages; the term, for example, "five-jointed antenna" is both inaccurate and misleading. The following derivative terms are defined in alphabetic position:
ANTENNAL SEGMENT   SUPERLINGUAL SEGMENT
INTERCALARY SEGMENT   SUPERNUMARY SEGMENT
SUBSEGMENT

segment 2 (*see also* segment 1) in the sense of a portion of a chromosome. The following derivative terms are defined in alphabetic position:
DIFFERENTIAL SEGMENT   PAIRING-SEGMENT
DISLOCATED SEGMENT

segmentation the division of an elongate organism into a number of similar parts. Usually, but inaccurately, considered to be synonymous with metamerism. The following derivative terms are defined in alphabetic position: (*for* the division of a fertilized egg, sometimes called segmentation, *see* cleavage)
METAMERIC S.   SUPERFICIAL S.

segregation the maintaining of closely connected things separate from each other. In genetics, the separation of pairs of alleles at meiosis

seiche [sāsh] a rhythmic, frequently sudden, rise

and fall at the surface level of a lake following strong winds or unusual barometric changes

-seio- *comb. form* meaning "shape" (cf. -seism-) frequently transliterated -sio-

-seiro- *comb. form* meaning a "rope"

-seismo- *comb. form* meaning "vibration'

Selachii [sĕl·ăk″·ē·ī] the order of elasmobranch Chondrichthyes that contains the sharks. It is distinguished from the rays (Batoidei) by the fact that the body is not significantly flattened. The genera *Carcharodon*, *Cetorhinus*, *Rhincodon*, *Sphyrna* and *Squalus* are the subjects of separate entries

Selaginella [sə·lăj′·ĭn·ĕl″·ə] a typical genus of Selaginellales, with prostrate, dichotomously branching stems. The strobilus bears the smaller microsporangia above the larger megasporangia

Selaginallales [sə·lăj′·ĭn·ăl·āl″·ēz] an order of lepidophyte plants distinguished from the Lycopodiales by the presence of ligules on the leaves and from the Isoetales by the absence of any secondary thickening on the stem. The genus *Selaginella* is defined above

-selen- *comb. form* meaning "moon"

self-sterility the condition of a hermaphrodite organism that cannot fertilize itself even though male and female gametes are mature at the same time

-sell- *comb. form* meaning "saddle"

sella turcica [sĕl″·ə tûrk″·ĭk·ə] literally, "Turkish saddle". Usually applied to the shallow depression in the basisphenoid bone into which the pituitary gland fits, but also, in some decapod crustacea, to a saddle shaped portion of the abdominal apodeme

-sem- *comb. form* meaning "sign" or "signal"

sematic color a color which is thought to serve a useful purpose to an organism (*see also* antiaposematic color, episematic color, pseudoepisematic color, aposematic color)

semen [sē″·mĕn] seed or spermatic fluid

-semi- properly meaning half, but widely used in biology to indicate "partly"

semicircular canal those tubes in the inner ear which are concerned with balancing

semilunar ganglion [sĕm′·ē·lōōn″·ər] a large ganglion at the base of the trigeminal nerve

semilunar valve a flat valve in the aorta

seminal vesicle [sĕm″·ĭn·əl] a vessel for the storage of sperm, variously situated on the side of or at the end of the vas deferens or redeptaculum seminalis

seminiferous tubule [sĕm′·ĭn·ĭf″·ər·əs] one of the tubules producing sperm in the testes

semipermiable membrane an unfortunate name for differentially permiable membrane (*see also* membrane and differentially permiable membrane)

semiplacenta [sĕm′·ē·plăs·ĕnt″·ə] one in which the embryonic and maternal tissue is not fused

-semper- *comb. form* meaning "always"

Semper's cell the cell of the crystalline eye cone in insects

sensillum [sĕn·sĭl″·əm] a sense organ, or a collection of neurosensory cells, not always of known function, in invertebrates

sensorium [sĕn·sôr·ē·əm] a hemispherical or flat shaped cavity covered by a membrane on the surface of some insects, probably a receptor of unknown function

sensory epithelium any of numerous types of epithelial cells modified to receive stimuli

sensory ganglion = cerebrospinal ganglion
sensory neuron a neuron which receives stimuli
sensu [sĕns″·ōō] in the sense of
sensu lato [lā″·tō] "in a broad sense", as when a species name is used in the sense of including subspecies
sensu stricto [strĭk″·tō] in a restricted sense in contrast to sensu latum
separation layer the region in the abscission zone where separation occurs
-sepes- comb. form meaning a hedge
Sepia [sēp″·ē·ə] a genus of cephalopod mollusks commonly called cuttle fish. They are distinguished from *Loligo* by the fact that the fins run the whole length of the body. *S. officinalis* [ŏf·ĭs′·ĭn·ăl″·ĭs] is the common "cuttle fish" of European waters
septomaxilla bone [sĕp′·tō·măks″·ĭl·ə] a membrane bone separating the two nasal passages in some reptiles and amphibia (see also maxilla bone, premaxilla bone)
septum [sĕp″·təm] a division or wall. In Coelenterata specifically one of the radial partitions projecting into the coelenteron. The following derivative terms are defined in alphabetic position:
MYOSEPTUM	SCLEROSEPTUM
NASAL SEPTUM	TRANSVERSE SEPTUM

Sequoia [sĕk·woi″·ə] a genus of Coniferales distinguished by its imbricated winter buds, principally known from the giant redwood (*S. sempervirens*) [sĕm′·pĕr·vĭr″·ĕnz] of the Pacific North West
-ser- comb. form confused from four Latin roots and therefore variously meaning "watery", "to plant", "to fasten or interweave", or "late" in the sense of a late season crop
seral [sĕr″·əl] pertaining to seres and therefore carrying the connotation "developmental"
serclimax [sĕr·klĭm″·ăks] an impermanent subclimax
sere [sĕr] one of a chain of successional ecological stages leading to a climax. The following derivative terms are defined in alphabetic position:
ADSERE	MESOSERE
ANGEOSERE	OXYSERE
AQUATOSERE	POSTCLISERE
CLISERE	PRECLISERE
CONSERE	PROSERE
COSERE	PSAMMOSERE
EOSERE	SUBSERE
GYMNOSERE	THALLOSERE
HYDROSERE	XEROSERE
LITHOSERE	

serial disposed in rows or ranks. Derivative terms are not separately defined since the meanings are obvious from the roots (e.g. biserial, in two ranks)
serial sections a series of successive sections from the same animal, or tissue, mounted in order on a slide
serine [sĕr″·ēn] 2-amino-3-hydroxy-propanoic acid (= β-Hydroxyalanine), $HOCH_2CH(NH_2)COOH$. A widely distributed amino acid not necessary for the nutrition of rats
serotinal [sĕr′·ō·tin″·əl] pertaining to the late summer and the early fall
serous [sĕr″·əs] watery, usually in distinction from mucous
serous gland one which secretes a clear, watery fluid
Serpentes [sûr·pĕn″·tēz] a sub-order of Squamata containing the snakes as distinct from the lizards (Sauria). No snakes have functional forelimbs (neither do many lizards) and no snakes have a moveable eyelid (also lacking in some lizards). The genera *Ancistrodon*, *Crotalus*, *Eunectes*, *Lachesis*, *Lampropeltis*, *Micrurus*, *Naja*, *Ophiosaurus*, *Pituophis*, *Sistrurus* and *Thamnophis* are the subjects of separate entries
Sertoli cells the supporting cell of the testicular epithelium
serum [sĕr″·əm] defibrinated blood plasma (see also antiserum)
sesamoid bone [sĕs″·ə·moid] one or more bones in the foot of some mammals lying below the metatarsal and phalanges
-sesqui- prefix meaning "one and one half"
sesquireciprocal hybrid [sĕs′·kwē·rĕs·ĭp″·rə·kəl] a reciprocal cross between an F_1 individual and one of its parents
sessile [sĕs·ĕl″, sĕs″·ĭl] lacking a stalk
sessile abdomen an insect abdomen that is not petiolate nor appreciably narrowed at its attachment with the thorax (see also abdomen, petiolated abdomen)
-sessor- comb. form meaning "seated", originally in the sense of having a "seat" or "house" in a community
-sesto- comb. form meaning "sieve"
seston [sĕs″·tən] the sum total of suspended matter in a given body of water (see bioseston, abioseston)
-set- comb. form meaning bristle
seta [sē″·tə] literally a bristle. Applied in botany to any bristle shaped body including the stalk of a moss sporangium. Confined in entomology to any monocellular cuticular outgrowth (cf. chaeta) (see also glandular seta)
Sewall Wright effect the result of random genetic drift
-sex- comb. form meaning six
sex a necessary attribute of a gamete producing organism. The term is definable only when anisogametes are produced in which case the parent of the larger gamete is referred to as having female sex and that of the smaller gamete as having male sex. The following derivative terms are defined in alphabetic position:
HETEROGAMETIC SEX	PHENOTYPIC INTERSEX
HOMOGAMETIC SEX	SUPER SEX
INTERSEX	

sex chromosome that chromosome which has no exact homologue among the chromosomes of the opposite sex (see W, X, Y and Z chromosomes)
sex cycle the periodic appearance and regression of sex specific characters, usually associated with the breeding season
sex-influenced character one which appears recessive in one sex and dominant in the other
sex-linked character one which is controlled by an allele on the sex chromosome and is therefore heterozygous in one sex and homozygous in the other
sexual character one pertaining to, or distinguishing between, male and female (see also primary sexual character, secondary sexual character, tertiary sexual character)
sexual communication the transfer of information between non-human animals in the course of mating, or in premating courtship
sexual reproduction the replication of an organism from the fusion of two, or parts of two, parents

Seymouria [sē·môr″·ē·ə] an extremely primitive lower Permian anapsid reptile sometimes regarded as an amphibian ancestral to reptiles

shadowing the device of projecting an oblique beam of vaporized metal across objects to be examined in the electron microscope. This adds a three dimensional aspect to the image

Sharpey's fibers fibers, thought by some to be nervous and by others to be connective tissue, running from the periostium through bone, or from the dentine of a tooth into the pulp, or from a bony scale or skin plate into the connective tissue below

shell gland any gland which secretes the outer coat of eggs (*see also* accessory shell gland)

Shigella [shĭg·ĕl″·ə] a large genus of eubacteriale Schizomycetes. Two members (*S. flexneri* [flĕks′·nə·ĭ] and *S. boydii* [boid″·ē·ĭ]) are the principal cause of bacillary dysentery

shikimic acid [shə·kĭm″·ĭk] $C_6H_6(OH)_3COOH$, a precursor of aromatic amino acids, first isolated from the Japanese star anise

shoot a plant stem together with the leaves and buds that it bears

shrub any low growing woody plant with many stems and no main trunk

shrub tundra that area of tundra in which shrubby growths are very noticeable or even predominant

sib [sĭb] variant of sibling but often misused as the adjectival form of this word

sib mating mating between brother and sister

Sibbaldus [sĭb·ăld″·əs] a genus of balaenopterid whales erected to contain the Blue Rorqual (*S. musculus*) [mus·kyū″·ləs]. It is the largest extant mammal (and also the largest extant animal) reaching a length of 110 feet and a weight of 120 tons. It is rapidly becoming extinct through the depredations of the whaling industry

sibling [sĭb″·lĭng] properly "kindred" but used in biology to indicate animals derived from a single birth, a single clutch of eggs or, in insects, a single mating: and in botany plants derived from the ovaries or the pollen of the same flower

-sicc- *comb. form* meaning "dry"

-siderip- *comb. form* meaning "magnet", but widely used for "iron"

sieve areas areas on the wall of sieve cells with clusters of perforations or pores, through which run protoplasmic connections (cf. sieve plate)

sieve cells sieve elements in the form of long and slender cells without clearly differentiated sieve areas

sieve plate a sieve area with larger holes and much larger callose cylinders than most, and usually placed on the terminal walls of elongated sieve plate elements

sigmoid colon that part of the colon, in man, which connects the descending colon to the rectum

silt rock particles less than 0.05 mm. in size (cf. sand, gravel)

Silurian period [si·lyōōr″·ē·ən] one of the Paleozoic periods extending from about 350 million years ago to about 300 million years ago. It was preceded by the Ordovician and followed by the Devonian. There were numerous invertebrates and ostracoderms. A few land plants had appeared

Simia see Pongo

Simioidea [sĭm″·ē·oid″·ē·ə] a sub-order of primate mammals containing the Old World monkeys as

distinct from the apes (Anthropoidea). They differ from the New World monkeys in having downwardly directed nostrils and from the apes in possessing tails, which are never, however, prehensile

simple epithelium an epithelium containing only one type of cell

simple gland in zoology any unbranched gland or, in botany, a unicellular gland

simple leaf one with a single blade

simple pit a plant pit not partially closed by an arch

simple protein one consisting of amino acids and which therefore lacks prosthetic groups

simplex uterus the type in which there is no trace left of the paired uteri

Sinanthropus [sin·ăn″·thrə·pŭs] the fossil "Peking Man" now thought to be a race of *Homo erectus* (*see Homo*)

sinus [si″·nəs] literally a "lake", "bay", or "fold of a garment", used mostly in the first two in zoology and the last sense in botany. The following derivative terms in this sense are defined in alphabetic position:

CARDINAL SINUS	SINUS OCEANICUS
CERVICAL SINUS	SINUS TERMINALIS
CUVERIAN SINUS	SINUS VENOSUS

sinus gland a gland in Crustacea located between the two basal optic ganglia and which secretes a hormone that increases the molting rate

sinus oceanicus = sinus terminalis

sinus terminalis [tûr′·mĭn·ăl″·ĭs] the circular, peripheral blood vessel, which delimits the area of the blastoderm in the development of a telolecithal egg

sinus venosus [vēn·ōs″·əs] the large sinus into which venous blood drains before reaching the atrium

-siph- *comb. form* meaning "tube"

siphon almost any type of tubular structure, plant or animal, through which a fluid flows; specifically the inhalent and exhalent tubes of some pelecypod mollusca, the similar tubes of urochordates, the sucking structures of sucking arthropods, the breathing tube on the larva of certain culicid insects, the gastrozooid of siphonophoran hydrozoa, and the elongated tube of Polysiphonia and similar algae (*see also* anal siphon)

Siphonaptera [sif″·ŏn·ăp″·tĕ·rə] the order of insects containing the fleas. They are characterized by the lack of wings, the mouth parts modified for biting, the laterally compressed shape of the body and the legs modified for jumping. The genus *Xenopsylla* is the subject of a separate entry

siphonoglyph [si″·fŏn·ō·glĭf′] a groove down one side of the pharynx of Anthozoan coelenterates

Siphonophora [*angl.* si′·fŏn·ŏf″·ər·ə, *orig.* si′·fŏn·ō·fôr″·ə] an order of hydrozoan coelenterates which exists as free-swimming colonies of polymorphic individuals attached to a common float. Specialized polyps are responsible for floatation, propulsion, food capture, food absorption and reproduction. The genera *Physalia* (Portuguese-man-of-war) and *Velella* are the subjects of separate entries

siphonostele [si″·fŏn·ō·stēl′] a tubular stele, which, therefore contains pith

siphuncle 1 [si″·fŭngk·əl] (*see also* siphuncle 2) in aphids, a hollow projection through which a defensive, sticky liquid is squirted

siphuncle 2 (*see also* siphuncle 1) the connective tissue cord which attaches the base of the body of argonautid mollusca to the shell in the first chamber

Sipunculoidea [si·pŭng″·kyūl·oid″·ē·ə]   a small phylum of the animal kingdom containing a group of unsegmented worms without bristles. The main characteristic is a crown of short hollow tentacles on the anterior end. This group has variously been regarded as a class of the Annelida, a class of the Gephyrea and as a separate Phylum. The genus *Dendrostomum* is the subject of a separate entry

Siren [sir″·ĕn]   a genus of completely neotenous urodele Amphibia, having only the anterior pair of limbs and lacking a pelvic girdle. There are three external gill slits. *S. lacertina* (the great siren) is found in the South East United States

Sirenia [sir·ēn″·ēə]   the order of placental mammals that contains the sea-cows, dugongs, and manatees. They are distinguished by their petal-like fins, forked tails, and pectoral mammary glands. The genera *Dugong* and *Trichechus* are the subjects of separate entries

Sirenodon   *see Ambystoma*

Sistrurus [sĭs·trûr″·əs]   a genus of snakes (Squamata) containing both the pygmy rattler (*S. miliaris* [mĭl′·ē·är″·ĭs]) and the larger *S. catenatus* [kăt′·ēn·ā″·təs] which is the rattlesnake of the North Central States (*see also Crotalus*)

skeletal muscle [skĕl″·ĭt·əl]   that portion of the muscular system which is attached to bony elements to which muscles are attached, and more loosely, supporting and protective structures such as the "skeleton" of some protozoa and other supporting structures such as the "skeleton" of leaves or insect wings. The following derivative terms are defined in alphabetic position:

AXIAL SKELETON   VISCERAL SKELETON
ENDOSKELETON
EXOSKELETON

-skia-   *comb. form* meaning "shade"

skin   loosely the outer coat of an organism, and specifically a covering of many animals consisting of both ectodermal and mesodermal elements

-skler-   *comb. form* meaning "hard" frequently transliterated -scler-

skolex-   *comb. form* meaning "worm", almost invariably transliterated -scolec-

skolio-   *comb. form* meaning "bent" or "curved", frequently transliterated -scolio-

skoto-   *comb. form* meaning "dark" or "dim", commonly transliterated -scoto-

skotoplankton [skō′·tō·plănk″·tən]   that which occurs below the 250 fathom mark

skull   the skeletal element of chordates that encloses the brain and articulates with the jaw or jaws

s.l.   abbreviation for *sensu lato* (q.v.)

slide   a 3″ ✕ 1″ strip of thin (usually about 1 mm) glass on which objects are mounted for microscopical examination (*see also* coverslip, grid)

small intestine   that portion of the intestine which runs from the stomach to the large intestine. It is the site of most nutrient absorption

smear   a preparation intended for microscopical examination prepared by smearing the material on a slide. Widely used in the examination of blood and of bacteria

smooth muscle   muscle which lacks striations and which is associated with uncontrolled (e.g. peristaltic) movement. In vertebrates smooth muscle (as also

cardiac muscle) is innervated by the autonomic nervous system

social parasite   an organism which lives on terms of amity with another organism from which it steals food or with which it competes for food

social symbiosis   the condition of one organism which feeds on food stored by another organism

sociation [sōsh′·ē·ā″·shŭn]   a stable plant community with one or more dominants at each level (*see also* association)

socies [sōsh″·ĭ·ēz]   a sere, or seral community, with one or more dominants (*see also* associes, consocies, isocies, subsocies)

society   a term used in so many various ways that it no longer has any specific meaning, as distinct from the generalized meaning of an assemblage of organisms. The following derivative terms are defined in alphabetic position:

ADOPTION SOCIETY   COMPLEMENTARY SOCIETY
CLOSED SOCIETY

sociohormone = pheromone

Solanaceae [sōl′·ăn·ās″·ē]   a large family of dicotyledons containing, in addition to the potato, eggplant, and tobacco the extremely poisonous nightshades and daturas. The numerous seeds and tubular plicate corolla are typical of the family. The genera *Atropa*, *Capsicum*, *Lycopersicum*, *Nicotiana* and *Solanum* are the subjects of separate entries

Solanum [sōl·ān″·əm]   the type genus of the family Solanaceae. Many are of great economic importance including *S. tuberosum* (potato) and *S. melongena* (eggplant)

Solaster [sŏl″·ăst·ə]   a genus of asteroid echinoderms. *S. endica* [ĕn·dĭk″·ə] is the common "purple sun star" of the Pacific

-sole-   *comb. form* meaning "sandal"

Solea [sōl″·ē·ə]   a genus of marine teleost fish with both eyes on the right side of the head. The European *S. solea* is the true Dover sole

-solen-   *comb. form* meaning a "tube"

solenocyte [sŏl·ēn″·ō·sit′]   an elongated flame cell in which one or more long flagella beat through the entire length of the neck

solenoglyph [sŏl·ēn″·ō·glĭf′]   said of snakes having hollow fangs which may be folded back

Solpugida [sŏl·pyōō″·gĭd·ə]   a small order of arachnid arthropods commonly called the sun-spiders. They are distinguished by the possession of a large carapace consisting of the head and first thoracic segment and the presence of peculiar stalked leaf-like organs (malleoli) on the fourth pair of legs

soluble protein   one in which chains of molecules are cross linked with weak internal hydrogen bonds and which therefore readily disperse in water

-soma-   *comb. form* meaning "body". The form -soma, -some and somite all enter into compounds, the last usually being used as synonymous with metamere

soma [sōm″·ə]   a body division (*see* podosoma, presoma and cf. -some-, somite)

somatic cell = propagatory cell

somatic hermaphrodite [sōm·ăt″·ĭk]   an organism having phenotypic characters not belonging to its genotype (*see also* primary somatic hermaphrodite, secondary somatic hermaphrodite)

somatic muscle   muscle associated with the body wall and its appendages

somatic nerve   a nerve carrying impulses to or from the outer layers of the body

somatic nucleus   a nucleus which is thought to play no part in reproduction, as the macronucleus of protozoans

somatic synapsis   the pairing of homologous chromosomes in mitosis

-some-   see -soma. The terminations -some and -soma are used interchangeably

-some 1   in the sense of a division of a body. The following derivative terms in this sense are defined in alphabetic position:

ACROSOME   NECTOSOME
CHOANOSOME   UROSOME

-some 2   in the sense of a discrete body, usually a cell inclusion. The following derivative terms in this sense are defined in alphabetic position:

ALLOSOME   EPISOME
AUTOSOME   IDIOSOME
CENTROSOME   KINETOSOME
CHROMOSOME   LYSOSOME
DESMOSOME   MICROSOME
DICTYOSOME   RIBOSOME
DIPLOSOME   ZYGOSOME
ECTOSOME

-somic   adjectival suffix pertaining to -soma or -some in any of its forms of means. -somal is synonymous. The following derivative terms are defined in alphabetic position:

DISOMIC   NULLISOMIC
MONOSOMIC

somite   [sōm"·ĭt] literally a "little body" but usually used for one segment of a metamerically segmented organism (cf. metamere) (see caudal somite)

sor   [sôr]   a shallow lagoon at the mouth of a large river

soredium   [sôr·ēd"·ē·əm, sər·ēd"·ē·əm]   an algal cell surrounded by hyphal tissues and which can, when detached, produce a lichen thallus

sorus   [sôr"·əs]   a cluster of sporangia in ferns or a mass of soredia on the surface of a lichen.

sp. nov.   abbreviation for *species nova* (new species) and appended to the species name (q.v.) in the first published description

sp., spps.   abbreviation for species (*sing.*) and species (plural)

spath-   *comb. form* meaning "a blade"

special transduction   the transfer through transduction of genes only known by other criteria

speciation   [spēsh'·ē·ā"·shun]   the process by which new species are formed. The following derivative terms are defined in alphabetic position:

ALLOCHRONIC S.   ALLOPATRIC S.   RING S.

species   an organism which is distinct in the sense that it does not interbreed with other organisms. This failure to interbreed may result from chromosomal incompatibility, differences in ecological or geographic habitat or from differences in the breeding season. Interspecific crosses of domestic animals that survive as distinct species are often sterile (as the mule) but sterility of offspring is not a measure of the validity of a species. The domestic cat is certainly, and the dog probably, the result of interspecific crosses from numerous domesticated species (*see also* species name). The following derivative terms are defined in alphabetic position:

ALLOCHRONIC SPECIES   INCIPIENT SPECIES
BUFFER SPECIES   INDEX SPECIES
CENOSPECIES   MICROSPECIES
CONDOMINANT SPECIES   PLASTOSPECIES
DIRECTIVE SPECIES   TYPE SPECIES
ECOSPECIES   VERSPECIES
EXCLUSIVE SPECIES

species immunity   complete immunity of one species to a pathogen affecting other species in the same genus

species name   a species name, such as the *proteus* in *Amoeba proteus* is always written in italics without a capital letter. The validity of the species name depends on the existence, or at least general acceptance of a type (q.v.). The name should be followed by the name of the author of the original description which is placed in parenthesis if the species has been transferred from one genus to another. Thus the example given should be *Amoeba proteus* (Leidy) since Leidy named *proteus* but put it in the genus *Chaos*

-speir-   *comb. form* meaning "twist" or "twisted", usually transliterated -spir-

Spemann's organiser   the dorsal lip of the blastopore in Amphibia. Its transplantation into another blastula induces a second embryo

-sperm-   *comb. form* meaning "seed". The following derivative terms are defined in alphabetic position:

CARPOSPERM   PERISPERM
ENDOSPERM   XENOENDOSPERM
PARTHENDOSPERM   ZOOSPERM

spermatangium   the male sex organ of Rhodophyta

spermatheca   a container of sperm

spermatic cord   [spûrm·ăt"·ĭk]   the vas deferens and its associated ligaments in the mammal

spermatid   [spûrm·ə·tĭd, spûrm·ăt"·ĭd]   the mother cell of an antherozoid or of a spermatozoon

spermatogonium   [spûrm'·ăt·ō·gōn"·ē·əm]   the male gametangium of plants or the spermatocyte of animals

Spermatophyta   that great division of the plant kingdom which contains the "flowering" or, more properly, "seed bearing" plants. It is divided into the Angiospermae, containing the great majority of extant plants, in which the seeds are contained in an ovary and the Gymnospermae (such as the conifers), in which they are not

spermatozoa   [spûrm'·ət·ō"·zō·ə, spûr·măt"·ō·zō·ə]   motile male gametes

spermatozooid   [spûrm'·ət·ō·zō"·id, spûr·măt"·ō·zōid]   a ciliated gamete within an antheridium

spermiducal vesicle   [spûrm'·ē·dūk"·əl]   a seminal vesicle consisting of an extension of the sperm duct; used by some as distinct from seminal vesicle which is then an outgrowth or appendage to the sperm duct

spermophore   [spûrm"·ō·fôr']   any portion of a plant bearing either seed-producing or gamete-producing structures

-sphac-   see sphak-

-sphaer-   *comb. form* meaning "sphere" and usually transliterated -spher-

Sphaeriales   [sfēr'·ē·āl"·ēz]   an order of Ascomycetes distinguished by the presence of discrete sessile globose perithecia. The genera *Ceratocystis*, *Giberella* and *Neurospera* are the subjects of separate entries

Sphagnobrya   [sfăg'·nō·bri"·ə]   a class of bryophytes containing the bogmosses or sphagnums. They are distinguished from all other mosses by the fact that the broadly thallose protonema produces a single gametophore

Sphagnum [sfăg"·nəm] the only genus of the Sphagnobrya. They are the dominant in sphagnum bogs, sharing these with a few ericaceous and sarraceniaceous plants that are equally tolerant of a low pH. The dried moss is widely used in horticulture

sphagnum bog = moss moor

-sphen- *comb. form* meaning "wedge" (*see* hemosphene, zygosphene)

Sphenodon [sfēn"·ō·dŏn'] a genus of rhynchocephalian reptiles of which only one, *S. punctatus* (the New Zealand "tuatara") is extant

sphenoid bone any of several membrane bones forming the base, and part of the lower sides, of the cranium (*see also* alisphenoid bone, basisphenoid bone, orbitosphenoid bone, parasphenoid bone, laterosphenoid bone)

Sphenophytina a division of vascular plants the only extant representatives of which are the Equisetales. There are numerous other divisions known from most geologic periods from the Devonian to the present. The genera *Hyenia* and *Equisetum* are the subjects of separate entries

Sphenopsida = Sphenophytina

spherical symmetry a pattern in which parts are arranged around the center of the sphere in which no definite poles are apparent

sphincter [sfĭng"·tə] a circular muscle capable of closing, by contraction, a tube or aperture (*see* Boyden's sphincter, Oddi's sphincter)

Sphyrna [sfûr"·nə, sfər·nə] a genus of selachian elasmobranchs containing the hammerhead sharks, a name adequately descriptive of their appearance

-spic- *comb. form* meaning a "point" or, by extension, any pointed object

spicule a small inorganic body, commonly calcareous or silicious, found embedded in the tissues of, or forming the skeleton of, certain invertebrates particularly sponges

spider cell = astrocyte

Spilogale [spil'·ō·gāl"·ē] a genus of mustelid carnivorous mammals containing the spotted skunks; the striped skunks are in the genus *Mephitis*. *S. putorius* [pyōō·tôr"·ē·əs] is the eastern spotted skunk

spinal nerve a nerve leaving the central nervous system posterior to the cranium. They are commonly identified by serial arabic numerals

spindle an active axis, particularly the fibrillar structure running from one cytocentrum to the other in sections of cells in mitosis

spindle fibers fibers found in smooth muscle

spine popular term for the vertebral column. Also used for any stiff process, not of necessity sharp, projecting from a plant or animal, particularly the calcareous, articulated, projections from the test of echinoid echinoderms and the sharp pointed, woody spikes protruding from plant stems

spinneret the fine tubular processes through which silk is spun. Also the adhesive tube at the posterior end of some nematodes

spinous cartilage one of a pair of ventral-lateral cartilages running backwards from each side of the annular cartilage in the chondrocranial complex of Marsippobranchii

-spir- *see* speir-

spiracle [spir"·əkl, spīr·əkl] a perforation in the arthropod exoskeleton through which air enters the tracheae; also used for the remnants of the hyoid gill slits in adult chondrichthyean fish

spiracular cartilage [spir·ăk·yūl·ə, spīr·ăk·yūl·ə] one of several small cartilages supporting the spiracular opening in Chondrichthian fish. They are usually considered to be a portion of the hyoid arch

spiral cleavage the type of cleavage in which, after the first few divisions, the daughter blastomeres lie at an angle to the longitudinal plane of the cell mass and thus tend to form a spiral pattern

spiral valve 1 (*see also* spiral valve 2) a valve in the aorta of certain fish, assisting in the separation of blood into the aortic arches

spiral valve 2 (*see also* spiral valve 1) a spiral structure in the intestines of some fish, formed by the central twisting of a detached portion of the wall

Spirobulus [spir·ŏb"·yōō·ləs] a genus of Diplopoda with legs on all anterior segments. *S. marginatus* [mär'·jĭn·ăt"·əs] is the large (3 inch) dark brown, red ringed, millipede of the eastern United States

Spirochaetales [spir'·ō·kēt·ăl"·ēz] an order of Schizomycetes consisting of spiral forms swimming freely by flexure of the body. Some reach 500 microns in length, and were at one time classed among the Protozoa. The genus *Treponema* is the subject of a separate entry

spirocyst [spir"·ō·sĭst', spīr"·ō·sĭst'] a type of nematocyst, or cell resembling a nematocyst, found in the Zoantharia. It differs from nematocyst proper in that the contained tube is unarmed and of uniform diameter

Spirogyra [spir'·ō·jir"·ə] a genus of filamentous green algae distinguished by the spiral bands of chloroplasts

Spirostomum [*angl.* spir·ŏst"·əm·əm, *orig.* spir'·ō·stōm"·əm] a genus of elongate spirotrichid ciliate protozoans. *S. ambiguum* [ăm·bĭg"·yōō·əm] may reach 4 mm in length

Spirotricha a subclass of ciliate protozoans with the cilia arranged in spiral rows. The genera *Euplotes, Nyctotherus, Spirostomum* and *Stylonychia* are the subjects of separate entries

-splachn- *comb. form* meaning "moss", frequently confused with -splanch-

-splanch- *comb. form* meaning "viscera"

splanchnic [splăngk"·nĭk] pertaining to the visceral mass as distinct from the body wall (cf. somatic)

splanchnocranium [splăngk'·nō·krān"·ē·əm] the jaws and visceral arches of a skull in contrast to the neurocranium

spleen a highly vascular ductless organ laying adjacent to the stomach in most vertebrates. It is concerned with the control of blood volume and of erythrocyte production. The adjective from spleen is lienic

splenial bone [splēn"·ē·əl] one of a pair of chondral bones lying along the medial face of the dentary in some reptiles (*see also* dentosplenial bone)

-spodo- *comb. form* meaning "ash"

spodogram [spō'·dō·grăm"] an ashed section used in the investigation of the localization of minerals

spondyl [spŏn"·dəl] a "spool" or by extension "vertebra". Usually used in biology in the adjectival form spondylous

spondylous [spŏn"·dəl·əs] (*see also* vertebra) pertaining to a spool or a spool-shaped articulation on a vertebra

Spongilla [spŭn·jĭl"·ə] a genus of freshwater sponges. They usually encrust submerged logs and twigs. The green *S. lacustris* [lăk·ŭs"·trĭs] is common all over the world

spongioblast [spŭn"·jē·ō·blăst'] either a cell in the epithelium of the embryonic neural tube which will subsequently become a neuroglia cell or a spongin producing scleroblast

spongocoel [spŭn'·jō·sēl"] the main cavity of a simple sponge

spongy bone lamellar bone lacking Haversian canals but with a system of interconnecting, blood-filled lacunae

spongy parenchyma that portion of the mesophyll which is not palisade parenchyma (*see under* parenchyma for other compound terms)

spontaneous generation the assumption, at one time prevalent, that living things sprang directly from dead things by an instantaneous mutation of inorganic matter into complex living forms

-spor- *comb. form* properly meaning "seed", but used today in compounds to mean either seed or spore, though -sperm- is more common for the former

sporangiophore [spôr·ănj"·ē·ō·fôr] an aerial hypha of a fungus which produces one or more sporangia

sporangium [spôr·ănj"·ē·əm] a hollow vessel in which "spores" are produced and thus applied both to the spore-bearing organ of lower plants and to the gamete producing organs of angiosperms. The following derivative terms are defined in alphabetic position: (*see also* megasporangium, microsporangium)

spore [spôr] used both in the sense of a resting stage or, more rarely, for any cell, or group of cells, capable of producing a new organism. The meanings of seed, sperm and spore are very confused. The following derivative terms are defined in alphabetic position:

AECIDIOSPORE	MICROSPORE
ANISOSPORE	OOSPORE
APLANOSPORE	PROTOSPORE
BASIDOSPORE	RESTING SPORE
BLASTOSPORE	SUMMER SPORE
CARPOSPORE	TELEUTOSPORE
ENDOGENOUS SPORE	TETRASPORE
ENDOSPORE	UREDOSPORE
ISOSPORE	ZOOSPORE
MEGASPORE	ZYGOSPORE

spore layer a layer of spore mother cells in certain fungi

sporocarp [spôr'·ō·kärp"] literally, a "spore bearing fruit"; that is a spore producing body derived from a zygote

sporocyst [spŏr"·ō·sĭst'] either any cyst containing spores, particularly in sporozooan protozoa *or* a stage in the life history of trematod worms (*see* next entry) or a cell from which spores are produced asexually (*see also* sporocarp)

sporocyst larva a hollow cyst developed from the miracidium in the intermediate host of a trematode. Germinal cells in the sporocysts produce either further sporocysts or redia

sporocyte = spore mother cell (*see also* megasporocyte, microsporocyte)

sporogamy [*angl.* spŏr·ŏg"·ə·mē, *orig.* spôr'·ō·

găm"·ē] the production of spores by an organism derived from a zygote

sporogenesis [spôr'·ō·jĕn"·əs·ĭs, spŏr·ō·jĕn"·əs·ĭs] the sum process of the production of spores or seeds

sporogonium [spŏr'·ō·gōn"·ē·əm, spŏr·ō·gōn"·ē·əm] the sporocarp of a moss

sporogony [*angl.* spŏr·ŏg"·ən·ē, *orig.* spôr'·ō·gōn"·ē] the process of multiple fission, following sexual fusion of gametes, and when the product is spores

sporont [spôr"·ŏnt] an individual that gives rise to spores

sporophore [spŏr"·ō·fôr', spŏr"·ō·fôr'] used in the sense of sporophyte in Pteridophyta and Bryophytae, but is also applied to any part of a plant which bears seeds or spores, though more usually used for the latter (*see also* microsporophore)

sporophyll [spŏr'·ō·fĭl", spŏr'·ō·fĭl"] a leaf, or leaf like structure, bearing spores

sporophyte [spôr'·ō·fit', spŏr'·ō·fit] in plants showing alternation of generations that generation which produces asexually; or in higher plants, that generation which produces the seed

sporosac [spŏr'·ō·săk", spŏr'·ō·săk"] an area on the blastostyle of some hydroid coelenterates in which sex cells develop directly, without the production of a medusoid form

Sporotrichum [spôr'·ō·trĭk"·əm] a genus of Deuteromycetes normally saprophytic. *S. schenki* [shĕnk"·i] can, however, cause serious lesions in human lymph vessels

Sporozoa [spŏr'·ō·zō"·ə, spŏr'·ō·zō"·ə] a class of Protozoa distinguished by their universally parasitic habit, lack of locomotor organelles and their reproduction by encysted zygotes (spores). The genus *Monocystis* is the subject of a separate entry

sporozoite [spŏr'·ō·zō"·it] the motile product of multiple fission, either of zygotes or spores

spray zone that area of an ocean shore which is close to high tide mark but wetted occasionally by salt spray

Squalus [skwā"·ləs] a large genus of rather small selachian elasmobranchs, better called dogfish than shark. *S. acanthias* [ăk'·ăn·pē·əs] is the common "shark" of elementary biology courses

-squam- *comb. form* meaning "scale"

Squamata [skwə·mät"·ə, skwäm·ät"·ə] an order of lepidosaurian reptiles, easily distinguished from the other extant order (Rhyncocephalia) by the moveable quadrate bone. The Squamata contains about 95% of living reptiles, including all those known as lizards (Sauria) and snakes (Serpentes). Some place the genus *Amphisbaena* in a separate order the Amphibaenea

squamosal bone [skwəm·ōs"·əl] one of a pair of membrane bones at the posterior side of the skull, immediately below the parietal, and from which rises the zygomatic arch which fuses with the jugal

squamose suture [skwăm"·ōz] one in which the bones overlap

squamous epithelium [skwăm"·əs] an epithelium which consists of flattened polygonal cells spread over the surface

squash a preparation intended for microscopical examination prepared by squashing tissues between a coverslip and a slide. The technique is widely used to demonstrate chromosomes, usually stained with orcein

s.s. abbreviation for *sensu stricto* (q.v.)

stab culture a microbial culture produced by stabbing the inoculating needle into the solid medium; descriptive terms for stab cultures are mostly self-explanatory (e.g. arborescent, filiform) though villose and plumose differ only in the thickness of the lateral branches and echinulate is used for a rather thick filiform stab having areas of prominences coming from it

stable association an association in a state of equilibrium

-stachy- *comb. form* meaning "spike" in the sense of the seed head of a grass

staining the device of applying various dyes to specimens to be examined under the optical microscope in order to increase the visibility of, or contrast in, the specimen (*see* vital stains, nuclear stains, plasma stains)

stamen [stā″·měn] the microsporophyll of seed plants

stapes bone [stā″·pēz] the outermost of three ossicles which conduct vibration from the tympanum to the inner ear (cf. malleus, incus)

starch a complex polysaccharide, usually a mixture of d-amylose and amylopectin, that is the principal food storage material of plants (*see also* floridean starch)

-stas- *comb. form* meaning "to be without movement" (cf. stat-)

stasis [stā″·sĭs] stability or cessation of movement, growth or change. The following derivative terms are defined in alphabetic position:
EPISTASIS   HOMEOSTASIS
GENEPISTASIS   HYPOSTASIS

-stat- *comb. form* meaning "stand", usually in the sense of "stand still" or "stabilize" (cf. -stas-)

statoblast [stăt″·ō·blăst′, stāt″·ō·blăst] a chitinous covered gemmule found in sponges and a somewhat similar structure, usually provided with hooks for attachment, found in some freshwater Ectoprocta. This term is sometimes most misleadingly used for statocyst

statoconium [stăt′·ō·kōn″·ē·əm, stāt′·ō·kōn″·ē·əm] one of numerous small granules acting on the hairs of neurosensory cells in statocysts (cf. statolith)

statocyst a sense organ, found in many invertebrates, consisting of a more or less spherical, fluid-filled, cavity with sensory hairs on the bottom and a contained free body (statolith), or bodies (statoconia), the movement of which orient the animal with regard to gravity

statolith [stăt″·ō·lith′] a mineral mass acting on the hairs of the neurosensory cells in statocysts (cf. statoconium)

statoreceptor [stăt′·ō·rē·sĕpt·ər] a receptor for perceiving the direction of gravitational pull, as in the semicircular canals of chordates on the statocysts of many other forms

-staur- *comb. form* meaning "cross"

Stauromedusae [stôr′·ō·měd·yūz″·ə] an order of sessile scyphozoan Coelenterata developing directly from a scyphostoma and remaining attached by an adoral stalk. The genus *Haliclystus* is the subject of a separate entry

-stego- *comb. form* meaning "roof"

stegochordal centrum [stĕj′·ō·kôrd″·əl] one in which only the dorsal arch of the perichordal sheath becomes ossified

Stegomyia = Aedes

Stegosaurus [stĕg′·ō·sôr″·əs] a genus of extinct diapsid reptiles from the early Jurassic. They were about twenty feet long with enlarged hindquarters and a very small brain. The tail carried four long bony spikes

-stel- *comb. form* meaning "pillar" or "column"

stele [stēl] a unit consisting of the vascular system and any tissue enclosed within it, which forms a cylinder running through the stem of vascular plants. The following derivative terms are defined in alphabetic position:
HAPLOSTELE   PROTOSTELE
MONARCH STELE   SIPHONOSTELE
PERIPHERAL STELE

-stell- *comb. form* meaning "star"

stenobath [stĕn″·ō·băþ′] an organism restricted to a narrow range of depth

stenohaline [stĕn′·ō·hāl″·ĭn] properly used of organisms capable of supporting only slight variations in salt concentrations but also used of those capable of supporting only low concentrations

stenohygric [stĕn′·ō·hig″·rĭk] pertaining to an organism tolerating only a narrow humidity range (*see also* euryhygric)

Steno's duct the duct of the parotid gland

Stenostomum [*angl.* stĕn·ŏst″·əm·əm, *orig.* stĕn′·ō·stōm″·əm] a large genus of rhabdocoel turbellarians with well developed anterior ciliated pits. Most are very small but *S. speciosum* [spēsh′·ē·ōs·əm] may exceed 2 mm in length. Some, such as *S. agile* [ăj″·ĭl·ē] have pigmented intestines. This form is also often found in long (up to 4 mm) chains of partially separated individuals

stenotherm [stĕn″·ō·þûrm′] an organism capable of tolerating only a narrow temperature range

stenozonal [stĕn′·ō·zōn″·əl] pertaining to an organism living at a restricted altitude or depth

Stentor [stĕn″·tôr, stĕn″·tər] a genus of trumpet-shaped spirotrichid ciliate protozoans. The bright blue *S. coeruleus* [sĕr·ōōl″·ē·əs] is one of the most beautiful protozoans in existence. *Stentor* is also remarkable for its very large elongate myofibrils

-stephanos- *comb. form* meaning "wreath"

steppe [stĕp] grasslands which lie in the rain-shadows of mountains in temperate latitudes with desert on the drier side and forest on the wetter side

stereoblastula [stĕr′·ē·ō·blăst″·yōō·lə] a blastula which fails to develop a central cavity (= morula in mammals)

stereogastrula larva = planula larva

stereospondylous vertebra [stĕr′·ē·ō·spŏnd″·əl·es] one consisting of a single body or, essentially, of the intercentrum

sterile 1 (*see also* sterile 2) in the sense of lacking the ability to breed or interbreed. The following derivative terms are defined in alphabetic position:
CHROMOSOMAL S.   HYBRID STERILITY
CYTOPLASMIC S.   SELF-STERILITY
GENIC STERILITY

sterile 2 (*see also* sterile 1) said of anything which is devoid of viable organisms

sternal [stûr″·nəl] pertaining to the sternum

sternite [stûr″·nit] the lower, or ventral, plate of the exoskeleton of a segmented animal, particularly an arthropod. The term is sometimes considered synonymous with sternum (q.v.), though most writers confine the latter term to chordates

sternum [stûrn″·əm] a cartilaginous or bony

structure of the vertebrate skeleton with which usually the ribs, but sometimes the pectoral girdle, are articulated in the ventral median plane

-stero- *comb. form* meaning "solid"

-sticho- *comb. form* meaning a "row"

stigma 1 [stĭg″·mə] almost any hole or spot. Usually, without modification the upper end of the style modified for the reception of pollen

stigma 2 (*see also* stigma 1, 3, 4) the eye spot of Protozoa

stigma 3 (*see also* stigma 1, 2, 4) the spiracle of arthropods

stigma 4 (*see also* stigma 1–3) the apertures that, in the larvacean Urochordates, open directly from the branchial net to the exterior)

stigmata [stĭg″·mə·tə] plural of stigma

stigmatic fluid [stĭg·măt″·ĭk] the sticky secretion on a stigma to which pollen grains adhere

stilt root large adventitious roots that support the trunk of a tree, such as the mangrove

sting an ovipositor, modified for the ejection of venom, found in some Hymenoptera. Also applied to other injectors of venom such as the calcareous projection at the base of a sting ray's, or the end of a scorpion's, tail

stinging cell = nematocyst

stink gland any gland producing a mephitic secretion, but more usually applied to such structures in insects than in vertebrates

stipe [stīp] the term can be applied to any stalk-like object but is most usually applied to the stalk of a capped basidiomycete

-stipul- *comb. form* meaning a "branch" or "long rod"

stipule [stĭp′·yōōl] a leaf-like appendage at the base of a leaf. They are sometimes modified as tendrils or spines, or may serve in place of leaves

stirn organ [stûrn] a small dermal body found in certain Salientia probably equivalent to the pineal

Stizostedium [stĭz′·ō·stĕd″·ē·əm] a genus of teleost fish, closely allied to *Perca*. S. *vitreum* [vĭt·rē″·em] is the wall-eye, often miscalled the wall-eyed pike or pickerel (*see* Lucius)

-stol- *comb. form* meaning "shoot"

stolon [stō′·lən] a lateral branch, frequently reproductive in function, produced either from the base of a stem of a plant or from the base of many sessile animals. In both cases it runs parallel to the substrate

-stom- *comb. form* meaning "mouth"

stoma [stōm·ə] an organ consisting of two guard cells, with a pore between them, on the surface of the leaf or stem of a higher plant. The expansion and contraction of the guard cells controls the flow of gas and vapor

stomach in vertebrates, that portion of the alimentary canal into which the esophagus leads and also any analogous structure in other animals. The following derivative terms are defined in alphabetic position:
CARDIAC STOMACH   PYLORIC STOMACH

stomata [stōm″·ə·tə] plural of stoma

-stome *adjectival suffix* meaning pertaining to mouths or to any other aperture or cavity which might conceivably be called a mouth. The following derivative terms using this suffix are defined in alphabetic position:
COELOMOSTOME   HYPOSTOME   PERISTOME
CYTOSTOME   NEPHROSTOME

stomium [stō″·mē·əm] any aperture that can conceivably be called a mouth. Specifically the opening through which pollen escapes from an anther. In most of its other meanings it is usually replaced by "stome"

stomochord [stŏm″·ō·kôrd] a diverticulum of the buccal tube which projects into the anterior cavity of hemichordates

stomodeum [stō′·mō·dē″·əm] an anterior opening into the archenteron or, in anthozoan coelenterates, a short inturned tube projecting into the coelenteron

stone canal the tube connecting the water-vascular system of echinoderms to the hydropore

-stracum- *comb. form* meaning "shell" (*see* periostracum)

-strat- *comb. form* originally meaning a "paved highway" but now used almost entirely in the sense of "layer"

stratified epithelium epithelium consisting of several layers of the same type of cell

streak culture a microbial culture produced by streaking the surface of a solid culture medium without breaking the surface; most of the descriptive terms are self-explanatory (e.g. filiform, rhizoid) but "effuse" is used for a streak which is diffuse around the edges

-streph- *comb. form* meaning "twisted" (cf. -strept-)

Strepsiptera [strĕps·ĭp″·tə·rə] a small order of minute insects, many parasitic, distinguished by the fact that the hind wings are functional but the fore wings are reduced to slender club-like appendages. At one time regarded as a family (Stylopidae) of Coleoptera

-strept- *comb. form* meaning "twisted"

Streptococcus [strĕp′·tō·kŏk″·əs] a genus of lactobacillaceous schizomycetes which form long chains. S. *pyogenes* [pī·ŏj″·ĕn·ēz] is a dangerous pathogen of man but S. *viridans* [vĭr″·ĭd·ănz′] lives as a commensal in the human mouth and throat as does S. *faecalis* [fē·kăl″·ĭs] in the intestine. S. *lactis* [lăk″·tĭs] is always found in milk which is allowed to curdle, through the formation of lactic acid, in the initial stages of making many cheeses

stretch receptor a mechanoreceptor stimulated by the stretching of peripheral nerve endings and therefore an organ of touch

stria [strī·ə] literally, a furrow; but, applied particularly to markings on the frustules of diatoms which have the appearance of lines until they are resolved into dots

striated muscle muscle, the fibers of which appear striated and which is associated with voluntary (e.g. limb) movement

-strobil- *comb. form* meaning a "cone" in the sense of a pine cone

strobila [strō′·bĭl·ə] an organism, or stage in the life history of an organism, which buds off successive parts from one end. In this meaning a tapeworm, as well as a larval stage of a jellyfish, may be called a strobila

strobila larva a scyphistoma larva of a scypozoan which has begun to split off larvas from its upper end

strobilocercus larva [strō·bĭl·ō·sûr″·kŭs] a terminal larval stage of a cestode in which the posterior portion of the cysticercus begins strobilization

strobilus a pine cone or anything resembling it; that is a terminal structure bearing sporangia. *See* megastrobilus, microstrobilus

-strom- *comb. form* meaning "mattress"

stroma [strŏm·ə] a spongelike framework usually produced by reticular cells. A section of a stroma has the appearance of a net

stromacenter [strŏm"·ə·sĕn'·tər] a proteinaceous aggregate of fibrils in the chloroplast of higher plants

-strot- *comb. form* meaning "to spread"

-strote *adjectival suffix* indicating an organism defined by its method of distribution. Derivative terms are not defined since the meanings are obvious from the roots (e.g. spermostrote, dispersed as seeds; zoostrote, dispersed by animals)

structural color colors produced by reflexion from surface structures, also called interference colors or schemochromes (*see also* Tyndall color)

-styl- *comb. form* meaning "column"

Stylaria [sti·lãr"·ē·ə] a genus of freshwater oligochaete worms easily distinguished from *Nais* by the long, proboscis-like, prostomium. *S. lacustris* [lăk·ŭs"·trïs] is very common and is occasionally found in "chains" since it reproduces by transverse fission

style literally a "column" but used in biology in a wide variety of senses

style 1 (*see also* style 2–6) the part of a carpel lying above the ovary. Ovary and style capped by the stigma make up the pistil

style 2 (*see also* style 1, 3–6) a cusp on a mammalian tooth

style 3 (*see also* style 1, 2, 4–6) a monaxonic sponge spicule

style 4 (*see also* style 1–3, 5, 6) a rod-like or column-like organ or skeletal element in a chordate. In this sense the following derivative terms are defined in alphabetic position:
ENDOSTYLE   PYGOSTYLE   UROSTYLE

style 5 (*see also* style 1–4, 6) a projection from, or rodlike structure in, an invertebrate (*see* crystalline style, odontostyle)

style 6 (*see also* style 1–5) a rodlike organelle. In this sense *see* parastyle

stylet diminutive of style in any of its meanings

stylohyal bone [sti'·lō·hī"·əl] that part of the ossified hyoid arch which, in mammals, is not fused to the tympanic bulla (*see also* hyal bone, tympanohyal bone)

styloid cartilage [stïl"·oid] the remnants of the hyoid arch in the visceral skeleton of Marsippobranchii

Stylonychia [stïl'·ō·nïk"·ē·ə] a genus of spirotrichid ciliate protozoans with two meganuclei which, however, fuse before division. There are long caudal setae and a number of ventral compound cilia

Stylotella [sti'·lō·tĕl"·ə] a genus of marine sponges of soft texture, but containing numerous megascleres, the bodies of which have finger like projections. *S. heliophila* [hēl'·ē·ō·fil"·ə] is an orange colored species common on the Atlantic coast

-sub- *comb. form* meaning "below", though used in biology compounds to mean "almost" as "subacute", meaning "not quite pointed"; most compounds of this type are not given in the present dictionary, since the meaning is self-evident

sub-imago a stage in some insects, particularly Ephemoptera, intercalated between the pupa and the perfect imago (cf. dun)

sub-phylum a taxon intermediate between phylum and class

subapical initial the initial from which the internal

tissue of some leaf axes is produced (*see also* apical initial)

subassociation an assemblage of organisms which do not, according to the particular usage of the individual writer, agree completely with an association

subcephalic pocket [sŭb'·sĕf·ăl"·ïk] the area underlying the head in an embryo

subclass a taxon intermediate between class and order

subclavian artery [sŭb·klāv"·ē·ən] the most anterior of the major arteries arising from the dorsal aorta, and which supplies blood to the forelimbs

subclavian vein brings blood from the forelimbs and enters the jugular vein just anterior to the junction of the posterior cardinal with the cuverian

subclimax either a climax which has not reached a stable level by reason of factors other than climatic or a stage which precedes a true climax but which persists for an unusually long time

subdominant an organism which becomes dominant in areas not controlled by the regular dominant

subdural space [sŭb·dyōōr"·əl] the space between the pia mater and the dura spinalis (*see also* peridural space)

suberin [sū·ber"·in] a fatty substance, the presence of which in cork cells, gives the peculiar consistency of the tissue cork

subfamily a division of a family containing several distinct but allied genera. In both botany and zoology the names of subfamilies usually terminate in -inae or -inaceae. For example, the dipteran family Culcidae is by some divided into the Culicinae, containing the blood sucking mosquitoes and the Corethrinae, the mouth parts of which are not adapted to blood sucking

subgenus a taxon immediately below genus, the members of which are thought to be associated by genetic rather than geographical factors

subinfluent an influent that is seasonal or transitory and therefore exercises very little effect

-subit- *comb. form* meaning "sudden"

subitaneous egg [sŭb'·ït·ān"·ē·əs] a thin-shelled egg, destined to hatch rapidly in contrast with a thick-shelled egg, intended to remain dormant for some time (cf. dormant egg)

sublingua [sŭb·lïng"·gwə] a horny pointed, or serrated, plate found beneath the tongue of, and peculiar to, the Lemuroidea

sublittoral pertaining to areas adjacent to the littoral zone. In freshwaters the sublittoral zone commences where rooted vegetation ceases

submarginal initial one of a line of initials lying immediately under the marginal initials, giving rise to the internal tissues of the lamina of the blade of a leaf

subnymph an intercalated stage between a nymph and a pupa

subocular arch one of a pair of arch-shaped cartilages, running downward and forwards from the chondrocranium of Cyclostomes

subopercular bone one of a pair of two large bones, lying immediately under the opercular bone (*see also* opercular bone, interopercular bone, preopercular bone)

suborder a taxon used to divide an order. Thus the insect order Hymenoptera is divided into the wasp-waisted Apocrita (ants, wasps and bees) and the thick-bodied Symphyta (sawflies)

subordinate association an association which is either progressive or regressive

subpleurodont a pleurodont in which the teeth are replaced by intercalation (*see also* eupleurodont)

subsegment a subdivided arthropod segment distinguished from a true segment by the lack of specific musculature

subsere [sŭb″·sēr] either a secondary sere or a partially developed climax

subsociation a portion of an association which is distinguished by the presence of a subdominant which is, however, itself clearly under the influence of the dominant

subsocies this term is to associes as subsociation is to association (*see also* socies, associes, consocies, isosocies)

substitute association a secondary formation which has replaced a stable association

subsuccession a sere beginning on a rock surface and ending in a mat-growth

subtemperate intermediate between temperate and extreme, properly used of the colder regions between the temperate and Arctic zones but occasionally also used of the regions between temperate and tropical zones

-subul- *comb. form* meaning "awl"

subumbrella the concave surface of the umbrella of a medusa

succession the event of one thing following another chronologically. Specifically, in ecology, the sequence of changes in the population of a given area. The following derivative terms are defined in alphabetic position:
AUTOGENIC SUCCESSION
ECOLOGICAL SUCCESSION
HALARCH SUCCESSION
HYDROSTATIC SUCCESSION
HYDROTROPIC SUCCESSION
SUBSUCCESSION

Suctoria [sŭk·tôr″·ē·ə] a group of Protozoa possessing tentacles and reproducing by ciliate young. They are variously regarded as a class, or as an order of the class Ciliata. The genus *Ephelota* is the subject of a separate entry

-sulca- *comb. form* meaning a "groove"

sulfatase [sŭlf″·ə·tāz] any of several enzymes that catalyze the hydrolysis of organic sulfates with the liberation of sulphuric acid

sulphur bacteria any bacteria driving its energy from sulphur or sulphur compounds

super literally "above", but also used in biology in the sense of "greater than" or "superior to" (cf. supra)

supersex an individual with an abnormal ration of sex chromosomes

superclass a taxon which is with great difficulty distinguished from subphylum

superfamily a biological taxon usually erected when a family contains so many genera that it must be divided into other families. The name of the original family is retained for the superfamily

superficial cleavage that type of cleavage of a telolecithal egg in which cleavage is gradually spread over the whole surface

superficial segmentation a form of metameric segmentation which involves only the outer layers of an organism

supermale the male genetic equivalent of a metafemale

superior ovary one beneath which all the floral parts are inserted

supernumary segment one which is intercolated between segments

superorder a taxon combining several orders of a class

superorganism an aggregate of organisms, or organisms and their environment, which may be studied as though they were a single organism. Oceans and forests are examples

suppressor gene a gene that suppresses a character

-supra- *comb. form* from super

supraangular bone one of a pair of dermal bones lying at the posterior end of the lower jaw, between the coracoid and angular bones in many vertebrates other than mammals (*see also* angular bone, multangular bone)

supracleithrum bone one of a pair of bony elements in the pectoral girdle of Crossopterygian fish, lying between the posttemporal and the postcleithrum (*see also* cleithrum bone, postcleithrum bone)

supracoracoid cartilage a cartilaginous extension of the coracoid bone (*see also* scapulocoracoid cartilage)

supraneuston [syū′·prə·nyōōs″·tŏn] organisms living on or in the surface film of water

supraoccipital bone one of a pair of chondral bones at the posterior end of the skull, lying immediately above the foramen magnum, adjacent to the postparietal, petrosal, and parietal bones (*see also* occipital bone, basioccipital bone, exoccipital bone)

suprarenal body cells corresponding to the cortex of the adrenal body in lower vertebrates

supratemporal bone one of a pair of membrane bones lying at the posterior junction of squamosal and parietal in many vertebrates other than mammals (*see also* temporal bone, intertemporal bone, posttemporal bone)

supratemporotabular bone [syū″·prə·tĕm′·pər·ō·tăb″·yōō·lə] one of a pair of membrane bones in the dermo-cranium of actinopterygian fish, corresponding to the supratemporal and tabular bones of Crossopterygian fish

sustentacular cell [sŭs′·tĕn·tăk·yū·lə] a cell in a taste bud having the form of a segmental slice of a thick walled hollow sphere

suture 1 [sōō″·chə, syū·tyōōr] (*see also* suture 2) in insects, the line of junction of sclerites

suture 2 (*see also* suture 1) a tight fibrous joint between bones. In this sense the following derivative terms are defined in alphabetic position:
BASAL SUTURE   SQUAMOSE SUTURE
FALSE SUTURE   TRUE SUTURE
FRONTAL SUTURE

swarm cells a term usually applied to the flagellated stage which results from the germination of the spores Myxomycetales

sweat gland any gland in the skin of mammals which is not a sebaceous gland (*see also* apocrine sweat gland and eccrine sweat gland)

swim bladder a thin walled, gas-filled bladder lying along the dorsal wall of the coelom in bony fish. It is developed as an outgrowth of the alimentary canal, and may retain, in some fish, an attachment either to the esophagus or to the stomach

Sycon [sī″·kŏn] a genus of calcareous sponges closely allied to *Grantia* but lacking flagellated chambers in the radial tubes

syconoid [sī″·kŏn·oid] a level of sponge structure in which the choanocytes are in simple pouch-like outfoldings from the central chamber (cf. asconoid, leuconoid, sylleibid)

sylleibid [sĭl″·ē·ə·bĭd] a grade of sponge structure intermediate between the syconoid and the leuconoid, in which each radial canal is subdivided into elongated, flagellated chambers grouped around a common excurrent channel (cf. asconoid, leuconoid, syconoid)

-sylv- *comb. form* meaning a "wood"

Sylvilagus [*angl.* sĭl·vĭl″·ə·gəs, *orig.* sĭl′·vē·lāg″·əs] a genus of lagomorph mammals containing the "cotton tails" or nonburrowing rabbits. There are many species all confined to the New World

Sylvius' fissure the deep, vertical, lateral fissure marking the posterior margin of the temporal lobe of the cerebral hemisphere

-sym- = **-syn-**

symbiosis [sĭm′·bē·ōs″·ĭs] the condition of two or more different organisms living together in close association. At one time, and occasionally today, the term is restricted to an association supposed to be to the mutual advantage of both organisms (*see also* antagonistic symbiosis, contingent symbiosis, endosymbiosis, conjunctive symbiosis, mutualistic symbiosis, parasymbiosis, social symbiosis, symphily)

symbiotic sapprophyte [sĭm′·bē·ŏt″·ĭk] a higher plant associated with a lower as in mycorhizae

symmetry the arrangement of similar parts on each side of a common axis or radially round a point. The following derivative terms are defined in alphabetic position:

ASYMMETRY
BILATERAL SYMMETRY
BIRADIAL SYMMETRY
MONAXIAL SYMMETRY
RADIAL SYMMETRY
RADIOBILATERAL S.
SPHERICAL SYMMETRY
TETRAMEROUS SYMMETRY
TRIAXIAL SYMMETRY

sympaedium [sĭm·pēd″·ē·əm] an assemblage of young animals which play together

sympathetic ganglion a ganglion which receives its primary impulses from the lateral horns of the spinal chord

sympathetic nervous system a series of ganglia, the principal of which are the chain ganglia and the prevertebral ganglia which are a part of the autonomic nervous system. The system contains mostly adrenergic fibers and acts principally on secretory structure and involuntary muscles (*see also* parasympathetic nervous system)

sympathetic system = sympathetic nervous system

sympathetic trunk the commissure connecting the chain ganglia of the sympathetic nervous system

sympatry [sĭm·păt″·rē] the condition of two populations occupying the same territory

symphile [sĭm″·fil] a pet of an ant or termite colony; the words guest or inquiline, frequently used, do not adequately describe the relationship

Symphyla [sĭm·fil″·ə] a small class of Arthropoda with a centipede-like body of fourteen segments of which the first twelve bear legs. They are easily distinguished from the true centipedes by their white translucent bodies. The genus *Scutigerella* is the subject of a separate entry

symphysis [sĭm″·fə·sĭs] literally "a growth together". Used generally for a joint in which a bone capped with cartilage is held together with dense fibrous connective tissue

Symphyta [sĭm·fit″·ə] a suborder of hymenopteran insects containing those forms such as sawflies and horntails in which the abdomen is broadly joined to the thorax and therefore lacks, the typical wasp-waisted appearance of the Apocrita

sympletic bone [sĭm·plĕt·ĭk] one of a pair of membrane bones derived from the hyoid arch of actinopterygian fish, lying immediately ventral to the hyomandibular bone

sympodial 1 [sĭm·pōd″·ē·əl] (*see also* sympodial 2) said of a method of branching in plant stems in which the main axis ceases growth and auxilliary buds near the tip assume the major role in shoot development

sympodial 2 (*see also* sympodial 1) said of the type of growth in a coelenterate colony in which temporary, terminal, hydranths are produced alternately on each side

-syn- *comb. form* meaning "with" or "combined with"; very frequently mis-transliterated as -sym-

synandria [sĭn·ăn″·drē·ə] a group of males living together (cf. syngynia)

synapse [sĭn″·ăps, sĭ·năps″] the surface of contact between nerve endings derived from separate cells

Synapsida [sĭn·ăps″·id·ə] a subclass of reptiles distinguished by having a single pair of lateral temporal openings in the skull. All are extinct, occurring from the Devonian to the Permian. They are the most mammal-like of the reptiles. The genera *Cynognathus*, *Dimetrodon*, *Lycaenops* and *Tyrannosaurus* are the subjects of separate entries

synapsis [sĭn·ăps″·ĭs] the pairing of homologous chromosomes in meiosis. The following derivative terms are defined in alphabetic position:

ALLOSYNAPSIS
AUTOSYNAPSIS
DESYNAPSIS
HOMOSYNAPSIS
PROCENTRIC SYNAPSIS
PROTERMINAL SYNAPSIS
SOMATIC SYNAPSIS

synarthrosis [sĭn′·är·thrō″·sĭs] an apparent joint, but one which is incapable of movement

synchondrosis [sĭn′·kŏn·drōs″·ĭs] a cartilaginous connection between two bones (*see also* chondrosis)

synchoropaedium [sĭn′·kôr·ō·pēd″·ē·əm] an aggregate of young animals of approximately the same age but different parentage

synchronogamy [*angl.* sĭn·krən·ŏg″·əm·ē, *orig.* sĭn′·krŏn·ō·găm″·ē] the condition of having the male and female sex organs mature at the same time

syncytium [sĭn·sĭt″·ē·əm] multi-nucleate animal tissues in which cell boundaries are not apparent (cf. coenocyte)

syndesmorchorial placenta [sĭn′·dĕz·mō·kôr″·ē·əl] one which the maternal uterine epithelium is eroded away, so that the embryonic membranes are in contact with the connective tissue of the uterus

syndiploid [sĭn·dĭp″·loid] a diploid produced through failure of chromosomes to separate in meiosis

Synechococcus [sĭn·ĕk′·ō·kŏk″·əs] a genus of unicellular, free-floating cyanophytes considered to be among the most primitive blue-green algae

synecology [sĭn′·ē·kŏl″·ō·jē] the relation between an association and its environment, or the ecology of communities (see also dynamic synecology, geographic synecology, morphological synecology)

synergid [sĭn″·ə′·jĭd, sĭn·ûrj″·id] one of the two nuclei of the upper end of the plant embryo sac

-synergo- comb. form meaning "assistant" or "assist"

syngamete = zygote

syngamy [sĭn″·gə·mē, sĭn·găm″·ē] the fusion of two gametes

syngen [sĭn″·jĕn″] a genetically isolated variety of ciliate protozoan which cannot interbreed fruitfully with another syngen of the same species

syngynia [sĭn·jĭn″·ē·ə] a group of females living together (cf. synandria)

synhaploid [sĭn·hăp″·loid] the condition arising from the fusion of two or more haploid nuclei (see also double haploid)

synhesmia [sĭn·hĕz″·mē·ə] a group of organisms gathered together in consequence of a reproductive drive (see also androsynhesmia, gynosynhesmia)

synkaryon [sĭn·kâr″·ē·ən] a nucleus formed by the fusion of two others, particularly the early fusion nucleus in a zygote

synovial fluid [sĭn·ōv″·ē·əl] the fluid of the synovial joint, consisting of tissue fluid enriched with mucin

synovial joint a joint between two bones which possesses a cavity and is specialized to permit more or less free movement

synovial membrane the inner layer of the joint capsule of a synovial joint

synsacrum [sĭn·sāk″·rəm] that part of the pelvic girdle which is derived from modified vertebrae (see also sacrum)

synthase [sĭn″·þāz] a term used to distinguish compound building lyase enzymes from compound-building ligase enzymes which are properly called synthetases

synthesis [sĭn″·thə·sĭs] the production of substances from other, usually simpler, compounds (see also biosynthesis, photosynthesis)

synthetase [sĭn″·thə·tāz′] a term used to distinguish compound-building ligase enzymes from compound-building lyase enzymes which are properly called synthases

synusium [sĭn·yōōz″·ē·əm] a group of organisms having the same ecological requirements and reacting in much the same way to their environment, but not taxonomically related

-syst- comb. form meaning "contract"

systemic circulation the circulation of blood through those parts of the body not concerned with respiration

systole [sĭst″·əl·ē] the contraction of a hollow body such as a contractile vacuole or heart

syzygy [sĭz″·ĭj·e] the apposition of two bodies particularly the apposition, but not union, of two sporozoan protozoa; or the apposition, and fusion, of two joints in the arm of a crinoid echinoderm

T

tabular bone one of a pair of membrane bones in the dermal cranium of Crossopterygian fish. They lie immediately on each side of the postparietals, immediately behind the supratemporal, and immediately in front of the extrascapulus

-tachy- *comb. form* meaning "quick"

tachyblastic egg [tăk'·ē·blăst"·ĭk] one which hatches rapidly and is not adapted to withstand adverse conditions

tachygenesis [tăk'·ē·jĕn"·əs·ĭs] abbreviated development, as when one or more larval stages are coalesced

Tachyglossidae [tăk'·ē·glŏs"·ĭd·ē] one of the two extant genera of monotremata containing the "spiny anteaters", a name descriptive of their appearance and habits. The type genus *Tachyglossus* differs from the only other genus (*Zaglossus*) in being adapted to desert, rather than forested, habitats

tachytelic [tăk'·ē·tēl"·ĭk] evolving at what appears to be a rapid rate

-tact- *comb. form* properly meaning "to touch" but frequently used as an adjectival form from -tax- (q.v.)

tactile corpuscle an end organ consisting of a single tactile cell in a terminal nerve cup

tadpole the free-swimming larva of Salientia (frogs and toads)

tadpole larva the larva produced by urochordates. It is essentially a miniature planktonic urochordate propelled by a tail similar, in transverse section, to that of a tadpole. The branchial basket, however, drains directly to the exterior and not into an atrial cavity

-taen- *comb. form* meaning a "band" or "ribbon"

Taenia [tēn"·ē·ə] a genus of large tapeworms. *T. solium* [sōl'·ē·əm], of which the cysticercus occurs in pigs, is a not uncommon human parasite. *T. saginata* [săj'·ĭn·ā"·tə], the cysticercus of which occurs in beef, is less common

taiga [ti"·gə] a zone of intermittent scattered trees between the tundra and the forest-tundra

tail fold that part of a posterior end of an embryo which rises above the blastoderm in the early development of a telolecithal egg

talon 1 (*see also* talon 2) a heavy recurved claw of a predator particularly of a bird

talon 2 (*see also* talon 1) an upper molar

tangoreceptor [tăng'·gō·rē·sĕp"·tər] a receptor perceiving contact stimuli

tannase [tăn"·āz] an enzyme that catalyzes the hydrolysis of ester links in tannins

tap-root the straight, long root, when it occurs, which derives directly from the continued growth of the radicle

-tapet- *comb. form* meaning a carpet

-taphr- *comb. form* meaning "ditch"

Taphrina [tăf·rĭn·ə] a genus of taphrinale Ascomycetes, that are leaf parasites of higher plants. *T. deformans* [dē·fôrm"·ănz] causes peach leaf curl

Taphrinales [tăf"·rĭn·āl·ēz] an order of ascomycete fungi. They are closely allied to the Endomycetales (yeasts) in that ascospores bud as do yeast cells but differ from yeast in that they produce hyphae in the plant tissues that they infest

Tapirus [tə·pir"·əs] a genus of perissodactyl mammals closely related to the Rhinoceros. These heavily built animals with a protruding upper lip were once widely distributed but are now represented by one Asiatic (*T. indicus* [ĭn·dĭk"·əs]), one Central American (*T. bairdii* [bârd"·ē·ī]) and one South American (*T. terrestris* [tĕr·rĕst"·rĭs]) species

Tarantula [tə·răn"·tyū·lə] a genus of Uropygi (*not* Aranaea) with the characteristics of the order. *T. whitei* [hwit"·ē·ī] is the western and *T. fuscimana* [fŭsk'·ē·män"·ə] the eastern United States species. The English word tarantula is applied also to numerous large spiders, particularly in U.S. those of the genus Eurypelma

Tardigrada [tär'·dē·grād"·ə] a small group of apparent arthropods commonly called the water-bears. They have variously been regarded as a separate phylum, a class of the Arthropoda, or an order of the Arachnida. They are small (less than 1 mm), unsegmented, cylindrical animals with four pairs of clawed legs. The genus *Echiniscus* is the subject of a separate entry

tarsal [tär"·səl] pertaining to the terminal joints of a limb or to the connective tissue in the eyelid

tarsal bone one of the bones in the vertebrate leg which lies between the metatarsal bones and the tibia. Commonly called ankle bones (*see also* metatarsal bone)

tarsal gland   a modified sweat gland on the edge of the eyelid

tarsal plate   a dense connective tissue plate, supporting the edge of the eyelid

tarsale bones   [tär·săl"·ē]   a group name for those tarsal bones (cuneiform and cuboid) which lie at the base of the metatarsals

Tarsius   [tär"·sē·əs]   a genus of lorisoid primates found in the East Indies. They have large external ears and enormous eyes. The English name tarsier is commonly used

taste bud   one of numerous types of sensory end organ in the tongue and palate

taste pore   a passageway through the epithelium reaching to a taste bud

taste receptor   a chemoreceptor activated by solutions

tautonym   [tôrt'·ō·nĭm"]   a binomial in which the genetic and specific names are the same as in *Gorilla gorilla*

Tawara's node   the heart node lying at the junction of the atrium and ventricle

-tax-   *comb. form* properly meaning "to arrange", but which has come to mean "to arrange, or move in the direction of", hence "to be attracted to" or "move toward"

taxis 1   [tăk"·sĭs]   (*see also* taxis 2) a movement of an animal oriented with respect to a source of stimulation. The adjective "negative" indicates movement away from, and the adjective "positive" indicates movement toward. The following derivative terms are defined in alphabetic position:

CHEMOTAXIS	PHOTOTAXIS
CYTOTAXIS	PHOTOHOROTAXIS
GALVANOTAXIS	POSITIVE CYTOTAXIS
GEOTAXIS	RHEOTAXIS
HETEROTAXIS	TELOTAXIS
KLINOTAXIS	THERMOTAXIS
MENOTAXIS	THIGMOTAXIS
NEGATIVE CYTOTAXIS	TOPOTAXIS
NEUROBIOTAXIS	TROPOTAXIS
PHAROTAXIS	

taxis 2   (*see also* taxis 1) in the sense of arrangement rather than movement (*see* cytotaxis, phyllotaxis)

Taxodiaceae   [tăks'·ōd·ē·ā·sē]   a family of coniferous Spermatophyta. It is distinguished by the flat, or peltate, cone scales each producing two to nine seeds. The genera *Metasequoia* and *Sequoia* are the subjects of separate entries

taxon   [tăks'·ŏn]   a group of organisms thought to be genetically related. The closeness of the relationship depends on the rank of the taxon. The smallest recognized taxon is the genus (e.g. *Culex*), a group of closely related species. Genera are grouped into families (e.g. *Culicidae*), families into orders (e.g. Diptera), orders into classes (e.g. Insecta) and classes into phyla (e.g. Arthropoda). There is little doubt as to the relatedness of individuals in most phyla, such as Chordata or Arthropoda, but widely varying views are held as to the relationship of phyla to each other

taxonomic character   [tăks'·ō·nŏm"·ĭk]   one on which classification (i.e. assignment to a taxon) is based

taxonomy   [tăks·ŏn"·əm·ē]   the science of arranging organisms in logical and natural groups, in such a

manner as to cast light on their evolution and affinities (*see also* cytotaxonomy)

Taxus   [tăks"·əs]   the genus of Coniferales containing the yews. They are distinguished by their spirally arranged leaves and by the fruit with a fleshy aril

tectin   [těk"·tĭn]   a glycoprotein, often referred to as pseudochitin. It is a frequent component of the skeleton of protozoa

tectum   [těk"·təm]   literally a roof, particularly parts of the "roof" of the brain

Tegiticula   [těg'·ē·tĭk·yū·lə]   a genus of moths containing the yucca moth *T. alba* [ăl"·bə] (*see Yucca*)

-tel-   *comb. form* meaning "far off"

-tele-   *comb. form* meaning "perfection" and by extension, "end" or termination. Frequently confused with -tel-

telegony   [*angl.* těl·ěg"·ən·ē, *orig.* těl'·ē·gōn"·ē]   the postulate, at one time widely accepted, that a female could transfer characters derived from one mate to the offspring of a subsequent mate

telencephalon   [těl'·ěn·sěf"·ə·lən]   the anterior portion of the forebrain, or the anterior division of the embryonic prosencephalon. It gives rise to the cerebral hemispheres

teleology   [těl'·ē·ŏl"·ō·gē]   the suggestion that something is shaped by a purpose or directed towards an end. It is an inexpiable heresy in the United States but it is viewed somewhat more leniently by biologists of other countries

Teleostei   [těl'·ē·ŏst"·ē·ī]   a superorder of actinopterygian gnathostomatous chordates containing the true boney fish. The group which contains the great majority of boney fishes differs from the Protospondyli in lacking ganoid scales and from the Chondrostei in possessing a completely ossified skeleton. The genera *Alosa, Amblyopsis, Carassus, Clupea, Coregonus, Cyprinus, Esox, Fundulus, Gadus, Gambusia, Hippoglossus, Lophius, Micropterus, Mola, Onchorrhynchus, Perca, Photocoryns, Remora, Salmo, Scomber, Scolea,* and *Stizostedium* are the subjects of separate entries

telepod   [těl'·ō·pŏd"]   a modified leg serving as an intromittent organ in Diplopoda

-teleut-   *comb. form* meaning "end"

teleutospore   [těl·ōōt"·ō·spôr']   a winter-resistant spore produced by the bikaryotic mycelium of a urediniale basidiomycete fungus. The nuclei divide within the spore before the promycelium is produced

teliospore = telentospore

-telic   *comb. suffix* of doubtful origin meaning "pertaining to evolution". The following terms using this suffix are defined in alphabetic position:

HOROTELIC	TACHYTELIC

teloblast   [tě'·lō·blăst']   a large mother cell that buds off columns of smaller cells at the growing tip of many invertebrate embryos

telocentric chromosome   [těl'·ō·sěnt"·rĭk]   one in which the centromere is terminal (*see also* acrocentric chromosome, metacentric chromosome)

teloceptor   [těl'·ō·sěp"·tər]   a sense organ, such as the eye or ear, which provides information about stimuli originating at a distance

telocoel   [těl'·ō·sēl"]   the cavity of the telencephalon

telogonic   [těl'·ō·gŏn"·ĭk]   said of a sac-shaped ovary in which germ cells are proliferated over the whole of the interior surface of the sac (cf. hologonic)

telolecithal [tēl′·ō·lĕs″·ĭth·əl]   said of an egg containing such large quantities of yolk that it is only capable of meroblastic cleavage

telomere [tē″·lō·mēr′]   the end portion of a chromosome

telomitic chromosome [*angl.* tē·lŏm″·ət·ĭk, *orig.* tēl′·ō·mĭt″·ĭk]   one having a terminal centromere

telophase [tēl″·ō·fāz]   the terminal phase of mitotic division in which the chromosomes reconstitute a resting nucleus

telopodite [*angl.* tēl·ŏp″·əd·ĭt, *orig.* tēl′·ō·pōd″·ĭt]   in insects, all those segments distal to the coxopodite

telotaxis [tēl′·ō·tăks″·ĭs]   the responses of an animal capable of moving towards a source using only one remaining, of previously paired, receptors (cf. tropotaxis)

telson 1 [tĕl″·sən]   (*see also* telson 2) the terminal segment of the crustacean abdomen

telson 2   (*see also* telson 1) the stinging segment at the end of the tail of a scorpion

-temn-   *comb. form* meaning "cut"

temnospondylous vertebra [tĕm′·nō·spŏnd″·əl·ŭs]   one consisting of several parts, or at least having the arch separated from the body

temporal bone [tĕm″·pər·əl]   a compound bone, or group of bones, in the cranium (*see also* intertemporal bone, posttemporal bone, supratemporal bone)

temporary climax   an apparent, but not actual, climax

tenaculum [tĕn·ăk″·yōō·ləm]   literally a hold fast and so used of Algae

tendon   a mass of white fibrous connective tissue attaching a muscle to a bone

-tene   a suffix used interchangeably with -nema (q.v.) in the description of chromosomes

tentacle   any flexible, as distinct from articulated appendage. Many are tactile

tentaculozooid [tĕn·tăk′·yōō·lō·zō″·ĭd]   a dactylozooid, usually on a siphonophoran, in the form of a single tentacle

-tentor-   *comb. form* meaning "tent"

tentorium 1 [tĕn·tôr″·ē·əm]   (*see also* tentorium 2, 3) the internal skeleton of the insect head

tentorium 2   (*see also* tentorium 1, 3) that transverse fold of the meninges which lies between the cerebral hemispheres and cerebellum

tentorium 3   (*see also* tentorium 1, 2) a secondary ossification, separating, in some mammals, the cerebral fossa from the cerebellar fossa

tepal [tĕp″·əl]   those parts of the perianth of a flower that cannot be distinguished either as petals or sepals

-tera-   *comb. form* meaning "prodigy" though usually used in the sense of "monster"

teratology [tĕr′·ə·tŏl″·ō·gē, tĕr′·ə·tŏl″·ō·gē]   the study of monsters and other abnormalities, particularly in embryos

Teredo [tə·rē″·dō]   a family of vermiform pelecypod mollusks that bore long tunnels in wood by rasping with their small anterior shells. Siphons extend the whole length of the body and are furnished with calcareous pallets that close the opening of the burrow

-terg-   *comb. form* meaning "back"

tergite [tûr″·jit, tûr″·git]   the upper, or dorsal, plate of the exoskeleton of a segmented animal, particularly an arthropod (*see also* sternite)

terminal intestine   a term used in place of "large intestine" for those fish in which the diameter of the posterior portion of the alimentary canal is less than that of the anterior portion

terminal lamina   the portion of the floor of the embryonic brain that lies anterior to the preoptic recess

-terr-   *comb. form* meaning "land"

terrarium   a small case in which terrestrial or amphibious animals are maintained for observation, study, or decorative purposes

territorial signal   one that establishes a breeding, or more rarely hunting, territory. Such signals are usually olfactory in mammals and auditory in birds

territorialism   the custom, particularly strong in birds and carnivorous mammals, of laying claim to exclusive rights to a territory

Tertiary period   the older period of the Cenozoic era extending from about 60 million years ago to 1 million years ago. It was preceded by the Cretaceous period and followed by the Quaternary period. It is divided into the Eocene, Oligocene, Miocene, Pliocene and Pleistocene epochs. Mammals, birds and insects dominated the well-forested land masses

tertiary sexual characters   those characters that visually, but not functionally, distinguish the sexes (i.e. plumage, etc.) (*see also* primary sexual character, secondary sexual character)

-test-   *comb. form* meaning "shell" in the sense of "hard covering"

test   a term applied to almost any hard outer covering of an organism or part of an organism

Testacea [tĕst·ās″·ē·ə]   an order of sarcodinous Protozoa distinguished by possessing a shell, usually chitinous in nature, containing a single aperture through which the pseudopodia protrude. The genera *Arcella* and *Difflugia* are the subjects of separate entries

testosterone [*angl.* tĕst·ŏst″·ə·rōn, *orig.* tĕst′·ō·stēr″·ōn]   the hormone secreted in the testes. It is necessary for the development and the maintenance of male secondary and tertiary sexual characters

Testudo [tĕst·yōō″·dō]   a large genus of land turtles (Chelonia) of wide distribution. *T. graecae* [grēk″·ē] is the common small European tortoise. *T. elephantopus* [ĕl′·ə·făn·tōp″·əs] is the huge, but nearly extinct, Galapagos tortoise

-tetra-   *comb. form* meaning "four" sometimes written -tetr-

tetrad [tĕt″·răd]   a group of four, particularly the four cells resulting from the division of a spore mother cell or the groups of four chromosomes resulting from the meiotic prophase

Tetrahymena [tĕt′·rə·hī′·mən·ə]   a genus of small holotrich ciliate protozoans. *T. pyriformis* [pir′·ē·fôrm″·ĭs] is usually ovoid and is so flexible that the shape constantly varies. This was the first animal successfully maintained in an axenic culture and an enormous number of physiological studies have therefore been made on it

tetramerous [*angl.* tĕt·răm″·ər·əs, *orig.* tĕt′·rə·mēr″·əs]   a variety of radial symmetry in which four major axes can be distinguished

tetraploid [tĕt′·rə·ploid″]   having twice (i.e. 4N) times the usual number of chromosomes (*see also* allotetraploid, diploid, double tetraploid, triploid, polyploid)

Tetraspora [*angl.* tĕt·răsp″·ər·ə, *orig.* tĕt′·rə·spôr·ə]   a genus of sessile, volvocale chlorophyte algae. The cells are enclosed in groups of four in a gelatinous matrix

tetraspore [tĕt'·rə·spôr] a spore produced by meiosis from a diploid generation descended from a diploid carpospore

temporal lobe that portion of the lower outer part of the cerebral hemisphere that lies below Sylvius' fissure

-thal- *comb. form* meaning a "twig" or, by extension, anything young or recently produced (cf. -thalam-)

-thalam- *comb. form* meaning "chamber" in the sense of enclosed space (cf. -thal-)

thalamencephalon [þăl'·ə·mĕn·sĕf"·ə·lŏn] the thickened portion of the roof of the diencephalon

-thalamic adjectival suffix used interchangeably as a derivative of thal- or -thalam-

thalamus that part of the diencephalon which lies on each side of the third ventricle (*see also* epithalamus, hypothalamus)

-thalass- *comb. form* meaning "sea"

Thalassicola [*angl*. þăl'·ăs·ĭk"·əl·ə, *orig*. þăl'·ăs·ē·kōl"·ə] a genus of radiolarian protozoans lacking a skeleton

thalassin [þăl·ə·sĭn] a toxin obtained by the alcohol extraction of nematocyst bearing tentacles (cf. congestin and hypotoxin)

thalassoplankton [þăl'·ə·sō·plănk"·tən] that which occurs in the open ocean

Thaliacea [þăl'·ē·ās"·ē·ə] a class of urochordata with the oral and atrial apertures at opposite ends of the body. All are free-swimming. Most occur in two forms; the solitary, produced sexually, and the aggregate produced asexually in chains budded off posteriorly. *Doliolum* has three asexually produced and one sexually produced form. The genera *Doliolum* and *Salpa* are the subjects of separate entries

-thall- *comb. form* meaning "branch" or "young shoot"

thallic pertaining to a thallus in any meaning of that word

thallophyte [þăl"·ō·fĭt] a plant which consists principally of a thallus not organized into root, stem, and leaf

thallus literally a branch, or young shoot, but most commonly used in reference to solid structures of non-vascular plants (*see also* prothallus, homothallus)

Thamnophis [þăm·nō"·fĭs] a genus of snakes (Squamata). *T. radix* is the common garter snake of the central United States

-thanat- *comb. form* meaning "death"

thanatosis [þăn'·ə·tōs"·ĭs] the condition of being dead or of feigning death

Thasmida [þăz"·mĭd·ə] an order of wingless insects containing the stick insects, at one time included in the Orthoptera. They are a standard textbook example of protective mimicry

-thec- *see* -thek-

theca [þē"·kə] the term has been applied to almost any structure which contains anything, such as the sporangium of a fern or the capsule of a moss, the loculus of an anther, the calcareous exoskeleton of echinoderms, and the wall as distinct from the basal plate, of a corallite. A more specific use is for a lorica which is an indirect contact with the cell membranes of some Algae and thus corresponds to the cellulose cell wall of higher forms. The following derivative terms are defined in

alphabetic position:

GASTROTHECA	OVOTHECA
GONOTHECA	SPERMATHECA
HYDROTHECA	

-thecium a suffix derived from -thek- (*see* perithecium)

thecodont [þē'·kō·dŏnt", þĕk'·ə·dŏnt'] said of a form in which the teeth are set in sockets

-thek- *comb. form* meaning "container", "case" or "cover", almost invariably transliterated -thec-. A Latin derivative of this form "apothecium", meaning "storehouse" or "warehouse", has caused a complete confusion in biological literature between -theca and -thecium a form which never existed save in the very restricted word cited

-thele- *comb. form* meaning "nipple", and by extension, sometimes used to mean "female". The origin of the word "epithelium", which actually means, "above the nipple", derives from this root

thelytoky [þĕl'·ē·tōk"·ē] the parthenogenic production of female offspring

theory a hypothesis which is sufficiently supported by a logical interpretation of experimental data to justify the supposition that it extends over a wide field, but which is not yet sufficiently universally verified to justify the use of the term "law" (q.v.). The following theories are given in alphabetic position:

AGE AND AREA THEORY	NEPHROCOEL THEORY
COSMOBIOTIC THEORY	POLYCLIMAX THEORY
FIELD THEORY	PREFORMATION THEORY
GENE THEORY	PSEUDOMETAMERIC T.
GONOCOEL THEORY	VITALIST THEORY
HATSCHEK'S THEORY	WAGNER'S SEPARATION T.
HOMOLOGOUS THEORY	WILLIS'S THEORY
IMBIBITION THEORY	ZIEGLER'S THEORY
LOCK AND KEY THEORY	
MECHANISTIC THEORY	

-ther- *comb. form* confused from three Greek roots and therefore variously meaning "summer", "mammal" and "to hunt", the last either in the sense of "seek", or of the activities of a predator

Theriodonta [þĕr'·ē·ō·dŏnt"·ə] a sub-order of sphenapsid reptiles probably ancestral to the mammals. Unlike other reptiles the teeth were specialized as are those of mammals. *Cynognathus* [*angl*. sĭn·ŏg"·nə·þəs, *orig*. sĭn'·ŏg·nāþ·əs] is typical

-therm- *comb. form* meaning "heat"

-therm *comb. suffix* used to describe organisms in relation to their reactions to heat. The derivatives ending in -thermic and -thermous are equally common but are not separately recorded. The following terms using this suffix are defined in alphabetic position:

ECTOTHERM	HOMOIOTHERM
ENDOTHERM	POIKELOTHERM
EURYTHERM	STENOTHERM
HEKISTOTHERM	XEROTHERM

thermocline [þûr"·mō·klĭn'] that layer of water in which the temperature changes 1°C with each meter increase in depth

thermoreceptor [þûr'·mō·rē·sĕp"·tə] a receptor sensitive to temperature change

thermotaxis [þûr'·mō·tăks·ĭs] orientation or movement in relation to temperature

thermotropism [þûr'·mō·trōp"·ĭzm] the condition of responding to heat

thigmocyte [þĭg"·mō·sĭt'] an insect blood cell

which aids in rapid clotting over a fracture in the exoskeleton

thigmomorphosis [thĭg'·mō·môrf·ōs"·ĭs] a change in structure due to contact stimulus

thigmonastic [*angl.* thĭg·mŏn"·əs·tĭk, *orig.* thĭg'·mō·nàst"·ĭk] response of a flattened plant organ to touch

thigmotaxis [thĭg'·mō·tăks"·ĭs] movement in response to tactile stimuli

thigmotropism [thĭg'·mō·trōp"·ĭzm] movement or, in plants, curvature stimulated by contact

Thiobacillus [thi'·ō·bās·ĭl"·əs] a genus of pseudomonodale Schizomycetes containing the "purple sulfur bacteria". They are chemosynthetic autotrophs that oxidize sulfur to sulphate

third ventricle the cavity of the diencephalon

Thompsonian coordinates coordinates drawn between dorso-ventral and anterior-posterior biologically determined points on the body of related organisms. Horizontal and vertical lines joining coordinates for one species differ in shape for those of another species and may indicate the existence and shape of an intermediate species

thoracic vertebrae [þôr·ăs"·ĭk] the vertebrae of the thorax. The transverse process has a flattened end for the articulation of the rib

thorax that section of the body which lies between the head and abdomen. The following derivative terms are defined in alphabetic position:

CEPHALOTHORAX	METATHORAX
ENDOTHORAX	PROTHORAX
MESOTHORAX	

Thos [thŏs, tŏs] the genus of canid carnivores containing the Old World jackals. The New World jackals are *Dusicyon*. The several species of *Thos* are nocturnal, omnivorous scavengers, only hunting when in large packs. They will interbreed with domestic dogs

thrombin [þrŏm"·bĭn] an enzyme catalyzing the conversion of fibrinogen into fibrin

thrombocyte [þrŏm"·bō·sĭt] a blood platelet

Thylacinus [thi'·lə·sĭn"·əs] an almost extinct genus of Tasmanian Metatherians containing the "pouched wolf" (*T. cynocephalus*) [sĭn'·ō·sĕf"·əl·əs], greatly resembling a wolf in size, shape and habits. This is the largest extant carnivorous marsupial

thylakoid [thi'·lə·koid"] the lamella of a chloroplast

thymus gland [thī"·məs] an endocrine gland arising from evaginations of the dorsal walls of the pharyngeal pouches. The function is not clearly established though it is thought to play some role in the development of immunities

Thyone [thi·ōn"·ē] a genus of holothurian echinoderms with tube feet over the entire surface. *T. briareus* [bri·âr"·ē·əs] is the common black holothurian of the Atlantic coast

-thyr- *comb. form* meaning either "door" or in the shape of an oblong "shield"

thyroglobulin [thir'·ō·glŏb"·yū·lĭn] a secretion of the thyroid gland, normally stored in thyroid follicles

thyroid cartilage [thir"·oid] the central of the three laryngeal cartilages

thyroid gland one of a pair of endocrine gland lying on each side of the thyroid cartilage in mammals, that secretes thyroxin and plays a major role in the establishment of the metabolic rate (*see also* parathyroid gland)

thyrotroph cell [thir'·ō·trôf] one of the basophilic cells found in the thyroid gland

thyrotropic hormone [thir'·ō·trŏp"·ĭk] a hormone produced in the adenohypophysis which stimulates secretion by the thyroid gland

thyroxine [thir·ŏks"·ĭn] β-[(3,5-diiodo-4-hydroxyphenoxy)-3,5-diiodophenyl]-alanine. An amino acid synthesized in the thyroid gland of higher forms but also known from many lower animals. Best known for its hormonal effects in growth and metamorphosis

Thysanoptera = Thysanura

Thysanura [this'·ə·nōōr"·ə, this'·ăn·nyōōr"·ə] a small order of wingless insects containing the thrips. The absence of the right mandible distinguish thrips from all other insects

tibia 1 [tĭb"·ē·ə] (*see also* tibia 2) that segment of the arthropod limb which precedes the tarsus

tibia 2 (*see also* tibia 1) the larger of the two bones lying between the femur and the ankle in the posterior limb of vertebrates. The other bone, with which it is often fused, is the fibula

-tiph- *comb. form* meaning "pool", and used by ecologists in the more specific sense of "pond"

tissue 1 (*see also* tissue 2) an aggregate of animal cells. The following derivative terms are defined in alphabetic position:

ADIPOSE TISSUE	MYELOID TISSUE
AREOLAR TISSUE	NERVOUS TISSUE
CONNECTIVE TISSUE	PERIPORTAL CONNECTIVE T.
LYMPHATIC TISSUE	

tissue 2 (*see also* tissue 1) an aggregation of plant cells. The following derivative terms are defined in alphabetic position:

CONNECTING TISSUE	SECONDARY TISSUE
FALSE TISSUE	VASCULAR TISSUE

tissue fluid a fluid that diffuses through the endothelial walls of capillaries

-tmes- *comb. form* meaning "cut"

Tmesipteris [mĕs·ĭp"·tər·ĭs] a genus of primitive leafless tracheophyte plants long thought to represent a survival of the extinct Rhiniophytina. They are now considered to be more closely allied to the true ferns

TMV *see* tobacco mosaic virus

tobacco mosaic virus a virus in the shape of an elongate cylinder (300 μ × 15 μ) with a helical arrangement of protein sub-units with the RNA molecules near the core

-tocous (*see also* -tok-) pertaining to the production of offspring (*see* polytocous)

-tok- *comb. form* meaning "offspring", or "production of offspring" sometimes transliterated -toc-

-toke *subs. suffix* indicating a part of an organism concerned with reproduction (*see* atoke, epitoke)

-toky *subs. suffix* indicating a method or condition of reproduction. The following derivative terms are defined in alphabetic position:

ARRHENOTOKY	EPITOKY
DEUTEROTOKY	THELYTOKY

Tolypothrix [*angl.* tŏl·ĭp"·ə·þrĭks, *orig.* tŏl'·ē·pōth"·rĭks] a genus of cyanophyte algae. *T. tenuis* is a nitrogen fixing species which has been widely used to increase the yield of rice by inoculating the rice fields

-tome- *comb. form* meaning "cut" or "slice" or, by inaccurate extension, "segment" for which -tomite is preferable. The following derivative terms are defined in alphabetic position :

MYOSCLEROTOME   NEPHROTOME
MYOTOME   SCLEROTOME

-tomite *subs. suffix* indicating "segment" (*see* sclerotomite)

-tomy *subs. suffix* indicating the state or condition of breaking or cutting. The following derivative terms are defined in alphabetic position :

ARCHITOMY   PARATOMY
AUTOTOMY   PLASMOTOMY
MICROTOMY   ULTRAMICROTOMY

tongue 1 (*see also* tongue 2) a more or less protrusible muscular organ arising from the floor of the mouth of many vertebrates (cf. rasping organ)

tongue 2 (*see also* tongue 1) an analogous structure in other forms. In botany the term is applied to the ligule and in insects to any structure that projects from the mouth part and is used in gathering food

tongue bar a downgrowth from the dorsal wall of the pharyngeal slit in *Branchiostoma* and hemichordates which divides the slit into anterior and posterior openings

-tonic *adjectival suffix* meaning "pertaining to stress". The following derivative terms are defined in alphabetic position :

HYPERTONIC   ISOTONIC

-tono- *comb. form* meaning a "chord", frequently confused with -ton-

tonsil a lymphoid gland in the roof of the mouth

tonus [tō″·nəs] the prolonged persistence of the contracted state in muscles

tooth 1 (*see also* tooth 2) the enamel-covered, dentine supported, biting, chewing, or grabbing structures produced from the jaw, and some other, bones, of vertebrates. In this sense the following derivative terms are defined in alphabetic position :

BISCUSPID TOOTH   INCISOR TOOTH
CANINE TOOTH   MILK TOOTH
CARNASSIAL TOOTH   MOLAR TOOTH
CUSPID TOOTH   PREMOLAR TOOTH
ETHMOID TOOTH

tooth 2 (*see also* tooth 1) any pointed structure arising from anything, particularly when the points are arranged serially, as on the jaws of many invertebrates

tooth cell one of the horny cells comprising the "teeth" of tadpoles

-top- *comb. form* meaning "place", usually in biology in the sense of environment

-topic *adjectival suffix* meaning "pertaining to place", usually in the sense of environment

topocline [tŏp″·ō·klīn′, tōp·ō·klīn] a geocline extending over an unusually long distance

topodeme [tŏp′·ō·dēm″, tōp′·ō·dēm″] a fraction of a population restricted to a specific area

topotype [tŏp″·ō·tīp′, tŏp″·ō·tīp] a specimen collected from the type locality. Usage varies as to whether this collection must be made by the original describer

tornaria larva [tôr·när″·ē·ə] the free-swimming larva of enteropneustans. Its general appearance, and arrangement of ciliated bands, is reminiscent of some echinoderm larva

Torpedo [tôr·pēd″·ō] a genus of batoid elasmobranchs containing the electric rays. The body is less flattened, and the tail relatively shorter, than in most rays. The powerful electric organs are in the lateral body projections, or "wings"

torpidity a condition found in some birds closely approximating hibernation

torsion twisting, as in the shells and some internal organs of gastropods

Torula see *Cryptoccus*

totipotency [tō′·tĭ·pŏt″·ən·sē] the ability of an egg to reproduce a whole organism and, particularly, the retention of this power by some early blastomeres

touch receptors usually a combination of tango- and stretch receptors

toxin a poison of organic origin (*see also* hypnotoxin)

trabecula [trə·bĕk″·yū·lə] literally a beam, and applied in botany to almost any transverse septa or transverse bars and to bars extending transversely across gymnosperm tracheids. In zoology cartilaginous rods in vertebrate embryos supporting the anterior end of the brain or the connective tissue framework of the spleen

trabecular bone lamellar bone in which the volume of the lacunae exceeds that of the boney substance

-trach- *comb. form* meaning "wind pipe"

trachea 1 [trāk″·ē·ə] (*see also* trachea 2, 3) water-conducting vessels in plant stems

trachea 2 (*see also* trachea 1, 3) a tube supported by cartilaginous rings conducting air from the mouth to the lungs

trachea 3 (*see also* trachea 1, 2) a tube, usually lined with rings or a spiral of chitin, which conducts air from the stigmata through the body cavity of arthropods

tracheal cartilages [trāk″·ē·əl] the cartilaginous rings which support the trachea of air breathing vertebrates

tracheid [trāk″·ē·ĭd] a long lignified cell in a plant stem both conducting water and serving as a support. It differs from trachea in having long tapering closed ends

tracheole [trāk″·ē·ōl] diminutive of trachea, particularly in the sense of trachea 2 and 3

-trachy- *comb. form* meaning "rough" or "wind pipe"

tract a specific area usually delimited within a larger one. The term is also used in the sense of "system". The following derivative terms are defined in alphabetic position :

ASCENDING TRACT   MECKEL'S TRACT
CORTICOSPINAL TRACT   OPTIC TRACT
DESCENDING TRACT

tractellum [trāk·tel″·əm] a flagellum used to pull, as distinct from push, a protozoan through water. The "grip" on the water is provided by backwardly directed mastigonemes

Tradescantia [trăd″·ĕs·kănt″·ē·ə] a genus of monocotyledenous plants of the family Commelinaceae often called spiderworts. *T. virginiana* [vûr′·jĭn·ē·än″·ə] is frequently used for class studies on chromosomes

tragus [trā″·gəs] the anterior flap which, in some mammals, can be closed to cover the ear hole

trans-configuration the condition in which two mutants at different sites within a cistron are located on different chromosomes

transcendentalism [trăn′·sĕn·dĕnt″·əl·ĭzm] in biology the doctrine that form precedes function. It is therefore, the antithesis of teleology

transduction a form of transfer of genetic information which depends on the ability of a bacteriophage to incorporate heritable characters of cells that they parasitize into their own genetic apparatus and then to transfer these characteristics to other cells that they subsequently invade (see also abortive transduction, special transduction)

transferase [trănz″·fer·āz′]   a group term for those enzymes that catalyze the transfer of atoms, or groups of atoms, between molecules

transformation   a term of bacterial genetics indicating an alteration in the characteristics of a bacterial clone

transfusion strand   the parenchymatous cells which lie between the xylem and phloem bundles

transhydrogenase [trănz′·hī·drŏj″·ĕn·āz]   an enzyme which catalyzes the transfer of hydrogen from one molecule to another

transitional association   one in the course of development

transitional epithelium   an epithelium composed of many types of cells

translocation 1   (see also translocation 2) the transfer of metabolites from one part of a plant to another

translocation 2   (see also translocation 1) a movement of groups of alleles within a chromosome or the transfer of this group to another chromosome. In this sense the following derivative terms are defined in alphabetic position:
ANEUCENTRIC T.   INSERTIONAL T.   RECIPROCAL T.

transpiration   the emission of water vapor by an organism

transverse axis   the axis at right angles to the sagittal axis

transverse colon   that portion of the colon which, in some mammals, passes from side to side of the abdominal cavity

transverse plane   a plane parallel to the transverse axis, that is at right angles to the longitudinal axis

transverse septum   the posterior wall of the pericardial cavity in vertebrates

transverse velum   the infolding of the dorsal floor of the primitive brain that divides the diencephalon from the telencephalon

-traum-   comb. form meaning "wound"

trehalase [trē″·he·lāz′]   an enzyme catalyzing the hydrolysis of trehalose into glucose

Treitz's ligament   the suspensor of the mammalian duodenum

-trem-   comb. form meaning "section", or "hole" (see peritrema)

Trematoda   a class of Platyhelminthes containing the parasitic forms called flukes. They are distinguished by the possession of one or two suckers at either end or sometimes at both ends. The eggs of most hatch into a free-swimming miracidium larva that parasitizes a molluscan host and produces a sporocyst, which may reproduce itself or produce redia that give rise to tailed, free-swimming cercaria larvas that are the infective stage. The genera Clonorchis, Fasciola, Fasciolopsis, Leucochloridium, Opisthorchis, Paragonimus and Schistosoma are the subjects of separate entries

Tremellales [trĕm′·ăl·āl″·ēz]   an order of Basidiomycetes distinguished by the fact that the hypobasidium becomes divided into two, three, or four cells

Treponema [trĕp′·en·ē″·mă]   a genus of spirochaetale Schyzomycetes T. *pallidum* [păl″·ĭd·em] is the causative agent of syphilis and T. *pertenue* [pûr·tĕn″·yōō·ē] of yaws

-tri-   comb. form meaning "three"

triangular bone   that carpal which lies, together with the pisiforme bone beneath it, opposite the head of the ulna

Triarthus [tri·är″·þres]   a genus of small (3–4 inch) Ordovician trilobites. Unlike most trilobites it is usually found flattened, not curled, and is therefore frequently used as a typical example

Triassic period   the oldest period of the Mesozoic era extending from about 200 million years ago to 150 million years ago. It was preceded by the Permian and followed by the Jurassic periods. It was an age of coniferous forests and reptiles. The first distinguishable mammals appeared during this period

Triatoma [angl. tri·ăt″·em·e, orig. tri·e·tōm·e]   a genus of hemipteran insects. T. *sanguisuga* [săng′·gwē·sōō″·ge] is the giant bedbug and its near ally T. *megista* [me·gĭst″·e] is the vector of *Trypanosoma cruzi* in South America

triaxial symmetry   a type of symmetry such as biradial symmetry or bilateral symmetry, in which there are three axes commonly known as the longitudinal, sagittal and transverse

tricarboxylic cycle [tri·kärb′·ŏks·ĭl″·ĭk]   the complex metabolic cycle starting with the production of citric acid from pyruvic acid through acetyl-CoA ultimately ending in the reproduction of citric acid through oxaloacetic acid

-trich-   comb. form meaning "hair", or by extension, "cilium"

Trichecus [tri·kĕk″·es]   one of the two extant genera of sirenian mammals. It contains the manatees, now confined to the Amazonian region. The genus is readily distinguished from the *Dugong* by its paddle-shaped tail and small head

Trichinella [trĭk′·ĭn·ĕl″·e]   a genus of parasitic nematodes. T. *spiralis* [spir·āl″·ĭs] is an intestinal parasite of many mammals including man, pig, cat, dog and rat. The larvas burrow through the intestinal mucosa and encyst in the muscles of the host. Heavily infected humans frequently die

trichite [trĭk″·ĭt]   hardened, rod-like structures in the bodies of some protozoa

trichium [trĭk″·ē·em]   a hair in any sense of that term

trichocyst [trĭk′·ō·sĭst″, trĭk·ō·sĭst″]   a rod, immediately under and at right angles to the pellicle of some ciliates. The function is obscure

Trichodesmium [trĭk′·ō·dĕz″·mē·em]   a "red" cyanophyte Alga. The prevalence of this genus in the Red Sea is said to be the origin of that name

trichogen [trĭk′·ō·jen″, trĭk′·ō·jen″]   a cell from which an insect seta is derived (cf. trichophore)

trichogyne [trĭk″·ō·jin′, trĭk″·ō·jin′]   either a filament arising from the ascogonium of ascomycete fungi or the receptive filament in the cross fertilization of some algae

trichome 1 [tri″·kōm]   (see also trichome 2) an epidermal appendage, which may be either unicellular or multicellular, of a plant; frequently called a plant hair

trichome 2   (see also trichome 1) a many-celled, frequently branched, filament of bacteria or, less usually, algae

Trichomonas [angl. trĭk·ŏm″·en·es, orig. trĭk″·ō·mōn″·ăs]   a genus of mastigophoran protozoans. They are pear-shaped organisms with from 3 to 5 anterior flagella. Most are internal parasites or commensals. *T. hominis* [hŏm"·ĭn·ĭs], *T. vaginalis* [vǎj'·ĭn·āl"·ĭs] and *T. tenax* [těn"·ăks] respectively inhabit the lower intestine, the vagina and the mouth of humans

Trichonympha [trĭk'·ō·nĭmf"·ə, trĭk'·ō·nĭmf·ə] a genus of zoomastigophoran Protozoa commensal in the gut of termites. The protozoan secretes a cellulase that the termite lacks and thus makes it possible for the termite to exist on the wood that it eats

trichophore [trĭk"·ō·fôr, trĭk'·ō·fôr] the structure from which the annelid chaeta is produced (cf. trichogen)

Trichoptera [*angl.* trĭk·ŏpt"·ər·ə, *orig.* trĭk'·ō·těr"·ə] the order of insects that contains the caddis flies. They are distinguished by two pairs of membranous wings with primitive venation usually densely hairy or scaly. The aquatic larvae either build portable cases of various materials or spin protective webs on rocks

-trichous *adjectival suffix* meaning "pertaining to hair" in any meaning of that word, including cilia

tricuspid valves [trī·kǔsp"·ĭd] the valves in the right hand atrio-ventricular canal of a four chambered heart

triennial [trī·ĕn"·ē·əl] lasting for three years

Trilobita [trī'·lōb·īt"·ə] the first order of arthropods of which there is a fossil record, since they occur in the lower Cambrian. The central segmented body had lobes on each side and there was an appendage on each segment. The genus *Triarthrus* is the subject of a separate entry

trilobite larva the trilobite-like larva of the Xiphisura

trimerism [trī·mēr"·ĭzm] the condition of those animal phyla (Echinodermata, Stomochorda, and Pogonophora) in which the coelom is divided into three parts

Trimerophytina [*angl.* trĭm'·ər·ō·fīt"·ĭn·ə, *orig.* trī'·mēr·ō·fīt"·ĭn·ə] a group of primitive vascular plants known only as Lower Devonian fossils. This group was once included with the Psilophyta. The genus *Psilophyton* is the subject of a separate entry

trimonoecism [*angl.* trī·mŏn"·ə·sĭzm, *orig.* trī'·mŏn·ō·ēs"·ĭzm] the condition of having male, female and perfect flowers on the same plant

Trionyx [trī"·ən·ĭks, trī·ŏn"·ĭks] a genus of Chelonia containing the soft-shelled turtles. *T. ferox* [fēr"·ŏks] is the large Florida form and *T. spinifera* [spĭn·ĭf"·ĕr·ə] is the smaller Mississippi basin species

-tripl- *comb. form* meaning "triple"

triplicate gene one of three non-allelic, noncumulative genes showing the same effect

triploblastic [trĭp'·lō·blăst"·ĭk] consisting of, or possessing, three developmental layers

triploid [trĭp"·loid] applied to a nucleus having one and half times the diploid number of chromosomes (*see also* allotriploid, diploid, tetraploid, polyploid)

tripton [trĭp·tən] the sum total of the suspended dead matter in water; it differs from leptopel in that the term is not restricted to minute particles (= abioseton)

-trit- *comb. form* meaning "third"

Triturus [trī·tyōōr"·əs] a genus once containing most of the aquatic, newt-like, urodelan Amphibia. The name is now confined to the European forms (previously called *Triton*) such as *T. vulgaris* [vōōl·găr·ĭs], the smooth newt. The best known American form (*T. viridescens*) is now *Notophthalmus viridescens*

trivial name the name by which an organism or chemical compound (e.g. "mouse" or "nitrate ester

reductase") is known as distinguished from its scientific name (e.g. *Mus musculus* or glutathione : polyolnitrate oxidoreductase)

trixenous [*angl.* trĭks"·ən·əs, *orig* trī·zēn"·əs] said of a parasite which requires three successive hosts

-troch- *comb. form* meaning "wheel". The term is particularly applied to rings of cilia

trochanter [trŏk"·ănt·ə] the segment of the arthropod limb, particularly in Insecta and Arachnida, that lies between the coxa and the femur

trochlea [trŏk"·lē·ə] literally a "pulley" and used for an articular surface on a large bone such as the humerus

trochlear nerve [trŏk"·lē·ə] cranial IV. Runs from the dorsal surface of the back end of the mesencephalon to the superior oblique muscle of the eye

trochophore larva [trŏk"·ō·fôr'] a top-shaped post gastrula larva, common to the Annelida and Mollusca. It is distinguished by the presence of an apical tuft of cilia and three preoral ciliated bands

-trog- *comb. form* meaning "to gnaw"

-troglo- *comb. form* properly meaning a "gnawed hole" but mostly supposed by biologists to mean "cave"

tropaxis [trŏp"·ăks·ĭs] that plane in a growing plant from which the epicotyl grows in one direction and the hypocotyl in another

-troph- *comb. form* meaning "nutrition", very frequently confused with -trop

-troph *substantive suffix* indicating an organism classified by its mode of nutrition (cf. -trophe, -trophy). The forms -troph and -trophe are frequently interchanged

trophallaxis [trŏf'·ə·lăks"·ĭs] the mutual exchange of food

trophi [trō"·fī] any of the numerous, cuticularized pieces within the mastax of a rotifer. Sometimes used as synonymous with mastax

-trophic *adjectival suffix* referring to food or nutrition. The following derivative terms with this suffix are defined in alphabetic position:

ALLOTROPHIC	METATROPHIC
AUTOTROPHIC	MIXOTROPHIC
AUXOTROPHIC	MONOTROPHIC
CHEMOLITHOTROPHIC	MYCOTROPHIC
DYSTROPHIC	MYXOTROPHIC
EUTROPHIC	OLIGOTROPHIC
HOLOTROPHIC	PHOTOLITHOTROPHIC
HYPOTROPHIC	POLYTROPHIC
MESOTROPHIC	

trophoderm [trŏf"·ō·dûrm'] the outermost layer of cells in the blastula stage of a placental mammal

trophodynamic [trō'·fō·dī·năm"·ĭk] pertaining to energy transfer between food levels in an ecosystem

trophozoite [trŏf'·ō·zō"·ĭt] the growing, vegetative stage of a sporozoan protozoan

-trophy *substantive suffix* indicating the condition or result, of a method or type of nutrition. Most forms given under -trophic can be converted to this form but are not listed separately unless there is a marked difference in meaning. The following derivative terms using this suffix are defined in alphabetic position:

ATROPHY	HETEROTROPHY
COMPENSATORY HYPERTROPHY	HYPERTROPHY
EMBRYOTROPHY	SYNTROPHY
HEMOTROPHY	

tropic either of two imaginary lines running parallel to the equator at a distance of 23°27' on each side of it. The "tropics" is the area contained between these lines
-tropic *adjectival suffix* from -trop in the sense of turning. References made to these compound words are under the *substantive form* -tropism
tropical in contemporary biology this adjective is confined to the meaning "pertaining to the tropics", in the ecological sense of that word, in distinction from "tropic" (*see* neotropical, palaeotropical)
tropical rain forest an equatorial rain forest (*see also* high rain forest)
tropism a growth response of plants or sessile animals, thus corresponding to taxis in free-living forms. A positive tropism means a response in the direction from which the stimulus is coming; negative tropism means the reverse. The following derivative terms using this suffix are defined in alphabetical position:

ANTHOTROPISM	HELIOTROPISM
DIAGEOTROPISM	NYCTOTROPISM
DIAPHOTOTROPISM	PHOTOTROPISM
DIATROPISM	THERMOTROPISM
GEODIATROPISM	THIGMOTROPISM
GEOTROPISM	

tropotaxis [trōp'·ō·tăks"·ĭs] the response of an animal using paired receptors without moving its head from side to side (cf. klinotaxis)
-trorse *adjectival suffix* indicating direction of movement. The following derivative terms with this suffix are defined in alphabetical position:

ANTRORSE	EXTRORSE
DETRORSE	INTRORSE
DEXTRORSE	RETRORSE

true rib a pleural rib which articulates directly with the sternum
true suture a synarthrosis formed by interlocking finger-like or tooth-like projections of the bone margin
trunk 1 (*see also* trunk 2) any portion of the body of an organism posterior to the neck (*see* alitrunk)
trunk 2 (*see also* trunk 1) a principal vessel or nerve to which other, smaller, vessels or nerves are joined (e.g. arterial trunk, sympathetic trunk)
trunk muscle a muscle associated with the body wall of a higher vertebrate
Trypanosoma [trĭp'·ăn·ō·sōm"·ə] a genus of trypanosomid protozoans all of which are blood parasites in most classes of vertebrates. *T. gambiense* [găm'·bē·ĕns"·ē] is the causative agent of human sleeping sickness but many other species are equally deadly to other mammals. *T. gambiense* is transmitted mostly by the bite of the tsetse fly (*Glossina morsitans*) but *T. equiperdum* [ĕk'·wē·pûrd"·əm] causes a venereal disease of horses. *T. cruzi* [krōōz"·ĭ], the vector of which is *Triatoma*, causes Chagas' disease
Trypanosomidae [trĭp'·ăn·ō·sōm"·ĭd·ē] a family of mastigophoran Protozoa often called the hemoflagellates. They are distinguished by the fact that the flagellum is joined to the body by an undulating membrane. The genera *Crithidia*, *Leishmania*, *Leptomonas* and *Trypanosoma* are the subjects of separate entries
trypsin [trĭp"·sĭn] an enzyme catalyzing the hydrolysis of numerous compounds having bonds with carboxyl groups of l-arginine or l-lysine. It is secreted in the pancreas as the precursor trypsinogen which becomes trypsin in the intestine
l-tryptophan [el·trĭp"·tō·făn'] *1-α-amino-3-indole-*

propionic acid. An amino acid essential in the nutrition of rats
Tsuga [sōō"·gə] the genus of Coniferales containing the Hemlocks. They are distinguished by drooping branches and reflexed cones
-tuba- *comb. form* meaning "trumpet", frequently confused with -tubi-
tube foot the extrusible portion of the water vascular system of echinoderms
tube nucleus one of the two nuclei derived from the primary division within a pollen cell (cf. generative nucleus)
tuber 1 (*see also* tuber 2) a swelling, usually for food storage, on the roots of a plant (cf. rhizome)
tuber 2 (*see also* tuber 1) a swelling on an animal or animal organ
tuber cinereum [tyū"·bə sĭn·ēr"·ē·əm] one of a pair of nerve centers in the hypothalamus, anterior to the mamillary bodies. The infundibulum is attached to its lower surface
tubercle diminutive of tuber. Without qualification specifically applied to a prominence from the test of an echinoderm on which spines fit with a ball-and-socket joint
tuberculum [tyū·bûrk"·yōō·ləm] the lower of the two articular processes of the vertebrate rib (cf. capitulum)
-tubi- *comb. form* meaning "pipe", frequently confused with -tuba-
Tubifex [tyū"·bē·fĕks'] a genus of freshwater oligochaete worms. *T. tubifex* is the familiar freshwater red wiggler, often so common as to form reddish patches on muddy bottoms
tubular gland a simple gland of which the secreting cells form a cylindrical tubule
Tubulidentata [tyū'·byōōl·ē·dĕnt"·ät·ə] a small order of placental mammals containing the forms commonly called aardvarks. They are distinguished, as the name indicates, by the possession of tube-like teeth
tundra [tŭn"·drə] an area of relatively luxuriant growth of scrub and heathland laying between the southern limit of the ice and the northern limit of the tree zone. The following derivative terms are defined in alphabetic position:

ALPINE TUNDRA	FOREST TUNDRA
ANTARCTIC TUNDRA	LICHEN TUNDRA
ARCTIC TUNDRA	SHRUB TUNDRA

tunic any covering. In zoology, without qualification, usually applies to the more or less thickened external covering of urochordates and in botany to the skin of a seed
Tunicata = Urochordate
tunicated bulb one with an outer coat like the onion
tunicin [tyū"·nĭs·ĭn] a polysaccharide, closely resembling cellullose, of which the tunic of urochordates is composed
Turbellaria [tûr'·bĕl·är"·ē·ə] the class of the phylum Platyhelminthes that contains the free-living flatworms. They are distinguished from the Cestoda and Trematoda by the presence of external cilia and by the absence of adaptations to parasitic life shown by these groups. The genera *Bdelloura*, *Convoluta*, *Crenobia*, *Dendrocoelum*, *Dugesia*, *Procotyla* and *Stenostomum* are the subjects of separate entries

turbin-   *comb. form* meaning "whirling around" and "top"

urbinal bone [tûr″·bĭn·əl] one of a pair of rolled heets of bone nearly filling the nasal passage

urgid [tûr″·jĭd] swollen with fluid

urgor [tûr″·gôr] the condition of being bloated, swollen, or inflated

usk a protuberant mammalian tooth, usually a canine. The tusk of the elephant, and the "horn" of the narwhale are, however, incisors

win one of two young produced at one birth by a viviparous animal (*see* parabiotic twin, Siamese twin)

ycholimnetic [ti′·kō·lĭm·nĕt″·ĭk] fortuitously limnetic; that is, normally dwelling in streams but found occasionally in lakes

ycholimnetic plankton that which is formed of algae which have broken away from the bottom and float in consequence of contained gas bubbles

ychopelagic= tycholimnetic

ychopotamic [ti′·kō·pŏt·ə·mĭk] fortuitously potamic; that is, normally dwelling in lakes but occasionally found in streams

tyl-   *comb. form* meaning a "knot" in the sense of lump

tympan-   *comb. form* meaning a drum or "sounding board"

ympanic bone [tĭm·păn″·ĭk] a dermal bone of the splanchnocranium which forms the outer part of the ympanic bulla

ympanic bulla the boney spheroid which surrounds the inner ear, formed in part from the tympanic bone

ympanic membrane a stretched vibrating membrane and thus the usual functional portion of a scolophore or phonoreceptor. In higher vertebrates a membrane separating the tympanic cavity from the outer ear and against which impinges the lower arm of the malleus

tympanohyal bone [tĭm′·pə·nō·hĭ″·əl] that part of the tympanic bulla which is formed from the hyoid arch (*see also* hyal bone, urohyal bone)

tympanum [tĭm″·pə·nəm] a cavity between the outer and inner ear, known as the "middle ear". The term is often applied by abbreviation to the tympanic membrane which separates it from the outer ear

Tyndall color one which is produced by a schemochrome blue over a pigment. Many green feathers, for example, are produced by a blue interference color over a yellow pigment

type 1 (*see also* type 2) specifically, in biology, the individual from which a species is described. The description must be published and the individuals are today almost invariably deposited in museums. Types from earlier times are acceptable from accurately illustrated written descriptions. The different varieties of type listed below, and defined in alphabetic position are all acceptable. Gray 1967 (Dictionary of the Biological Sciences) lists 38 additional types, to which reference is sometimes found in the literature, but which are no longer acceptable to the International Commission on Nomenclature (*see also* species name)

ALLOTYPE   MEROTYPE
ECOPHENOTYPE   NEALLOTYPE
ECOTYPE   NEOTYPE
GENERITYPE   PARATYPE
HOLOTYPE   PLESIOTYPE
ISOHOLOTYPE   TOPOTYPE

type 2 (*see also* type 1) in the more generalized sense of a kind of something. The following derivative terms are defined in alphabetic position:

AGROTYPE   GENOTYPE
BIOTYPE   KARYOTYPE
CLIMATIC ECOTYPE   MIMOTYPE
ECOTYPE   PHENOTYPE
EDAPHIC ECOTYPE   WILD TYPE
GENERITYPE

type species the species from which the characteristics of a genus are derived

typhlosole [tĭf″·lō·sōl′] a longitudinal fold projecting into the intestine in many invertebrates

-typlo-   *comb. form* meaning "blind" and hence "caecum" in the sense of a blindly ending tube

typonym [tĭp′·ō·nĭm″] a name having priority but which is based on the same type as the name currently in use

Tyrannosaurus [tĭr′·ăn·ō·sôr″·əs] a genus of giant synapsid reptiles of the Cretaceous period. *T. rex* [rĕks] was about forty feet long with a semi-upright stance that raised its head about twenty feet above the ground. The forelimbs were greatly reduced but there was a huge skull armed with immense, dagger-like teeth

l-tyrosine [el·tĭr″·ō·sēn′] β-(*p-hydroxyphenyl*)*alanine*. $HOC_6H_4CH_2CH(NH_2)COOH$. An amino acid necessary to the nutrition of rats

U

Uca [yū″·kə, ōō·kə] a genus of decapod Crustacea containing the "fiddler crabs". They are so called because one cheliped is enormously more developed than the other

UDP = uridine diphosphate

UDPG usually means UDP glucose but has also been used for UDP glucuronate

ulnare bone = triangular bone

Ulothrix [yū″·lō·thrĭks′] a genus of Chlorophyceae. They have unbranched filaments produced by the repeated division of cells in a straight line. Reproduction is by zoospores and isogametes

ultimobranchial body [ŭlt′·ĭm·ō·brănk″·ē·əl] a glandular structure of unknown function derived from the posterior wall of the fifth branchial pouch

ultramicroscopic invisible under an optical microscope but visible under an electron microscope (*see also* microscopic)

ultramicrotomy microtomy as applied to the electron microscope. Sections are cut of objects embedded in plastic at a thickness of about 200 Å

ultrastructure those features of cellular structure which are disclosed by electron microscopy

umbel [ŭm″·bəl] a type of influorescence in which the pedicels appear to spring from a common point usually resulting in a flat flower cluster

Umbelliferae a very large family of dicotyledons including numerous edible plants such as parsley, cumin, fennel, the carrots, etc. The arrangement of the simple flowers in umbels is characteristic. The genus *Conium* is the subject of a separate entry

umbilical cord [ŭm·bĭl″·ək·əl] the connection between a mammalian embryo and the placenta

Umbilicaria [ŭm′·bĭl·ĭk·är″·ē·ə] a genus of large foliose lichens of circumpolar distribution. They are sometimes used for human food under the name of rock tripe

umbilicus [ŭm·bĭl″·ĭk·əs] literally, the navel and applied to so many depressions of this shape in various organisms as to have no specific meaning

umbo [ŭm″·bō] literally, the boss in the center of a round shield, and therefore applied to any prominence in the center of a round mass such as the beak of a brachiopod shell or the protuberance on the surface of many pelecypod mollusk shells

umbones [ŭm·bōn″·ēz] plural of umbo

-umbracul- comb. form meaning "umbrella" or "parasol"

umbrella the main mass of a medusoid coelenterate. It consists of mesogloea bounded by a one cell thick coat of ectoderm. The coelenteron penetrates the mesogloea from a mouth on the end of a manubrum as radial canals terminating in a circular canal. The edge of the umbrella carries nematocyst-bearing tentacles (*see also* exumbrella subumbrella)

UMP uridine 5′-phosphate

un- comb. prefix meaning "not"; words beginning with this prefix are not given in the present dictionary, since the meaning is always self-evident

-unci- comb. form meaning "hook"

uncinate [ŭn″·sĭn·āt] hooked, or having the form of a hook, or possessing an unusually prominent uncus in those forms to which this term is applied

uncinate process a dorso-lateral projection from the central portion of the rib of some reptiles

uncus [ŭng″·kəs] literally, a hook. Specifically, one of the hook shaped pair of trophi in the mastax of a rotifer, a hooked process on the intromittent organ of some insects and the anterior end of the hippocampus

underwing the posterior pair of wings of butterflies and moths

undulating membrane a delicate membrane supported by several rows of cilia in the cytopharynx of some protozoa. Also the membrane that attaches the flagellum to the body in Trypanosomidae

-ungui- comb. form meaning "claw", "talon", "nail"

-ungulat- comb. form sometimes used for "claw" but more properly meaning "hoof"

Ungulata [ŭng′·gyū·lāt″·ə] at one time an order of placental mammals referred to as the hoofed animals. Now usually regarded as consisting of two separate orders the Perissodactyla and the Artiodactyla

uni- comb. prefix meaning "one" or "single". Compound words beginning with uni- are only given in this dictionary, when the meaning is not self-evident

unipolar ingression the production of endoderm from vegetal pole cells which detach themselves and rearrange themselves under the ectoderm

184

nit membrane a membrane consisting of a lipid layer between two layers of protein

nivalent [yū'·nē·vāl"·ənt] the odd chromosome which does not pair in an aneuploid

nivoltine reproducing only once a season

nstable association one which is just beginning

Q = ubiquinone

ur- *comb. form* confused from three roots and therefore variously meaning "urine", "tail" and "mountain"

rachus [yūr·ə·kəs] the connection between the mbryonic bladder and the allantois which later becomes a ligament supporting the bladder

rea cycle the cyclic formation of urea by the hydrolysis of arginine, yielding argenine and ornithine, following the synthesis of arginine from ornithine

rease [yūr"·ē·āz'] an enzyme catalyzing the hydrolysis of urea to carbon dioxide and ammonia

Jrediniales the order of Basidiomycetes that contains the rusts. They are all obligate parasites of vascular plants echnically distinguished by producing basidiospores hat develop directly with a hypha. The genus *Puccinea* s the subject of a separate entry

redospore [yūr·ē"·dō·spôr] a spore which germinates immediately to produce a mycelium which itself splits into other uredospores or sometimes, teleutospores

reter [yūr·ē"·tə] the entire kidney duct of lower vertebrates and, in higher forms, that portion of the kidney duct which transports the products of excretion from he kidney to the bladder

rethra [yūr·ēth"·rə] that portion of the kidney duct which connects the bladder to the exterior

rinary bladder [yūr"·ĭn·rē] an outgrowth of the indgut, or derivative of the allantois, in amniotic vertebrates into which the ureters open. The term is often improperly applied to the water storing bladder of Amphibia

rinary papilla the papilla through which the ureter opens into a cloaca

robilin [yūr'·ō·bil"·ĭn] the product to which bilirubin is converted in the large intestine

Urochordata [yūr'·ō·kôrd·ā"·tə] a subphylum of craniate chordates containing the sea squirts or tunicates. Sexual reproduction involves a tadpole larva which resembles an adult larvacean (*see* Larvacea). The group is divided into the Larvacea, the Ascidiacea and he Thaliaceae, each of which is the subject of a separate entry. They are distinguished by the presence of an outer covering, or test, a greatly reduced central nervous system, and a greatly enlarged branchial region

Urocyon [yūr'·ō·si"·ən] the genus of canid carnivores containing the American gray fox. It is longer legged than the red fox (*Canis vulpes*), originally lived on savannahs and hunted in packs

Urodela [yūr'·ō·dēl"·ə] an order of Amphibia containing those forms that retain the tail in the adult. The legs, absent in a few forms, are of equal size. Sometimes known as Caudata. The genera *Ambystoma, Amphiuma, Cryptobranchus, Eurycea, Megalobatrachus, Necturus, Notophthalmus, Plethodon, Siren* and *Triturus* are the subjects of separate entries

urodeum [yūr'·ō·dē"·əm] the posterior portion of the cloaca into which open the ureters and sex ducts

urogastrone [*angl.* yōōr·og"·əs·trōn, *orig.* yōōr'·ō·gast"·rōn] a hormone secreted in the gastrointestinal tract. It is active in controlling the functions of the stomach (*see also* enterogastrone)

urohyal bone [yūr'·ō·hi"·əl] a derivative of the hyoid arch, lying anterior to and below the copular in some fish (*see also* hyal bone, tympanohyal bone, stylohyal bone)

uropore [yūr'·ō·pôr"] the opening of an excretory organ particularly that of an arthropod

uropygeal gland [yūr·ō·pĭg"·ē·əl, yūr·ō·pĭj·ē·əl] that integumentary gland of birds found immediately above the tail which produces a secretion used by the bird for dressing, or "preening", its feathers. The gland is very prominent in aquatic birds (= preen gland, oil gland)

Uropygi [yūr'·ō·pĭj"·ē] a small order of arachnid arthropods at one time united with the Amblypygi into the Pedipalpi. The Uropygi are characterized by the possession of a multi-articulate tail giving them the name whip-scorpion. The genus *Tarantula* is the subject of a separate entry

urosacral vertebra [yūr'·ō·sāk"·rəl] one of those anterior caudal vertebrae which are fused with the sacrum (*see also* sacral vertebra and pseudosacral vertebra)

Urosalpinx [yūr'·ō·săl"·pĭngks] a genus of predatory gasteropod mollusks. *U. cincerareus* [sĭn'·ə·râr"·ē·əs] (the oyster drill) is a major pest of oysters, destroying large numbers by drilling holes through the shell

urostyle [yūr'·ō·stīl"] an unsegmented, posterior prolongation of the vertebral column. It is a thin, rod-like structure in Amphibia salientia, a group of fused vertebrae in certain Chelonia, and a terminal group of hypural bones in actinopterygian fish

Ustilaginales [ŭst'·ĭl·ăj'·ĭn·āl"·ēz] the order of basidiomycetes that contains the smuts. They are technically distinguished by the fact that the teleutospores are formed from intercalary cells of a dikaryotic mycelium. The genus *Ustilago* is the subject of a separate entry

Ustilago [ŭst'·ĭl·ā"·gō] a typical genus of ustilaginale basidiomycete fungi. *U. maydis* [mā"·dĭs] is corn smut, one of the most destructive pests of that crop

uterus 1 [yū"·tə·əs] (*see also* uterus 2) that part of the invertebrate female reproductive system in which eggs for embryos are stored

uterus 2 (*see also* uterus 1) that portion of the female mammalian reproductive system in which the embryo develops. The following derivative terms are defined in alphabetic position:

BICORNUATE UTERUS	DUPLEX UTERUS
BIPARTITE UTERUS	SIMPLEX UTERUS

UTP = uridine triphosphate

-utricul- *comb. form* meaning a "small bladder"

utriculus [yū·trĭk"·yōō·ləs] the dorsal of the two divisions into which the auditory vesicle becomes permanently divided (cf. sacculus)

-uv- *comb. form* meaning "grape"

V

vacuole [văk″·yū·ōl] a small cavity, usually one within a cell (*see* contractile vacuole)

-vag- *comb. form* meaning "wander"

-vagin- *comb. form* meaning "sheath"

vagina that part of the female reproductive system designed for the reception of the intromittent organ

vagus nerve [vā″·gəs] cranial X. Arising from several roots in a dorso-lateral wall of the medulla. In gilled animals, it sends branches to the gillslits as well as to the viscera and the lateral line

-val- *comb. form* meaning "strong". The word "equivalent" (equally strong) has led to the use of -valent in the sense of value (*see* bivalent, univalent)

valine [văl″·ēn, vāl″·ēn] α-aminoisovaleric acid. $CH_2CHCHNH_2COOH$. An amino acid necessary for the nutrition of rats

Valisneria [văl′·is·nēr″·ē·ə] a genus of aquatic dicotyledons, mostly of freshwaters. *V. spiralis*, a common aquarium plant, is often called eel grass, a term, however, also applied to *Zostera*

-vall- *comb. form* meaning "rampart" or "valley"

-valv- *comb. form* properly meaning "the part of a door which swings open and shut" and thus, by extension, to any structure which opens or closes an aperture

valve 1 (*see also* valve 2) in the original sense of a folding shutter, mostly applied to the shells of pelecypod mollusks, brachiopods, and diatoms and to that portion of a seed capsule which opens in dehiscence. In this sense the following derivative terms are defined in alphabetic position:

SPATHE VALVE   VIEUSSENS' VALVE

valve 2 (*see also* valve 1) in the derivative sense of a device permitting one-way flow. In this sense the following derivative terms are defined in alphabetic position:

EUSTACHIAN VALVE   PYLORIC VALVE
KERCKRING'S VALVES   SEMILUNAR VALVE
MITRAL VALVE   SPIRAL VALVE
TRICUSPID VALVES

-var- *comb. form* meaning "ingrown", frequently confused with -vari-

Varanus [və·ăn″·əs] a genus of lizards (Squamata) containing numerous small forms and also the so-called komodo dragon (*V. komodoensis*) [kō·mō′·dō·ĕns″·is] which reaches a length of ten feet and a weight of 250 pounds

-vari- *comb. form* meaning "diverse"

Variolus′ bridge (pons Varolii) a transverse band of fibers running across the lower surface of the brain between the two sides of the cerebellum

-vas- *comb. form* meaning a "vessel", "dish" o "utensil"

vas deferens [văs. dĕf″·ər·ĕnz] literally a duct that "carries down" or "away" and specifically one of the ducts that conveys sperm out of the body

vas efferens [văs ĕf″·ər·ĕnz] literally a duct that "carries to" or "brings out" and specifically one of the ducts that carries sperm from the testis to the vas deferens

vasa [väz″·ə] plural of vas

vascular [văs″·kyū·lər] pertaining to, or containing vessels

vascular cambium that meristem which produces secondary growth (= vascular meristem)

vascular cylinder the conducting tissue of plants when it forms a cylinder separating the outer cortex from the inner pith

vascular meristem the meristem that produces secondary growth (= vascular cambium)

vascular ray the phloem and xylem rays taken together

vascular system 1 (*see also* vascular system 2) the sum total of fluid conducting vessels in an organism and therefore including veins, arteries, and lymph vessels

vascular system 2 (*see also* vascular system 1) the conducting tissues of a plant

vascular tissue conducting vessels of either plants or, less frequently, animals

vasomotor [văz′·ō·mō″·tər] concerned with the movement—actually wall contraction—of blood vessels

vasopressin [văz′·ō·prĕs″·ĭn] a hormone secreted by the neurohypophysis which acts on the capillaries and arterioles

Vater's ampulla a swelling in the pancreatic duct at the point of junction with the intestine

Vater's corpuscle an end organ sensing pressure, consisting of a central elongated granular mass containing the nerve ending, surrounded by many layers of thin connective tissue

Vater's papilla a papilla around the mouth of the pancreatic duct within the intestine

Vaucheria [vō·shēr"·ē·ə]   one of the very few genera of siphonale green algae to occur in freshwater. *V. repens* [rĕp"·ĕnz], typical of the genus, forms a felt-like mass of aseptate filaments. In sexual reproduction an oogonium and an antheridium develop side by side on the same septum

vector [vĕk·tə]   an organism which carries and transmits a parasite

-veget-   *comb. form* meaning "lively", "active" or "vigorous". Most modern derivative words mean the reverse of these terms

vegetal [vĕj"·ə·təl']   a word synonymous with vegetable but preferred by most in non-botanical usage

vegetal pole   the lower, yolk encumbered, pole of a medialecithal egg

vegetative [vĕj·ə·tā"·tiv]   used in the sense of asexual, in contrast to sexual, reproduction

vegetative cell   that cell in the pollen granule which produces the pollen tube

vegetative nucleus   any of those nuclei of the pollen tube which do not take part in fertilization

vein 1   (*see also* vein 2, 3) a vessel carrying blood to the heart. Names of veins that are self-explanatory (e.g. epigastric) are not given in this dictionary. However, the following derivative terms are defined in alphabetic position:

ADVEHENT VEINS   OMPHALO-MESENTERIC V.
ANTERIOR ABDOMINAL V.   PORTAL VEIN
CUTANEOUS VEIN   POSTERIOR CARDINAL V.
ILIAC VEIN   REVEHENT VEIN
JUGULAR VEIN   SUBCLAVIAN VEIN
LATERAL ABDOMINAL V.   VITELLINE VEINS

vein 2   (*see also* vein 1, 3) the vascular strands in a leaf. Confused with nerve 3 but properly applying to those smaller vascular strands which do not cause an elevation of the leaf surface

vein 3   (*see also* vein 1, 2) the strands of chitin that stiffen the wings of insects. Names of these veins are not entered in this dictionary

veldt [vĕlt]   a type of grassland habitat found in South Africa in which a luxuriant growth of grass is augmented by many patches of low shrubs and occasional trees

Velella [vĕl·ĕl"·ə]   a common genus of siphonophoran hydroids with an elliptical disk-shaped pneumatophore bearing a longitudinal crest which acts as a sail

velum [vē"·ləm]   literally an "awning", but applied in biology to many membranous structures which do not have a supporting function. Specifically in botany used of the envelope within which the whole fruiting body of some Basidomycetes are produced. In zoology a delicate annular membrane projecting inwards from the edge of the bell of medusae, a skirt-like development from the posterior end of the proglottid of some tapeworms which envelopes the anterior region of the proglottid next behind, the swimming organ of marine molluscan larvae which is developed from the preoral ciliated rings of the trochophore, and the membranes surrounding the pseudostome of many ciliate protozoa. It is, rarely, also used for the soft palate. In *Branchiostoma*, it is a membrane that separates the cavity of the oral funnel from the pharynx and bears the mouth in its center

velumen   literally a fleece and specifically the epidermis on the roots of epiphytic orchids

venous system [vēn·əs]   the sum total of those

blood vessels which return blood under low pressure to the heart

ventilating pits   analogues of stomata found in some ferns

ventral   in zoology pertaining to the underside or belly. In botany the ventral face of a leaf is the upper surface which was originally the inner face in the bud (cf. dorsal)

ventral aorta   the aorta that runs directly forward from the heart along the floor of the pharynx and from which the aortic arches arise (*see also* aorta, dorsal aorta)

ventral canal cell   a cell formed in the developing archegonium of Bryophyta intermediate in position between the egg and the ventral canal cells

ventral cartilage   one of a series of cartilages precursor to the ventral portions of the vertebrae

ventral root   the metameric motor ganglia (inside the chord) which form the ventral roots of spinal nerves

ventricle 1   [vĕn"·trĭk·əl]   (*see also* ventricle 2) that muscular division of the heart which pumps blood from the heart

ventricle 2   (*see also* ventricle 1) the cavities of the brain. In this sense the following derivative terms are defined in alphabetic position:

COMMON VENTRICLE   LATERAL VENTRICLE
FOURTH VENTRICLE   THIRD VENTRICLE

-verg-   *comb. form* meaning "twig"

-verm-   *comb. form* meaning "worm"

vermiform appendix   [vûrm"·ē·fôrm']   a blind tube which terminates the caecum of mammals

vermis [vûrm"·ĭs]   the unpaired, dorsal, central portion of the cerebellum

vernal [vûrn"·əl]   pertaining to spring

vernalization [vûr'·năl·ĭz·ā"·shŭn]   the treatment of seeds in such a manner as to shorten the maturation period of the plant developed from them

-vers-   *comb. form* meaning "turn"

verspecies [və·spĕsh"·ēz]   a term applied to a species, as such, when it is desired to indicate that it is neither a super species nor a subspecies

vertebra [vûr"·tĕb·rə]   one of the bones which together make the vertebral column. The following derivative terms are defined in alphabetic position:

ADELOSPONDYLOUS V.   OPISTHOCOELOUS V.
AMPHICOELOUS V.   PHYLLOSPONDYLOUS V
AMPHIPLATYAN V.   PROCOELOUS V.
ASPIDOSPONDYLOUS V   PSEUDOSACRAL V.
CAUDAL V.   SACRAL V.
COELOUS V.   SPONDYLOUS V.
HOLOSPONDYLOUS V.   STEROSPONDYLOUS V.
LEPTOSPONDYLOUS V   TEMNOSPONDYLOUS V
LUMBAR V.   THORACIC V.
NOTOCENTROUS V.   UROSACRAL V.
NOTOCHORDAL V.

vertebral artery   one of a pair of derivative arteries of the subclavian which run forward to unite as the basilar artery running forward under the medulla

vertebral lamina   [vûr"·tĕb·rəl]   one, of the two, sides of the neural arch

vertebral ring   those cartilaginous elements of a vertebra which lie between the prospondylous and opisthospondylous rings

vertebrate [vûr·tĕb·brāt']   a loose term applying to those chordates that have vertebral columns

vesicle [vĕs"·ĭk·əl]   a small bladder. The term was at one time used for "cell" but is applied to almost any cavity, particularly in zoology those filled with fluid but in botany also those filled with gas. The following derivative terms are defined in alphabetic position :

AUDITORY VESICLE   POLIAN VESICLE
GERMINAL VESICLE   SEMINAL VESICLE
MADREPORIC VESICLE   SPERMIDUCAL VESICLE

-vesp-   comb. form meaning "wasp"

-vesper-   comb. form meaning "evening"

vessel member   a xylem element differing from a tracheid in being hollow

-vexill   comb. form meaning "flag"

Vibrio [vĭb″·rē·ō] a genus of pseudomonodale schizomycetes. *V. cholerae* [kŏl″·ə·rē] is the causative agent of cholera

vibrissa [vĭ·brĭs″·ə] one of the large, laterally projecting, bristles on the snouts of some mammals. The term is also applied to sensitive trichomes on plants

vicariant [vĭ·kâr″·ē·ənt] one species which, in any given ecological situation, represents a speces found in a similar environment in another geographical location

Vicia [vĭs″·ē·ə] a large genus of leguminous plants of great commercial value. Many, such as *L. sativa* [săt·ĕv″·ə] are used as fodder under the general name of vetches while *V. faber* [fā″·bə] is the broad bean, widely eaten in Europe and the Near East. Favism is an acute hemolytic anemia resulting from an excess of *Vicia faba* in the diet

Vieussen's valve   the transverse velum that separates the medulla from the cerebellum

-vill-   comb. form meaning "hairy" in the sense of "shaggy"

villikinin [vĭl″·ē·kĭn″·ĭn] a hormone liberated in the intestine which increases the rate of movement of the villi

villus [vĭl″·əs] in zoology a small finger-like process, particularly those arising from the intestinal epithelium (*see also* microvillus)

vinculum [vĭn″·kyū·ləm] plural vinculi; a chain, band or strap and used for so many structures of this general designation as to have no valid specific meaning

virgin 1 (*see also* virgin 2) in humans, an individual who has never engaged in sexual intercourse; in other animals, particularly insects, an unfertilized female

virgin 2 (*see also* virgin 1) anything untouched by human hands (e.g. virgin forest), or in a natural state (e.g. virgin wool), or produced without artifice (e.g. virgin oil)

virus [vĭr″·əs] the smallest known living entity. They are distinguished by their small size, inability to exhibit metabolic activity outside a host cell, and their simple structure consisting almost invariably of a core of nucleic acid in a protein sheath. Tobacco mosaic virus is the subject of a separate entry

viscera [vĭs″·ə·rə] the contents of the abdominal cavity

visceral arch [vĭs·ə·rəl] one of a series of U-shaped aggregations of cartilages or bones supporting the pharyngeal cavity and forming the jaw of vertebrates. In all bony forms, the maxillary arch, or upper jaw, is fused to the cranium. The mandibular arch forms the lower jaw in all vertebrates. The hyoid arch, which lies immediately behind or below the mandibular, varies greatly in structure and function in different classes. Behind the hyoid there are from five (normal) to seven (the shark *Heptanchus*) branchial arches which support the pharynx in water-dwelling forms, but are modified or lost in terres-

trial forms. The components of a typical branchial arch are, from the top down, pharyngobranchial, epibranchial, ceratobranchial, and basibranchial cartilages or bones, as the case may be (cf. branchial arch)

visceral muscle   muscles associated with the walls of the alimentary canal

visceral nerve   a nerve carrying impulses to or from viscera

visceral skeleton   the visceral arches of fish and their derivatives or equivalents in other groups

viscus   singular of viscera

visual pigment   photosensitive pigments in the eye. Rhodopsin and porphyropsin occur in the rod cells while iodopsin and cyanopsin occur in the cone cells

visual purple = rhodopsin

vitalism   the doctrine that development, or indeed all metabolic processes, are controlled by a force (entelechy) which has not yet been detected, and quite possibly cannot be detected by physical equipment

vitalist theory   originally the postulate that fermentation and putrefaction were due to organisms. Often nowadays transferred to the proponents of vitalism (q.v.)

vitamin A   3,7-Dimethyl-9-(2,-6,6-trimethyl-l-cyclohexen-l-yl)-2,4-6,8-nonatetraen. An oil soluble vitamin required by most species of animal

vitamin B₁   thiamine. 3-(4-Amino-2-methylpyrimidyl-5-methyl)-4-methyl-5-β-hydroxyethylthiazolium chloride hydrochloride. A water soluble vitamin required in the nutrition of almost all organisms

vitamin B₂   riboflavin. 6,7-Dimethyl-9,(D-l′-ribityl)-isoaloxazine. A water soluble vitamin forming part of the electron transfer system in the cellular metabolism of most living forms

vitamin B₆   pyridoxine. 5-Hydroxy-6-methyl-3,4-pyridine dimethanol hydrochloride. A water soluble nutrient thought by many to be a vitamin though no human requirements have been established and the evidence is doubtful for most other species

vitamin B₁₂   cobalamin. 5,6-Dimethylbenzimi-dazolyl cyanocobamide. A water soluble vitamin acting as a hemopoietic agent in many animals though its specific human need has not been determined

vitamin B_c = folic acid

vitamin C   ascorbic acid. A water soluble vitamin synthesized by most mammals except man and the guinea pig. It was originally regarded only as an anti-scorbutic reagent but is now known to enter into the respiratory metabolic cycle of cells

vitamin D₂   activated ergosterol 9,10-secoergosta-5,7,10(19),22-tetra-en-3-ol. A fat soluble vitamin required by most animals for calcium metabolism

vitamin D₃   activated 7-dehy-drocholesterol. 22,23-dihydro-24-demethylcalciferol. An analog of vitamin D₂ about 50 times as active in birds as in D₂ but without apparent difference in activity in most other forms

vitamin E   α-tocopherol, 5,7,8-trimethyltocol. An oil soluble vitamin playing an ill defined role in the metabolism of some animals

vitamin G = vitamin B₂

vitamin H   biotin. cis-hexahydro-2-oxo-l-H-thieno-[3,4] imidazole-4-valeric acid. A water soluble vitamin the necessity of which in human nutrition has not been illustrated. It protects many animals against egg white injury

vitamin K₁   2-methyl-3-phytyl-l,4-naphthoquinone. An oil soluble vitamin controlling production of prothrombin and therefore essential for blood clotting in all vertebrates

vitamin K₂ 2-methyl-3-difarnesyl-1,4-naphthoquinone. Essentially similar in action to vitamin K₁

vitamin L a so-called lactation factor of doubtful existence

vitamin M = folic acid

vitamin P a substance of unknown composition derived from citrus fruits supposed to affect the permeability and fragility of capillary blood vessels

-vitell- *comb. form* meaning "yolk"

vitelline cell [vĭt·ĕl″·ĭn, vĭt·ĕl″·īn] a cell that stores or produces yolk

vitelline membrane the external surface of an egg (*see also* pervitelline membrane)

vitelline veins extraembryonic veins collecting blood from the surface of the yolk sac and carrying it back to the embryo

vitellophage [vĭt·ĕl″·ō·fāj′] a wandering cell in a yolk mass which is alleged to assist in its digestion

-viti- *comb. form* meaning "vine"

Vitis [vī″·tĭs] the genus of vines that produce grapes. The cultivation of the Old World *V. vinifera* [vĭn·ĭf″·ər·ə] precedes written history and is the only extant Old World species. There are numerous American species

vitreous [vĭt″·rē·əs] glassy, either in regard to the transparency, surface texture, or greenish yellow color

vitreous humor the transparent colloid filling the posterior chamber of an eye

viviparous [*angl.* vĭv·ĭp″·ər·əs, *orig.* vĭv′·ē·pär·əs] said of females in which the young are maintained for some time internally before birth

vocal cord tendons stretched across the glottis, or in some vertebrates within the laryngeal cavity, used in the production of speech and song. The term is also applied, in insects, to modified thoracic spiracles which produce the typical hum of a mosquito and similar forms

Volkmann's canal a transverse canal in bones joining Haversian canals to each other, and to the periostium

-voltine *adjectival suffix* meaning "pertaining to the seasons of the year". The following terms using this suffix are defined in alphabetic position:
MULTIVOLTINE   UNIVOLTINE

voluntary muscle = striated muscle

volutin [vŏl·yōō·tĭn] protein granules with a high phosphate content found in the cells of many lower animals

volva [vŏl″·və] a membrane totally enclosing some growing basidomycete fungi, such as *Amanita* and its allied forms, and which remains until the cap has emerged as a veil round the bottom of the stalk

Volvocales [vŏl′·vō·kāl″·ēz] an order of Chlorophytae distinguished by the fact that the vegetative cells are flagellated and actively motile. The genera *Eudorina, Platydorina, Pleodorina* and *Volvox* are the subjects of separate entries

Volvox a typical genus of Volvocales. The hollow spherical reticulated colonies are unmistakeable. *V. perglobator* [pûr′·glōb·āt″·ər], the common United States species often reaches 1 mm in diameter but the European *V. globator* [glōb·āt″·ər] is usually half this size

vomer bone [vō″·mər] one of a pair of cranial bones lying in the floor of the cranium immediately anterior to the basisphenoid

Von Ebner's gland albuminous glands associated with the circumvallate papillae in the mammalian tongue

-vor- *comb. form* meaning to "devour"

-vorous *adjectival suffix* meaning eating (cf. -phagous). The *substantive form* -vore is more common than its equivalent -phage. Derivative compounds are not listed since the meanings are apparent from the roots; e.g. omnivorous (eating anything or everything), insectivorous (insect eating)

Vorticella [vôr′·tē·sĕl″·ə] the commonest of the freshwater peritrichan ciliate protozoans. The bell-shaped body, containing a U-shaped nucleus, is carried on a contractile stalk. It is interesting that *V. convallaria* [kŏn′·văl·âr″·ē·ə] was the first protozoan described by Leeuwenhoek

Vulpes [vŭl″·pēz] the genus of canid carnivore containing most of the foxes (*see also Alopex, Urocyon*). The common red fox of the Northern Hemisphere is *V. vulpes* but there are several other species

vulva [vŭl″·və] female external genitalia

W chromosome a term used by some biologists for a sex chromosome which is heterozygous in the female (ZW) and absent in the male (ZZ)

Wagner's separation theory a new race can only be formed when there is marked geographic separation from the parent species

Wallace's line the imaginary line that separates the Australasian from the Asiatic faunas; the important part of the line runs north-northwest between the islands of Bali and Lombok and then between Borneo and Celebes; Bali and Borneo thus have an Asiatic fauna while Celebes and Lombok represent the limits of the Australasian

walzing syndrome a mouse behavior pattern, controlled by many genes at different locations, which involves a rambling, circling movement with head shaking

water cells suberized, water-retaining cells in the palisade tissue of succulents

water leaf the submerged leaf in those aquatic plants having two kinds of leaf

water parasite an epiphyte, such as the mistletoe, which derives only water from its host

weather migrant an animal which migrates in response to changes, usually seasonal, in the climate

Weberian apparatus a series of small bones connecting the swim bladder with the inner ear found in some Osteichthyes. These bones are modifications of the anterior vertebra

Weber's organ the vestigeal uterus in the male mammal. This term is sometimes also applied to lateral glands of the tongue

Wharton's duct the duct of the submaxillary gland

white fibrous connective tissue the principal connective tissue of animals. It consists of collagen strands in a gelatinous matrix of mucopolysaccharides

white matter that part of the central nervous system which consists principally of medullated fibers (*see also* grey matter)

wholemount a whole organism, or part of an organism, mounted for microscopical examination (*see also* section, smear, squash)

wild type the stock or population from which mutants are derived

wild type gene the allele normally found at any given locus on a chromosome

Willis' circle the arterial ring formed where the basilar artery splits to pass round on each side of the hypophysis and reunites on the other side

Willis's theory the more widely distributed an organism is in a given area, the longer it has been resident in that area

winter egg one produced by some invertebrates, particularly Cladocera, which are adapted to resist freezing temperatures and which usually will not hatch until after they have been frozen

winter rigidity a term used to describe the hibernation of poikilothermic animals in distinction to the dormancy of homoiothermic animals

Wirsung's duct = pancreatic duct

Wolffia [*angl.* wōōl″·fē·ə, *orig.* vŏl″·fē·ə] a genus of Lemnaceae containing the smallest known flowering plants, some species being less than 1 mm when fully grown

Wolffian body = mesonephoros

wood the hard part of the stem of a tree consisting principally of xylem. Wood in the standing tree is called timber and, when sawn or split for use, lumber. The following derivative terms are defined in alphabetic position:

DIFFUSE POROUS WOOD   RING POROUS WOOD

worker (*see also* ergate) a female ant, who, being unable to reproduce, busies herself with other things

Wuchereria [*angl.* wōō′·shĕr·ēr″·ē·ə] a genus of nematodes parasitic in humans. The young of *W. bancrofti* [băn·krŏft″·ī], transmitted by various mosquitoes (principally *Culex fatigans*), choke the lymph vessels and produce elaphantiasis

X

X bone = squamosal bone

X chromosome that sex chromosome which is commonly homozygous (XX) in the female and heterozygous (XY) in the male

-xanth- *comb. form* meaning "yellow"

xanthin any of numerous yellow carotenoid pigments found in plants and animals (*see also* astaxanthin)

xanthine a yellow nitrogenous pigment found in animals, and a few plants, formed by the hydrolysis of guanine (*see also* hypoxanthine)

Xenarthra [zĕn·ärþ″·rə] the order of placental mammals which contains the anteaters and sloths. They are distinguished by the lack of true teeth and by the large recurved claws on the feet. The armadillos are sometimes included in this group and sometimes placed in a separate order the Loricata. The genera *Choloepus* and *Myrmecophaga* are the subjects of separate entries

-xeno- *comb. form* meaning "a host" in the opposite sense to "guest" or "foreigner"

xenoparasite [zĕn′·ō·păr″·ə·sĭt] either one which infests the host not normal to it or one only capable of invading a host after the latter is injured adventitiously

Xenopsylla [zĕn′·ŏp·sĭl″·ə] a genus of siphonapteran insects. *X. cheopsis* the Asiatic rat flea is the transmittor of *Pasteurella pestis* the causative agent of plague. Originally found on rats in Egypt, is now of world wide distribution on a very large variety of rodents

Xenopus [*angl.* zĕn″·ə·pəs, *orig.* zĕn·ōp″·əs] a genus of tongueless frogs known as the clawed toads since the three inner toes bear claws. *X. laevis* [lē″·vĭs] was once used in the first test perfected for human pregnancy

-xenous *comb. suffix* meaning pertaining to a "host" or "guest". The following terms using this suffix are defined in alphabetic position:

LIPOXENOUS   TRIXENOUS

-xer- *comb. form* meaning "dry"

xerophyte [zēr″·ō·fit′] a plant specifically adapted to life in dry places

xerosere [zēr″·ō·sēr′] a succession developing in virtue of a scanty supply of water

xerotherm [zēr″·ō·thûrm′] an organism capable of withstanding both drought and heat

-xiph- *comb. form* meaning "sword"

xiphoid cartilage [zĭf″·oid] that which terminates the sternum

Xiphosura [zĭf″·ō·sûr″·ə] an order of arthropods commonly called horseshoe crabs or king crabs. They are distinguished by the horseshoe-shaped body, long pointed telson, and the presence of 5 pairs of gill books on the opisthoma. They were at one time, but are no longer, thought to be related to the Trilobita through the superficial resemblance of the larva of *Limulus* to this group. The genus *Limulus* is the subject of a separate entry (cf. Merostomata)

-xyl- *comb. form* meaning "wood"

xylanase [zĭl″·ə·nāz′] an enzyme hydrolyzing β-1, 4-xylan links in carbohydrates

xylem [zi″·ləm] woody tissue or cells. The following derivative terms are defined in alphabetic position:

ENDARCH XYLEM   METAXYLEM
EXARCH XYLEM   PRIMARY XYLEM
LEPTOXYLEM   PROTOXYLEM
SECONDARY XYLEM

Y chromosome the sex chromosome that is commonly heterozygous (XY) in the male and absent from the female

yellow elastic connective tissue the principal constituent of ligaments in which the collagen fibers are interspersed with brownish fibers of elastin

yield a portion of the productivity of a population removed per unit time, for example by a predator

yolk the intracellular food reserve of an egg

yolk plug the heavily yolked unpigmented endodermal cells which are visible in the blastopore of the amphibian gastrula

Yucca [yŭk″·ə] a genus of American liliaceous plants best known to biologists for the complete mutual dependance of *Y. filamentosa* [fĭl′·ə·mĕnt·ōz″·ə] and the incurvariid moth *Tegiticula alba.* The plant depends for its fertilization on the insect and the insect larva depends for its food on the seeds produced in consequence of the fertilization

Z

Z-band the light area in the center of the A band of striped muscle. It represents the portion in which the actin filaments extending into the A-disc do not meet *see also* I-disc, A-disc)

Z chromosome a term used for the sex chromosome of those forms in which the male is homozygous (ZZ) and the female heterozygous (ZW)

Z-line a line which appears to bisect an I-disc

Zaglossus *see* Tachyglossidae

Zamia [zām″·ē·ə] a genus of extant Cycadales, with short, thick "trunk" and long pinnate leaves. The tip of the "trunk" bears either numerous male strobili, or a single female strobilus

Zea [zē′·ə] a monotypic genus of Graminaceae. *Z. mais* [mā″·is] (corn, maize) is known only as a cultigen

zeugite [zyōō″·gīt] a spore, such as a teleutospore, containing two fused nuclei

Ziegler's theory = nephrocoel theory

Zinjanthropus *see* Australopithecine

zo- *comb. form* meaning "animal"

zoa *comb. suffix* indicating "animal" in the plural

zoaea larva [zō″·ē·ə] that stage in the development of higher crustacea with compound eyes, a carapace overlapping the thorax and a segmented abdomen. It is usually the last planktonic instar

Zoantharia [zō′·ǎn·þâr″·ē·ə] a subclass of antho-zoan Coelenterata distinguished by the fact that the tentacles and septa are usually in multiples of six but never eight as in the Alcyonaria. This subclass contains the sea anemones and the true corals. The genus *Antipatharia* is the subject of a separate entry

zoanthella larva [zō′·ǎn·thĕl″·ə] a larva of a zoan-thid coral, in the shape of an inverted pegtop with a girdle of very long cilia near the oral pole

zoarium [zō·âr″·ē·əm] a colony of animals. Most commonly applied to Ectoprocta

-zoic *adjectival suffix* meaning "pertaining to an animal", or animals. The alternative form -zoous, though occasionally seen in the literature, is ignored in this dictionary. The following terms using this suffix are defined in alphabetic position:

COPROZOIC	EPIZOIC
ENDOZOIC	HOLENDOZOIC

-zoite *adjectival suffix* indicating a part of an animal produced by fission (cf. -zooid). The following terms

using this suffix are defined in alphabetic position:

MEROZOITE	SPOROZOITE
TROPHOZOITE	

-zon- *comb. form* meaning "girdle", which has come in contemporary writing to be extended into the meaning "zone"

-zonal *comb. suffix* pertaining to a zone, particularly in biology in the sense of zone 1 (below) (*see* euryzonal, stenozonal)

zonary placenta one having a band of villi encircling the membranous cover

zone 1 (*see also* zone 2–4) a horizontal layer either of water or air (cf. horizon). In this sense the following derivative terms are defined in alphabetic position:

APHOTIC ZONE	DYSPHOTIC ZONE
APHYTAL ZONE	EUPHOTIC ZONE

zone 2 (*see also* zone 1, 3, 4) a specific limited area of a land mass, particularly one in which the population is limited by physical conditions. In this sense the following derivative terms are defined in alphabetic position:

CLIMAX ZONE	LIFE ZONE
EULITTORAL ZONE	SPRAY ZONE

zone 3 (*see also* zone 1, 2, 4) in the original sense of girdle (*see* adoral zone)

zone 4 (*see also* zone 1–3) in the sense of a portion of a solid. In this sense the following derivative terms are defined in alphabetic position:

ALARY ZONE	CAMBIAL ZONE
BASAL ZONE	

-zonic *adjectival suffix* meaning pertaining to zone, particularly in the sense of zone 2 (*see* heterozonic, hormozonic)

-zoo- *comb. form* derived from -zo-, usually meaning animal, but used in botany to denote motile in contrast to immobile

zoochlorella [zō′·ō·klōr·ĕl″·ə] a general term for green algae living in or among animal cells

zooecium [zō·ēs″·ē·əm] a case containing an animal particularly applied to the zooids of Ectoprocta

zoogamete = planogamete

zoogamy [zō″·găm·ē] the condition of plants having motile sex cells

zooid 1 [zō″·id] (*see also* zooid 2) a motile plant gamete. In this sense, the following derivative terms are defined in alphabetic position:

ANTHEROZOOID   MICROZOOID
MEGAZOOID   SPERMATOZOOID

zooid 2 (*see also* zooid 1) an ill-defined term applied for convenience sake to individual members or distinct functional portions of aggregates of animals like Ectoprocta, or colonial or compound animals like Hydrozoa. Frequently, but unfortunately, used in place of -zoite. In this sense the following derivative terms are defined in alphabetical position:

ACANTHOZOOID   GASTROZOOID
AUTOZOOID   GONOZOOID
BLASTOZOOID   HETEROZOOID
DACTYLOZOOID   SIPHONOZOOID
TENTACULOZOOID

Zoomastigophora [zō'·ō·măst·ĭg·ŏf"·ər·ə] a subclass of mastigophorous Protozoa distinguished from the Phytomastigophora by their lack of chlorophyll. The genera *Codosiga*, *Giardia*, *Oikomonas*, *Trichomonas* and *Trichonympha*, as well as the family Trypanosomidae, are the subjects of separate entries

-zoon [zō"·ŏn] *adjectival suffix* meaning "an animal". The distinction from -zooid 1 is not always clear

zooplankton [zō·ō·plănk"·tən] the animal constituent of the planktonic population

zoosperm = zoospore

zoosphere [zō'·ō·sfēr"] a ciliated swarm cell of an alga which subsequently forms an oosphere

zoospore [zō"·ō·spôr'] a motile spore found in both Fungi and Algae

Zostera [zŏs'·tər·ə] a genus of potamogaceous angiosperms and one of the very few marine representatives of that group. There are six species called "eel grass", though this is misleading since the same name is also applied in freshwaters *Valisneria spiralis*

Zosterophyllophytina [zŏs'·tə·rō·fĭl'·ō·fi·tĭn"·ə] a group of primitive vascular plants known only as fossils from the earliest Devonian strata. This group was once included with the Psilophyta

-zyg- *comb. form* meaning "yoke"

zygantra [zĭ·găn"·trə] an articular process, analogous to a zygapophysis but arising from the neural arch, in snakes

zygapophysis [zĭg'·ə·pŏf"·əs·ĭs] processes that articulate vertebrae. They are paired, flattened processes on which the anterior pair of one vertebra slide under the posterior pair of the next vertebra (*see also* apophysis, anterior apophysis, gonapophysis, parapophysis)

zygoid = diploid

zygoid parthenogenesis [zĭ"·goid] asexual reproduction from an egg that remains diploid or becomes diploid again in the course of its development

zygomatic arch [zi'·gō·măt"·ĭk] a prominence rising from the squamosal bone and articulating in front with the jugal bone

zygomatic process either of the processes from the squamosal or maxilla, which articulate to form the zygomatic arch

Zygomycetes [zĭ'·gō·mĭ·sēt"·ēz] a class of fungi distinguished by the production of a zygospore from the fusion of two gametangia. The genera *Mucor* and *Rhizopus* are the subjects of separate entries

zygonema [zĭ'·gō·nēm"·ə] that stage in the prophase of meiosis in which the synapsis of chromosomes occurs

Zygophyllaceae [zĭ'·gō·fĭl·ās"·ē] that family of dicotyledons which contains the lignum vitae tree. The family is readily distinguished by its prickly fruits. The genus *Larrea* is the subject of a separate entry

zygosphene [zĭ"·gō·sfēn'] a median articular process on the neural arch of the vertebrae of snakes

zygospore 1 [zĭ"·gō·spôr] (*see also* zygospore 2) a thick walled spore resulting in zygomycete fungi from the fusion of the tips of two hyphae

zygospore 2 (*see also* zygospore 1) the product of the fusion of two isogametes

zygote [zĭ"·gōt] the cell that results from the fusion of two gametes (*see* homozygote, heterozygote)

zygotic sterility [zĭ·gŏt"·ĭk] the condition in which gametes can fuse but the zygote is not viable

zygous *comb. suffix* meaning "pertaining to zygote"

-zym- *comb. form* meaning "yeast"

zymogen [zĭ"·mō·jen"] an inactive enzyme, or enzyme precursor, commonly observed in the form of granules